Photon-Atom Interactions

Photon-Atom Interactions

Mitchel Weissbluth

Department of Applied Physics
Stanford University
Stanford, California

ACADEMIC PRESS, INC.

Harcourt Brace Jovanovich, Publishers

Boston San Diego New York
Berkeley London Sydney
Tokyo Toronto

United Kingdom Edition published by
ACADEMIC PRESS INC. (LONDON) LTD.
24–28 Oval Road, London NW1 7DX

ACADEMIC PRESS, INC.
1250 Sixth Avenue, San Diego, CA 92101

Designed by Joni Hopkins

Library of Congress Cataloging-in-Publication Data
Weissbluth, Mitchel.
Photon-atom interactions / Mitchel Weissbluth.
p. cm.
Bibliography: p.
Includes index.
ISBN 0-12-743660-X
1. Photonuclear reactions. 2. Quantum theory. 3. Statistical physics. I. Title.
QC794.8.P4W45 1988
530.1'2—dc 19 88-12642
CIP

Printed in the United States of America
89 90 91 92 9 8 7 6 5 4 3 2 1

In Memory
of my Parents

Contents

Preface

A substantial part of the history of quantum mechanics is associated with efforts directed toward an understanding of the interactions between light and matter on the atomic and molecular level. For the first several decades of this century, activity in this area remained at a high level, and the continuing stream of advances in spectroscopy, both in theory and experiment, contributed enormously to fundamental physics and to various applied sciences. By modern standards, the light sources available during that early period produced light over a broad spectrum and at relatively low intensities, so that on the whole, perturbative treatments to low orders were sufficient to deal with experimental observations. A new era of light-matter interactions began in the 1960s with the invention of the laser. Its unique properties of high intensity, monochromaticity, directionality, and coherence led to the disclosure of new optical phenomena, the development of novel forms of high-resolution spectroscopy, and the invention of numerous optical devices. As applications proliferated, new subfields under various titles—Nonlinear Optics, Quantum Electronics, Laser Physics, and Quantum Optics—came into existence.

Though each subfield is more or less unique in content, whether it be spectroscopy, chemical analysis, medical application, communication, or any of the myriad applications, there exists a significant body of theory shared by all. This body of theory consists of a mixture of classical electromagnetism, statistical physics, and quantum mechanics. These are, of course, well-established branches of physics. Nevertheless, the manner in which they are applied and the particular combinations found to be appropriate in modern optics are of more recent origin. The purpose of this book is to provide an introduction to some of the new concepts and formulations with emphasis on the quantum and statistical aspects.

The first chapter introduces the nomenclature, definitions, and certain basic formulae associated with the mathematics of stochastic processes. Included is a description of Brownian motion to illuminate the significance of the Langevin and Fokker–Planck approaches. Chapter II is devoted to the density matrix, evolution (time-development) operator, time-dependent perturbation theory, correlation functions, Green's functions, and an introduction to two-sided Feynman diagrams.

Not infrequently in the history of physics, a new field, when examined more closely, turns out to bear a close kinship to an older, well-established field. Such is the case with modern optics in relation to magnetic resonance, best exemplified by the close formal analogy between a spin-1/2 system in a time-varying magnetic field and a two-level atom (or molecule) in a radiation field. Indeed, evidence of this analogy is found in some of the optical terminology as well as in the methods employed in certain types of experiments. Portions of the theory of magnetic resonance are therefore included in Chapter III to serve as a background for understanding these fruitful connections.

Quantization of the radiation field and the harmonic oscillator formalism are treated in Chapter IV. Several types of states and their statistical properties are discussed, including photon number, coherent, and squeezed states. In Chapter V, the radiation field is coupled to an atomic system and the resulting processes—absorption, emission, and scattering—are formulated in both the semiclassical and quantized versions. Coherence functions in first and higher orders are defined as well as the connection with light beams exhibiting bunching, antibunching, and random statistics.

Chapter VI discusses damping and the master equation derived on the basis of the interactions of a dynamical system with a reservoir (heat bath). Both Langevin and density matrix methods are employed with applications to the damped oscillator, the optical Bloch equations, photon echoes, spontaneous emission from the standpoint of vacuum fluctuations, and several kinds of line shapes. Finally, in Chapter VII, a number of nonlinear and multiphoton processes—two-photon absorption and emission, stimulated Raman processes, three- and four-wave mixing, dressed states—are discussed. A prominent role is assigned to the susceptibility function and its representation in terms of two-sided Feynman diagrams.

The laws of nature are drawn from experience, but to express them one needs a special language: for, ordinary language is too poor and too vague to express relations so subtle, so rich, so precise. Here then is the first reason why a physicist cannot dispense with mathematics: it provides him with the one language he can speak ... Who has taught us the true analogies, the profound analogies which the eyes do not see, but which reason can divine? It is the mathematical mind, which scorns content and clings to pure form.

Henri Poincaré, *Analysis and Physics*

I Stochastic Processes

It is a matter of general experience that all physical measurements are subject to fluctuations. Random perturbations, which may originate in molecular collisions, spontaneous emission, lattice vibrations, and various other processes, manifest themselves in phenomena such as spectral line broadening and relaxation effects. We find, for example, that light beams may have different statistical properties depending on how they are generated and that such differences have an important bearing on optical nonlinear interactions. Considerations of this sort are relevant in both classical and quantum mechanical formulations; it will therefore be necessary to treat events that can be described only in probabilistic language. The present chapter is devoted to a summary of a number of properties of stochastic processes. For readers interested in more extensive treatments, numerous sources exist, some of which are listed in the general references at the end of this book.

1.1 Discrete and Continuous Random Variables

Consider a simple coin-tossing experiment. For each experiment there are two possible outcomes: heads(H) or tails(T). If we let ξ represent the outcome,

then

$$\xi = \begin{cases} \text{H}, \\ \text{T}. \end{cases} \tag{1.1}$$

In general it is preferable, for the purpose of further mathematical manipulation, to represent outcomes by numerical values. We therefore might construct a function $X(\xi)$ such that

$$X(\xi) = \begin{cases} 1 & \text{when} \quad \xi = \text{H}, \\ 0 & \text{when} \quad \xi = \text{T}. \end{cases} \tag{1.2}$$

The function $X(\xi)$ is known as a random or stochastic variable, defined as a variable whose value depends on the outcome of a random experiment. The probability of an outcome $p(\xi)$ must satisfy

$$0 \le p(\xi) \le 1, \qquad \sum_{\xi} p(\xi) = 1. \tag{1.3}$$

Beyond these statements, the assignment of numerical values to $p(\xi)$ lies outside the purview of the mathematical theory of probability which is primarily concerned with the manipulation of probabilities and ultimately rests on an axiomatic foundation. From a physical standpoint, one proceeds by assigning values to $p(\xi)$ based on a mixture of available knowledge (or lack of it) concerning the system, physical reasoning, and possibly other considerations. In the final analysis, it is only through experiment that one can determine whether the assignments are justifiable. If the coin is tossed N times and $n(\text{H})$ is the number of times the outcome is H, it is reasonable to suppose that the probability $p(\text{H})$ is given by the ratio $n(\text{H})/N$ when N is large. This is merely an assumption, however, since $n(\text{H})/N$ has no limit as N approaches infinity.

$X(\xi)$ in the coin-tossing experiment is a discrete, random variable with just two values: 1 and 0. In many experiments, the random variable X is continuous, that is, the outcome of an experiment may lie anywhere on a continuum. Furthermore, because much of our work will involve temporal changes, the random variable will be regarded as a function of time and will be written $X(t)$.

To illustrate these ideas, consider the case of a fluctuating voltage. One may obtain a continuous record of the voltage taken over a specified time interval. A record, $V(t)$, is a curve of voltage *vs.* time and is regarded as the outcome of a random experiment. If there are many replicas of the system that generate the voltage, each replica would produce its own record $V(t, \xi_i)$ where ξ_i is a label to keep track of individual records. $V(t, \xi)$, where $\xi = \xi_1, \xi_2, \ldots$, represents a family or ensemble of curves.

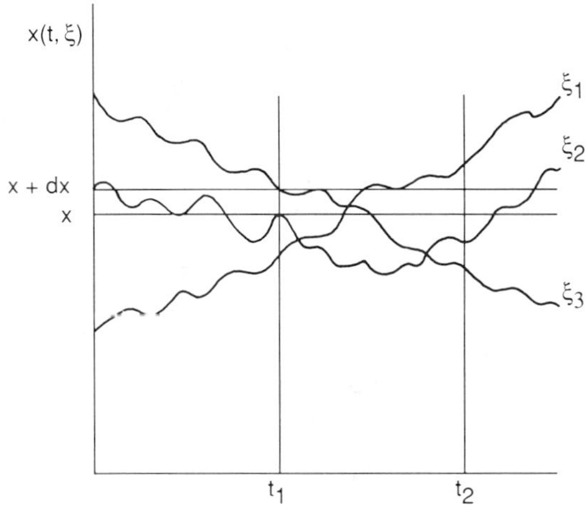

FIGURE 1.1 The set of curves $X(t, \xi)$ is a stochastic process; $X(t, \xi_i)$ with ξ_i constant is a function of time; $X(t_i, \xi)$ with t_i constant is a random variable. The quantity of interest is the probability that the random variable lies in the interval $(x, x + dx)$ at various times t_i.

Now, in place of a voltage, we may generalize to a physical quantity X (e.g., position, velocity, phase) whose value x is subject to fluctuations. The ensemble of records $X(t, \xi)$ (Fig 1.1) is known as a *stochastic process*. When ξ is kept constant, say $\xi = \xi_i$, the function $X(t, \xi_i)$ is simply a function of time and represents the outcome of an individual experiment (as in the case of a single record of the voltage fluctuations). On the other hand, when the value of t is fixed at $t = t_i$, the function $X(t_i, \xi)$ is the *random variable*. If both t and ξ are fixed at $t = t_i$ and $\xi = \xi_i$ then $X(t_i, \xi_i)$ is simply a number (x). As in the coin-tossing experiment, we shall be interested in the probability of a particular outcome, but since the variables are continuous the statements refer to a probability that the random variable $X(t_i, \xi)$ has a value that lies between x and $x + dx$ at the time t_i. It is customary to suppress the dependence on ξ unless it is explicitly required. The random variable $X(t_i, \xi)$ is then written $X(t_i)$ but since t may be varied, $t = t_1, t_2, \ldots$ (Fig 1.1), the random variable is usually written $X(t)$.

There is a fundamental difference between a random variable, $X(t)$, associated with a stochastic process and a deterministic function, $f(t)$. For the latter, the value of f is completely specified at every value of the time t but for the random variable there is no functional relation between the value of X and the value of t. In fact, for any given t the value of X can be anything within its range of variation and all we can say is that X has a certain probability of lying in a particular interval.

1.2 Probability Densities

For the continuous random variable $X(t)$ we define a function $W_1(x,t)$ known as the first-order *probability density* or *probability distribution* function such that $W_1(x,t)\,dx$ is the probability that the value of $X(t)$ lies in the interval $(x, x + dx)$ at the time t (Fig. 1.2):

$$W_1(x,t)\,dx = p\{x < X(t) \le x + dx\}. \tag{1.4}$$

This definition is illustrated in Fig. 1.1 for two values of the time, t_1 and t_2. Under special circumstances, the probabilities at t_1, t_2 and other values of the time may all be the same, but for a general definition such an assumption is not required. Since probabilities are inherently positive,

$$W_1(x,t) \ge 0. \tag{1.5}$$

We may also include discrete random processes in which the random variable $X(t)$ is defined only for integral values s. For this case,

$$\int_{-\infty}^{\infty} W_1(x,t)\,\delta(x - s)\,dx = p\{X(t) - s\} \equiv p(s). \tag{1.6}$$

When $W_1(x,t)$ is integrated with respect to x over the interval (a,b), we obtain the probability that the random variable $X(t)$ acquires values lying in (a,b). Thus,

$$\int_a^b W_1(x,t)\,dx = p\{a < X(t) \le b\}. \tag{1.7}$$

and if the limits are extended to $\pm\infty$ to encompass the full range of x, the probability achieves its maximum value:

$$\int_{-\infty}^{\infty} W_1(x,t)\,dx = p\{-\infty < X(t) < \infty\} = 1. \tag{1.8}$$

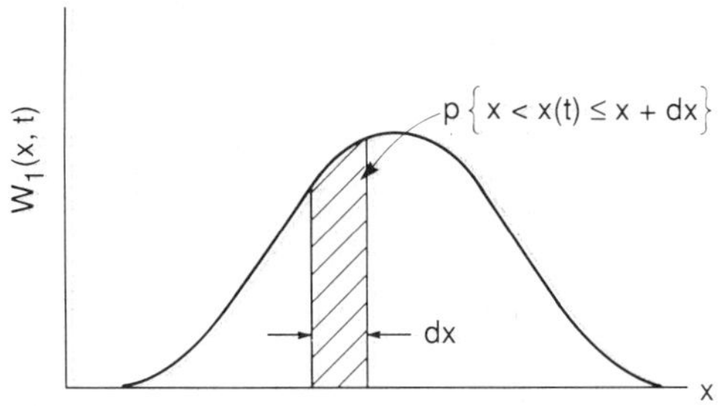

FIGURE 1.2 The curve $W_1(x,t)$ as a function of x is the first-order probability density. The area $W_1(x,t)\,dx$ is the probability that $X(t)$ lies in the interval $(x, x + dx)$ at the time t.

For two values of the time, t_1 and t_2, the second-order or *joint* probability density $W_2(x_1t_1;x_2t_2)$ is defined by the statement that $W_2(x_1t_1;x_2t_2)\,dx_1\,dx_2$ is the probability that, at $t = t_1$, the random variable $X(t_1)$ lies in the interval $(x_1, x_1 + dx_1)$ and that at $t = t_2$, $X(t_2)$ is located within $(x_2, x_2 + dx_2)$:

$$W_2(x_1t_1;\, x_2t_2)\,dx_1\,dx_2$$
$$= p\{x_1 < X(t_1) \le x_1 + dx_1;\, x_2 < X(t_2) \le x_2 + dx_2\}. \tag{1.9}$$

Integrating $W_2(x_1t_1;\, x_2t_2)$ over the entire range of x_2 gives the probability of finding $X(t_1)$ in the interval $(x_1, x_1 + dx_1)$. Thus,

$$\int_{-\infty}^{\infty} W_2(x_1t_1;\, x_2t_2)\,dx_2 = W_1(x_1, t_1). \tag{1.10a}$$

Similarly,

$$\int_{-\infty}^{\infty} W_2(x_1t_1;\, x_2t_2)\,dx_1 = W_1(x_2, t_2). \tag{1.10b}$$

Also, as in Eq. (1.8)

$$\int_{-\infty}^{\infty} W_2(x_1t_1;\, x_2t_2)\,dx_1\,dx_2 = 1. \tag{1.11}$$

Evidently, the definitions may be extended to an nth order probability density $W_n(x_1t_1;\, x_2t_2;\ldots;\, x_nt_n)$ where

$$W_n(x_1t_1;\ldots;\, x_nt_n)\,dx_1\cdots dx_n$$
$$= p\{x_1 < X(t_1) \le x_1 + dx_1;\ldots;\, x_n < X(t_n) \le x_n + dx_n\}. \tag{1.12}$$

The quantity $W_n(x_1t_1;\ldots;\, x_nt_n)\,dx_1\cdots dx_n$ is analogously interpreted as the probability that at $t = t_1, \ldots t_n$ the random variables $X(t_1), \ldots, X(t_n)$ are found in the intervals $(x_1, x_1 + dx_1), \ldots, (x_n, x_n + dx_n)$, respectively. The probability densities satisfy the following conditions:

1. $W_n(x_1t_1;\ldots;\, x_nt_n)$ is symmetric with respect to an interchange of a pair (x_it_i) with a pair (x_kt_k).
2. $W_n(x_1t_1;\ldots;\, x_nt_n) \ge 0.$ (1.13)
3. $\displaystyle\int_{-\infty}^{\infty} W_n(x_1t_1;\ldots;\, x_nt_n)\,dx_n = W_{n-1}(x_1t_1;\ldots;\, x_{n-1}t_{n-1}).$ (1.14)
4. $\displaystyle\int_{-\infty}^{\infty} W_n(x_1t_1;\ldots;\, x_nt_n)\,dx_1\cdots dx_n = 1.$ (1.15)

Assuming $X(t_1) = x_1$ (that is, $X(t_1)$ lies in the interval $(x_1, x_1 + dx_1)$), the probability of finding $X(t_2)$ in the interval $(x_2, x_2 + dx_2)$ at the time t_2

is defined as the *conditional* or *transition* probability and is written $P_2(x_1t_1 \,|\, x_2t_2)\,dx_2$. In more physical language, $P_2(x_1t_1 \,|\, x_2t_2)\,dx_2$ may be regarded as the probability of a transition from x_1 to x_2 in a time $t_2 - t_1$. Therefore, if $P_2(x_1t_1 \,|\, x_2t_2)\,dx_2$ is multiplied by $W_1(x_1t_1)\,dx_1$, the latter being the probability of finding the random variable $X(t_1)$ in the interval $(x_1, x_1 + dx_1)$, we will obtain the joint probability $W_2(x_1t_1; x_2t_2)\,dx_1\,dx_2$ of finding $X(t_1)$ in $(x_1, x_1 + dx_1)$ and $X(t_2)$ in $(x_2, x_2 + dx_2)$. One may therefore define the conditional probability density by the relation

$$P_2(x_1t_1 \,|\, x_2t_2) = \frac{W_2(x_1t_1; x_2t_2)}{W_1(x_1t_1)} \geq 0. \tag{1.16}$$

In the general case, P_2 may depend on t_1 and t_2 separately; in special cases, P_2 depends only on the difference $t_2 - t_1$. From Eq. (1.16) and Eq. (1.10a)

$$\int_{-\infty}^{\infty} P_2(x_1t_1 \,|\, x_2t_2)\,dx_2 = \frac{\int_{-\infty}^{\infty} W_2(x_1t_1; x_2t_2)\,dx_2}{W_1(x_1t_1)} = 1, \tag{1.17}$$

and

$$\int_{-\infty}^{\infty} W_1(x_1t_1) P_2(x_1t_1 \,|\, x_2t_2)\,dx_1 = \int_{-\infty}^{\infty} W_2(x_1t_1; x_2t_2)\,dx_1 = W_1(x_2t_2). \tag{1.18}$$

By extension to nth order, the probability of finding $X(t_k), \ldots, X(t_n)$ in the respective intervals $(x_k, x_k + dx_k), \ldots, (x_n, x_n + dx_n)$, given that $X(t_1), \ldots, X(t_{k-1})$ have preassigned values $x_1, \ldots, x_{k-1}$, is determined by the conditional probability density

$$P_n(x_1t_1; \ldots; x_{k-1}t_{k-1} \,|\, x_kt_k; \ldots; x_nt_n) = \frac{W_n(x_1t_1; \ldots; x_nt_n)}{W_{k-1}(x_1t_1; \ldots; x_{k-1}t_{k-1})} \geq 0, \tag{1.19}$$

with

$$t_1 < t_2 < t_3 < \cdots < t_n \tag{1.20}$$

and

$$\int_{-\infty}^{\infty} P_n(x_1t_1; \ldots; x_{k-1}t_{k-1} \,|\, x_kt_k; \ldots; x_nt_n)\,dx_k \cdots dx_n = 1. \tag{1.21}$$

In many physical situations, the statistical properties of the fluctuations are invariant under a displacement in time, i.e., the stochastic properties of the system are independent of the origin of time. Such a random process is said to be *stationary*. If t_0 is an arbitrary time, a stationary process is characterized by

$$W_n(x_1, t_1 + t_0; \ldots; x_n, t_n + t_0) = W_n(x_1, t_1; \ldots; x_n, t_n). \qquad (1.22)$$

Thus,

$$W_1(x, t + t_0) - W_1(x, t), \qquad (1.23)$$

and if one chooses $t_0 = -t$, then

$$W_1(x, t + t_0) = W_1(x, 0) \equiv W_1(x), \qquad (1.24)$$

which shows that the first-order stationary probability density is independent of time. In second order,

$$W_2(x_1, t_1 + t_0; x_2, t_2 + t_0) = W_2(x_1, t_1; x_2, t_2). \qquad (1.25)$$

Again we may choose $t_0 = -t_1$, in which case

$$W_2(x_1, t_1 + t_0; x_2, t_2 + t_0) = W_2(x_1, 0; x_2, t_2 - t_1) \equiv W_2(x_1; x_2, \tau) \qquad (1.26)$$

where

$$\tau = t_2 - t_1. \qquad (1.27)$$

The second-order stationary probability density therefore depends only on the *difference* in time. Similar considerations apply to stationary conditional probability densities as, for example,

$$P_2(x_1, t_1 \mid x_2, t_2) = P_2(x_1 \mid x_2, \tau). \qquad (1.28)$$

One should note, as these examples indicate, that a process that is stationary in the statistical sense need not be one whose parameters are independent of time.

If $W_n(x_1 t_1; \ldots; x_n t_n)$ can be expressed in the form

$$W_n(x_1 t_1; \ldots; x_n t_n) = W_1(x_1 t_1) \cdots W_1(x_n t_n), \qquad (1.29)$$

the random variables $X(t_1), \ldots, X(t_n)$ are said to be *statistically independent*. $X(t_i)$ is then completely independent of $X(t_j)$ and there is no relation between the values of X at two different times. Such a process is also said to be *purely* random and is characterized completely by the first-order probability densities.

1.3 Statistical Averages and Ergodicity

Statistical or ensemble averages of random variables and several closely related quantities are the following:

1. Mean

$$\langle X(t)\rangle = \eta(t) = \int_{-\infty}^{\infty} x W_1(x,t)\,dx. \tag{1.30}$$

2. Variance (also known as the mean-square deviation or dispersion)

$$\begin{aligned}\sigma^2(t) &\equiv \mathrm{var}\{X(t)\} \equiv (\Delta X)^2 \\ &= \langle [X(t) - \langle X(t)\rangle]^2\rangle = \langle X^2(t)\rangle - \langle X(t)\rangle^2\end{aligned} \tag{1.31}$$

in which

$$\langle X^2(t)\rangle = \int_{-\infty}^{\infty} x^2 W_1(x,t)\,dx. \tag{1.32}$$

The variance is a measure of the degree to which the value of $X(t)$ deviates from the mean value $\langle X(t)\rangle$. The square root of the variance $\sigma(t)$ is the *standard deviation.* Higher moments of the probability distribution are defined by

$$M_k \equiv \langle X^k(t)\rangle = \int_{-\infty}^{\infty} x^k W_1(x,t)\,dx. \tag{1.33}$$

3. Correlation function

$$\begin{aligned}G(t_1,t_2) &\equiv \langle X(t_1)X(t_2)\rangle \\ &= \int_{-\infty}^{\infty} x_1 x_2 W_2(x_1 t_1; x_2 t_2)\,dx_1\,dx_2.\end{aligned} \tag{1.34}$$

The correlation function provides a measure of the influence exerted by a value of X at time t_1 on the value of X at t_2.

4. Covariance

$$\begin{aligned}\sigma_{12}(t_1,t_2) &= \langle [X(t_1) - \langle X(t_1)\rangle][X(t_2) - \langle X(t_2)\rangle]\rangle \\ &= G(t_1,t_2) - \langle X(t_1)\rangle\langle X(t_2)\rangle.\end{aligned} \tag{1.35}$$

5. Correlation coefficient

$$\rho_{12}(t_1,t_2) = \frac{\sigma_{12}(t_1,t_2)}{\sqrt{\sigma^2(t_1)\sigma^2(t_2)}}. \tag{1.36}$$

It may be shown that

$$0 \le |\rho_{12}(t_1, t_2)| \le 1. \tag{1.37}$$

When $\rho_{12}(t_1, t_2) = 1$, the random variables $X(t_1)$ and $X(t_2)$ are said to be *correlated* and when $\rho_{12}(t_1, t_2) = -1$, $X(t_1)$ and $X(t_2)$ are *anticorrelated.* When $\rho_{12}(t_1, t_2) = 0$, the fluctuations in the two random variables are *uncorrelated.*

We saw that for a stationary process the first-order probability density is independent of time. Therefore, from Eqs. (1.30) and (1.32),

$$\langle X(t)\rangle = \langle X(0)\rangle \equiv \langle X\rangle = \text{a constant}, \tag{1.38}$$

$$\langle X^2(t)\rangle = \langle X^2(0)\rangle \equiv \langle X^2\rangle = \text{a constant}, \tag{1.39}$$

independent of time. The second-order probability density depends only on the time difference; hence, setting $t_2 - t_1 = \tau$,

$$\langle X(t_1)X(t_2)\rangle = \langle X(t_1)X(t_1 + \tau)\rangle = G(\tau) \tag{1.40}$$

which indicates that the correlation function, too, depends only on the time difference. The invariance of $G(\tau)$ under time translation permits one to write

$$G(\tau) = \langle X(0)X(\tau)\rangle. \tag{1.41}$$

If $X(t_1)$ and $X(t_2)$ are statistically independent,

$$G(t_1, t_2) = \langle X(t_1)X(t_2)\rangle = \langle X(t_1)\rangle\langle X(t_2)\rangle. \tag{1.42}$$

In that case, the covariance $\sigma_{12}(t_1, t_2)$ and the correlation coefficient $\rho_{12}(t_1, t_2)$ both vanish, indicating that $X(t_1)$ and $X(t_2)$ are uncorrelated. Two uncorrelated random variables are not necessarily statistically independent, however.

In addition to ensemble averages, one also may compute time averages. When the two are equal the stochastic process is said to be *ergodic*. In that case, the mean of the random variable $X(t)$ and its correlation function satisfy

$$\langle X\rangle = \lim_{T\to\infty} \frac{1}{2T}\int_{-T}^{T} X(t)\,dt = \int_{-\infty}^{\infty} xW_1(x)\,dx, \tag{1.43}$$

$$\begin{aligned} G(\tau) = \langle X(t)X(t+\tau)\rangle &= \lim_{T\to\infty} \frac{1}{2T}\int_{-T}^{T} X(t)X(t+\tau)\,dt \\ &= \int_{-\infty}^{\infty} x_1x_2W_2(x_1; x_2, \tau)\,dx_1\,dx_2. \end{aligned} \tag{1.44}$$

Since the time has been averaged out, $\langle X\rangle$ and $G(\tau)$ are independent of time; hence, the ergodic process is stationary, but not all stationary processes are ergodic.

For counting experiments in which the counts are statistically independent as in the case of radioactive disintegrations, for example, the resulting probability distribution is the *Poisson distribution*

$$p(k; \lambda t) = e^{-\lambda t}\frac{(\lambda t)^k}{k!}. \tag{1.45}$$

Here, $p(k; \lambda t)$ is the probability of recording k counts in a counting interval of t seconds when the average counting rate is λ particles per second. More generally, the Poisson distribution is written (Fig. 1.3)

$$p(s) \equiv p\{X = s\} = \frac{e^{-m}m^s}{s!}, \qquad 0 < m < \infty, \qquad s = 0, 1, 2, \ldots. \tag{1.46}$$

It is evident from the power series expansion of e^{-m} that

$$\sum_{s=0}^{\infty} p(s) = 1. \tag{1.47}$$

Also,

$$\langle s \rangle = e^{-m}\sum_{s=0}^{\infty}\frac{m^s}{s!}s = e^{-m}\sum_{s=1}^{\infty}\frac{m^s}{(s-1)!} = me^{-m}\sum_{s=1}^{\infty}\frac{m^{s-1}}{(s-1)!} = m, \tag{1.48a}$$

$$\langle s^2 \rangle = e^{-m}\sum_{s=0}^{\infty}\frac{m^s}{s!}s^2 = e^{-m}\left[m^2\sum_{s=2}^{\infty}\frac{m^{s-2}}{(s-2)!} + m\sum_{s=1}^{\infty}\frac{m^{s-1}}{(s-1)!}\right] = m(m+1). \tag{1.48b}$$

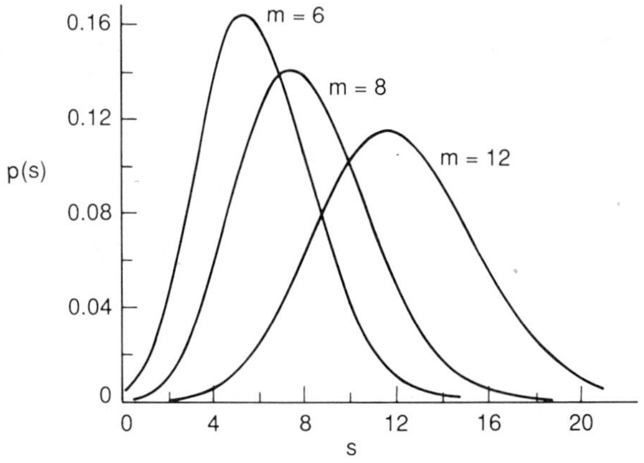

FIGURE 1.3 The Poisson distribution $p(s) = e^{-m}m^s/s!$ for $m = 6, 8$, and 12. The mean η and the variance σ^2 are both equal to m.

Thus,

$$\sigma^2 = \langle s^2 \rangle - \langle s \rangle^2 = m, \tag{1.49}$$

that is, for the Poisson distribution (Eq. (1.46)), the variance σ^2 and mean η are both equal to m.

If m is a large number and $(s - m)/m \ll 1$, the application of Stirling's asymptotic formula,

$$x! \simeq x^x e^{-x}(2\pi x)^{1/2} \tag{1.50}$$

to the Poisson distribution leads to the (normalized) *Gaussian* distribution (Fig. 1.4)

$$p(s) = \frac{1}{(2\pi m)^{1/2}} \exp\left\{-\frac{1}{2}\frac{(s-m)^2}{m}\right\}. \tag{1.51}$$

For a continuous random variable, the first-order Gaussian probability density is

$$W_1(x, t) = \frac{1}{\sigma(t)\sqrt{2\pi}} \exp\left\{-\frac{1}{2}\left[\frac{x - \eta(t)}{\sigma}\right]^2\right\}, \tag{1.52}$$

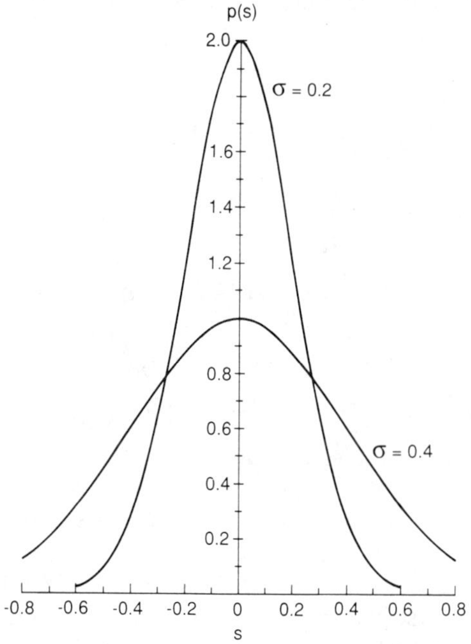

FIGURE 1.4 The Gaussian distribution $p(s) = (1/2\pi\sqrt{\sigma})\exp[-\frac{1}{2}(\frac{s}{\sigma})^2]$ for $\sigma = 0.2$ and 0.4.

where $\eta(t)$ and $\sigma^2(t)$ are the mean and variance, respectively. The second order density is

$$W_2(x_1t_1; x_2t_2) = \frac{1}{2\pi\sigma_1\sigma_2\sqrt{1-\rho_{12}^2}} \exp\left\{-\frac{1}{2(1-\rho_{12}^2)} \cdot \left[\left(\frac{x_1-\eta_1}{\sigma_1}\right)^2 - 2\rho_{12}\frac{(x_1-\eta_1)(x_2-\eta_2)}{\sigma_1\sigma_2} + \left(\frac{x_2-\eta_2}{\sigma_2}\right)^2\right]\right\}, \tag{1.53}$$

in which

$$\sigma_1 = \sigma(t_1), \qquad \sigma_2 = \sigma(t_2); \qquad \eta_1 = \eta(t_1), \qquad \eta_2(t_2); \qquad \rho_{12} = \rho_{12}(t_1, t_2). \tag{1.54}$$

When the correlation coefficient $\rho_{12} = 0$,

$$W_2(x_1t_1; x_2t_2) = \frac{1}{2\pi\sigma_1\sigma_2} \exp\left\{-\frac{1}{2}\left[\left(\frac{x_1-\eta_1}{\sigma_1}\right)^2 + \left(\frac{x_2-\eta_2}{\sigma_2}\right)^2\right]\right\}$$
$$= W_1(x_1t_1)W_1(x_2t_2) \tag{1.55}$$

as required for statistical independence. This is an important property of Gaussian random variables, namely, lack of correlation does imply statistical independence which, as indicated previously, is not true in general. If $\sigma_1 = \sigma_2 = \sigma$ and $\eta_1 = \eta_2 = 0$,

$$W_2(x_1t_1; x_2t_2) = \frac{1}{2\pi\sigma^2\sqrt{1-\rho_{12}^2}} \exp\left\{-\frac{x_1^2 + x_2^2 - 2\rho_{12}x_1x_2}{2(1-\rho_{12}^2)\sigma^2}\right\}. \tag{1.56}$$

1.4 Markov Processes, Chapman-Kolmogorov Equation

A random process is known as a *Markov* process if, and only if

$$P_n(x_1t_1; \ldots; x_{n-1}t_{n-1} \,|\, x_nt_n) = P_2(x_{n-1}t_{n-1} \,|\, x_nt_n); \tag{1.57}$$

that is, the probability that $X(t_n)$ is found in the interval $(x_n, x_n + dx_n)$ at $t = t_n$ depends only on the value of the random variable at $t = t_{n-1}$. Other information that we might have concerning the values of the random variables at earlier times $t = t_1, \ldots, t_{n-2}$ is irrelevant. This means that the future development of a Markovian system is determined entirely by the most recent state whereas the past history of the system has no influence on its future behavior. It is often said that a Markovian system is one that has suffered a complete loss of memory of its past.

A completely random process is fully characterized by the first-order probability densities, as has been indicated by Eq. (1.29). A Markov process is, in a sense, less random, to the extent that all the probability densities W_n may be expressed in terms of W_1 and W_2. As an illustration, we have, according to Eq. (1.19)

$$W_3(x_1t_1; x_2t_2; x_3t_3) = W_2(x_1t_1; x_2t_2)P_3(x_1t_1; x_2t_2 \,|\, x_3t_3). \tag{1.58}$$

Using the Markov condition (Eq. (1.57)),

$$P_3(x_1t_1; x_2t_2 \,|\, x_3t_3) = P_2(x_2t_2 \,|\, x_3t_3); \tag{1.59}$$

but

$$P_2(x_2t_2 \,|\, x_3t_3) = \frac{W_2(x_2t_2; x_3t_3)}{W_1(x_2, t_2)}. \tag{1.60}$$

Hence, Eq. (1.58) may be written in the form

$$W_3(x_1t_1; x_2t_2; x_3t_3) = \frac{W_2(x_1t_1; x_2t_2)W_2(x_2t_2; x_3t_3)}{W_1(x_2, t_2)} \tag{1.61}$$

We now derive an important relation which may also be regarded as a defining property of a Markov process. Referring to Eqs. (1.14) and (1.16),

$$\begin{aligned}\int_{-\infty}^{\infty} W_3(x_1t_1; x_2t_2; x_3t_3)\,dx_2 &= W_2(x_1t_1; x_3t_3)\\ &= W_1(x_1, t_1)P_2(x_1t_1 \,|\, x_3t_3).\end{aligned} \tag{1.62}$$

But from Eq. (1.19) and the Markov condition,

$$\begin{aligned}&\int_{-\infty}^{\infty} W_3(x_1t_1; x_2t_2; x_3t_3)\,dx_2\\ &\quad= \int_{-\infty}^{\infty} W_2(x_1t_1; x_2t_2)P_3(x_1t_1; x_2t_2 \,|\, x_3t_3)\,dx_2,\\ &\quad= W_1(x_1, t_1)\int_{-\infty}^{\infty} P_2(x_1t_1 \,|\, x_2t_2)P_3(x_1t_1; x_2t_2 \,|\, x_3t_3)\,dx_2,\\ &\quad= W_1(x_1, t_1)\int_{-\infty}^{\infty} P_2(x_1t_1 \,|\, x_2t_2)P_2(x_2t_2 \,|\, x_3t_3)\,dx_2.\end{aligned} \tag{1.63}$$

Comparing Eqs. (1.62) and (1.63), we obtain the *Smoluchowski* or *Chapman-Kolmogorov* equation for a Markov process:

$$P_2(x_1t_1 \,|\, x_3t_3) = \int_{-\infty}^{\infty} P_2(x_1t_1 \,|\, x_2t_2)P_2(x_2t_2 \,|\, x_3t_3)\,dx_2. \tag{1.64}$$

In other words, the conditional or transition probability density $P_2(x_1t_1 \mid x_3t_3)$ for $X(t)$ to be found in the interval $(x_3, x_3 + dx_3)$ at $t = t_3$ given that $X(t_1) = x_1$ may be computed by integrating the product of $P_2(x_1t_1 \mid x_2t_2)$ and $P_2(x_2t_2 \mid x_3t_3)$ over the range of x_2. In more physical language, Eq. (1.64) states that the evolution of a system during the time $t_3 - t_1$ can be described in terms of the evolution during the times $t_3 - t_2$ and $t_2 - t_1$, bearing in mind that $t_1 < t_2 < t_3$.

If the Markov process is stationary, P_2 will depend only on the difference between the two values of the time as in (Eq. (1.28)). Letting

$$\tau = t_3 - t_1, \qquad t = t_2 - t_1, \tag{1.65}$$

the Smoluchowski equation for a stationary Markov process becomes

$$P_2(x_1 \mid x_3, \tau) = \int_{-\infty}^{\infty} P_2(x_1 \mid x_2, t) P_2(x_2 \mid x_3, \tau - t)\, dx_2, \tag{1.66}$$

or, in a somewhat more general notation, with

$$x_1 = x, \qquad x_3 = y, \qquad x_2 = z, \qquad \Delta t = \tau - t, \tag{1.67}$$

we write

$$P(x \mid y, t + \Delta t) = \int_{-\infty}^{\infty} P(x \mid z, t) P(z \mid y, \Delta t)\, dz, \tag{1.68}$$

in which the now superfluous subscript on P_2 has been omitted.

1.5 Fokker-Planck Equation

The Fokker-Planck (F-P) equation is a differential equation that governs the time development of the conditional probability density for a stationary Markov process. In the notation of the previous section, we write $P(x \mid y, t)$ for the conditional probability density. Thus, for example, x may represent the initial, known position of a Brownian particle at $t = 0$ while y is the position at the time t. In that case $P(x \mid y, t)$ is interpreted as the probability for the Brownian particle to be found in the space interval $(y, y + dy)$ at the time t, given the position x at $t = 0$. We now give an elementary derivation of the F-P equation.

Consider a time interval Δt of sufficiently short duration that one may write

$$\frac{\partial P(x \mid y, t)}{\partial t} \Delta t = P(x \mid y, t + \Delta t) - P(x \mid y, t), \tag{1.69}$$

or in terms of the Smoluchowski equation (Eq. (1.68)) for a stationary Markov process,

$$\frac{\partial P(x\,|\,y,t)}{\partial t}\Delta t = -P(x\,|\,y,t) + \int P(x\,|\,z,t)P(z\,|\,y,\Delta t)\,dz. \tag{1.70}$$

It is further assumed that during Δt the transition probability density $P(z\,|\,y,\Delta t)$ has appreciable values only when z and y differ by a small quantity, ε. In terms of the motion of the Brownian particle, this assumption implies that during Δt only small changes in the position of the particle have a significant probability whereas large changes in position have a negligible probability. In that case,

$$P(z\,|\,y,\Delta t) = P(y-\varepsilon\,|\,y,\Delta t) = P(y\,|\,y+\varepsilon,\Delta t), \tag{1.71}$$

and the integrand in Eq. (1.70) becomes

$$P(x\,|\,z,t)P(z\,|\,y,\Delta t) = P(x\,|\,y-\varepsilon,t)P(y\,|\,y+\varepsilon,\Delta t). \tag{1.72}$$

Expanding the right-hand side in a Taylor series about $P(x\,|\,y,t)P(y\,|\,y+\varepsilon,\Delta t)$ one obtains, to second order,

$$\begin{aligned} P(x\,|\,y-\varepsilon,t)P(y\,|\,y+\varepsilon,\Delta t) &= P(x\,|\,y,t)P(y\,|\,y+\varepsilon,\Delta t) \\ &\quad - \varepsilon\frac{\partial}{\partial y}[P(x\,|\,y,t)P(y\,|\,y+\varepsilon,\Delta t)] \\ &\quad + \frac{\varepsilon^2}{2}\frac{\partial^2}{\partial y^2}[P(x\,|\,y,t)P(y\,|\,y+\varepsilon,\Delta t)]. \end{aligned} \tag{1.73}$$

The integration with respect to z in Eq. (1.70) is now replaced by an integration with respect to ε. In view of the normalization condition (Eq. (1.21)),

$$\begin{aligned} \int P(x\,|\,y,t)P(y\,|\,y+\varepsilon,\Delta t)\,d\varepsilon &= P(x\,|\,y,t)\int P(y\,|\,y+\varepsilon,\Delta t)\,d\varepsilon \\ &= P(x\,|\,y,t). \end{aligned} \tag{1.74}$$

Writing

$$M_1 = \frac{1}{\Delta t}\int \varepsilon P(y\,|\,y+\varepsilon,\Delta t)\,d\varepsilon, \tag{1.75a}$$

$$M_2 = \frac{1}{\Delta t}\int \varepsilon^2 P(y\,|\,y+\varepsilon,\Delta t)\,d\varepsilon, \tag{1.75b}$$

and integrating Eq. (1.73)

$$\int P(x\,|\,y-\varepsilon, t)P(y\,|\,y+\varepsilon, \Delta t)\,d\varepsilon$$

$$= P(x\,|\,y,t) - \frac{\partial}{\partial y}[M_1 P(x\,|\,y,t)]\,\Delta t + \frac{1}{2}\frac{\partial^2}{\partial y^2}[M_2 P(x\,|\,y,t)]\,\Delta t. \quad (1.76)$$

Upon substituting Eqs. (1.72) and (1.76) into Eq. (1.70) we obtain the *Fokker-Planck* equation

$$\frac{\partial P(x\,|\,y,t)}{\partial t} = -\frac{\partial}{\partial y}[M_1 P(x\,|\,y,t)] + \frac{1}{2}\frac{\partial^2}{\partial y^2}[M_2 P(x\,|\,y,t)]. \quad (1.77)$$

If $M_1 = 0$ and M_2 is independent of y, the F-P equation reduces to the diffusion equation

$$\frac{\partial P}{\partial t} = \frac{1}{2} M_2 \frac{\partial^2 P}{\partial y^2}. \quad (1.78)$$

Further insight into the Fokker-Planck equation will be gained when we return to it in connection with Brownian motion (Section 1.9) and, later, in connection with the damping of a radiation mode coupled to an atomic reservoir (Section 6.8).

1.6 Correlation Functions, Wiener-Khinchine Theorem

Correlation functions were defined in Section 1.3. In view of their central importance in understanding the response of physical systems to external stimuli, we now examine some of their properties [1, 2], but before doing so, a change of notation is in order. Previously $X(t)$ represented a random variable and x the value of the random variable at time t. But in the physical literature it is common practice not to distinguish between a random variable and its value. We therefore shall let $y(t)$ represent a real, random variable as well as its value. The correlation function, then, is written

$$G(t, t') = \langle y(t)y(t')\rangle. \quad (1.79)$$

If $y(t)$ and $y(t')$ commute, $G(t, t') = G(t', t)$, and when the stochastic process is stationary,

$$G(t, t') = G(\tau) = \langle y(t)y(t+\tau)$$
$$= \langle y(0)y(\tau) = \langle y(-\tau)y(0), \quad (1.80)$$

in which $\tau = t' - t$. Thus,

$$G(\tau) = G(-\tau). \tag{1.81}$$

Since $G(0) = \langle y^2(t) \rangle = \langle y^2(0) \rangle$ and $\langle y^2(t) \rangle$ is positive definite we have

$$G(0) \geq 0. \tag{1.82}$$

Noting that

$$\langle [y(t + \tau) \pm y(t)]^2 \rangle = \langle y^2(t + \tau) \rangle + \langle y^2(t) \rangle \pm 2\langle y(t)y(t + \tau) \rangle \geq 0, \tag{1.83}$$

and that for a stationary process

$$\langle y^2(t + \tau) \rangle = \langle y^2(t) \rangle = G(0), \tag{1.84}$$

$$\langle y(t)y(t + \tau) \rangle = G(\tau), \tag{1.85}$$

the inequality becomes

$$2[G(0) \pm G(\tau)] \geq 0, \tag{1.86}$$

or

$$G(0) \geq |G(\tau)|. \tag{1.87}$$

Thus it is seen that the correlation function $G(\tau)$ is an even (or symmetric) function of τ with its maximum value at $\tau = 0$.

Quite often, in cases of physical interest, the probability that y assumes a certain value at the time $t + \tau$ becomes less and less dependent on the value of y at the time t, as the delay τ increases. This means that $G(\tau)$, the correlation between the value of y at t and the value of y at $t + \tau$, ultimately vanishes as $\tau \to \infty$. When $G(\tau)$ decreases exponentially from its maximum value at $\tau = 0$, the decay constant τ_c is known as the correlation time. More generally, the notion of a correlation time may be extended to situations in which $G(\tau)$ does not necessarily decay in an exponential fashion, in which case τ_c serves as a measure of the effective time over which the system retains a memory of its past. Sketches of several correlation functions are shown in Fig. 1.5.

Complex correlation functions are of the form

$$G(t, t') = \langle y^*(t)y(t') \rangle \tag{1.88}$$

which becomes

$$G(\tau) = \langle y^*(t)y(t + \tau) \rangle = \langle y^*(0)y(\tau) \rangle \tag{1.89}$$

for a stationary process. Since

$$G^*(\tau) = \langle y(0)y^*(\tau) \rangle, \tag{1.90a}$$

$$G^*(-\tau) = \langle y(0)y^*(-\tau) \rangle = \langle y(\tau)y^*(0) \rangle, \tag{1.90b}$$

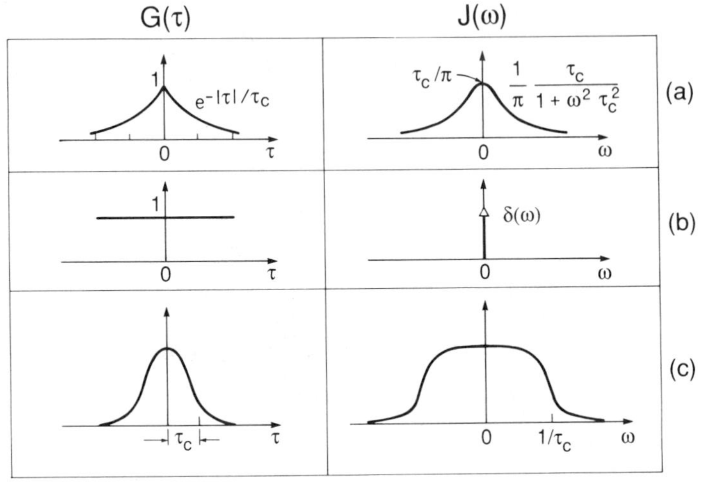

FIGURE 1.5 Examples of correlation functions and their power spectra. (a) Exponentially decaying correlation function and the associated Lorentzian power spectrum. (b) Constant correlation function and its δ-function power spectrum. (c) A general bell-shaped correlation function corresponding to irregular noise yields a power spectrum that is approximately constant up to a frequency $\omega \simeq 1/\tau_c$ where τ_c is the correlation time.

we have

$$G^*(-\tau) = G(\tau), \tag{1.91}$$

or

$$\begin{aligned} \operatorname{Re} G(\tau) &= \operatorname{Re} G(-\tau), \\ \operatorname{Im} G(\tau) &= -\operatorname{Im} G(-\tau). \end{aligned} \tag{1.92}$$

We shall now derive a theorem [3,4] that establishes the connection between the frequency spectrum of a random variable and its correlation function for a stationary ergodic process. Let

$$y_T(t) = \begin{cases} y(t) & -T \leq t \leq T \\ 0 & \text{otherwise} \end{cases} \tag{1.93}$$

and let

$$G_T(\tau) = \frac{1}{2T} \int_{-T}^{T} y_T(t) y(t+\tau)\, dt. \tag{1.94}$$

In view of the fact that $y_T(t)$ is confined to the interval $\pm T$ and is zero elsewhere, the limits of the integral may be extended to $\pm\infty$. We now take the

Fourier transform of $G_T(\tau)$:

$$\frac{1}{2\pi}\int_{-\infty}^{\infty} G_T(\tau)e^{i\omega t}\,d\tau = \frac{1}{2\pi}\frac{1}{2T}\int_{-\infty}^{\infty} d\tau e^{i\omega\tau}\int_{-\infty}^{\infty} y_T(t)y_T(t+\tau)\,dt$$
$$= \frac{1}{2\pi}\frac{1}{2T}\int_{-\infty}^{\infty} y_T(t)e^{-i\omega t}\,dt\int_{-\infty}^{\infty} y_T(t+\tau)e^{i\omega(t+\tau)}\,d\tau. \tag{1.95}$$

Replacing τ by $t' - t$,

$$\frac{1}{2\pi}\int_{-\infty}^{\infty} G_T(\tau)e^{i\omega\tau}\,d\tau = \frac{\pi}{T}\left\{\frac{1}{2\pi}\int_{-\infty}^{\infty} y_T(t)e^{-i\omega t}\,dt\,\frac{1}{2\pi}\int_{-\infty}^{\infty} y_T(t')e^{i\omega t'}\,dt'\right\}$$
$$= \frac{\pi}{T}Y(-\omega)Y(\omega), \tag{1.96}$$

in which $Y(\omega)$ is the Fourier transform of $y_T(t)$. Provided $y_T(t)$ is a real function,

$$Y(-\omega) = Y^*(\omega) \tag{1.97}$$

and

$$\frac{1}{2\pi}\int_{-\infty}^{\infty} G_T(\tau)e^{i\omega\tau}\,d\tau = \frac{\pi}{T}|Y(\omega)|^2. \tag{1.98}$$

The function $G_T(\tau)$, defined by Eq. (1.94), is the time average of the product $y_T(t)y_T(t+\tau)$ over the interval $2T$. We shall now let $T \to \infty$ and assume the stochastic process to be ergodic. Then,

$$\lim_{T\to\infty} G_T(\tau) = \lim_{T\to\infty}\frac{1}{2T}\int_{-T}^{T} y_T(t)y_T(t+\tau)\,dt$$
$$= \langle y(t)y(t+\tau)\rangle = \langle y(0)y(\tau)\rangle = G(\tau). \tag{1.99}$$

The *spectral density* or *power spectrum* is defined as the real function

$$J(\omega) = \lim_{T\to\infty}\frac{\pi}{T}|Y(\omega)|^2. \tag{1.100}$$

Combining Eqs. (1.98), (1.99) and (1.100),

$$J(\omega) = \frac{1}{2\pi}\int_{-\infty}^{\infty} G(\tau)e^{i\omega\tau}\,d\tau, \tag{1.101a}$$

and

$$G(\tau) = \int_{-\infty}^{\infty} J(\omega)e^{-i\omega\tau}\,d\omega. \tag{1.101b}$$

If $G(\tau)$ is real it is also an even function of τ; expressions (1.101) may then be replaced by

$$J(\omega) = \frac{1}{\pi}\int_0^\infty G(\tau)\cos\omega\tau\,d\tau; \qquad G(\tau) = 2\int_0^\infty J(\omega)\cos\omega\tau\,d\tau. \quad (1.102)$$

The relations constitute the *Wiener-Khinchine theorem* which states that the power spectrum and the correlation function for a real random function of the time associated with a stationary ergodic process are Fourier transforms of one another.

If $J(\omega)$ is integrated over all frequencies one obtains

$$J \equiv \int_{-\infty}^{\infty} J(\omega)\,d\omega = \frac{1}{2\pi}\int_{-\infty}^{\infty} G(\tau)\,d\tau \int_{-\infty}^{\infty} e^{i\omega\tau}\,d\omega$$

$$= \int_{-\infty}^{\infty} G(\tau)\,\delta(\tau)\,d\tau = G(0). \quad (1.103)$$

As an example of these relations let

$$G(\tau) = e^{-|\tau|/\tau_c}. \quad (1.104)$$

The power spectrum is

$$J(\omega) = \frac{1}{\pi}\int_0^\infty e^{-\tau/\tau_c}\cos\omega\tau\,d\tau = \frac{1}{\pi}\frac{\tau_c}{1+\omega^2{\tau_c}^2} \equiv L(\omega). \quad (1.105)$$

Since $L(\omega)$ is a normalized Lorentzian function, its integral over all frequencies is equal to 1, which is precisely the value of $G(0)$ as required by Eq. (1.104). It is seen, then, that the power spectrum of an exponentially decaying correlation function is a Lorentzian as illustrated in Fig. 1.5a. It may be shown [5] that a stationary Gaussian process with a correlation function of the type shown in Eq. (1.104) is Markovian or, alternatively, a stationary Gaussian Markov process has an exponentially decaying correlation function (Doob's Theorem).

As another example, let

$$G(\tau) = e^{-i\omega_0\tau - |\tau|/\tau_c}. \quad (1.106)$$

In this case $G(\tau)$ is complex; we shall need, therefore, the more general form of Eq. (1.101a)) to compute the power spectrum

$$J(\omega) = \frac{1}{2\pi}\int_{-\infty}^{\infty} G(\tau)e^{i\omega\tau}\,d\tau$$

$$= \frac{1}{2\pi}\int_{-\infty}^{0} G(\tau)e^{i\omega\tau}\,d\tau + \frac{1}{2\pi}\int_0^\infty G(\tau)e^{i\omega\tau}\,d\tau. \quad (1.107)$$

In the integral from $-\infty$ to 0, we replace τ by $-\tau$; then

$$\int_{-\infty}^{0} G(\tau)e^{i\omega\tau}\,d\tau = \int_{0}^{\infty} G(-\tau)e^{-i\omega\tau}\,d\tau = \int_{0}^{\infty} G^*(\tau)e^{-i\omega\tau}\,d\tau \tag{1.108}$$

and

$$J(\omega) = \frac{1}{\pi}\,\mathrm{Re}\int_{0}^{\infty} G(\tau)e^{i\omega\tau}\,d\tau = \frac{1}{\pi}\int_{0}^{\infty} e^{-\tau/\tau_c}\cos(\omega_0 - \omega)\tau\,d\tau$$
$$= \frac{1}{\pi}\,\frac{\tau_c}{1 + (\omega_0 - \omega)^2\tau_c^2}. \tag{1.109}$$

For the Gaussian power spectrum,

$$J(\omega) = \frac{1}{\delta\sqrt{2\pi}}\exp\left[-\frac{(\omega_0 - \omega)^2}{2\delta^2}\right]; \tag{1.110}$$

the correlation function is

$$G(\tau) = \int_{-\infty}^{\infty} J(\omega)e^{-i\omega\tau}\,d\omega = e^{-i\omega_0\tau}\exp\left[-\frac{\delta^2\tau^2}{2}\right]. \tag{1.111}$$

If $E(t)$ is a complex, time-varying electric field associated with a light field in vacuum, the intensity $I(t)$ is given by

$$I(t) = c\varepsilon_0 E^*(t)E(t) \tag{1.112}$$

and its long-time average is

$$\langle I(t)\rangle = \lim_{T\to\infty}\frac{1}{2T}\int_{-T}^{T} I(t)\,dt. \tag{1.113}$$

Assuming that the fluctuations of the field are stationary and ergodic, $\langle I(t)\rangle$ may also be expressed in terms of the correlation function $G(\tau) = \langle E^*(t)E(t+\tau)\rangle$:

$$\langle I(t)\rangle = c\varepsilon_0\langle E^*(t)E(t)\rangle = c\varepsilon_0 G(0). \tag{1.114}$$

In view of the Wiener-Khinchine theorem the power spectrum is given by

$$I(\omega) = \frac{1}{2\pi}\int_{-\infty}^{\infty} G(\tau)e^{i\omega\tau}\,d\tau \tag{1.115}$$

where $I(\omega)\,d\omega$ is the intensity within the frequency interval $(\omega, \omega + d\omega)$. Hence, $I(\omega)$ may be regarded as the line shape. When $I(\omega)$ is integrated over all frequencies, one obtains,

$$\int_{-\infty}^{\infty} I(\omega)\,d\omega = \frac{1}{2\pi}\int_{-\infty}^{\infty} G(\tau)\,d\tau \int_{-\infty}^{\infty} e^{i\omega\tau}\,d\omega$$

$$= \int_{-\infty}^{\infty} G(\tau)\,\delta(\tau)\,d\tau = G(0). \tag{1.116}$$

Clearly, this result is not unexpected since the intensity $\langle I(t)\rangle$ can be calculated either by averaging over time or by integrating over all Fourier components.

1.7 Random Walk

A good example of a stochastic process is the random walk. In its simplest one-dimensional version it consists of N random steps of equal length l, taken at equal time intervals T and with equal probabilities for steps in the positive and negative directions. It n_1 and n_2 are the number of steps in the positive and negative directions, respectively, the total number of steps is $N = n_1 + n_2$ and the total distanced traveled from the starting point in a time $t = NT$ is

$$L = l(n_1 - n_2) = l(2n_1 - N). \tag{1.117}$$

Since both n_1 and n_2 may have values lying between zero and N, the distance L lies between $-Nl$ and $+Nl$.

Each step can occur in one of two ways; hence, the number of ways in which N steps can occur is 2^N. The total distance L depends only on the number of positive and negative steps and not on the order of their occurrence. Therefore, the number of ways in which a given value of L (i.e., a given value of n_1 and n_2) can occur is given by the binomial coefficient

$$\binom{N}{n_1} = \frac{N!}{n_1!(N - n_1)!}, \tag{1.118}$$

and the probability for the occurrence of a particular value of L is

$$P_L(n_1) = \binom{N}{n_1}\frac{1}{2^N}. \tag{1.119}$$

The average values of L and L^2 are then given by

$$\langle L\rangle = \sum_{n_1} LP_L = \frac{1}{2^N}\sum_{n_1}(2n_1 - N)\binom{N}{n_1}, \tag{1.120}$$

$$\langle L^2\rangle = \sum_{n_1} L^2P_L = \frac{l^2}{2^N}\sum_{n_1}(2n_1 - N)^2\binom{N}{n_1}. \tag{1.121}$$

With the aid of the binomial theorem

$$(1 + x)^n = \sum_{k=0}^{n} \binom{n}{k} x^k, \tag{1.122}$$

one readily finds

$$\langle L \rangle = 0, \tag{1.123}$$

$$\langle L^2 \rangle = Nl^2 = \frac{l^2}{T} t. \tag{1.124}$$

That is, the average distance $\langle L \rangle$ covered by the random walk is zero, as is to be expected from the equal probabilities for positive and negative steps. $\langle L^2 \rangle$ is not zero, however, but is proportional to the first power of the time.

One also might be interested in the probability of the random walk arriving at a particular location after a specified time. Let the positions of the various steps be labeled $q = 0, \pm 1, \pm 2, \ldots$ and let the probability of reaching a particular position at the time t be $p(q, t)$. Transitions from adjacent positions $q \pm 1$ into q will increase $p(q, t)$ while transitions in the reverse direction will reduce $p(q, t)$. If the number of steps during the time t is large, the time for a single step may be regarded as infinitesimal. One may then define $p(q, q \pm 1)\,dt$ as the probability for a transition $q \pm 1 \to q$ in the time dt; similarly, $p(q \pm 1, q)\,dt$ is the corresponding quantity for a transition $q \to q \pm 1$. To obtain the net change in $p(q, t)$ we must multiply each transition probability into and out of q by the probability that the random walk is located at the initial position. As an example, $p(q - 1, t)p(q, q - 1)\,dt$ is the product of $p(q - 1, t)$, the probability that the random walk is at $q - 1$ at the time t, and $p(q, q - 1)\,dt$, the probability for a transition from $q - 1$ to q in the time dt. A product of this type is proportional to the rate at which $p(q, t)$ increases. The net change $dp(q, t)$ may now be written

$$\begin{aligned} dp(q, t) = {} & p(q, q - 1)p(q - 1, t)\,dt + p(q, q + 1)p(q + 1, t)\,dt \\ & - p(q + 1, q)p(q, t)\,dt - p(q - 1, q)p(q, t)\,dt. \end{aligned} \tag{1.125}$$

The first two terms increase $p(q, t)$ as a result of transitions $q \pm 1 \to q$; the third and fourth terms produce the opposite effect due to transitions $q \to q \pm 1$ (Fig. 1.6). An equation of this type is known as a *master* or *rate* equation.

The conditions

$$\begin{aligned} p(q, q + 1)p(q + 1, t) &= p(q + 1, q)p(q, t), \\ p(q, q - 1)p(q - 1, t) &= p(q - 1, q)p(q, t), \end{aligned} \tag{1.126}$$

are sufficient to guarantee equilibrium, by which we mean $dp(q, t)/dt = 0$. But more than that, they require the probability for a transition $q \to q + 1$

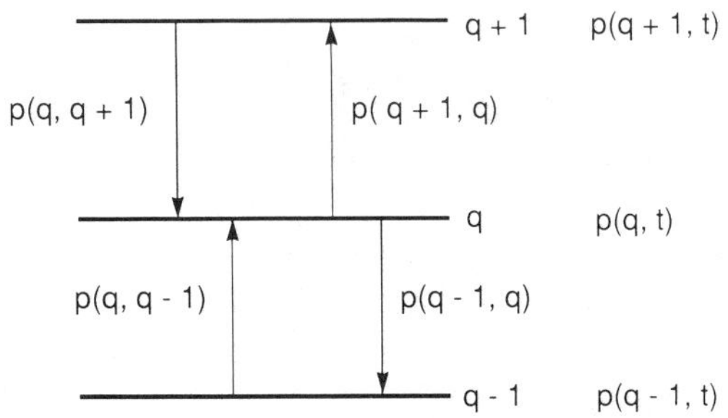

FIGURE 1.6 $P(q \pm 1, t)$ and $p(q, t)$ are the probabilities that the random walk is located at $q \pm 1$ and q, respectively, at time t. Transitions $q \pm 1 \rightarrow q$ with probabilities $p(q, q \pm 1)$ enhance $p(q, t)$; transitions $q \rightarrow q \pm 1$ with probabilities $p(q \pm 1, q)$ reduce $p(q, t)$. When Eqs. (1.126) are satisfied $p(q, t)$ remains constant, independent of time.

to be equal to the probability for the reverse transition $q + 1 \rightarrow q$, similarly, for transitions $q \rightarrow q - 1$ and $q - 1 \rightarrow q$. This is an illustration of the important principle of *detailed balance*. Equations (1.125) and (1.126) may be generalized to

$$\frac{dp(q,t)}{dt} = \sum_{q'} p(q,q')p(q',t) - p(q,t)\sum_{q'} p(q',q) \tag{1.127}$$

and

$$p(q',t)p(q,q') = p(q,t)p(q',q). \tag{1.128}$$

Returning to Eq. (1.125), let us regard q as a continuous variable and expand $p(q + 1, t)$ and $p(q - 1, t)$ to second order:

$$p(q \pm 1, t) = p(q,t) \pm \frac{\partial p(q,t)}{\partial q}\Delta q + \frac{1}{2}\frac{\partial^2 p(q,t)}{\partial q^2}(\Delta q)^2. \tag{1.129}$$

Assuming

$$\begin{aligned} p(q+1,q) &= p(q,q-1), \\ p(q-1,q) &= p(q,q+1), \end{aligned} \tag{1.130}$$

the differential equation (1.129) becomes the Fokker-Planck equation

$$\frac{\partial p(q,t)}{\partial t} = -M_1\frac{\partial p(q,t)}{\partial q} + \frac{1}{2}M_2\frac{\partial^2 p(q,t)}{\partial q^2}, \tag{1.131}$$

with

$$\begin{aligned} M_1 &= \Delta q[p(q,q-1) - p(q,q+1)], \\ M_2 &= (\Delta q)^2[p(q,q-1) + p(q,q+1)]. \end{aligned} \tag{1.132}$$

1.8 Brownian Motion and the Langevin Equation

In 1828 Robert Brown, a Scottish botanist, was engaged in microscopic investigations of pollen grains suspended in water [6, 7]. He observed that, contrary to intuition, the grains were never stationary but were jumping constantly about in an irregular fashion without ever coming to rest. This type of motion, observable with any small particles ($\sim 10^{-3}$ mm radius or smaller) suspended in a fluid medium, has come to be known as *Brownian motion.* It was not until 1905 that the correct explanation was given by Einstein who postulated that the irregular motion was due to bombardments of the particle by the molecules of the fluid [8–10]. Using the diffusion equation, Einstein demonstrated that the mean square displacement of a Brownian particle is proportional to the first power of the time—a result reminiscent of the random walk problem. Although Einstein's work laid to rest the essential physics of Brownian motion, it nevertheless continues to be of interest to physicists because it serves as a prototype for numerous stochastic processes [10, 11], including those that are of importance in radiation interactions.

Consider a Brownian particle after it has been projected with an initial velocity into a fluid medium where it is buffeted by random collisions with the molecules of the fluid. Because of its forward motion, the particle will suffer, on the average, more collisions in front than in back, causing it to slow down. For motion in one dimension we might write

$$m\frac{dv}{dt} = -\alpha v \tag{1.133}$$

for the equation of motion of a particle of mass m. The velocity v would then decay exponentially with a damping constant α/m where the constant α is the coefficient of friction. Clearly, Eq. (1.133) does not describe Brownian motion since it predicts that the particle ultimately comes to rest, in clear contradiction to the observation of persistent erratic motion. To account for the latter, Langevin augmented Eq. (1.133) with and additional term $F(t)$ assumed to be a random, rapidly fluctuating force independent of the velocity:

$$m\frac{dv}{dt} = -\alpha v + F(t). \tag{1.134}$$

The origin of $F(t)$ is attributed to the numerous microscopic collisons but is not defined explicitly, although certain assumptions are made concerning its statistical properties.

The presence of the random force $F(t)$ precludes the possibility of treating the Langevin equation as an ordinary differential equation. We are confronted with a nondeterministic or stochastic differential equation in which the initial conditions do not determine the future behavior of the system, as is the case with ordinary differential equations. A knowledge of the position and velocity of the particle at one instant of time does not yield the precise position (or velocity) at a later time; only statistical predictions are possible. Nevertheless, it is precisely the statistical features that are the most relevant for experimental observations.

There are several time scales associated with Brownian motion. The shortest is the mean collison time, τ_c, between the Brownian particle and the molecules of the medium—typically about 10^{-21} seconds in a liquid. The average period of the fluctuating force $F(t)$ must then be of the same order of magnitude. We shall therefore assume that $\langle F(t)\rangle$, the average value of $F(t)$, rapidly goes to zero in a time not too much longer than τ_c. Since the mass of a fluid molecule is orders of magnitude smaller than the mass of a Brownian particle, the effect of a single collision on the latter will be very small. It is only after a very large number of collisions that their cumulative action will produce a perceptible effect in the position and velocity of a particle; such changes, therefore, must occur on a much longer time scale. This means that there exist time intervals Δt sufficiently short for the variations in position and velocity to be very small but long enough for $F(t)$ to have undergone many fluctuations, so that $\langle F(t)\rangle = 0$. A third time scale is associated with the resolution of the detecting instrument.

As a first step in the analysis of the Langevin Eq. (1.134) we compute the mean square displacement. One-dimensional motion already contains the essential physics and is sufficient for our purpose. Then

$$m\dot{v} = m\ddot{x} = -\alpha\dot{x} + F(t), \tag{1.135}$$

or, upon multiplying through by x—the distance the Brownian particle moves from its starting point at $t = 0$,

$$m\ddot{x}x \equiv m\frac{d}{dt}(x\dot{x}) - m\dot{x}^2 = -\alpha x\dot{x} + xF(t). \tag{1.136}$$

The corresponding equation for average values is

$$m\frac{d}{dt}\langle x\dot{x}\rangle - m\langle\dot{x}^2\rangle + \alpha\langle x\dot{x}\rangle = \langle xF(t)\rangle. \tag{1.137}$$

A crucial assumption at this stage is to set

$$\langle xF(t)\rangle = x\langle F(t)\rangle = 0. \tag{1.138}$$

The justification is based on the fact that any time interval $t_2 - t_1$ that is large enough for the position or velocity to have changed appreciably and that is the interval employed to obtain the average $\langle xf(t)\rangle$, can be subdivided into a large number of the aforementioned small time intervals Δt. During these small time intervals, $\langle F(t) \rangle = 0$ while x and v are essentially constant. Also, assuming the system is in thermal equilibrium, the equipartition law in one dimension gives

$$m\langle \dot{x}^2 \rangle = kT. \tag{1.139}$$

Equation (1.137) then simplifies to

$$m\frac{d}{dt}\langle x\dot{x} \rangle + \alpha\langle x\dot{x} \rangle = kT, \tag{1.140}$$

and its solution, with the initial condition $x = 0$ at $t = 0$, is

$$\langle x\dot{x} \rangle = \frac{1}{2}\frac{d}{dt}\langle x^2 \rangle = \frac{kT}{\alpha}(1 - e^{-\gamma t}), \tag{1.141}$$

in which γ is a relaxation constant equal to α/m. Hence, the mean square displacement is

$$\begin{aligned}\langle x^2 \rangle &= \frac{2kT}{\alpha}\left[t - \int_0^t e^{-\gamma t'}\,dt'\right] \\ &= \frac{2kT}{\alpha}\left[t - \frac{1}{\gamma}(1 - e^{-\gamma t})\right].\end{aligned} \tag{1.142}$$

Depending on whether t is much smaller or much greater than the relaxation time γ^{-1}, one obtains the two relations

$$\langle x^2 \rangle = \frac{kT}{m}t^2 \qquad \text{for} \quad t \ll \frac{1}{\gamma}, \tag{1.143}$$

$$\langle x^2 \rangle = \frac{2kT}{\alpha}t \qquad \text{for} \quad t \gg \frac{1}{\gamma}. \tag{1.144}$$

Equation (1.144) is known as the *Einstein equation*; its most important feature is the proportionality between $\langle x^2 \rangle$ and the first power of the time—a result previously obtained for the random walk problem (Eq. (1.124)). For three-dimensional motion the mean kinetic energy is

$$\frac{1}{2}m\langle v^2 \rangle = \frac{3}{2}kT, \tag{1.145}$$

and the mean square displacement is

$$\langle r^2 \rangle = \frac{6kT}{\alpha}\left[t - \frac{1}{\gamma}(1 - e^{-\gamma t})\right]. \tag{1.146}$$

We illustrate the foregoing with a numerical example. Assume that the Brownian particles are of unit density and have radii $a = 10^{-4}$ cm. The friction coefficient in water may be calculated by means of the Stokes relation

$$\alpha = 6\pi\eta a \tag{1.147}$$

where η is the viscosity of water (10^{-3} Pas). These values give a relaxation time $1/\gamma \sim 10^{-7}$ seconds—a time much longer than a period of $F(t)$. Evidently, $F(t)$ goes through many oscillations before there is a significant change in the displacement, x.

Returning to the Langevin equation, we now compute the mean square velocity. With the relation

$$\frac{d}{dt}\left[e^{-\gamma t}\int_0^t e^{\gamma t'}F(t')\,dt'\right] = F(t) - \gamma e^{-\gamma t}\int_0^t e^{\gamma t'}F(t')\,dt', \tag{1.148}$$

the solution to the Langevin equation, with the initial condition $v = v_0$ at $t = 0$, may be expressed formally by

$$v(t) = v_0 e^{-\gamma t} + \frac{1}{m}e^{-\gamma t}\int_0^t e^{\gamma t'}F(t')\,dt'. \tag{1.149}$$

Since $\langle F(t)\rangle = 0$,

$$\langle v(t)\rangle = v_0 e^{-\gamma t}, \tag{1.150}$$

indicating that the average velocity of a Brownian particle decays to zero.

Proceeding from Eq. (1.149),

$$\langle v^2(t)\rangle = v_0^2 e^{-2\gamma t} + \frac{e^{-2\gamma t}}{m^2}\int_0^t\int_0^t e^{\gamma(t'+t'')}\langle F(t')\rangle\,dt'\,dt'' \tag{1.151}$$

in which the cross term has been omitted in view of the condition $\langle F(t)\rangle = 0$. With $\tau = t' - t''$

$$\int_0^t\int_0^t e^{\gamma(t'+t'')}\langle F(t')F(t'')\rangle\,dt'\,dt''$$

$$= \int_0^t e^{2\gamma t''}\,dt''\int_{-t''}^{t-t''} e^{\gamma\tau}\langle F(t'')F(t''+\tau)\rangle\,d\tau. \tag{1.152}$$

Two assumptions are now introduced. The first is that the correlation function $\langle F(t'')F(t''+\tau)\rangle$ depends only on τ, i.e., Brownian motion is assumed to be a stationary process. It is permissible then to write

$$G(\tau) = \langle F(t'')F(t''+\tau)\rangle = \langle F(0)F(\tau)\rangle. \tag{1.153}$$

The second assumption is that the correlation time associated with $G(\tau)$ must be of the same order of magnitude as τ_c, the average fluctuation period of

$F(t)$ (or average collision time). This means that $G(\tau)$ peaks sharply at $\tau = 0$ and drops to zero after a few collision times, in a manner approximating a δ-function.

With these assumptions we now may evaluate the integral in Eq. (1.151). Following Reif [13], the domain of integration, shown in Fig. 1.7, is separated into two parts as defined by the two functions

$$\tau = -t'', \qquad \tau = t' - t''. \tag{1.154}$$

The t'' integration may be carried out in two steps corresponding to the upper and lower shaded areas. In the upper area,

$$0 \leq t'' \leq t - \tau, \qquad 0 \leq \tau \leq t, \tag{1.155a}$$

and in the lower area,

$$-\tau \leq t'' \leq t, \qquad -t \leq \tau \leq 0. \tag{1.155b}$$

The right-hand side of Eq. (1.152) then becomes

$$\int_0^t e^{\gamma\tau} G(\tau)\, d\tau \int_0^{t-\tau} e^{2\gamma t''}\, dt'' + \int_{-t}^0 e^{\gamma\tau} G(\tau)\, d\tau \int_{-\tau}^t e^{2\gamma t''}\, dt'', \tag{1.156}$$

or, after performing the t'' integration,

$$\frac{1}{2\gamma}\int_0^t e^{\gamma\tau} G(\tau)[e^{2\gamma(t-\tau)} - 1]\, d\tau + \frac{1}{2\gamma}\int_{-t}^0 e^{\gamma\tau} G(\tau)[e^{2\gamma t} - e^{-2\gamma\tau}]\, d\tau. \tag{1.157}$$

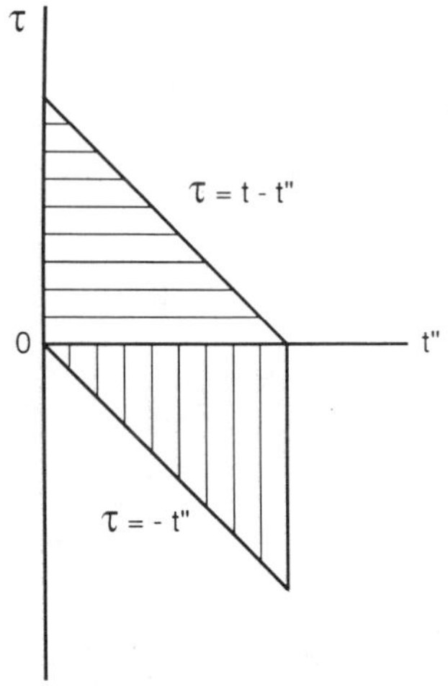

FIGURE 1.7 The domains of integration for the right-hand integral in Eqs. (1.152) (after Reif [13]).

Based on the assumption that $G(\tau) \sim 0$ when $\tau \neq 0$, the integrands will be effectively zero at all times except for $\tau \sim 0$ or $e^{\gamma\tau} \sim 1$. The last two integrals then reduce to

$$\frac{1}{2\gamma}\int_0^t G(\tau)[e^{2\gamma t} - 1]\,d\tau + \frac{1}{2\gamma}\int_{-t}^0 G(\tau)[e^{2\gamma t} - 1]\,d\tau$$

$$= \frac{e^{2\gamma t} - 1}{2\gamma}\int_{-t}^{t} G(\tau)\,d\tau. \tag{1.158}$$

The limits may be extended to $\pm\infty$ by virtue of the restriction on $G(\tau)$ to the region $\tau \sim 0$. The final result for the mean square velocity is then given by

$$\langle v^2(t)\rangle = v_0^2 e^{-2\gamma t} + \frac{1 - e^{-2\gamma t}}{2\gamma m^2}\int_{-\infty}^{\infty} G(\tau)\,d\tau. \tag{1.159}$$

For times that are long compared to $1/\gamma$, the system reaches thermal equilibrium, in which case, for one dimension,

$$\langle v^2(t)\rangle = \langle \dot{x}^2(t)\rangle = \frac{kT}{m} = \frac{1}{2\gamma m^2}\int_{-\infty}^{\infty} G(\tau)\,d\tau, \tag{1.160}$$

or

$$\alpha = m\gamma = \frac{1}{2kT}\int_{-\infty}^{\infty} G(\tau)\,d\tau. \tag{1.161}$$

This is an important expression that relates the friction coefficient to the correlation function of the fluctuation force; it is a special case of the general *fluctuation-dissipation* theorem which connects the dissipative characteristics of a physical system with its statistical properties.

Combining Eqs. (1.159) and (1.160),

$$\langle v^2(t)\rangle = v_0^2 e^{-2\gamma t} + \frac{kT}{m}(1 - e^{-2\gamma t}) \tag{1.162}$$

$$= v_0^2 \qquad \text{for} \quad t \ll 1/\gamma, \tag{1.163}$$

$$= \frac{kT}{m} \qquad \text{for} \quad t \gg 1/\gamma. \tag{1.164}$$

For three-dimensional motion kT is replaced by $3kT$.

A more direct derivation of Eq. (1.162) may be obtained by means of the approximation $\langle F(t')F(t'')\rangle = K\delta(t' - t'')$ in which K is a constant to be evaluated. Then

$$\begin{aligned}\int_0^t \int_0^t & e^{\gamma(t'+t'')} \langle F(t')F(t'')\rangle \, dt' \, dt'' \\ &= K \int_0^t \int_0^t e^{\gamma(t'+t'')} \, \delta(t' - t'') \, dt' \, dt'' \\ &= K \int_0^t e^{2\gamma t'} \, dt' = \frac{K}{2\gamma}(e^{2\gamma t} - 1),\end{aligned} \tag{1.165}$$

and the mean square velocity, from Eq. (1.151), is

$$\begin{aligned}\langle v^2(t)\rangle &= v_0^2 e^{-2\gamma t} + \frac{e^{-2\gamma t}}{m^2} \frac{K}{2\gamma}(e^{2\gamma t} - 1) \\ &= v_0^2 e^{-2\gamma t} + \frac{K}{2\gamma m^2}(1 - e^{-2\gamma t}).\end{aligned} \tag{1.166}$$

For times that are long compared to $1/\gamma$

$$\langle v^2(t)\rangle = \frac{K}{2\gamma m^2} = \frac{kT}{m}. \tag{1.167}$$

Hence,

$$\langle v^2(t)\rangle = v_0^2 e^{-2\gamma t} + \frac{kT}{m}(1 - e^{-2\gamma t}), \tag{1.168}$$

as in Eq. (1.162). Clearly, the approximation $\langle F(t')F(t'')\rangle = K\,\delta(t' - t'')$ achieves the same result more expeditiously. It does not disclose, however, the important connection between the friction coefficient and the correlation function.

1.9 Brownian Motion and the Fokker-Planck Equation

Further insight into the motion of Brownian particle, beyond that provided by the mean square displacement and velocity, is obtained by computing the probability for a transition from an initial position x_0 at $t = 0$ to a position x at the time t and the analogous transition probability for the velocity. Both the displacement and velocity are assumed to be stationary Markov processes; that is, the motion of the Brownian particle is invariant under time translation and is determined by the instantaneous values of the physical parameters without reference to the previous history of the motion. The transition probability densities, therefore, may be computed from the Fokker-Planck Eq. (1.177).

For the displacement, let

$$P(x \mid y, t) = P(x_0 \mid x, t) \tag{1.169}$$

represent the probability density that a Brownian particle is found at the position x at the time t, given that the particle had been at the position x_0 at $t = 0$. The quantity M_1, defined by Eq. (1.75a), becomes

$$\begin{aligned} M_1 &= \frac{1}{\Delta t} \int \varepsilon P(y \mid y + \varepsilon, \Delta t)\, d\varepsilon \\ &= \frac{1}{\Delta t} \int \Delta x P(x \mid x + \Delta x, \Delta t)\, d(\Delta x) = \frac{\langle \Delta x \rangle}{\Delta t}, \end{aligned} \tag{1.170}$$

in which $\langle \Delta x \rangle$ is the average displacement in the time Δt. For the same basic reason, as in the case of the random walk, $\langle \Delta x \rangle = 0$ so that $M_1 = 0$. For M_2 (Eq. (1.75b) we have

$$\begin{aligned} M_2 &= \frac{1}{\Delta t} \int \varepsilon^2 P(y \mid y + \varepsilon, \Delta t)\, d\varepsilon \\ &= \frac{1}{\Delta t} \int (\Delta x)^2 P(x \mid x + \Delta x, \Delta t)\, d(\Delta x) = \frac{\langle (\Delta x)^2 \rangle}{\Delta t}. \end{aligned} \tag{1.171}$$

The displacement $\langle (\Delta x)^2 \rangle$ is obtained from the Langevin equation. Assuming $\Delta t \gg 1/\gamma$, we refer to Eq. (1.144) which then yields

$$M_2 = \frac{2kT}{\alpha}. \tag{1.172}$$

The Fokker-Planck equation (Eq. (1.77)), in the notation of Eq. (1.169), now acquires the form

$$\frac{\partial P(x_0 \mid x, t)}{\partial t} = \frac{kT}{\alpha} \frac{\partial^2 P(x_0 \mid x, t)}{\partial x^2}, \qquad (t \gg 1/\gamma), \tag{1.173}$$

which is just the one-dimensional diffusion equation with diffusion constant $D = kT/\alpha$. The solution is the Gaussian

$$P(x_0 \mid x, t) = \frac{1}{\sqrt{4\pi \mathrm{D}t}} \exp\left[-\frac{(x - x_0)^2}{4\mathrm{D}t} \right]. \tag{1.174}$$

As time goes on, the probability distribution gradually broadens (Fig. 1.8); that is, if initially a group of Brownian particles is clustered around x_0, the particles eventually spread out. This is a diffusion-type motion; it is irreversible and is directly attributable to the random, fluctuating forces

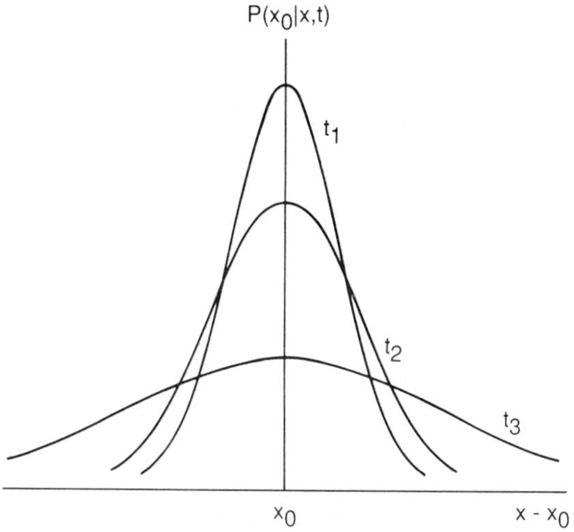

FIGURE 1.8 The diffusion of Brownian particles initially clustered around x_0 as a function of time. At a given instant the distribution of positions is a Gaussian that broadens as time progresses.

exerted by the fluid molecules on the Brownian particles. The Fokker-Planck equation establishes a relation between the diffusion constant D and the friction coefficient α; it also predicts that the probability for a displacement falls off rapidly as the displacement increases. Hence, a Brownian particle tends, most often, to undergo only small changes in position.

For three-dimensional Brownian motion,

$$P(r_0 \mid r, t) = \left[\frac{1}{4\pi \mathrm{D}t}\right]^{3/2} \exp\left[-\frac{(\Delta r)^2}{4\mathrm{D}t}\right] \tag{1.175}$$

in which

$$\Delta r = r - r_0, \qquad \langle(\Delta r)^2\rangle = 6\mathrm{D}t = \left(\frac{6kT}{\alpha}\right)t. \tag{1.176}$$

A similar analysis may be carried out for the velocities. For this case,the transition probability density is $P(v_0 \mid v, t)$, where v_0 is the initial velocity and v is the velocity after a time t. The quantities M_1 and M_2 are

$$M_1 = \frac{1}{\Delta t}\int \Delta v P(v \mid v + \Delta v, \Delta t)\, d(\Delta v) = \frac{\langle(\Delta v)\rangle}{\Delta t}, \tag{1.177a}$$

$$M_2 = \frac{1}{\Delta t}\int (\Delta v)^2 P(v \mid v + \Delta v, \Delta t)\, d(\Delta v) = \frac{\langle(\Delta v)^2\rangle}{\Delta t}. \tag{1.177b}$$

For small time intervals Δt, the Langevin equation (1.134) may be written

$$m\left(\frac{\Delta v}{\Delta t}\right) = -\alpha v + F(t). \tag{1.178}$$

Upon taking averages with $\langle F(t)\rangle = 0$ and $\langle v\rangle \sim v$, we have

$$M_1 = \frac{\langle \Delta v\rangle}{\Delta t} = -\left(\frac{\alpha}{m}\right)v = -\gamma v. \tag{1.179}$$

M_2 may also be obtained from Eq. (1.179). Starting with

$$\Delta v = -\frac{\alpha v}{m}\Delta t + \frac{1}{m}\int_t^{t+\Delta t} F(t')\,dt', \tag{1.180}$$

the mean square is

$$\langle(\Delta v)^2\rangle = \frac{\alpha^2 v^2}{m^2}(\Delta t)^2 - \frac{2\alpha v}{m^2}\Delta t \int_t^{t+\Delta t} \langle F(t')\rangle\,dt' + \frac{1}{m^2}\int_t^{t+\Delta t} dt' \int_t^{t+\Delta t} \langle F(t')F(t'')\rangle\,dt''. \tag{1.181}$$

To first order in Δt, the first term on the right containing $(\Delta t)^2$ may be eliminated; the second term is eliminated also because $\langle F(t)\rangle = 0$. Setting $\tau = t'' - t'$,

$$\langle(\Delta v)^2\rangle = \frac{1}{m^2}\int_t^{t+\Delta t} dt' \int_{t-t'}^{t-t'+\Delta t} \langle F(t')F(t'+\tau)\rangle\,d\tau. \tag{1.182}$$

As in the previous section, the condition that $G(\tau) = \langle F(t')F(t'+\tau)\rangle$ is a sharply peaked function of τ centered at $\tau = 0$ permits the extension of the limits on the second integral to $\pm\infty$. We may then use Eq. (1.161) to write

$$M_2 = \frac{\langle(\Delta v)^2\rangle}{\Delta t} = \frac{1}{m^2\,\Delta t}\int_t^{t+\Delta t} dt' \int_{-\infty}^{\infty} G(\tau)\,d\tau = \frac{2kT\gamma}{m}. \tag{1.183}$$

With M_1 and M_2 given by Eqs. (1.179) and (1.183), respectively, the Fokker-Planck equation for $P(v_0\,|\,v,t)$ is

$$\frac{\partial P}{\partial t} = \gamma\frac{\partial}{\partial v}(vP) + \frac{\gamma kT}{m}\frac{\partial^2 P}{\partial v^2}. \tag{1.184}$$

The quantities γv and $\gamma kT/m$ are known as *drift* and *diffusion* coefficients, respectively.

We now introduce the transformation [13]

$$P(v_0\,|\,v,t) = e^{\gamma t}Q(u_0\,|\,u,t) \tag{1.185}$$

where

$$u_0 = v_0, \qquad u = ve^{\gamma t}. \tag{1.186}$$

In terms of the new variables, Eq. (1.184) transforms to

$$\frac{\partial Q}{\partial t} = \frac{\gamma kT}{m} e^{2\gamma t} \frac{\partial^2 Q}{\partial u^2}. \tag{1.187}$$

Letting

$$\Theta = \frac{1}{2\gamma}(e^{2\gamma t} - 1), \tag{1.188}$$

we obtain the diffusion equation

$$\frac{\partial Q}{\partial \Theta} = \frac{\gamma kT}{m} \frac{\partial^2 Q}{\partial u^2}, \tag{1.189}$$

whose solution is

$$Q = \sqrt{4\pi c\Theta} \exp\left[-\frac{(u-u_0)^2}{4c\Theta}\right], \tag{1.190}$$

with $c = \gamma kT/m$. Finally, reverting to the original variables

$$P(v_0 \,|\, v, t) = \sqrt{\frac{m}{2\pi kT(1 - e^{-2\gamma t})}} \exp\left[-\frac{m(v - v_0 e^{-\gamma t})^2}{2kT(1 - e^{-2\gamma t})}\right]. \tag{1.191}$$

The transition probability density for the velocity is a Gaussian, as we found to be the case for the displacement. For small values of t, the probability density is sharply peaked at $v = v_0$. As time progresses, the distribution spreads and as $t \to \infty$, the one-dimensional transition probability density becomes

$$P(v_0 \,|\, v, t \to \infty) \equiv P(v) = \sqrt{\frac{m}{2\pi kT}} \exp\left[-\frac{mv^2}{2kT}\right], \tag{1.192}$$

independent of both t and v_0. In other words, the system has forgotten its initial velocity consistent with the previous result (Eq. (1.164)) for $t \gg 1/\gamma$. The distribution is now a stationary Maxwellian distribution at the temperature T of the medium.

Let us now recapitulate some of the basic assumptions and conclusions concerning Brownian motion:

1. Brownian motion is a stationary Markov process.

2. The force $F(t)$ which appears in the Langevin equation fluctuates very rapidly; its correlation time τ_c is on the order of the mean time between

collisions ($\sim 10^{-21}$ s in water). During so short a time interval, the change in the position or velocity of a Brownian particle will be very small. A second characteristic time is $1/\gamma = m/\alpha\,(\sim 10^{-7} - 10^{-9})$; on this time scale changes in position and velocity are observable. Between τ_c and $1/\gamma$ there are time intervals Δt during which $F(t)$ has undergone many fluctuations but the position and velocity of the particle have remained essentially unchanged.

3. The motions of a Brownian particle in different time intervals are independent processes.

4. In the Einstein equation, Eq. (1.144), the mean square displacement is proportional to the first power of the time—a result identical with that obtained for the random walk in Eq. (1.124).

5. The Einstein equation (1.144) contains a friction coefficient which, as shown by Eq. (1.161), is proportional to the integral of the correlation function of the fluctuating force $F(t)$. Thus, the basic source of friction (or viscosity) is intimately related to the fluctuating forces exerted by the molecules of the medium on the Brownian particles. Since frictional forces are associated with energy dissipation, the motion of the Brownian particles is irreversible.

6. The mean square velocity, Eq. (1.164), is independent of the initial velocity—a feature characteristic of diffusion-type motion in which the system loses memory of its initial values.

7. The Fokker-Planck equation for the displacement (Eq. (1.173)) is a diffusion equation with a diffusion constant related to the friction coefficient.

8. The Fokker-Planck equation for the velocity leads to a probability density (Eq. (1.191)) that ultimately becomes independent of the initial velocity.

Since Brownian motion is a direct consequence of molecular collisions and since $F(t)$ represents the force term associated with such collisions, one conceivably might write the equation of motion of a Brownian particle in the form

$$\frac{mdv}{dt} = F(t), \tag{1.193}$$

that is, without the friction term which, in principle, should ultimately be derivable from $F(t)$. Such an approach would require a detailed specification of $F(t)$ based on microscopic collision dynamics. The computations would become more complex by many orders of magnitude; nor would the resulting information be particularly useful. But, on the other hand, if one adopts a purely stochastic approach and sets $\langle F(t)\rangle = 0$, the integral of Eq. (1.193) then leads to the conclusion that the average velocity is independent of time and remains equal to the initial velocity. In other words, there is no mechanism for slowing down the particle. To avoid such an unphysical result it is necessary to include a friction term. We see, then, that Brownian motion is neither a truly microscopic nor a completely macroscopic phenomenon. It has features characteristic of both and therefore requires both frictional and fluctuating forces.

A point of view often found to be useful is one in which the Brownian particles and the medium are regarded as two separate systems with a coupling between them that allows for energy exchange. The medium acts as a bath or reservoir with many degrees of freedom whose detailed description is neither feasible nor relevant. The friction coefficient is a measure of the coupling strength between the system (Brownian particles) and the reservoir (medium), the latter serving as a sink for the energy dissipated by the system. We shall find a similar situation in connection with photon-atom interactions (Chapter VI), where the Langevin and Fokker-Planck formulations appear in close analogy with Brownian motion.

References

[1] R. G. Gordon, *Adv. Mag. Res.* **3**, 1(1968).
[2] B. J. Berne and G. D. Harp, *Adv. Chem. Phys.* **17**, 63(1970).
[3] N. Wiener, *Acta Math.* **55**, 117(1930).
[4] A. Khinchine, *Math. Annalen* **105**, 604(1934).
[5] J. L. Doob, *Ann. Math.* **43**, 351(1942).
[6] R. Brown, *Phil. Mag.* **4**, 161(1828).
[7] R. Brown, *Ann. Phys. Chem.* **14**, 294(1828).
[8] A. Einstein, *Ann. Phys.* **17**, 549(1905).
[9] A. Einstein, *Ann. Phys.* **19**, 289(1906).
[10] A. Einstein, *Ann. Phys.* **19**, 371(1906).
[11] S. Chandrasekhar, *Rev. Mod. Phys.* **15**, 1(1943).
[12] M. C. Wang and G. E. Uhlenbeck, *Rev. Mod. Phys.* **17**, 323(1945).
[13] F. Reif, *Fundamentals of Statistical and Thermal Physics*. McGraw-Hill, New York, 1965.

II Density Matrices and Perturbation Theory

The basic quantities that appear in quantum mechanics are wave functions and observables. The former are solutions to the Schrödinger equation for the physical system and the latter are Hermitian operators corresponding to a measurable property of the system, e.g., energy, momentum, electric or magnetic dipole moment, and so forth. Once the wave functions and their dependence on time are known, it becomes feasible, at least in principle, to compute matrix elements of the operators and it is the matrix elements that serve the function of a bridge between theory and experiment. There are numerous cases, however, where, for various reasons—practical and fundamental—wave functions are not available for the system under consideration. Common examples are encountered in dealing with collections or ensembles of interacting systems, e.g., atoms, molecules, simple harmonic oscillators, modes of a radiation field, and so forth. Under such circumstances, a powerful formalism based on the density operator and its matrix representation provides the maximum possible information [1–3]. Although density operators appear in both classical and quantum statistical mechanics, our primary emphasis will be on the latter.

2.1 Definitions and General Properties

Consider a collection of identical atoms in thermal equilibrium. The fraction of atoms found in the various atomic states is determined by the Boltzmann distribution. Such a system is said to be in a *mixed state*; the characteristic feature of a mixed state is that it is impossible to assign a wave function to any given atom in the system. All that can be said is that the atom has a certain probability of residing in one state or another. Should an atom be selected at random, the distribution assigns a probability P_i for the atom to be found in the particular atomic state $|\psi_i\rangle$. The *density operator* ρ is defined then by

$$\rho = \sum_i P_i|\psi_i\rangle\langle\psi_i|, \tag{2.1}$$

where the combination $|\psi_i\rangle\langle\psi_i|$ is to be understood as an operator and not as a product of a wave function and its complex conjugate. The definition is quite general; it is applicable not only to systems in thermal equilibrium but also to systems with arbitrary probability distributions. The probabilities P_i, by their very nature as probabilities, evidently satisfy

$$0 \le P_i \le 1, \qquad \sum_i P_i = 1, \qquad \sum_i P_i^2 \le 1. \tag{2.2}$$

If all the atoms are in the same quantum state $|\psi_j\rangle$, there is no way to distinguish one atom from another—each atom is definitely in a state with a wave function $|\psi_j\rangle$. In that case, $P_j = 1$; all other probabilities are zero and the density operator simplifies to

$$\rho = |\psi_j\rangle\langle\psi_j|. \tag{2.3}$$

Such a system is said to be in a *pure* state. Whether a system is in a pure state or a mixed state is determined, therefore, by the existence of a wave function in the former case and the absence of one in the latter.

If the set $\{|\psi_i\rangle\}$ consists of orthonormal functions, the definition (2.1) leads to the relation

$$\rho|\psi_i\rangle = P_i|\psi_i\rangle. \tag{2.4}$$

Hence, the density operator could be defined equally well as an operator with eigenfunctions $|\psi_i\rangle$ and corresponding eigenvalues P_i. An immediate consequence of Eq. (2.4) is that the diagonal matrix element $\langle\psi_i|\rho|\psi_i\rangle$ is simply P_i, the probability of finding the system in the state $|\psi_i\rangle$. In an equivalent interpretation, $\langle\psi_i|\rho|\psi_i\rangle$ is the fraction of the total number of systems (population) in $|\psi_i\rangle$. In view of the assumed orthonormality of the $\{|\psi_i\rangle\}$, off-diagonal matrix elements $\langle\psi_j|\rho|\psi_i\rangle$ obviously vanish.

The density operator may be given a matrix representation by adopting a complete orthonormal basis set $\{|n\rangle\}$ satisfying the *closure* relation

$$\sum_n |n\rangle\langle n| = 1. \tag{2.5}$$

Then

$$|\psi_i\rangle = \sum_n |n\rangle\langle n|\psi_i\rangle \equiv \sum_n c_{ni}|n\rangle, \tag{2.6}$$

where the constants $c_{ni} = \langle n|\psi_i\rangle$ are expansion coefficients in the basis $\{|n\rangle\}$. Inserting the closure relation twice into the definition for ρ,

$$\begin{aligned} \rho = \sum_i P_i|\psi_i\rangle\langle\psi_i| &= \sum_i P_i \sum_{nm} |n\rangle\langle n|\psi_i\rangle\langle\psi_i|m\rangle\langle m|, \\ &= \sum_i P_i \sum_{nm} c_{ni}c_{mi}^*|n\rangle\langle m|, \end{aligned} \tag{2.7}$$

and

$$\langle n|\rho|m\rangle = \sum_i P_i c_{ni} c_{mi}^*. \tag{2.8}$$

The diagonal matrix elements

$$\langle n|\rho|n\rangle = \sum_i P_i |c_{ni}|^2 \tag{2.9}$$

are real and positive; also, since

$$\langle m|\rho|n\rangle = \sum_i P_i c_{mi} c_{ni}^* = \langle n|\rho|m\rangle^*, \tag{2.10}$$

the density operator ρ is Hermitian. For a pure state, Eqs. (2.7), (2.8) and (2.9) reduce to

$$\rho = \sum_{nm} c_{ni}c_{mi}^*|n\rangle\langle m|, \qquad \langle n|\rho|m\rangle = c_{ni}c_{mi}^*, \qquad \langle n|\rho|n\rangle = |c_{ni}|^2. \tag{2.11}$$

Note the distinction between matrix elements of ρ with respect to a particular basis set $\{|n\rangle\}$ used in the expansion of the wave function $|\psi_i\rangle$ and the matrix elements of ρ with respect to the various functions $|\psi_i\rangle$.

For the trace (sum of diagonal elements) of ρ we have

$$\begin{aligned} \mathrm{Tr}\,\rho \equiv \sum_n \langle n|\rho|n\rangle &= \sum_{ni} P_i\langle n|\psi_i\rangle\langle\psi_i|n\rangle = \sum_{ni} P_i\langle\psi_i|n\rangle\langle n|\psi_i\rangle \\ &= \sum_i P_i\langle\psi_i|\psi_i\rangle = \sum_i P_i = 1. \end{aligned} \tag{2.12}$$

This is a fundamental property of density operators; it is valid in any basis set because the trace is invariant under a change in basis. It may be concluded further from Eq. (2.12) that, since $\langle n|\rho|n\rangle$ is real and positive,

$$0 \le \langle n|\rho|n\rangle \le 1. \tag{2.13}$$

The density operator has been shown to be Hermitian; hence, there must exist a basis set $\{|q\rangle\}$ in which ρ is diagonal. Let us then write, in the diagonal basis,

$$\begin{aligned} \mathrm{Tr}\{\rho^2\} &= \sum_q \langle q|\rho^2|q\rangle = \sum_{qr} \langle q|\rho|r\rangle\langle r|\rho|q\rangle\,\delta_{qr} \\ &\le \left(\sum_q \langle q|\rho|q\rangle\right)\left(\sum_q \langle q|\rho|q\rangle\right) \le (\mathrm{Tr}\,\rho)^2 = 1. \end{aligned} \tag{2.14}$$

Therefore, irrespective of basis,

$$\mathrm{Tr}\{\rho^2\} \le 1. \tag{2.15}$$

For a pure state $\rho^2 = |\psi\rangle\langle\psi|\psi\rangle\langle\psi| = |\psi\rangle\langle\psi| = \rho$ and

$$\mathrm{Tr}\{\rho^2\} = \mathrm{Tr}\,\rho = 1. \tag{2.16}$$

The property $\rho^2 = \rho$ for a pure state characterizes ρ as a projection operator. It may be shown that the relation $\rho^2 = \rho$ is also sufficient to ensure that the state in question is a pure state. Once again we choose a basis in which ρ is diagonal. Then, if $\rho^2 = \rho$,

$$\langle q|\rho^2|q\rangle = \sum_r \langle q|\rho|r\rangle\langle r|\rho|q\rangle\,\delta_{qr} = \langle q|\rho|q\rangle\langle q|\rho|q\rangle = \langle q|\rho|q\rangle, \tag{2.17}$$

which can only be satisfied when $\langle q|\rho|q\rangle = 0$ or 1. But $\mathrm{Tr}\,\rho^2 = \mathrm{Tr}\,\rho = 1$; hence, only one diagonal element, say, $\langle 0|\rho|0\rangle = 1$ and all others are zero. Referring to Eq. (2.9), we have in this case

$$\langle 0|\rho|0\rangle = \sum_i P_i|c_{oi}|^2 = 1, \qquad \langle q|\rho|q\rangle = \sum_i P_i|c_{qi}|^2 = 0. \tag{2.18}$$

But according to Eq. (2.2) the probabilities P_i cannot all be zero; therefore, $c_{qi} = 0$ $(q \ne 0)$ and $c_{oi} \ne 0$. The expansion of an arbitrary normalized wave function $|\psi_i\rangle$ then reduces to

$$|\psi_i\rangle = c_{oi}|0\rangle = e^{i\alpha}|0\rangle. \tag{2.19}$$

Thus, all wave functions $|\psi_i\rangle$ correspond to the same function $|0\rangle$, apart from an arbitrary phase factor. The system is therefore in a pure state proving, thereby, that $\rho^2 = \rho$ is a sufficient condition.

Since $\rho^2 = \rho$ is both a necessary and sufficient condition for a pure state, it may be concluded that, for a mixed state, $\rho^2 \ne \rho$ and $\mathrm{Tr}\{\rho^2\} < 1$.

If A is an observable the ensemble average of A is defined by the averaged expectation value

$$\langle A\rangle = \sum_i P_i\langle\psi_i|A|\psi_i\rangle \tag{2.20}$$

where $\langle\psi_i|A|\psi_i\rangle$ is the expectation value of A in the state $|\psi_i\rangle$ and the sum is taken over all the states the system may occupy. For a pure state, the ensemble average is just the expectation value $\langle A\rangle = \langle\psi|A|\psi\rangle$. We now show that the ensemble average $\langle A\rangle$ may be expressed in terms of the density operator. By definition,

$$\rho A = \sum_i P_i|\psi_i\rangle\langle\psi_i|A, \qquad \langle n|\rho A|m\rangle = \sum_i P_i\langle n|\psi_i\rangle\langle\psi_i|A|m>. \tag{2.21}$$

Therefore, using the closure relation,

$$\begin{aligned}\mathrm{Tr}\{\rho A\} &= \sum_n \langle n|\rho A|n\rangle = \sum_{in} P_i\langle n|\psi_i\rangle\langle\psi_i|A|n\rangle \\ &= \sum_{in} P_i\langle\psi_i|A|n\rangle\langle n|\psi_i\rangle = \sum_i P_i\langle\psi_i|A|\psi_i\rangle \equiv \langle A\rangle. \end{aligned} \tag{2.22}$$

This is an important application of density matrices; it shows that the ensemble average of an operator is calculable once the matrix elements of the density operator are known in some basis set. For a pure state $|\psi\rangle$, $\mathrm{Tr}\{\rho A\}$ is nothing more than the expectation value $\langle\psi|A|\psi\rangle$.

According to a general theorem, which follows directly from the definition of the trace operation, $\mathrm{Tr}\{\mathrm{ABC}\cdots\}$ is invariant under a cyclic permutation of the operators $\mathrm{ABC}\cdots$. We then may write

$$\mathrm{Tr}\{\rho A\} = \mathrm{Tr}\{A\rho\} = \langle A\rangle, \tag{2.23}$$

independent of basis. For the product $A^\dagger A$,

$$\begin{aligned}\langle A^\dagger A\rangle = \mathrm{Tr}\{\rho A^\dagger A\} &= \sum_i P_i\langle\psi_i|A^\dagger A|\psi_i\rangle = \sum_{in} P_i\langle\psi_i|A^\dagger|n\rangle\langle n|A|\psi_i\rangle \\ &= \sum_{in} P_i|\langle n|A|\psi_i\rangle|^2 \geq 0. \end{aligned} \tag{2.24}$$

Further, it is noted that for a general operator A,

$$\mathrm{Tr}\, A^\dagger = [\mathrm{Tr}\, A]^*. \tag{2.25}$$

2.2 Spin-1/2 System

We now shall derive an explicit expression for the density operator associated with a system of spin-1/2 particles. This is a particularly revealing exercise because, as will be shown in Chapter V, there is a close formal connection between a spin-1/2 system and an atomic two-level system. The Pauli spin operators in their usual matrix representation are

$$\sigma_x = \begin{pmatrix} 0 & 1 \\ 1 & 0 \end{pmatrix}, \quad \sigma_y = \begin{pmatrix} 0 & -i \\ i & 0 \end{pmatrix}, \quad \sigma_z = \begin{pmatrix} 1 & 0 \\ 0 & -1 \end{pmatrix}, \tag{2.26}$$

with

$$\begin{aligned}\mathrm{Tr}\,\sigma_i\sigma_j &= 2\delta_{ij}, \\ \sigma_x^2 = \sigma_y^2 &= \sigma_z^2 = I, \end{aligned} \tag{2.27}$$

where I is the 2×2 identity matrix. The three spin operators, collectively represented by the vector operator $\boldsymbol{\sigma}$, are Hermitian and obey the commutation rules

$$[\sigma_x, \sigma_y] = 2i\sigma_z, \qquad [\sigma_y, \sigma_z] = 2i\sigma_x, \qquad [\sigma_z, \sigma_x] = 2i\sigma_y, \tag{2.28a}$$

or, written more compactly,

$$[\sigma_i, \sigma_y] = 2i\sigma_k, \qquad (i, j, k \text{ cyclic}). \tag{2.28b}$$

The spin states

$$|\alpha\rangle = \begin{pmatrix} 1 \\ 0 \end{pmatrix}, \qquad |\beta\rangle = \begin{pmatrix} 0 \\ 1 \end{pmatrix}, \tag{2.29}$$

are simultaneous eigenstates of σ_z and $\sigma^2 (= \sigma_x^2 + \sigma_y^2 + \sigma_z^2)$:

$$\begin{aligned} \sigma_z|\alpha\rangle &= |\alpha\rangle, & \sigma^2|\alpha\rangle &= 3|\alpha\rangle, \\ \sigma_z|\beta\rangle &= -|\beta\rangle, & \sigma^2|\beta\rangle &= 3|\beta\rangle, \end{aligned} \tag{2.30}$$

and satisfy

$$\langle\alpha|\alpha\rangle = \langle\beta|\beta\rangle = 1, \qquad \langle\alpha|\beta\rangle = 0. \tag{2.31}$$

The spin states $|\alpha\rangle$ and $|\beta\rangle$ form a complete set in the spin-1/2 system; the corresponding density matrix then may be written

$$\begin{aligned} \rho &= \begin{pmatrix} \langle\alpha|\rho|\alpha\rangle & \langle\alpha|\rho|\beta\rangle \\ \langle\beta|\rho|\alpha\rangle & \langle\beta|\rho|\beta\rangle \end{pmatrix} \equiv \begin{pmatrix} \rho_{\alpha\alpha} & \rho_{\alpha\beta} \\ \rho_{\beta\alpha} & \rho_{\beta\beta} \end{pmatrix} \\ &= c_0 I + c_1\sigma_x + c_2\sigma_y + c_3\sigma_z, \end{aligned} \tag{2.32}$$

the last expression arising from the fact that any 2 × 2 matrix can be expressed in terms of the Pauli spin operators together with the 2 × 2 identity matrix, I. Evaluating the matrix elements in Eq. (2.32),

$$\rho = \begin{pmatrix} c_0 + c_3 & c_1 - ic_2 \\ c_1 + ic_2 & c_0 - c_3 \end{pmatrix}. \tag{2.33}$$

But

$$\begin{aligned} \operatorname{Tr}\rho = 1 = 2c_0, \qquad \operatorname{Tr}\{\rho\sigma_x\} \equiv \langle\sigma_x\rangle = 2c_1, \\ \operatorname{Tr}\{\rho\sigma_y\} \equiv \langle\sigma_y\rangle = 2c_2, \qquad \operatorname{Tr}\{\rho\sigma_z\} \equiv \langle\sigma_z\rangle = 2c_3. \end{aligned} \tag{2.34}$$

Thus,

$$\rho = \frac{1}{2}\begin{pmatrix} 1 + \langle\sigma_z\rangle & \langle\sigma_x\rangle - i\langle\sigma_y\rangle \\ \langle\sigma_x\rangle + i\langle\sigma_y\rangle & 1 - \langle\sigma_z\rangle \end{pmatrix} = \frac{1}{2}(I + \langle\boldsymbol{\sigma}\rangle \cdot \boldsymbol{\sigma}). \tag{2.35}$$

If $|\psi\rangle$ is a general, normalized one-particle spin state, it may be expanded in the (complete) basis set consisting of $|\alpha\rangle$ and $|\beta\rangle$.

$$|\psi\rangle = a|\alpha\rangle + b|\beta\rangle, \qquad |a|^2 + |b|^2 = 1. \tag{2.36}$$

We note that the state described by the wave function $|\psi\rangle$ is, by definition, a pure state. It is then readily verified that the density operator acquires the form

$$\rho = |\psi\rangle\langle\psi| = \begin{pmatrix} |a|^2 & ab^* \\ a^*b & |b|^2 \end{pmatrix}. \tag{2.37}$$

Equation (2.36) is a particularly simple example of the general expansion (2.6) in an orthonormal basis. Such expansions are regarded as *coherent* superpositions of basis states and are characterized by nonvanishing off-diagonal matrix elements of the density operator. The notion of coherence, in the sense used here, is not an obvious one; further clarification will be postponed until Section 2.8 where the coherent superposition will be compared with the thermal distribution.

Comparing Eqs. (2.32), (2.35), and (2.37), one finds

$$\begin{aligned} \langle\sigma_x\rangle &= \mathrm{Tr}\{\rho\sigma_x\} = \rho_{\alpha\beta} + \rho_{\beta\alpha} = b^*a + a^*b, \\ \langle\sigma_y\rangle &= \mathrm{Tr}\{\rho\sigma_y\} = i[\rho_{\alpha\beta} - \rho_{\beta\alpha}] = i[b^*a - a^*b], \\ \langle\sigma_z\rangle &= \mathrm{Tr}\{\rho\sigma_z\} = \rho_{\alpha\alpha} - \rho_{\beta\beta} = |a|^2 - |b|^2, \\ \langle\sigma_x\rangle^2 &+ \langle\sigma_y\rangle^2 + \langle\sigma_z\rangle^2 = (|a|^2 + |b|^2)^2 = 1, \\ \det\rho &= \rho_{\alpha\alpha}\rho_{\beta\beta} - \rho_{\alpha\beta}\rho_{\beta\alpha} = 0. \end{aligned} \tag{2.38}$$

It is observed that $\langle\sigma_z\rangle$—the difference between the two diagonal matrix elements of ρ—corresponds to the difference in the occupation probabilities of the two spin states.

We define two new operators:

$$\begin{aligned} \sigma^+ &= \frac{1}{2}(\sigma_x + i\sigma_y) = \begin{pmatrix} 0 & 1 \\ 0 & 0 \end{pmatrix} = \begin{pmatrix} 1 \\ 0 \end{pmatrix}(0 \quad 1) = |\alpha\rangle\langle\beta|, \\ \sigma^- &= \frac{1}{2}(\sigma_x - i\sigma_y) = \begin{pmatrix} 0 & 0 \\ 1 & 0 \end{pmatrix} = \begin{pmatrix} 0 \\ 1 \end{pmatrix}(1 \quad 0) = |\beta\rangle\langle\alpha|. \end{aligned} \tag{2.39}$$

In terms of σ^+ and σ^-, the Pauli spin operators are

$$\begin{aligned} \sigma_x &= \sigma^+ + \sigma^-, \qquad \sigma_y = i(\sigma^- - \sigma^+), \\ \sigma_z &= 1 - 2\sigma^-\sigma^+ = 2\sigma^+\sigma^- - 1, \end{aligned} \tag{2.40}$$

and the density operator is

$$\begin{aligned} \rho &= \begin{pmatrix} \langle\sigma^+\sigma^-\rangle & \langle\sigma^-\rangle \\ \langle\sigma^+\rangle & \langle\sigma^-\sigma^+\rangle \end{pmatrix} \\ &= \frac{1}{2}I + \left[\langle\sigma^+\sigma^-\rangle - \frac{1}{2}\right]\sigma_z + \langle\sigma^+\rangle\sigma^- + \langle\sigma^-\rangle\sigma^+, \end{aligned} \tag{2.41}$$

with

$$\langle\sigma^+\sigma^-\rangle = \mathrm{Tr}\{\rho\sigma^+\sigma^-\}, \qquad \langle\sigma^+\rangle = \mathrm{Tr}\{\rho\sigma^+\}, \qquad \langle\sigma^-\rangle = \mathrm{Tr}\{\rho\sigma^-\}. \quad (2.42)$$

For future reference we list a number of additional relations among the spin operators:

$$\begin{aligned}
&(\sigma^+)^2 = (\sigma^-)^2 = 0,\\
&\sigma^+\sigma^- = \begin{pmatrix}1 & 0\\ 0 & 0\end{pmatrix} = |\alpha\rangle\langle\alpha|, \qquad \sigma^-\sigma^+ = \begin{pmatrix}0 & 0\\ 0 & 1\end{pmatrix} = |\beta\rangle\langle\beta|,\\
&\sigma^+\sigma^- + \sigma^-\sigma^+ = \begin{pmatrix}1 & 0\\ 0 & 1\end{pmatrix} = |\alpha\rangle\langle\alpha| + |\beta\rangle\langle\beta| = I,\\
&\sigma^+\sigma^- + \sigma^-\sigma^+ \equiv [\sigma^+, \sigma^-] = \begin{pmatrix}1 & 0\\ 0 & -1\end{pmatrix} = \sigma_z = 2(\sigma^+\sigma^- - \tfrac{1}{2}).
\end{aligned} \quad (2.43)$$

For the commutators one gets

$$\begin{aligned}
&[\sigma^+, \sigma_x] = \sigma_z, && [\sigma^-, \sigma_x] = -\sigma_z,\\
&[\sigma^+, \sigma_y] = i\sigma_z, && [\sigma^-, \sigma_y] = i\sigma_z,\\
&[\sigma^+, \sigma_z] = -2\sigma^+, && [\sigma^-, \sigma_z] = 2\sigma^-.
\end{aligned} \quad (2.44)$$

The spin operators acting on the eigenstates $|\alpha\rangle$ and $|\beta\rangle$ yield:

$$\begin{aligned}
&\sigma_x|\alpha\rangle = |\beta\rangle, && \sigma_x|\beta\rangle = |\alpha\rangle,\\
&\sigma_y|\alpha\rangle = i|\beta\rangle, && \sigma_y|\beta\rangle = -i|\alpha\rangle,\\
&\sigma_z|\alpha\rangle = |\alpha\rangle, && \sigma_z|\beta\rangle = -|\beta\rangle,\\
&\sigma^+|\alpha\rangle = 0, && \sigma^+|\beta\rangle = |\alpha\rangle,\\
&\sigma^-|\alpha\rangle = |\beta\rangle, && \sigma^-|\beta\rangle = 0.
\end{aligned} \quad (2.45)$$

In place of the Pauli spin operators in Eq. (2.26), it is often desirable to employ the Hermitian operators defined by

$$\mathbf{s} = \frac{\boldsymbol{\sigma}}{2} \quad (2.46)$$

that satisfy the angular momentum commutation rules

$$[s_x, s_y] = is_z, \qquad [s_y, s_z] = is_x, \qquad [s_z, s_x] = is_y, \quad (2.47)$$

or

$$\mathbf{s} \times \mathbf{s} = i\mathbf{s}. \quad (2.48)$$

Since the spin states $|\alpha\rangle$ and $|\beta\rangle$ are simultaneous eigenstates of σ_z and σ^2, as shown by Eq. (2.30), they are also simultaneous eigenstates of s_z and s^2:

$$s_z|\alpha\rangle = \frac{1}{2}|\alpha\rangle, \qquad s^2|\alpha\rangle = \frac{3}{4}|\alpha\rangle,$$
$$s_z|\beta\rangle = -\frac{1}{2}|\beta\rangle, \qquad s^2|\beta\rangle = \frac{3}{4}|\beta\rangle. \tag{2.49}$$

In a more general notation, the spin states are written $|sm\rangle$ where

$$|\alpha\rangle \equiv \left|\frac{1}{2}\,\frac{1}{2}\right\rangle, \qquad |\beta\rangle \equiv \left|\frac{1}{2}\,-\frac{1}{2}\right\rangle. \tag{2.50}$$

The four equations in Eq. (2.49) are then represented by

$$s_z|sm\rangle = m|sm\rangle, \qquad s^2|sm\rangle = s(s+1)|sm\rangle \tag{2.51}$$

with $s = 1/2$ and $m = \pm 1/2$.

This formalism may be extended to systems involving N spins. We define

$$\mathbf{S} = \sum_{i=1}^{N} \mathbf{s}_i, \tag{2.52}$$

where $\mathbf{s}_i$ is the spin operator for the ith spin. In view of the commutation rules of Eq. (2.47), we also have

$$\mathbf{S} \times \mathbf{S} = i\mathbf{S}. \tag{2.53}$$

Employing the rules for the coupling of angular momenta, one may construct simultaneous eigenstates of S_z and S^2:

$$S_z|SM\rangle = M|SM\rangle, \qquad S^2|SM\rangle = S(S+1)|SM\rangle \tag{2.54}$$

where

$$S = 0, 1, 2, \ldots, \frac{N}{2}, \qquad N \text{ even},$$
$$= \frac{1}{2}, \frac{3}{2}, \ldots, \frac{N}{2}, \qquad N \text{ odd}, \tag{2.55}$$
$$M = S, S-1, \ldots, -S+1, -S.$$

Thus, for two spins, $S = 0$ (singlet) or 1 (triplet). The normalized eigenstates—obtained by the rules for the addition of angular momenta—are

$$|00\rangle = \frac{1}{\sqrt{2}}[|\alpha_1\beta_2\rangle - |\alpha_2\beta_1\rangle] \tag{2.56a}$$

for the singlet and

$$|11\rangle = |\alpha_1\alpha_2\rangle, \qquad |1\,-1\rangle = |\beta_1\beta_2\rangle,$$
$$|10\rangle = \frac{1}{\sqrt{2}}[|\alpha_1\beta_2\rangle + |\alpha_2\beta_1\rangle] \tag{2.56b}$$

for the triplet. It is observed that the states corresponding to $S = 1$ are symmetric under an interchange of spins (i.e., indices 1 and 2) while the state corresponding to $S = 0$ is antisymmetric. For N greater than 2, the construction of spin eigenstates generally requires reference to the properties of the permutation group. The extension of these rules to atomic coherent states is given by Arecchi *et al.* [4] and will be illustrated in Section 5.1.

2.3 Schrödinger Representation

The evolution in time, also known as the dynamical behavior, of a quantum system is governed by the Hamiltonian operator, $\mathscr{H}$. There are several possibilities, however, whereby the system can be "driven" by the Hamiltonian. Its influence may be exerted entirely on the wave functions, entirely on the operators representing physical quantities (observables), or on both wave functions and observables. These are known, respectively, as the *Schrödinger*, *Heisenberg*, and *interaction* representations or pictures. An observable may have an intrinsic time-dependence which then needs to be taken into account in addition to the dynamical behavior. We shall have occasion to use all three representations; a summary of their properties is therefore included for reference.

The Schrödinger representation, as the name implies, is based on the time-dependent Schrödinger equation

$$i\hbar\frac{\partial}{\partial t}|\psi(\mathbf{r},t)\rangle = \mathscr{H}|\psi(\mathbf{r},t)\rangle, \tag{2.57}$$

which governs the temporal evolution of the wave function. The Hamiltonian may or may not be time-dependent. In the discussion that follows, the spatial dependence of the wave function is of no concern and will be suppressed in the notation.

In a purely formal manner, one may write

$$|\psi(t)\rangle = U(t,t_0)|\psi(t_0)\rangle, \tag{2.58a}$$
$$\langle\psi(t)| = \langle\psi(t_0)|U^\dagger(t,t_0), \tag{2.58b}$$

with

$$U(t_0,t_0) = 1, \tag{2.59}$$

in which $U(t, t_0)$ is known as the *evolution* or *time-development* operator. Clearly, the wave function $|\psi(t)\rangle$ may be sought either by attempting to solve the time-dependent Schrödinger equation or by seeking the specific form of $U(t, t_0)$, since both methods will give $|\psi(t)\rangle$ at an arbitrary time t when $|\psi(t_0)\rangle$ at an initial time t_0 is known. It will become apparent as we proceed that $U(t, t_0)$ appears prominently in numerous applications involving the temporal development of a physical system.

Let us first examine some general properties of $U(t, t_0)$. A system in an eigenstate $|\psi_j(t_0)\rangle$ at $t = t_0$ will evolve into a state $U(t, t_0)|\psi_j(t_0)\rangle$ at the time t. Assuming the eigenstates of the system form an orthonormal basis, one may expand $U(t, t_0)|\psi_j(t_0)\rangle$ in the form

$$U(t, t_0)|\psi_j(t_0)\rangle = \sum_k a_k|\psi_k(t)\rangle \tag{2.60}$$

in which a typical coefficient, a_i, is given by

$$a_i = \langle\psi_i(t)|U(t, t_0)|\psi_j(t_0)\rangle. \tag{2.61}$$

But a_i is just the probability amplitude for the system to be found in the state $|\psi_i(t)\rangle$ at the time t, given that the system had been in the state $|\psi_j(t)\rangle$ at $t = t_0$. Hence, the transition probability is

$$P_{ij} \equiv P(j \to i) = |a_i|^2 = |\langle\psi_i(t)|U(t, t_0)|\psi_j(t_0)\rangle|^2. \tag{2.62}$$

Since

$$\begin{aligned}|\psi(t_2)\rangle &= U(t_2, t_1)|\psi(t_1)\rangle = U(t_2, t_1)U(t_1, t_0)|\psi(t_0)\rangle \\ &= U(t_2, t_0)|\psi(t_0)\rangle,\end{aligned} \tag{2.63}$$

we have the group property

$$U(t_2, t_0) = U(t_2, t_1)U(t_1, t_0). \tag{2.64}$$

An interchange of t and t_0 in Eq. (2.58a) gives

$$|\psi(t_0)\rangle = U(t_0, t)|\psi(t)\rangle, \tag{2.65}$$

but upon multiplying Eq. (2.58a) on the left by $U^{-1}(t, t_0)$,

$$|\psi(t_0)\rangle = U^{-1}(t, t_0)|\psi(t)\rangle. \tag{2.66}$$

Thus,

$$U^{-1}(t, t_0) = U(t_0, t). \tag{2.67}$$

We now show that $U(t, t_0)$ is a unitary operator. Substitution of Eq. (2.58a) into Eq. (2.57) yields the operator equation

$$i\hbar\frac{\partial U(t, t_0)}{\partial t} = \mathscr{H}U(t, t_0), \tag{2.68}$$

and its Hermitian conjugate

$$-i\hbar\frac{\partial U^{\dagger}(t,t_0)}{\partial t} = U^{\dagger}(t,t_0)\mathscr{H}, \tag{2.69}$$

under the assumption that $\mathscr{H}^{\dagger} = \mathscr{H}$. This is the only restriction on the Hamiltonian, which in some cases may be a function of time. If Eq. (2.68) is multiplied on the left by $U^{\dagger}(t,t_0)$ and Eq. (2.69) is multiplied on the right by $U(t,t_0)$, the combination of the two equations yields

$$i\hbar\left[U^{\dagger}\frac{\partial U}{\partial t} + \frac{\partial U^{\dagger}}{\partial t}U\right] = i\hbar\frac{\partial}{\partial t}U^{\dagger}U = 0. \tag{2.70}$$

Thus, $U^{\dagger}U$ is a constant independent of time. To satisfy the normalization condition of Eq. (2.59), we must set

$$U^{\dagger}U = 1 \tag{2.71}$$

or

$$U^{\dagger}(t,t_0) = U^{-1}(t,t_0) = U(t_0,t). \tag{2.72}$$

The unitary property (Eq. (2.72)) leads to a useful interpretation of $U(t,t_0)$. If $|\psi(t_0)\rangle$ is regarded as a state vector pointing in a certain direction in Hilbert space, the effect of $U(t,t_0)$ on $|\psi(t_0)\rangle$ is to rotate the vector in a continuous fashion as time progresses from t_0 to t. In view of Eq. (2.72), $U^{\dagger}(t,t_0)$ rotates the state vector in the opposite sense corresponding to propagation of the wave function from t to t_0.

The differential equation (2.68) may be converted to an integral equation whose chief virtue is that it lends itself to an iterative solution. With the initial condition of Eq. (2.59),

$$U(t,t_0) = 1 - \frac{i}{\hbar}\int_{t_0}^{t}\mathscr{H}(t')U(t',t_0)\,dt': \tag{2.73}$$

we have allowed the Hamiltonian to be a function of time. By successive substitutions, $U(t,t_0)$ is expanded into the series

$$U(t,t_0) = 1 - \frac{i}{\hbar}\int_{t_0}^{t}dt_1\mathscr{H}(t_1) + \left(\frac{-i}{\hbar}\right)^2\int_{t_0}^{t}dt_1\int_{t_0}^{t_1}dt_2\mathscr{H}(t_1)\mathscr{H}(t_2) + \cdots. \tag{2.74}$$

When the Hamiltonian is independent of time, integration of Eq. (2.57) with respect to time gives

$$|\psi(t)\rangle = e^{-i\mathscr{H}(t-t_0)/\hbar}|\psi(t_0)\rangle, \tag{2.75}$$

which permits the identification

$$U(t,t_0) \equiv e^{-i\mathscr{H}(t-t_0)/\hbar}$$
$$= 1 + \frac{1}{i\hbar}\mathscr{H}(t-t_0) + \frac{1}{2!}\left(\frac{1}{i\hbar}\right)^2 \mathscr{H}^2(t-t_0)^2 + \cdots. \tag{2.76}$$

Since $\mathscr{H}$ commutes with each term in the series, it is evident that

$$[e^{\pm i\mathscr{H}(t-t_0)/\hbar}, \mathscr{H}] = 0. \tag{2.77}$$

If $|\psi_i\rangle$ is an eigenstate of the Hamiltonian $\mathscr{H}$, that is, $\mathscr{H}|\psi_i\rangle = \mathrm{E}_i|\psi_i\rangle$, the power series expansion in Eq. (2.76) leads to the following relations:

$$\begin{aligned} U(t,t_0)|\psi_i\rangle &= e^{-i\mathscr{H}(t-t_0)/\hbar}|\psi_i\rangle = e^{-i\mathrm{E}_i(t-t_0)/\hbar}|\psi_i\rangle, \\ U^\dagger(t,t_0)|\psi_i\rangle &= e^{i\mathscr{H}(t-t_0)/\hbar}|\psi_i\rangle = e^{i\mathrm{E}_i(t-t_0)/\hbar}|\psi_i\rangle, \\ \langle\psi_i|U(t,t_0) &= [U^\dagger(t,t_0)|\psi_i\rangle]^\dagger = \langle\psi_i|e^{-i\mathrm{E}_i(t-t_0)/\hbar}, \\ \langle\psi_i|U^\dagger(t,t_0) &= [U(t,t_0)|\psi_i\rangle]^\dagger = \langle\psi_i|e^{i\mathrm{E}_i(t-t_0)/\hbar}. \end{aligned} \tag{2.78b}$$

When the Hamiltonian is of the form $\mathscr{H} = \mathscr{H}_0 + V(t)$, solutions to the time-dependent Schrödinger equation (2.57) may be sought as an expansion in the complete set of orthonormal eigenstates of the unperturbed Hamiltonian, $\mathscr{H}_0$. Letting

$$\mathscr{H}_0|\phi_k(\mathbf{r})\rangle = \mathrm{E}_k|\phi_k(\mathbf{r})\rangle, \tag{2.79}$$

$$|\psi(\mathbf{r},t)\rangle = \sum_k a_k(t)|\phi_k(\mathbf{r})\rangle, \tag{2.80}$$

the substitution of Eq. (2.80) into Eq. (2.57) leads to

$$i\hbar\sum_k \dot{a}_k(t)|\phi_k(\mathbf{r})\rangle = \sum_k \mathrm{E}_k a_k(t)|\phi_k(\mathbf{r})\rangle + \sum_k a_k(t)V(t)|\phi_k(\mathbf{r})\rangle. \tag{2.81}$$

When this equation is multiplied on the left by $\langle\phi_l(\mathbf{r})|$ and integrated over $\mathbf{r}$, the result is

$$i\hbar\dot{a}_l(t) = \mathrm{E}_l a_l(t) + \sum_k a_k(t)\langle\phi_l(\mathbf{r})|V(t)|\phi_k(\mathbf{r})\rangle. \tag{2.82}$$

To obtain the coefficients in the expansion (Eq. (2.80)), it is necessary, therefore, to solve the system of coupled differential equations exemplified by Eq. (2.82).

In place of Eq. (2.80), the expansion of $|\psi(\mathbf{r},t)\rangle$ also may be written

$$|\psi(\mathbf{r},t)\rangle = \sum_k c_k(t)|\phi_k(\mathbf{r})\rangle e^{-i\mathrm{E}_k t/\hbar}, \tag{2.83}$$

in which case the resulting differential equations analogous to Eq. (2.82) are

$$i\hbar\dot{c}_l(t) = \sum_k c_k(t)\langle\phi_l(\mathbf{r})|V(t)|\phi_k(\mathbf{r})\rangle e^{i\omega_{lk}t} \tag{2.84}$$

where $\hbar\omega_{lk} = \mathrm{E}_l - \mathrm{E}_k$. Clearly, the relation between the expansion coefficients is

$$a_k(t) = c_k(t)e^{-i\mathrm{E}_k t/\hbar}. \tag{2.85}$$

To illustrate how one obtains successive approximations to the coefficients, suppose that initially a single coefficient $c_m(t_0) = 1$ whereas all other coefficients $c_k(t_0) = 0$, i.e.,

$$c_m(t_0) \equiv c_k^{(0)} = \delta_{km}. \tag{2.86}$$

The general iterative solution then will be of the form

$$c_l^{(r+1)}(t) = -\frac{i}{\hbar}\sum_k \int_{t_0}^{t} c_k^{(r)}(t')\langle\phi_l|V(t')|\phi_k\rangle e^{i\omega_{lk}t'}\,dt' \tag{2.87}$$

which is an approximate solution to Eq. (2.84) to order $(r + 1)$. Thus,

$$c_l^{(1)}(t) = -\frac{i}{\hbar}\sum_k \int_{t0}^{t} c_k^{(0)}(t')\langle\phi_l|V(t')|\phi_k\rangle e^{i\omega_{lk}t'}\,dt', \tag{2.88}$$

$$c_l^{(2)}(t) = -\frac{i}{\hbar}\sum_k \int_{t_0}^{t} c_k^{(1)}(t')\langle\phi_l|V(t')|\phi_k\rangle e^{i\omega_{lk}t'}\,dt', \tag{2.89}$$

to first and second order. If the perturbation is independent of time, the first- and second-order solutions, with $c_k^{(0)} = \delta_{km}$, $t_0 = 0$, $V_{lm} = \langle\phi_l|V|\phi_m\rangle$ and $V_{km} = \langle\phi_k|V|\phi_m\rangle$ are

$$c_l^{(1)}(t) = \frac{V_{lm}}{\hbar\omega_{lm}}(1 - e^{i\omega_{lm}t}), \tag{2.90}$$

$$c_l^{(2)}(t) = \sum_k \frac{V_{lk}V_{km}}{\hbar\omega_{km}}\left(\frac{1 - e^{i\omega_{lk}t}}{\hbar\omega_{lk}} - \frac{1 - e^{i\omega_{lm}t}}{\hbar\omega_{lm}}\right). \tag{2.91}$$

Initially, the state $|\phi_m\rangle$ is occupied and all other states, including $|\phi_l\rangle$, are vacant. At time progresses, the probability amplitude c_l increases from its initial value of zero. We may therefore regard $|c_l(t)|^2$ as the probability for a transition from $|\phi_m\rangle$ to $|\phi_l\rangle$ in the time t. To first and second order,

$$\begin{aligned}
|c_l^{(1)}(t)|^2 &= \frac{2|V_{lm}|^2}{(\hbar\omega_{lm})^2}(1 - \cos\omega_{lm}t) \\
&= \frac{2\pi t}{\hbar^2}|V_{lm}|^2\frac{(1 - \cos\omega_{lm}t)}{\pi t\omega_{lm}^2} \\
&= \frac{2\pi t}{\hbar^2}|V_{lm}|^2\,\delta(\omega_{lm}) \qquad (t \to \infty) \\
&= \frac{2\pi t}{\hbar}|V_{lm}|^2\,\delta(\mathrm{E}_l - \mathrm{E}_m).
\end{aligned} \tag{2.92}$$

$$|c_1^{(2)}(t)|^2 = \frac{2\pi t}{\hbar}\left|\sum_k \frac{V_{lk}V_{km}}{\hbar\omega_{km}}\right|^2 \delta(\mathrm{E}_l - \mathrm{E}_m). \tag{2.93}$$

Equation (2.92) is the most common form of the *Fermi Golden Rule.*

2.4 Heisenberg Representation

The transition to the Heisenberg representation is accomplished by defining a new wave function

$$|\psi_{\mathrm{H}}\rangle = U^{\dagger}(t, t_0)|\psi(\mathbf{r}, t)\rangle \tag{2.94}$$

where $|\psi(\mathbf{r}, t)\rangle$ is a Schrödinger wave function, that is, a function that satisfies the time-dependent Schrödinger equation. Differentiating with respect to time and noting that $U^{\dagger}(t, t_0)$ and $|\psi(t)\rangle$ satisfy Eqs. (2.69) and (2.57), respectively,

$$\frac{\partial}{\partial t}|\psi_{\mathrm{H}}\rangle = 0 \tag{2.95a}$$

or

$$|\psi_{\mathrm{H}}(\mathbf{r})\rangle = |\psi(\mathbf{r}, t_0)\rangle \tag{2.95b}$$

for all time. Since the spatial coordinates are not involved in the transformation, there is no need to show them explicitly. For an observable represented by an operator A, the expectation value $\langle\psi(t)|A|\psi(t)\rangle$ in the Schrödinger representation may be converted to the Heisenberg representation by transforming the wave function according to Eq. (2.94)

$$\begin{aligned}\langle\psi(t)|A|\psi(t)\rangle &= \langle U(t, t_0)\psi_{\mathrm{H}}|A|U(t, t_0)\psi_{\mathrm{H}}\rangle \\ &= \langle\psi_{\mathrm{H}}|U^{\dagger}(t, t_0)AU(t, t_0)|\psi_{\mathrm{H}}\rangle \\ &= \langle\psi_{\mathrm{H}}|A_{\mathrm{H}}(t)|\psi_{\mathrm{H}}\rangle = \langle A\rangle,\end{aligned} \tag{2.96}$$

in which the Heisenberg operator $A_{\mathrm{H}}(t)$ is defined by

$$A_{\mathrm{H}}(t) = U^{\dagger}(t, t_0)AU(t, t_0). \tag{2.97}$$

It is seen that it makes no difference whether the expectation value of an operator is evaluated in the Schrödinger or Heisenberg representation, provided the wave function and operator are transformed in accordance with Eqs. (2.94) and (2.97), respectively. At $t = t_0$, Heisenberg wave functions and operators evidently merge with the corresponding Schrödinger wave functions and operators.

The product of two Schrödinger operators A and B transforms to the Heisenberg representation according to

$$(AB)_{\mathrm{H}} = U^{\dagger}(t, t_0)ABU(t, t_0). \tag{2.98}$$

But since $U(t, t_0)$ is unitary,

$$(AB)_{\mathrm{H}} = U^{\dagger}(t, t_0)AU(t, t_0)U^{\dagger}(t, t_0)BU(t, t_0) = A_{\mathrm{H}}B_{\mathrm{H}}. \tag{2.99}$$

It then follows that a commutator relation of the form

$$[A, B] = C \tag{2.100}$$

in the Schrödinger representation implies the corresponding relation

$$[A_{\mathrm{H}}, B_{\mathrm{H}}] = C_{\mathrm{H}} \tag{2.101}$$

in the Heisenberg representation.

The time rate of change of a Heisenberg operator is obtained directly from Eqs. (2.97), (2.68), and (2.69):

$$\begin{aligned} i\hbar\frac{d}{dt}A_{\mathrm{H}}(t) &= i\hbar\frac{d}{dt}\left[U^{\dagger}(t, t_0)AU(t, t_0)\right] \\ &= U^{\dagger}(t, t_0)A\mathscr{H}U(t, t_0) - U^{\dagger}(t, t_0)\mathscr{H}U(t, t_0) \\ &= A_{\mathrm{H}}(t)\mathscr{H}_{\mathrm{H}}(t) \quad \mathscr{H}_{\mathrm{H}}(t)A_{\mathrm{H}}(t) = [A_{\mathrm{H}}(t), \mathscr{H}_{\mathrm{H}}(t)]. \end{aligned} \tag{2.102}$$

This is the Heisenberg equation. If an operator A in the Schrödinger representation commutes with $\mathscr{H}$, i.e., $[A, \mathscr{H}] = 0$, then $[A_{\mathrm{H}}, \mathscr{H}_{\mathrm{H}}] = 0$ and $A_{\mathrm{H}}(t) = A$ independent of time. One also may compute the time rate of change of the expectation value. From Eq. (2.102) and the relation $\langle A_{\mathrm{H}}(t)\rangle = \langle A\rangle$,

$$\begin{aligned} i\hbar\frac{d}{dt}\langle\psi_{\mathrm{H}}|A_{\mathrm{H}}(t)|\psi_{\mathrm{H}}\rangle \equiv i\hbar\frac{d}{dt}\langle A_{\mathrm{H}}(t)\rangle &= \langle[A_{\mathrm{H}}(t), \mathscr{H}_{\mathrm{H}}]\rangle \\ &= i\hbar\frac{d}{dt}\langle A\rangle = \langle[A, \mathscr{H}]\rangle. \end{aligned} \tag{2.103a}$$

In other words, if

$$i\hbar\frac{d}{dt}A_{\mathrm{H}}(t) = [A_{\mathrm{H}}(t), \mathscr{H}_{\mathrm{H}}(t)] \tag{2.103b}$$

in the Heisenberg representation, then

$$i\hbar\frac{d}{dt}\langle A\rangle = \langle[A, \mathscr{H}]\rangle. \tag{2.103c}$$

Ignoring possible intrinsic time-dependence, an operator A in the Schrödinger representation is independent of time; hence, $dA/dt = 0$. But $d\langle A\rangle/dt$ is not automatically zero owing to the time-dependent wave function contained in $\langle A\rangle$.

In the event that $\mathscr{H}$ is independent of time, we write, in place of Eqs. (2.94) and (2.97),

$$|\psi_{\mathrm{H}}\rangle = e^{i\mathscr{H}(t-t_0)/\hbar}|\psi(t)\rangle, \tag{2.104}$$

$$A_{\mathrm{H}}(t) = e^{i\mathscr{H}(t-t_0)/\hbar}Ae^{-i\mathscr{H}(t-t_0)/\hbar}, \tag{2.105}$$

and

$$\mathscr{H}_{\mathrm{H}} = \mathscr{H}. \tag{2.106}$$

The equation of motion (Eq. (2.102)) then becomes

$$i\hbar\frac{d}{dt}A_{\mathrm{H}}(t) = [A_{\mathrm{H}}(t), \mathscr{H}]. \tag{2.107}$$

When $|k\rangle$ and $|l\rangle$ are eigenstates of $\mathscr{H}$ with eigenvalues E_k and E_l, we have, with the help of Eqs. (2.78),

$$\begin{aligned}\langle k|A_{\mathrm{H}}(t)|l\rangle &= \langle k|e^{i\mathscr{H}(t-t_0)/\hbar}Ae^{-i\mathscr{H}(t-t_0)/\hbar}|l\rangle \\ &= \langle k|A|l\rangle e^{i(E_k-E_l)(t-t_0)/\hbar},\end{aligned} \tag{2.108a}$$

$$\langle k|A_{\mathrm{H}}(t)|k\rangle = \langle k|A|k\rangle. \tag{2.108b}$$

For an operator $A(t)$ with an intrinsic time-dependence, the transformation to the Heisenberg representation may still be written as in Eq. (2.105) provided $\mathscr{H}$ is independent of time. But the time derivative of $A_{\mathrm{H}}(t)$ now contains an additional term due to $\partial A(t)/\partial t$, so that the equation of motion (Eq. (2.107)) must be modified to

$$i\hbar\,\dot{A}_{\mathrm{H}}(t) = [A_{\mathrm{H}}(t), \mathscr{H}] + i\hbar\left[\frac{\partial}{\partial t}A(t)\right]_{\mathrm{H}} \tag{2.109}$$

where in the last term

$$\left[\frac{\partial}{\partial t}A(t)\right]_{\mathrm{H}} = U^{\dagger}(t,t_0)\left[\frac{\partial}{\partial t}A(t)\right]U(t,t_0). \tag{2.110}$$

2.5 Interaction Representation

The interaction representation is particularly suited to the solution of the time-dependent Schrödinger equation when the Hamiltonian is of the form

$$\mathscr{H} = \mathscr{H}_0 + V \tag{2.111}$$

where V is regarded as a perturbation. In problems of this type it is generally assumed that $\mathscr{H}_0$ is independent of time and that its eigenfunctions and eigenvalues are known. The perturbation, on the other hand, may be time-dependent. If $\psi(t)$ is a wave function and A an observable (with no intrinsic time-dependence) in the Schrödinger representation, we define $|\tilde{\psi}(t)\rangle$ and $\tilde{A}(t)$ in the *interaction* representation by the transformations

$$|\tilde{\psi}(t)\rangle = e^{i\mathscr{H}_0(t-t_0)/\hbar}|\psi(t)\rangle, \tag{2.112}$$

$$\tilde{A}(t) = e^{i\mathscr{H}_0(t-t_0)/\hbar}Ae^{-i\mathscr{H}_0(t-t_0)/\hbar}. \tag{2.113}$$

These definitions are similar in form to the corresponding definitions in the Heisenberg representation, except for the important distinction that the Heisenberg representation is based on the total Hamiltonian $\mathscr{H}$ while the interaction representation is based on the unperturbed Hamiltonian $\mathscr{H}_0$.

The equation for the time-dependence of $|\psi(t)\rangle$ is obtained from the definition (2.112) and the time-dependent Schrödinger equation

$$i\hbar\frac{\partial}{\partial t}|\psi(t)\rangle = (\mathscr{H}_0 + V)|\psi(t)\rangle. \tag{2.114}$$

The result is

$$i\hbar\frac{\partial}{\partial t}|\tilde{\psi}(t)\rangle = \tilde{V}(t)|\tilde{\psi}(t)\rangle. \tag{2.115}$$

Similarly, when $\tilde{A}(t)$, defined by Eq. (2.113), is differentiated with respect to time, the result is

$$i\hbar\frac{\partial\tilde{A}(t)}{\partial t} = [\tilde{A}(t), \mathscr{H}_0]. \tag{2.116}$$

We see, then, that in the interaction representation both the wave function $|\tilde{\psi}(t)\rangle$ and the operator $\tilde{A}(t)$ are functions of time. The time-dependence of $|\tilde{\psi}(t)\rangle$, however, is governed by the perturbation term V while the time-dependence of $\tilde{A}(t)$ is controlled by the unperturbed part of the Hamiltonian $\mathscr{H}_0$.

Solutions to Eq. (2.115) may be obtained in a form analogous to Eq. (2.80). Writing

$$|\tilde{\psi}(\mathbf{r}, t)\rangle = \sum_k c_k(t)|\phi_k(\mathbf{r})\rangle, \tag{2.117}$$

where $\mathscr{H}_0|\varphi_k(\mathbf{r})\rangle$, the equations for the coefficients resulting from substitution in Eq. (2.115) are

$$\begin{aligned} i\hbar\dot{c}_1(t) &= \sum_k c_k(t) < \phi_1(\mathbf{r})|\tilde{V}(t)|\phi_k(\mathbf{r})\rangle \\ &= \sum_k c_k(t)\langle\phi_1(\mathbf{r})|V|\phi_k(\mathbf{r})\rangle e^{i\omega_{1k}t} \end{aligned} \tag{2.118}$$

as in Eq. (2.84).

The expectation value of an operator in the interaction representation is defined by

$$\langle \tilde{A}(t) \rangle = \langle \tilde{\psi}(t) | \tilde{A}(t) | \tilde{\psi}(t) \rangle. \tag{2.119}$$

The right side of the equation may be converted to the Schrödinger representation by means of Eqs. (2.112) and (2.113). Then

$$\langle \tilde{A}(t) \rangle = \langle \psi(t) | A | \psi(t) \rangle \equiv \langle A \rangle. \tag{2.120}$$

This relation, as in Eq. (2.96), demonstrates that the expectation value of an operator is independent of the representation in which it is expressed. This is an expected result since expectation values of observables are simply numbers associated with physical measurements and therefore cannot depend on the choice of representation which, after all, is merely a computational device.

If $|k\rangle$ and $|l\rangle$ are eigenstates of $\mathscr{H}_0$ with eigenvalues E_k and E_1, respectively,

$$\begin{aligned} \langle k | \tilde{A}(t) | l \rangle &= \langle k | e^{i\mathscr{H}_0(t-t_0)/\hbar} A e^{-i\mathscr{H}_0(t-t_0)/\hbar} | l \rangle \\ &= \langle k | A | l \rangle e^{i(E_k - E_l)(t-t_0)/\hbar}, \end{aligned} \tag{2.121}$$

which reduces to

$$\langle k | \tilde{A}(t) | k \rangle = \langle k | A | k \rangle \tag{2.122}$$

for diagonal matrix elements. These relations are similar to Eqs. (2.108a) and (2.108b) in the Heisenberg representation with the important difference that, in the latter, the wave functions $|k\rangle$ and $|l\rangle$ are eigenstates of the total Hamiltonian $\mathscr{H}$, whereas in Eqs. (2.121) and (2.122), they are eigenstates of the unperturbed Hamiltonian $\mathscr{H}_0$.

In summary, the principal characteristics of the three representations are the following:

- Schrödinger representation. The wave functions are time-dependent; they obey the Schrödinger equation and are the source of information on the time-development of the system. Schrödinger operators (observables) do not evolve in time under the influence of the Hamiltonian but may have an intrinsic time-dependence. The expectation value of an operator, evaluated at any time t, is the same in all representations and is time-dependent unless the operator commutes with the Hamiltonian.

- Heisenberg representation. The wave functions are independent of time, but operators are time-dependent and obey the Heisenberg equation of motion. Information on the time-development of the system is based on the operators.

- Interaction representation. Both wave functions and operators are time-dependent. When $\mathscr{H} = \mathscr{H}_0 + V$, the wave functions owe their time-dependence to V and the operators to $\mathscr{H}_0$. At the initial time t_0 (which may be zero) the three representations are indistinguishable, i.e.,

$$|\tilde{\psi}(t_0)\rangle = |\psi_H\rangle = |\psi(t_0)\rangle, \tag{2.123a}$$

$$\tilde{A}(t_0) = A_H(t_0) = A. \tag{2.123b}$$

The time-development of a Schrödinger wave function was described in terms of the evolution operator $U(t, t_0)$. We should like to construct an analogous unitary operator $\tilde{U}(t, t_0)$ to describe the time-development of a wave function in the interaction representation. Let

$$|\tilde{\psi}(t_2)\rangle = \tilde{U}(t_2, t_1)|\tilde{\psi}(t_1)\rangle = \tilde{U}(t_2, t_1)e^{i\mathscr{H}_0(t_1 - t_0)/\hbar}|\psi(t_1)\rangle, \tag{2.124}$$

but

$$\begin{aligned} |\tilde{\psi}(t_2)\rangle &= e^{i\mathscr{H}_0(t_2 - t_0)/\hbar}|\psi(t_2)\rangle \\ &= e^{i\mathscr{H}_0(t_2 - t_0)/\hbar}U(t_2, t_1)|\psi(t_1)\rangle. \end{aligned} \tag{2.125}$$

Comparing the two expressions, we have

$$\tilde{U}(t_2, t_1) = e^{i\mathscr{H}_0(t_2 - t_0)/\hbar}U(t_2, t_1)e^{-i\mathscr{H}_0(t_1 - t_0)/\hbar}, \tag{2.126a}$$

but since the initial time t_0 is arbitrary, we may set $t_0 = 0$. In this case,

$$\tilde{U}(t_2, t_1) = e^{i\mathscr{H}_0 t_2/\hbar}U(t_2, t_1)e^{-i\mathscr{H}_0 t_1/\hbar}. \tag{2.126b}$$

In a more convenient notation, for two arbitrary times t and t_0,

$$\begin{aligned} \tilde{U}(t, t_0) &= e^{i\mathscr{H}_0 t/\hbar}U(t, t_0)e^{-i\mathscr{H}_0 t_0/\hbar} \\ &= e^{i\mathscr{H}_0 t/\hbar}e^{-i\mathscr{H}(t - t_0)/\hbar}e^{-i\mathscr{H}_0 t_0/\hbar} \end{aligned} \tag{2.127}$$

and

$$|\tilde{\psi}(t)\rangle = \tilde{U}(t, t_0)|\psi(t_0)\rangle, \tag{2.128}$$

$$\tilde{U}(t_0, t_0) = U(t_0, t_0) = 1. \tag{2.129}$$

We also shall find it useful to define

$$U_0(t, t_0) = e^{-i\mathscr{H}_0(t - t_0)/\hbar}. \tag{2.130}$$

Equation (2.62) relates the transition probability $P_{ij} \equiv P(j \to i)$ to the evolution operator $U(t, t_0)$. Converting to the interaction representation, we obtain an alternative expression:

$$\begin{aligned} P_{ij} &= |\langle e^{-i\mathscr{H}_0(t - t_0)/\hbar}\tilde{\psi}_i(t)|U(t, t_0)|\psi_j(t_0)\rangle|^2 \\ &= |\langle\tilde{\psi}_i(t)|e^{i\mathscr{H}_0(t - t_0)/\hbar}U(t, t_0)|\psi_j(t_0)\rangle|^2 \\ &= |\tilde{\psi}_i(t)|\tilde{U}(t, t_0)|\psi_j(t_0)\rangle|^2. \end{aligned} \tag{2.131}$$

The insertion of Eq. (2.127) into Eq. (2.115) yields a differential equation analogous to Eq. (2.68),

$$i\hbar\frac{\partial\tilde{U}(t,t_0)}{\partial t} = \tilde{V}(t)\tilde{U}(t,t_0), \tag{2.132}$$

with the important difference that Eq. (2.68) is based on the total Hamiltonian $\mathscr{H}$, whereas Eq. (2.132) is based on the perturbation V alone. Here, as in Eq. (2.73), an equivalent integral equation containing the initial condition of Eq. (2.75) may be written

$$\tilde{U}(t,t_0) = 1 - \frac{i}{\hbar}\int_{t_0}^{t} \tilde{V}(t')\tilde{U}(t',t_0)\,dt'. \tag{2.133}$$

As mentioned previously, the advantage of the integral equation is that it may be iterated to arbitrary order:

$$\begin{aligned}\tilde{U}(t,t_0) &= \tilde{U}^{(0)}(t,t_0) + \tilde{U}^{(1)}(t,t_0) + \tilde{U}^{(2)}(t,t_0) + \cdots,\\ &= 1 - \frac{i}{\hbar}\int_{t_0}^{t}\tilde{V}(t_1)\,dt_1 + \left(\frac{-i}{\hbar}\right)^2\int_{t_0}^{t}dt_1\int_{t_0}^{t_1}dt_2\,\tilde{V}(t_1)\tilde{V}(t_2) + \cdots,\end{aligned} \tag{2.134}$$

where

$$t > t_1 > t_2 > \cdots > t_0. \tag{2.135}$$

A further useful step is to convert Eq. (2.134) to the Schrödinger representation with the help of Eq. (2.113) and (2.128):

$$U(t,t_0) = U^{(0)}(t,t_0) + U^{(1)}(t,t_0) + U^{(2)}(t,t_0) + \cdots \tag{2.136}$$

where

$$\begin{aligned}U^{(0)}(t,t_0) &= e^{-i\mathscr{H}_0(t-t_0)/\hbar}\tilde{U}^{(0)}(t,t_0) = e^{-i\mathscr{H}_0(t-t_0)/\hbar}\\ &= U_0(t,t_0),\end{aligned} \tag{2.137a}$$

$$\begin{aligned}U^{(1)}(t,t_0) &= e^{-i\mathscr{H}_0(t-t_0)/\hbar}\tilde{U}^{(1)}(t,t_0)\\ &= -\frac{i}{\hbar}\int_{t_0}^{t}dt_1 e^{-i\mathscr{H}_0(t-t_0)/\hbar}\tilde{V}(t_1)\\ &= -\frac{i}{\hbar}\int_{t_0}^{t}dt_1 e^{-i\mathscr{H}_0(t-t_1)/\hbar}V(t_1)e^{-i\mathscr{H}_0(t_1-t_0)/\hbar},\\ &= -\frac{i}{\hbar}\int_{t_0}^{t}dt_1 U_0(t,t_1)V(t_1)U_0(t_1,t_0),\end{aligned} \tag{2.137b}$$

$$\begin{aligned}
U^{(2)}(t,t_0) &= e^{-i\mathscr{H}_0(t-t_0)/\hbar}\tilde{U}^{(2)}(t,t_0) \\
&= \left(\frac{-i}{\hbar}\right)^2 \int_{t_0}^{t} dt_1 \int_{t_0}^{t_1} dt_2 e^{-i\mathscr{H}_0(t-t_0)/\hbar}\tilde{V}(t_1)\tilde{V}(t_2) \\
&= \left(\frac{-i}{\hbar}\right)^2 \int_{t_0}^{t} dt_1 \int_{t_0}^{t_1} dt_2 e^{-i\mathscr{H}_0(t-t_1)/\hbar}V(t_1) \\
&\quad \times e^{-i\mathscr{H}_0(t_1-t_2)/\hbar}V(t_2)e^{-i\mathscr{H}_0(t_2-t_0)/\hbar}, \\
&= \left(\frac{-i}{\hbar}\right)^2 \int_{t_0}^{t} dt_1 \int_{t0}^{t_1} dt_2 U_0(t,t_1)V(t_1)U_0(t_1,t_2) \\
&\quad \times V(t_2)U_0(t_2,t_0).
\end{aligned} \tag{2.137c}$$

The general form is

$$\begin{aligned}
U^{(n)}(t,t_0) = \left(\frac{-i}{\hbar}\right)^n \int_{t_0}^{t} dt_1 \int_{t_0}^{t_1} dt_2 \cdots \int_{t_0}^{t_{n-1}} dt_n U_0(t,t_1)V(t_1)U_0(t_1,t_2) \\
\times V(t_2)\cdots U_0(t_{n-1},t_n)V(t_n)U_0(t_n,t_0).
\end{aligned} \tag{2.138}$$

With this expansion, a solution to the time-dependent Schrödinger equation may be written

$$|\psi(t)\rangle = |\psi^{(0)}(t)\rangle + |\psi^{(1)}(t)\rangle + \cdots + |\psi^{(n)}(t)\rangle + \cdots \tag{2.139}$$

where

$$|\psi^{(n)}(t)\rangle = U^{(n)}(t,t_0)|\psi(t_0)\rangle. \tag{2.140}$$

Following the same procedure,

$$U^{\dagger}(t,t_0) = U^{(0)\dagger}(t,t_0) + U^{(1)\dagger}(t,t_0) + U^{(2)\dagger}(t,t_0) + \cdots, \tag{2.141}$$

where

$$U^{(0)\dagger}(t,t_0) = e^{i\mathscr{H}_0(t-t_0)/\hbar} = U_0^{\dagger}(t,t_0), \tag{2.142a}$$

$$\begin{aligned}
U^{(1)\dagger}(t,t_0) &= \frac{i}{\hbar}\int_{t_0}^{t} dt'_1 e^{i\mathscr{H}_0(t'_1-t_0)/\hbar}V(t'_1)e^{i\mathscr{H}_0(t-t'_1)/\hbar} \\
&= \frac{i}{\hbar}\int_{t_0}^{t} dt'_1 U_0^{\dagger}(t'_1,t_0)V(t'_1)U_0^{\dagger}(t,t'_1),
\end{aligned} \tag{2.142b}$$

$$\begin{aligned}
U^{(2)\dagger}(t,t_0) &= \left(\frac{i}{\hbar}\right)^2 \int_{t_0}^{t} dt'_1 \int_{t_0}^{t'_1} dt'_2 e^{i\mathscr{H}_0(t'_2-t_0)/\hbar}V(t'_2) \times e^{i\mathscr{H}_0(t'_1-t'_2)/\hbar}V(t'_1)e^{i\mathscr{H}_0(t-t'_1)/\hbar}, \\
&= \left(\frac{i}{\hbar}\right)^2 \int_{t_0}^{t} dt'_1 \int_{t_0}^{t'_1} dt'_2 U_0^{\dagger}(t'_2,t_0)V(t'_2) \times U_0^{\dagger}(t'_1,t'_2)V(t'_1)U_0^{\dagger}(t,t'_1),
\end{aligned} \tag{2.142c}$$

$$U^{(n)\dagger}(t, t_0) = \left(\frac{i}{\hbar}\right)^n \int_{t_0}^{t} dt'_1 \int_{t_0}^{t'_1} dt'_2 \cdots \int_{t_0}^{t'_{n-1}} dt'_n U_0^\dagger(t'_n, t_0)$$
$$\times V(t'_n) U_0^\dagger(t'_{n-1}, t'_n) V(t'_{n-1}) \cdots V(t'_2)$$
$$\times U_0^\dagger(t'_1, t'_2) V(t'_1) U_0^\dagger(t, t'_1), \tag{2.143}$$

with

$$t > t'_1 > t'_2 > \cdots > t_0. \tag{2.144}$$

Corresponding to Eq. (2.140) we have

$$\langle \psi^{(n)}(t)| = \langle \psi(t_0)| U^{(n)\dagger}(t, t_0). \tag{2.145}$$

Primes have been inserted on the time variables in Eqs. (2.142) and (2.143) in order to distinguish between the temporal development of the evolution operator and that of its Hermitian conjugate. It should be noted that, apart from t_0 being the earliest time and t the latest, there is no constraint on the relative ordering of primed and unprimed time variables; for example, t_1 may be earlier or later than t'_1 or any other primed time variabled. This feature becomes relevant in the successive approximation of density matrix elements and their representations by two-sided Feynman diagrams.

2.6 Equations of Motion

The temporal behavior of the density operator is of central importance in the time-development of a physical system under the influence of various interactions. If $|\psi_i(t)\rangle$ is a function of time, the density operator, defined by Eq. (2.1), also will depend on the time. Referring to Eq. (2.58),

$$\rho(t) \sum_i P_i |\psi_i(t)\rangle\langle\psi_i(t)|$$
$$= \sum_i P_i |U(t, t_0)\psi_i(t_0)\rangle\langle U(t, t_0)\psi_i(t_0)| \tag{2.146}$$
$$= \sum_i P_i U(t, t_0)|\psi_i(t_0)\rangle\langle\psi_i(t_0)| U^\dagger(t, t_0),$$

and writing

$$\rho(t_0) = \sum_i P_i |\psi_i(t_0)\rangle\langle\psi_i(t_0)|, \tag{2.147}$$

the time-dependence of ρ is given by

$$\rho(t) = U(t, t_0)\rho(t_0)U^\dagger(t, t_0). \tag{2.148}$$

$\mathrm{Tr}\{\rho(t)\}$ is independent of time, however, since under all circumstances $\mathrm{Tr}\,\rho = 1$. For a time-independent Hamiltonian,

$$\rho(t) = e^{-i\mathscr{H}(t-t_0)/\hbar}\rho(t_0)e^{i\mathscr{H}(t-t_0)/\hbar}. \tag{2.149}$$

Eq. (2.149) is equivalent to Eq. (2.148).

The differential equations (2.68) and (2.69) for $U(t, t_0)$ and $U^\dagger(t, t_0)$ now may be used to obtain the equation of motion for the density operator:

$$\begin{aligned} i\hbar\dot{\rho}(t) &= i\hbar\left[U\rho(t_0)\frac{\partial U^\dagger}{\partial t} + \frac{\partial U}{\partial t}\rho(t_0)U^\dagger\right] \\ &= -U\rho(t_0)U^\dagger\mathscr{H} + \mathscr{H}U\rho(t_0)U^\dagger = -\rho(t)\mathscr{H} + \mathscr{H}\rho(t) \end{aligned} \tag{2.150}$$

or

$$i\hbar\dot{\rho}(t) = [\mathscr{H}, \rho(t)] \equiv L\rho(t). \tag{2.151}$$

The operator L is known as the *Liouville operator* and Eq. (2.151) is the quantum mechanical Liouville equation. It should be noted that Eq. (2.151), despite its close resemblance to the Heisenberg equation (Eq. (2.107)—apart from the difference in sign—is, in fact, in the Schrödinger representation. That is, the density operator, in contrast to observables, may be a time-dependent operator in the Schrödinger representation unless ρ commutes with $\mathscr{H}$.

The Liouville equation may be converted to the equivalent integral equation

$$\rho(t) = \rho(t_0) - \frac{i}{\hbar}\int_{t_0}^{t}[\mathscr{H}(t'), \rho(t')]\,dt', \tag{2.152}$$

which may be inserted in Eq. (2.151) to obtain a variation of the Liouville equation:

$$i\hbar\dot{\rho}(t) = [\mathscr{H}, p(t_0)] - \frac{i}{\hbar}\int_{t_0}^{t}[\mathscr{H}(t'), [\mathscr{H}(t'), \rho(t')]]\,dt'. \tag{2.153}$$

If A is a Schrödinger operator with no intrinsic time-dependence, then

$$\begin{aligned} i\hbar\frac{d}{dt}\langle A\rangle &= i\hbar\frac{d}{dt}(\mathrm{Tr}\{\rho A\}) \\ &= i\hbar\,\mathrm{Tr}\left(\frac{d\rho}{dt}A\right) = \mathrm{Tr}\{[\mathscr{H}, \rho(t)]A\}. \end{aligned} \tag{2.154}$$

But due to the invariance of the trace under cyclic permutations,

$$\begin{aligned} \mathrm{Tr}\{\mathscr{H}\rho(t)A\} &= \mathrm{Tr}\{\rho(t)A\mathscr{H}\} = \langle A\mathscr{H}\rangle, \\ \mathrm{Tr}\{\rho(t)\mathscr{H}A\} &= \langle\mathscr{H}A\rangle. \end{aligned} \tag{2.155}$$

Therefore,

$$i\hbar \frac{d}{dt}\langle A\rangle = \langle [A, \mathscr{H}]\rangle, \tag{2.156}$$

which is identical in form to Eq. (2.103). Since the present derivation is based on a general density operator, however, we interpret $\langle A\rangle$ to be an ensemble average; it is only for a pure state that $\langle A\rangle$ reduces to a quantum mechanical expectation value. It is in the latter, more restricted, sense that Eq. (2.103) was derived.

The iterated solution to Eq. (2.152) gives rise to the series

$$\rho(t) = \rho^{(0)}(t) + \rho^{(1)}(t) + \rho^{(3)}(t) + \cdots \tag{2.157}$$

where

$$\rho^{(0)}(t) = \rho(t_0) \tag{2.158a}$$

$$\begin{aligned}\rho^{(1)}(t) &= -\frac{i}{\hbar}\int_{t_0}^{t} dt_1[\mathscr{H}(t_1), \rho^{(0)}(t_1)] \\ &= -\frac{i}{\hbar}\int_{t_0}^{t} dt_1[\mathscr{H}(t_1), \rho(t_0)],\end{aligned} \tag{2.158b}$$

$$\begin{aligned}\rho^{(2)}(t) &= -\frac{i}{\hbar}\int_{t_0}^{t} dt_1[\mathscr{H}(t_1), \rho^{(1)}(t_1)] \\ &= \left(-\frac{i}{\hbar}\right)^2\int_{t_0}^{t} dt_1\int_{t_0}^{t_1} dt_2[\mathscr{H}(t_1), [\mathscr{H}(t_2), \rho(t_0)]],\end{aligned} \tag{2.158c}$$

$$\begin{aligned}\rho^{(n)}(t) &= \left(-\frac{i}{\hbar}\right)^n\int_{t_0}^{t} dt_1\int_{t_0}^{t_1} dt_2\cdots\int_{t_0}^{t_{n-1}} dt_n \\ &\quad \times [\mathscr{H}(t_1), [\mathscr{H}(t_2), \cdots, [\mathscr{H}(t_n), \rho(t_0)]\cdots],\end{aligned} \tag{2.158d}$$

with

$$t > t_1 > t_2 > \cdots > t_0. \tag{2.159}$$

These equations enable us to make successive approximations to $\rho(t)$ when $\rho(t_0)$ is known. They have a wide range of applicability in time-dependent perturbation theory and play a fundamental role in the description of linear and nonlinear radiative interactions.

In the Heisenberg representation, the density operator follows the rule of Eq. (2.97),

$$\rho_{\mathrm{H}}(t) = U^{\dagger}(t, t_0)\rho(t)U(t, t_0), \tag{2.160}$$

which yields the result

$$\dot{\rho}_{\mathrm{H}}(t) \equiv 0, \tag{2.161}$$

as a consequence of Eqs. (2.68), (2.69), and (2.151).

Although the density operator is independent of time in the Heisenberg representation, such is not the case in the interaction representation. When the Hamiltonian is in the form $\mathscr{H} = \mathscr{H}_0 + V$, with $\mathscr{H}_0$ independent of time, the density operator in the interaction representation becomes

$$\tilde{\rho}(t) = e^{i\mathscr{H}_0(t-t_0)/\hbar}\rho(t)e^{-i\mathscr{H}_0(t-t_0)/\hbar}, \tag{2.162}$$

with

$$\tilde{\rho}(t_0) = \rho(t_0). \tag{2.163}$$

Evaluating the time derivative and inserting the Liouville equation,

$$\begin{aligned} i\hbar\frac{d}{dt}\tilde{\rho}(t) &= e^{i\mathscr{H}_0(t-t_0)/\hbar}[V, \rho(t)]e^{-i\mathscr{H}_0(t-t_0)/\hbar} \\ &= [\tilde{V}(t), \tilde{\rho}(t)]. \end{aligned} \tag{2.164}$$

Of particular importance here is the appearance in the commutator of the perturbation Hamiltonian V, rather than the total Hamiltonian $\mathscr{H}$ as in Eq. (2.151), in the Schrödinger representation. Because of this feature, computations involving the temporal behavior of the density operator are often more conveniently performed in the interaction representation. The relation between Eq. (2.151) and Eq. (2.164) may be obtained by writing

$$\begin{aligned} i\hbar\dot{\rho}(t) &= [\mathscr{H}_0, \rho(t)] + [V, \rho(t)] \\ &= [\mathscr{H}_0, \rho(t)] + e^{-i\mathscr{H}_0(t-t_0)/\hbar}[\tilde{V}(t), \tilde{\rho}(t)]e^{i\mathscr{H}_0(t-t_0)/\hbar} \\ &= [\mathscr{H}_0, \rho(t)] + i\hbar e^{-i\mathscr{H}_0(t-t_0)/\hbar}\frac{d\tilde{\rho}(t)}{dt}e^{i\mathscr{H}_0(t-t_0)/\hbar} \end{aligned} \tag{2.165}$$

or

$$i\hbar\frac{d\tilde{\rho}(t)}{dt} = i\hbar e^{i\mathscr{H}_0(t-t_0)/\hbar}\dot{\rho}(t)e^{-i\mathscr{H}_0(t-t_0)/\hbar} - [\mathscr{H}_0, \tilde{\rho}(t)]. \tag{2.166}$$

Again, with the initial condition of Eq. (2.163), the differential equation (2.164) is converted to the integral equation

$$\tilde{\rho}(t) = \rho(t_0) - \frac{i}{\hbar}\int_{t_0}^{t}[\tilde{V}(t'), \tilde{\rho}(t')]\,dt'. \tag{2.167}$$

Eq. (2.167) is then inserted into the commutator of Eq. (2.164) to obtain the alternative form of the equation of motion

$$i\hbar\frac{d\tilde{\rho}(t)}{dt} = [\tilde{V}(t), \rho(t_0)] - \frac{i}{\hbar}\int_{t_0}^{t} [\tilde{V}(t), [\tilde{V}(t'), \tilde{\rho}(t')]]\, dt' \qquad (2.168)$$

which, clearly, has the same structure as Eq. (2.153).

Successive iterations of Eq. (2.167) produce the series

$$\tilde{\rho}(t) = \tilde{\rho}^{(0)}(t) + \tilde{\rho}^{(1)}(t) + \tilde{\rho}^{(2)}(t) + \cdots, \qquad (2.169)$$

where

$$\tilde{\rho}^{(0)}(t) = \rho(t_0). \qquad (2.170a)$$

$$\begin{aligned}\tilde{\rho}^{(1)}(t) &= -\frac{i}{\hbar}\int_{t_0}^{t} dt_1 [\tilde{V}(t_1), \tilde{\rho}^{(0)}(t_1)] \\ &= -\frac{i}{\hbar}\int_{t_0}^{t} dt_1 [\tilde{V}(t_1), \rho(t_0)]. \end{aligned} \qquad (2.170b)$$

$$\begin{aligned}\tilde{\rho}^{(2)}(t) &= -\frac{i}{\hbar}\int_{t_0}^{t} dt_1 [\tilde{V}(t_1), \tilde{\rho}^{(1)}(t_1)] \\ &= \left(-\frac{i}{\hbar}\right)^2 \int_{t_0}^{t} dt_1 [\tilde{V}(t_1), [\tilde{V}(t_2), \rho(t_0)]], \end{aligned} \qquad (2.170c)$$

$$\begin{aligned}\tilde{\rho}^{(n)}(t) &= \left(-\frac{i}{\hbar}\right)^n \int_{t_0}^{t} dt_1 \int_{t_0}^{t_1} dt_2 \cdots \int_{t_0}^{t_{n-1}} dt_n \\ &\quad \times [\tilde{V}(t_1), [\tilde{V}(t_2), \ldots, [\tilde{V}(t_n), \rho(t_0)] \ldots], \end{aligned} \qquad (2.170d)$$

with

$$t > t_1 > t_2 > \cdots > t_0. \qquad (2.171)$$

Again we see the close resemblance between the expansion of the density operator in the interaction representation and the corresponding expansion given by Eq. (2.158) in the Schrödinger representation.

As an example we compute the time-development of the ensemble average of an operator A:

$$\begin{aligned}\langle A(t)\rangle &= \mathrm{Tr}\{\tilde{\rho}(t)\tilde{A}(t)\} \\ &= \mathrm{Tr}\{\tilde{\rho}^{(0)}(t)\tilde{A}(t)\} + \mathrm{Tr}\{\tilde{\rho}^{(1)}(t)\tilde{A}(t)\} + \mathrm{Tr}\{\tilde{\rho}^{(2)}(t)\tilde{A}(t)\} + \cdots \\ &= \mathrm{Tr}\{\rho(t_0)\tilde{A}(t)\} - \frac{i}{\hbar}\int_{t_0}^{t} dt_1\, \mathrm{Tr}\{[\tilde{V}(t_1), \rho(t_0)]\tilde{A}(t)\} \\ &\quad + \left(-\frac{i}{\hbar}\right)^2 \int_{t_0}^{t} dt_1 \int_{t_0}^{t_1} dt_2\, \mathrm{Tr}\{[\tilde{V}(t_1), [\tilde{V}(t_2), \rho(t_0)]]\tilde{A}(t)\} + \cdots. \end{aligned} \qquad (2.172)$$

We may choose a basis such that $\rho(t_0)$ commutes with $\mathscr{H}_0$; upon invoking the cyclic property of the trace, one finds

$$\begin{aligned}\mathrm{Tr}\{\rho(t_0)\tilde{A}(t)\} &= \mathrm{Tr}\{\rho(t_0)e^{i\mathscr{H}_0(t-t_0)/\hbar}Ae^{-i\mathscr{H}_0(t-t_0)/\hbar}\}\\ &= \mathrm{Tr}\{\rho(t_0)A\} = \langle A(t_0)\rangle, \end{aligned}\tag{2.173}$$

$$\begin{aligned}\mathrm{Tr}\{[\tilde{V}(t_1),\rho(t_0)]\tilde{A}(t)\} &= \mathrm{Tr}\{[\tilde{A}(t),\tilde{V}(t_1)]\rho(t_0)\}\\ &= \langle[\tilde{A}(t),\tilde{V}(t_1)]\rangle,\end{aligned}\tag{2.174}$$

$$\mathrm{Tr}\{[\tilde{V}(t_1),[\tilde{V}(t_2),\rho(t_0)]]\tilde{A}(t)\} = \langle[[A(t),V(t_1)],V(t_2)]\rangle. \tag{2.175}$$

Thus,

$$\begin{aligned}\langle A(t)\rangle &= \langle A(t_0)\rangle - \frac{i}{\hbar}\int_{t_0}^{t} dt_1\langle[\tilde{A}(t),\tilde{V}(t_1)]\rangle\\ &\quad + \left(-\frac{i}{\hbar}\right)^2\int_{t_0}^{t} dt_1\int_{t_0}^{t_1} dt_2\langle[[\tilde{A}(t),\tilde{V}(t_1)],\tilde{V}(t_2)]\rangle + \cdots.\end{aligned}\tag{2.176}$$

An equivalent expression is obtained by converting $\tilde{A}(t)$ to the Schrödinger representation. We note that with $t_0 = 0$,

$$\begin{aligned}\langle\tilde{A}(t)\tilde{V}(t_1)\rangle &= \mathrm{Tr}\{\rho(t_0)e^{i\mathscr{H}_0t/\hbar}Ae^{-i\mathscr{H}_0t/\hbar}e^{i\mathscr{H}_0t_1/\hbar}Ve^{-i\mathscr{H}_0t_1/\hbar}\}\\ &= \mathrm{Tr}\{\rho(t_0)Ae^{i\mathscr{H}_0(t_1-t)/\hbar}Ve^{-i\mathscr{H}_0(t_1-t)/\hbar}\}\\ &= \langle A\tilde{V}(t_1-t)\rangle.\end{aligned}\tag{2.177}$$

Continuing with this procedure, the result is

$$\begin{aligned}\langle A(t)\rangle &= \langle A(0)\rangle - \frac{i}{\hbar}\int_0^t dt_1\langle[A,\tilde{V}(t_1-t)]\rangle\\ &\quad + \left(-\frac{i}{\hbar}\right)^2\int_0^t dt_1\int_0^{t_1} dt_2\langle[[A,\tilde{V}(t_1-t)],\tilde{V}(t_2-t)]\rangle + \cdots.\end{aligned}\tag{2.178}$$

In Eq. (2.158d) an expression for $\rho^{(n)}(t)$ in the Schrödinger representation was given in terms of the total Hamiltonian $\mathscr{H}$. When $\mathscr{H} = \mathscr{H}_0 + V(t)$, it is useful to write $\rho^{(n)}(t)$ explicitly in terms of $\mathscr{H}_0$ and $V(t)$. This is accomplished by transforming Eq. (2.170) to the Schrödinger representation. Employing the definition (2.130),

$$\tilde{\rho}^{(0)}(t) = U_0^\dagger(t,t_0)\rho^{(0)}U_0(t,t_0) \tag{2.179}$$

or

$$\rho^{(0)}(t) = U_0(t,t_0)\rho(t_0)U_0^\dagger(t,t_0). \tag{2.180}$$

For the two terms associated with the commutator in Eq. (2.170b), let

$$\tilde{\rho}^{(1)}(t) = \tilde{\rho}_a^{(1)}(t) + \tilde{\rho}_b^{(1)}(t) \tag{2.181}$$

where

$$\tilde{\rho}_a^{(1)}(t) = -\frac{i}{\hbar}\int_{t_0}^{t} dt_1 \tilde{V}(t_1)\rho(t_0), \tag{2.182a}$$

$$\tilde{\rho}_b^{(1)}(t) = \frac{i}{\hbar}\int_{t_0}^{t} dt_1 \rho(t_0)\tilde{V}(t_1). \tag{2.182b}$$

To transform Eq. (2.182a) to the Schrödinger representation, we write

$$\begin{aligned}\tilde{\rho}_a^{(1)}(t) &= U_0^\dagger(t,t_0)\rho_a^{(1)}(t)U_0(t,t_0)\\ &= -\frac{i}{\hbar}\int_{t_0}^{t} dt_1 U_0^\dagger(t_1,t_0)V(t_1)U_0(t_1,t_0)\rho(t_0)\end{aligned} \tag{2.183a}$$

or

$$\rho_a^{(1)}(t) = -\frac{i}{\hbar}\int_{t_0}^{t} dt_1 U_0(t,t_0)U_0^\dagger(t_1,t_0)V(t_1)U_0(t_1,t_0)\rho(t_0)U_0^\dagger(t,t_0). \tag{2.183b}$$

But

$$U_0(t,t_0)U_0^\dagger(t_1,t_0) = U_0(t,t_0)U_0(t_0,t_1) = U_0(t,t_1). \tag{2.184}$$

Thus,

$$\rho_a^{(1)}(t) = -\frac{i}{h}\int_{t_0}^{t} dt_1 U_0(t,t_1)V(t_1)U_0(t_1,t_0)\rho(t_0)U_0^\dagger(t,t_0), \tag{2.185a}$$

$$\rho_b^{(1)}(t) = \frac{i}{\hbar}\int_{t_0}^{t} dt_1 U_0(t,t_0)\rho(t_0)U_0^\dagger(t_1,t_0)V(t_1)U_0^\dagger(t,t_1). \tag{2.185b}$$

The four terms in the commutator of Eq. (2.170c) lead to

$$\rho^{(2)}(t) = \rho_a^{(2)}(t) + \rho_b^{(2)}(t) + \rho_c^{(2)}(t) + \rho_d^{(2)}(t) \tag{2.186}$$

$$\begin{aligned}\rho_a^{(2)}(t) = \left(-\frac{i}{\hbar}\right)^2 \int_{t_0}^{t} dt_1 \int_{t_0}^{t_1} dt_2 U_0(t,t_1)V(t_1)U_0(t_1,t_2)\\ \times V(t_2)U_0(t_2,t_0)\rho(t_0)U_0^\dagger(t,t_0),\end{aligned} \tag{2.187a}$$

$$\begin{aligned}\rho_b^{(2)}(t) = \left(\frac{i}{\hbar}\right)^2 \int_{t_0}^{t} dt_1 \int_{t_0}^{t_1} dt_2 U_0(t,t_1)V(t_1)U_0(t_1,t_0)\\ \times \rho(t_0)U_0^\dagger(t_2,t_0)V(t_2)U_0^\dagger(t,t_2),\end{aligned} \tag{2.187b}$$

$$\rho_c^{(2)}(t) = \left(\frac{i}{\hbar}\right)^2 \int_{t_0}^{t} dt_1 \int_{t_0}^{t_1} dt_2 U_0(t, t_2)V(t_2)U_0(t_2, t_0)$$
$$\times \rho(t_0)U_0^\dagger(t_1, t_0)V(t_1)U_0^\dagger(t, t_1), \tag{2.187c}$$
$$\rho_d^{(2)}(t) = \left(-\frac{i}{\hbar}\right)^2 \int_{t_0}^{t} dt_1 \int_{t_0}^{t_1} dt_2 U_0(t, t_0)\rho(t_0)U_0^\dagger(t_2, t_0)$$
$$\times V(t_2)U_0^\dagger(t_1, t_2)V(t_1)U_0^\dagger(t, t_1). \tag{2.187d}$$

In this fashion, we obtain the time-development of the density operator in a series in which the role played by $\mathscr{H}_0$ and $V(t)$ is explicitly displayed in each term.

2.7 Matrix Elements

The translation of quantum mechanical theory into experimental observations is accomplished through the matrix elements of the relevant observables. In view of Eq. (2.22), which relates the ensemble average (or expectation value in the case of a pure state) of an operator A to the trace of ρA, one needs the matrix elements of the density operator ρ. More often, it is the *time-development* of the density matrix elements that is of primary interest. As an example, the optical Bloch equations, which provide the basic theory for coherent optical phenomena, are temporal differential equations of quantities that are nothing more than certain combinations of density matrix elements. Another important example occurs in the theory of nonlinear optics based on the computation of a nonlinear polarization induced in the medium by propagating electromagnetic waves. The induced polarization depends on the susceptibility and the latter is described by quantum mechanical expressions obtained from the density matrix elements as a function of time.

When the Hamiltonian is $\mathscr{H} = \mathscr{H}_0 + V$, the main problem is finding how the system evolves under the influence of the perturbation V, the presumption being that the Schrödinger equation in the absence of the perturbation has been solved. If $|k\rangle, |l\rangle, \ldots$ are eigenstates of $\mathscr{H}_0$ with eigenvalues $\mathrm{E}_k, \mathrm{E}_l, \ldots$, the general matrix element of Eq. (2.151) is

$$\begin{aligned} i\hbar\langle k|\dot{\rho}(t)|l\rangle &= \langle k|[\mathscr{H}, \rho(t)]|l\rangle \\ &= \langle k|[\mathscr{H}_0, \rho(t)]|l\rangle + \langle k|[V, \rho(t)]|l\rangle \\ &= (\mathrm{E}_k - \mathrm{E}_l)\langle k|\rho(t)|l\rangle + \langle k|[V, \rho(t)]|l\rangle. \end{aligned} \tag{2.188}$$

Employing the closure relation,

$$\sum_r |r\rangle\langle r| = 1, \tag{2.189}$$

and the matrix element of the commutator is

$$\langle k|[V, \rho(t)]|l\rangle = \sum_r (\langle k|V|r\rangle\langle r|\rho(t)|l\rangle - \langle k|\rho(t)|r\rangle\langle r|V|l\rangle),$$
$$= \sum_r (V_{kr}\rho_{rl}(t) - \rho_{kr}(t)V_{rl}). \tag{2.190}$$

Let us now suppose that the system has been prepared so that at $t = t_0 = 0$ only one eigenstate of $\mathscr{H}_0$, say $|l\rangle$, is occupied and all other states are vacant. If $|k\rangle$ is one of the eigenstates that is vacant at $t = 0$, it may no longer be vacant after the application of the perturbation V. As indicated in Section 2.1, the diagonal matrix element $\langle k|\rho(t)|k\rangle$ is the probability for the system to be found in the state $|k\rangle$ after a time t has elapsed. Since, initially, $|l\rangle$ was occupied and $|k\rangle$ was vacant, $\langle k|\rho(t)|k\rangle$ also represents the transition probability for a transition from $|l\rangle$ to $|k\rangle$. In the same sense, $\langle k|\dot{\rho}(t)|k\rangle$ is the probability per unit time for a transition $|l\rangle \to |k\rangle$, or the rate of growth of the population in $|k\rangle$.

Of particular interest are the equations for the matrix elements in systems that contain only two levels with eigenstates $|g\rangle$ and $|e\rangle$, eigenvalues E_g and E_e, respectively, and $\mathrm{E}_e - \mathrm{E}_g = \hbar\omega_0$. Owing to the high monochromaticity of modern lasers, the two-level system is often a good model for a medium interacting with a light beam. Since the two eigenstates $|g\rangle$ and $|e\rangle$ form a complete set, a matrix element such as $\langle g|V\rho|e\rangle$ may be written

$$\langle g|V\rho|e\rangle = \langle g|V|g\rangle\langle g|\rho|e\rangle + \langle g|V|e\rangle\langle e|\rho|e\rangle$$
$$\equiv V_{gg}\rho_{ge} + V_{ge}\rho_{ee}. \tag{2.191}$$

If the diagonal matrix elements of V are zero (as is the case for radiative interactions in the dipole approximation), the two-level equations corresponding to Eq. (2.188) are

$$\begin{aligned} i\hbar\dot{\rho}_{gg}(t) &= V_{ge}\rho_{eg} - V_{eg}\rho_{ge}, \\ i\hbar\dot{\rho}_{ee}(t) &= V_{eg}\rho_{ge} - V_{ge}\rho_{eg} = -i\hbar\dot{\rho}_{gg}(t), \\ i\hbar\dot{\rho}_{ge}(t) &= -\hbar\omega_0\rho_{ge} + V_{ge}(\rho_{ee} - \rho_{gg}), \\ i\hbar\dot{\rho}_{eg}(t) &= \hbar\omega_0\rho_{eg} - V_{eg}(\rho_{ee} - \rho_{gg}). \end{aligned} \tag{2.192}$$

The same equations may be derived by forming the superposition state

$$|\psi\rangle = a_g|g\rangle + a_e|e\rangle \tag{2.193}$$

which is a special case of Eq. (2.80). The coefficients a_g and a_e are subject then to equations of the type of Eq. (2.82):

$$\begin{aligned} i\hbar\dot{a}_g &= \mathrm{E}_g a_g + a_e V_{ge}, \\ i\hbar\dot{a}_e &= \mathrm{E}_e a_e + a_g V_{eg}. \end{aligned} \tag{2.194}$$

The density operator is

$$\rho = |\psi\rangle\langle\psi| = [a_g|g\rangle + a_e|e\rangle][a_g^*\langle g| + a_e^*\langle e|], \tag{2.195}$$

and the density matrix is

$$\rho = \begin{pmatrix} \rho_{ee} & \rho_{eg} \\ \rho_{ge} & \rho_{gg} \end{pmatrix} = \begin{pmatrix} a_e a_e^* & a_e a_g^* \\ a_g a_e^* & a_g a_g^* \end{pmatrix}. \tag{2.196}$$

We then have

$$\begin{aligned} i\hbar\dot{\rho}_{gg} &= i\hbar(\dot{a}_g a_g^* + a_g \dot{a}_g^*) \\ &= (\mathrm{E}_g a_g + a_e V_{ge})a_g^* + a_g(-\mathrm{E}_g a_g^* - a_e^* V_{eg}) \\ &= a_e a_g^* V_{ge} - a_g a_e^* V_{eg} = V_{ge}\rho_{eg} - V_{eg}\rho_{ge}. \end{aligned} \tag{2.197}$$

The other relations in Eq. (2.192) are evaluated in a similar fashion.

Equations (2.192) may be written in matrix form:

$$i\hbar\begin{pmatrix} \dot{\rho}_{gg} \\ \dot{\rho}_{ge} \\ \dot{\rho}_{eg} \\ \dot{\rho}_{ee} \end{pmatrix} = \begin{pmatrix} 0 & -V_{eg} & V_{ge} & 0 \\ -V_{ge} & -\hbar\omega_0 & 0 & V_{ge} \\ V_{eg} & 0 & \hbar\omega_0 & -V_{eg} \\ 0 & V_{eg} & -V_{ge} & 0 \end{pmatrix}\begin{pmatrix} \rho_{gg} \\ \rho_{ge} \\ \rho_{eg} \\ \rho_{ee} \end{pmatrix}, \tag{2.198}$$

or in vector form:

$$\dot{\mathbf{v}} = -\mathbf{v} \times \mathbf{\Omega} \tag{2.199}$$

where

$$\mathbf{v} = (\rho_{eg} + \rho_{ge})\hat{\mathbf{i}} + i(\rho_{eg} - \rho_{ge})\hat{\mathbf{j}} + (\rho_{ee} - \rho_{gg})\hat{\mathbf{k}}, \tag{2.200}$$

$$\mathbf{\Omega} = \frac{1}{\hbar}[(V_{eg} + V_{ge})\hat{\mathbf{i}} + i(V_{eg} - V_{ge})\hat{\mathbf{j}} + \hbar\omega_0\hat{\mathbf{k}}]. \tag{2.201}$$

Specifically,

$$\begin{aligned} \dot{v}_x &= -[v_y\Omega_z - v_z\Omega_y] \\ &= -\frac{i}{\hbar}[\hbar\omega_0(\rho_{eg} - \rho_{ge}) + (\rho_{ee} - \rho_{gg})(V_{ge} - V_{eg})], \\ \dot{v}_y &= -[v_z\Omega_x - v_x\Omega_z] \\ &= \frac{1}{\hbar}[\hbar\omega_0(\rho_{eg} + \rho_{ge}) - (\rho_{ee} - \rho_{gg})(V_{eg} + V_{ge})], \\ \dot{v}_z &= -[v_x\Omega_y - v_y\Omega_x] = -2\frac{i}{\hbar}[V_{ge}\rho_{ge} - V_{ge}\Omega_{ge}]. \end{aligned} \tag{2.202}$$

In the interaction representation, the equations analogous to Eq. (2.192) are derived from Eq. (2.164) whose general matrix element is

$$i\hbar\left\langle k\frac{d}{dt}\tilde{\rho}(t)|l\right\rangle = \langle k|[\tilde{V}(t), \tilde{\rho}(t)]|l\rangle$$
$$= \sum_r (\tilde{V}_{kr}(t)\tilde{\rho}_{rl}(t) - \tilde{\rho}_{kr}(t)\tilde{V}_{rl}(t)). \tag{2.203}$$

For the two-level system, one obtains

$$\begin{aligned} i\hbar\frac{d}{dt}\tilde{\rho}_{gg}(t) &= \tilde{V}_{eg}\tilde{\rho}_{eg} - \tilde{V}_{eg}\tilde{\rho}_{ge}, \\ i\hbar\frac{d}{dt}\tilde{\rho}_{ee}(t) &= \tilde{V}_{eg}\tilde{\rho}_{ge} - \tilde{V}_{ge}\tilde{\rho}_{eg} = -i\hbar\frac{d}{dt}\tilde{\rho}_{gg}, \\ i\hbar\frac{d}{dt}\tilde{\rho}_{ge}(t) &= \tilde{V}_{ge}(\tilde{\rho}_{ee} - \tilde{\rho}_{gg}), \\ i\hbar\frac{d}{dt}\tilde{\rho}_{eg}(t) &= -\tilde{V}_{eg}(\tilde{\rho}_{ee} - \tilde{\rho}_{gg}), \end{aligned} \tag{2.204a}$$

which, just as in Eq. (2.192), also may be written in matrix form

$$i\hbar\frac{d}{dt}\begin{pmatrix}\tilde{\rho}_{gg}\\ \tilde{\rho}_{ge}\\ \tilde{\rho}_{eg}\\ \tilde{\rho}_{ee}\end{pmatrix} = \begin{pmatrix} 0 & -\tilde{V}_{eg} & \tilde{V}_{ge} & 0 \\ -\tilde{V}_{ge} & 0 & 0 & \tilde{V}_{ge} \\ \tilde{V}_{eg} & 0 & 0 & -\tilde{V}_{eg} \\ 0 & \tilde{V}_{eg} & -\tilde{V}_{ge} & 0 \end{pmatrix}\begin{pmatrix}\tilde{\rho}_{gg}\\ \tilde{\rho}_{ge}\\ \tilde{\rho}_{eg}\\ \tilde{\rho}_{ee}\end{pmatrix}, \tag{2.204b}$$

or in vector form,

$$\frac{d}{dt}\tilde{\mathbf{v}} = -\tilde{\mathbf{v}} \times \tilde{\mathbf{\Omega}}, \tag{2.205}$$

where

$$\tilde{\mathbf{v}} = (\tilde{\rho}_{eg} + \tilde{\rho}_{ge})\hat{\mathbf{i}} + i(\tilde{\rho}_{eg} - \tilde{\rho}_{ge})\hat{\mathbf{j}} + (\tilde{\rho}_{ee} - \tilde{\rho}_{gg})\hat{\mathbf{k}}, \tag{2.206}$$

$$\tilde{\mathbf{\Omega}} = \frac{1}{\hbar}[(\tilde{V}_{eg} + \tilde{V}_{ge})\hat{\mathbf{i}} + i(\tilde{V}_{eg} - \tilde{V}_{ge})\hat{\mathbf{j}}]. \tag{2.207}$$

Matrix elements of $\tilde{\rho}(t)$ based on the expansion in Eq. (2.169) also may be evaluated. Assuming $t_0 = 0$ and $\rho(0)$ to be diagonal, the zero-order term is

$$\langle k|\tilde{\rho}^{(0)}(t)|l\rangle \equiv \tilde{\rho}_{kl}^{(0)}(t) = \rho_{kl}(0)\,\delta_{kl}. \tag{2.208}$$

To obtain the first-order matrix element, we have

$$\langle k|[\tilde{V}(t_1), \rho(0)]|l\rangle = \sum_r (\tilde{V}_{kr}(t_1)\rho_{rl}(0) - \rho_{kr}(0)\tilde{V}_{rl}(t_1))$$
$$= \tilde{V}_{kl}(t_1)[\rho_{ll}(0) - \rho_{kk}(0)]. \tag{2.209}$$

Converting the right side to the Schrödinger representation and substituting in Eq. (2.170b), we obtain

$$\tilde{\rho}_{kl}^{(1)}(t) = -\frac{i}{\hbar}\int_0^t dt_1\, V_{kl}[\rho_{ll}(0) - \rho_{kk}(0)]e^{i\omega_{kl}t_1} \tag{2.210}$$

in which $\hbar\omega_{kl} = E_k - E_l$. Similarly, the second-order term is

$$\begin{aligned}\tilde{\rho}_{kl}^{(2)}(t) = \left(-\frac{i}{\hbar}\right)^2 \int_0^t dt_1 \int_0^{t_1} dt_2 \sum_r \{[\rho_{ll}(0) - \rho_{rr}(0)]\\ \times V_{kr}(t_1)V_{rl}(t_2)e^{i\omega_{kr}t_1}e^{i\omega_{rl}t_2}\\ - [\rho_{rr}(0) - \rho_{kk}(0)]V_{kr}(t_2)V_{rl}(t_1)e^{i\omega_{kr}t_2}e^{i\omega_{rl}t_1}\}.\end{aligned} \tag{2.211}$$

Equation (2.168) also provides an iterative expansion. Setting $t_0 = 0$, the expansion to first order is obtained by replacing $\tilde{\rho}(t')$ by $\rho(0)$ in the integrand:

$$i\hbar\frac{d}{dt}\tilde{\rho}(t) = [\tilde{V}(t), \rho(0)] - \frac{i}{\hbar}\int_0^t dt'[\tilde{V}(t), [\tilde{V}(t'), \rho(0)]]. \tag{2.212}$$

Assuming only $|l\rangle$ is occupied initially, $\rho_{ll}(0) = 1$ and all other matrix elements of $\rho(0)$—diagonal or off-diagonal—are zero. Then

$$[\tilde{V}(t), \rho(0)]_{kk} = \sum_r [\tilde{V}_{kr}(t)\rho_{rk}(0) - \rho_{kr}(0)\tilde{V}_{rk}(t)] = 0. \tag{2.213}$$

The commutator in the integral consists of four terms:

$$\begin{aligned}\langle k|\tilde{V}(t)\tilde{V}(t')\rho(0)|k\rangle &= \sum_{rs} \tilde{V}_{kr}(t)\tilde{V}_{rs}(t')\rho_{sk}(0) = 0,\\ \langle k|\rho(0)\tilde{V}(t')\tilde{V}(t)|k\rangle &= \sum_{rs} \rho_{kr}(0)\tilde{V}_{rs}(t')\tilde{V}_{sk}(t) = 0,\\ \langle k|\tilde{V}(t)\rho(0)\tilde{V}(t')|k\rangle &= \sum_{rs} \tilde{V}_{kr}(t)\rho_{rs}(0)\tilde{V}_{sk}(t')\\ &= \tilde{V}_{kl}(t)\rho_{ll}(0)\tilde{V}_{lk}(t'') = \tilde{V}_{kl}(t)\tilde{V}_{lk}(t'),\\ \langle k|\tilde{V}(t')\rho(0)\tilde{V}(t)|k\rangle &= \sum_{rs} \tilde{V}_{kr}(t')\rho_{rs}(0)\tilde{V}_{sk}(t)\\ &= \tilde{V}_{kl}(t')\rho_{ll}(0)\tilde{V}_{lk}(t) = \tilde{V}_{kl}(t')\tilde{V}_{lk}(t).\end{aligned} \tag{2.214}$$

Inserting these results in Eq. (2.212), we have, to first order,

$$\frac{d\tilde{\rho}_{kk}(t)}{dt} = \frac{1}{\hbar^2}\int_0^t dt'[\tilde{V}_{kl}(t)\tilde{V}_{lk}(t') + \tilde{V}_{kl}(t')\tilde{V}_{lk}(t)]. \tag{2.215}$$

To convert to the Schrödinger representation, it is noted that Eq. (2.166) implies

$$\frac{d\tilde{\rho}_{kk}(t)}{dt} = \frac{d\rho_{kk}(t)}{dt}. \tag{2.216}$$

Therefore,

$$\frac{d\rho_{kk}(t)}{dt} = \frac{1}{\hbar^2}\int_0^t dt'[V_{kl}(t)V_{lk}(t')e^{i\omega_{kl}(t-t')} + V_{kl}(t')V_{lk}(t)e^{-i\omega_{kl}(t-t')}] \quad (2.217)$$

In the Schrödinger representation, matrix elements of $\rho^{(0)}(t)$, $\rho^{(1)}(t)$, and $\rho^{(2)}(t)$ are derived from Eqs. (2.180), (2.185) and (2.187). With the help of relations analogous to those shown in Eq. (2.78) and in the notation

$$\mathscr{H}_0|k\rangle = \mathrm{E}_k|k\rangle, \qquad \langle k|l\rangle = \delta_{kl},$$
$$\mathrm{E}_k \equiv \hbar\omega_k, \qquad \mathrm{E}_k - \mathrm{E}_l \equiv \hbar\omega_{kl}, \quad (2.218)$$

$$U_0(t,t_0)|k\rangle = e^{-i\mathscr{H}_0(t-t_0)/\hbar}|k\rangle = e^{-i\omega_k(t-t_0)}|k\rangle,$$
$$\langle k|U_0(t,t_0) = (U_0^\dagger(t,t_0)|k\rangle)^\dagger = \langle k|e^{-i\omega_k(t-t_0)}, \quad (2.219)$$

we have

$$\begin{aligned}\langle k|\rho^{(0)}(t)|l\rangle &= \langle k|U_0(t,t_0)\rho(t_0)U_0^\dagger(t,t_0)|l\rangle \\ &= e^{-i\omega_k(t-t_0)}\langle k|\rho(t_0)|l\rangle e^{i\omega_l(t-t_0)} \\ &= \rho_{kl}(t_0)e^{i\omega_{lk}(t-t_0)},\end{aligned} \quad (2.220)$$

$$\begin{aligned}\langle k|\rho_a^{(1)}(t)|l\rangle &= -\frac{i}{\hbar}\int_{t_0}^t dt_1\langle k|U_0(t,t_1)V(t_1)U_0(t_1,t_0)\rho(t_0)U_0^\dagger(t,t_0)|l\rangle \\ &= -\frac{i}{\hbar}\int_{t_0}^t dt_1 e^{-i\omega_k(t-t_1)}\langle k|V(t_1)U_0(t_1,t_0)\rho(t_0)|l\rangle e^{i\omega_l(t-t_0)} \\ &= -\frac{i}{\hbar}\sum_r e^{-i\omega_{kl}t}\int_{t_0}^t dt_1 e^{i\omega_k t_1}\langle k|V(t_1)U_0(t_1,t_0)|r\rangle \\ &\quad \times \langle r|\rho(t_0)|l\rangle e^{-i\omega_l t_0} \\ &= -\frac{i}{\hbar}\sum_r e^{-i\omega_{kl}t}\int_{t_0}^t dt_1 e^{i\omega_k t_1}V_{kr}(t_1)e^{-i\omega_r(t_1-t_0)} \\ &\quad \times \rho_{rl}(t_0)e^{-i\omega_l t_0} \\ &= -\frac{i}{\hbar}\sum_r e^{-i\omega_{kl}t}\int_{t_0}^t dt_1 V_{kr}(t_1)e^{i\omega_{kr}t_1}\rho_{rl}(t_0)e^{i\omega_{rl}t_0}.\end{aligned} \quad (2.221a)$$

Following the same procedure,

$$\langle k|\rho_b^{(1)}(t)|l\rangle = \frac{i}{\hbar}\sum_r e^{-i\omega_{kl}t}\int_{t_0}^t dt_1 \rho_{kr}(t_0)e^{i\omega_{kr}t_0}V_{rl}(t_1)e^{i\omega_{rl}t_1}. \quad (2.221b)$$

In second order, the results are

$$\langle k|\rho_a^{(2)}(t)|l\rangle = \left(-\frac{i}{\hbar}\right)^2 \sum_{rs} e^{-i\omega_{kl}t} \int_{t_0}^{t} dt_1 \int_{t_0}^{t_1} dt_2\, V_{kr}(t_1) e^{i\omega_{kr}t_1} \times V_{rs}(t_2) e^{i\omega_{rs}t_2} \rho_{sl}(t_0) e^{i\omega_{sl}t_0}, \tag{2.222a}$$

$$\langle k|\rho_b^{(2)}(t)|l\rangle = \left(\frac{1}{\hbar}\right)^2 \sum_{rs} e^{-i\omega_{kl}t} \int_{t_0}^{t} dt_1 \int_{t_0}^{t_1} dt_2\, V_{kr}(t_1) e^{i\omega_{kr}t_1} \times \rho_{rs}(t_0) e^{i\omega_{rs}t_0} V_{sl}(t_2) e^{i\omega_{sl}t_2}, \tag{2.222b}$$

$$\langle k|\rho_c^{(2)}(t)|l\rangle = \left(\frac{1}{\hbar}\right)^2 \sum_{rs} e^{-i\omega_{kl}t} \int_{t_0}^{t} dt_1 \int_{t_0}^{t_1} dt_2\, V_{kr}(t_2) e^{i\omega_{kr}t_2} \times \rho_{rs}(t_0) e^{i\omega_{rs}t_0} V_{sl}(t_1) e^{i\omega_{sl}t_1},$$

$$\langle k|\rho_d^{(2)}(t)|l\rangle = \left(-\frac{i}{\hbar}\right)^2 \sum_{rs} e^{-i\omega_{kl}t} \int_{t_0}^{t} dt_1 \int_{t_0}^{t_1} dt_2\, \rho_{kr}(t_0) e^{i\omega_{kr}t_0} \times V_{rs}(t_2) e^{i\omega_{rs}t_2} V_{sl}(t_1) e^{i\omega_{sl}t_1}. \tag{2.222c}$$

It often is found that the rate of change of the density matrix elements in the interaction representation is governed by two types of terms—one type varying slowly and the other varying rapidly. In the *secular* approximation, the rapidly varying terms are ignored. Thus, if

$$\frac{d}{dt}\tilde{\rho}_{kl} = A_{kl} + B_{kl} e^{i\omega t}, \tag{2.223}$$

then the secular approximation consists of the elimination of the second term on the right if it oscillates much more rapidly compared to variations in A_{kl}.

2.8 Thermal Equilibrium, Correlation Functions

One of the most important situations, because of its frequent occurrence in physical problems, arises when a system is in contact with a heat bath (or reservoir) under conditions where the random exchange of energy between the system and the heat bath results in thermal equilibrium. Under these circumstances, the probability P_i, which appears in the definition of the density operator (Eq. (2.1)), is given by an explicit expression known as the *Boltzmann* distribution:

$$P_i = \frac{1}{Z} e^{-\beta E_i} \tag{2.224a}$$

where

$$Z = \sum_i e^{-\beta \mathrm{E}_i}, \qquad \beta = \frac{1}{\mathrm{k}T}. \tag{2.224b}$$

The quantity Z is known as the *partition function* and k is the Boltzmann constant. The sum is taken over all accessible states of the system. If E_i is the energy of the state $|\psi_i\rangle$, the thermal density operator ρ_0, i.e., the density operator under thermal equilibrium, becomes

$$\rho_0 = \frac{1}{Z}\sum_i e^{-\beta \mathrm{E}_i}|\psi_i\rangle\langle\psi_i|. \tag{2.225}$$

A more general definition that avoids reference to any particular basis is written

$$\rho_0 = \frac{1}{Z}e^{-\beta\mathscr{H}}, \qquad Z = \mathrm{Tr}\, e^{-\beta\mathscr{H}}. \tag{2.226}$$

The two expressions for ρ_0 merge when the states $|\psi_i\rangle$ obey the closure relation

$$\sum_i |\psi_i\rangle\langle\psi_i| = 1 \tag{2.227}$$

and are eigenstates of the Hamiltonian with eigenvalues E_i. Then

$$Z = \mathrm{Tr}\, e^{-\beta\mathscr{H}} = \sum_i \langle\psi_i|e^{-\beta\mathscr{H}}|\psi_i\rangle = \sum_i e^{-\beta \mathrm{E}_i}, \tag{2.228}$$

and

$$\rho_0 = \frac{1}{Z}e^{-\beta\mathscr{H}} = \frac{1}{Z}\sum_i e^{-\beta\mathscr{H}}|\psi_i\rangle\langle\psi_i| = \frac{1}{Z}\sum_i e^{-\beta \mathrm{E}_i}|\psi_i\rangle\langle\psi_i|. \tag{2.229}$$

It is evident that in this (orthonormal) basis only diagonal matrix elements of ρ_0 are nonvanishing, i.e.,

$$\langle\psi_i|\rho_0|\psi_j\rangle = \langle\psi_i|\rho_0|\psi_i\rangle\,\delta_{ij} \equiv \rho_{ii}^0\,\delta_{ij}. \tag{2.230}$$

The definition (Eq. (2.226) of ρ_0 implies that ρ_0 commutes with $\mathscr{H}$; hence, the Liouville Eq. (2.151) gives

$$i\hbar\dot{\rho}_0 = [\mathscr{H}, \rho_0] = \frac{1}{Z}[\mathscr{H}, e^{-\beta\mathscr{H}}] = 0, \tag{2.231}$$

and

$$(\rho_0)_{\mathrm{H}} = e^{i\mathscr{H}(t-t_0)/\hbar}\rho_0 e^{-i\mathscr{H}(t-t_0)/\hbar} = \rho_0. \tag{2.232}$$

A system, initially in thermal equilibrium, subsequently may be subjected to an interaction that disturbs the equilibrium; that is, $\mathscr{H} = \mathscr{H}_0$ for $t \le t_0$ and

$\mathcal{H} = \mathcal{H}_0 + V$ for $t > t_0$. During the time that the system is in thermal equilibrium ($t \leq t_0$), the density operator satisfies

$$\rho_0 = \frac{1}{Z} e^{-\beta \mathcal{H}_0}, \qquad Z = \operatorname{Tr} e^{-\beta \mathcal{H}_0}, \tag{2.233}$$

$$\dot{\rho}_0 = 0, \tag{2.234}$$

or

$$\rho_0 = \frac{1}{Z} \sum_i e^{-\beta \mathrm{E}_i} |i\rangle\langle i|, \tag{2.235}$$

in which the eigenstate $|i\rangle$ and eigenvalue E_i belong to $\mathcal{H}_0$. For $t > t_0$, the system evolves in response to the perturbation and the density opertor $\rho(t)$, as well as its matrix elements, which are now functions of time, are calculated by the methods described in the previous sections.

It is important to observe that a system in thermal equilibrium is in a mixed state. To prove this, it is sufficient to consider a two-level system with states $|\psi_1\rangle, |\psi_2\rangle$ and energics E_1, E_2. Employing the definitions

$$P_1 = |a_1|^2 = \frac{1}{Z} e^{-\beta \mathrm{E}_1}, \qquad P_2 = |a_2|^2 = \frac{1}{Z} e^{-\beta \mathrm{E}_2}, \qquad P_1 + P_2 = 1, \tag{2.236}$$

in which $|a_1|^2$ and $|a_2|^2$ are the probabilities for the system to be found in the states $|\psi_1\rangle$ and $|\psi_2\rangle$, respectively, the thermal density operator is

$$\rho_0 = |a_1|^2 |\psi_1\rangle\langle\psi_1| + |a_2|^2 |\psi_2\rangle\langle\psi_2|. \tag{2.237}$$

In matrix form,

$$\rho_0 = \begin{pmatrix} |a_1|^2 & 0 \\ 0 & |a_2|^2 \end{pmatrix}, \qquad \rho_0^2 = \begin{pmatrix} |a_1|^4 & 0 \\ 0 & |a_2|^4 \end{pmatrix}. \tag{2.238}$$

Since

$$\operatorname{Tr} \rho_0^2 = |a_1|^4 + |a_2|^4 < (|a_1|^2 + |a_2|^2)^2 = 1, \tag{2.239}$$

the system is evidently in a mixed state.

For comparison let us consider the superposition state similar to Eq. (2.36) in which all atoms of the ensemble have the same wave function

$$|\psi\rangle = a_1|\psi_1\rangle + a_2|\psi_2\rangle, \qquad |a_1|^2 + |a_2|^2 = 1. \tag{2.240}$$

Here, too, the occupation probabilities for the two states are $|a_1|^2$ and $|a_2|^2$ as in the thermal equilibrium case. But this system is in a pure state since

$$\rho = |\psi\rangle\langle\psi|, \qquad \rho^2 = \rho, \qquad \operatorname{Tr} \rho^2 = 1. \tag{2.241}$$

Note, also, that the density matrix

$$\rho = \begin{pmatrix} |a_1|^2 & a_1 a_2^* \\ a_1^* a_2 & |a_2|^2 \end{pmatrix} \tag{2.242}$$

has off-diagonal elements whereas the density matrix in Eq. (2.238) for the thermal case is diagonal. As noted in Section 2.2, the absence of off-diagonal matrix elements in the latter indicates that $|\psi_1\rangle$ and $|\psi_2\rangle$ are incoherent whereas the nonvanishing off-diagonal elements in Eq. (2.242) reveal a coherence between $|\psi_1\rangle$ and $|\psi_2\rangle$.

To see what this means, let us assume that the two-level system described by the superposition state $|\psi\rangle$ is brought in contact with a heat bath. The random exchanges of energy between the system and the heat bath will induce random fluctuations in the magnitudes and phases of the complex coefficients a_1 and a_2. The phase fluctuations will have no effect on $|a_1|^2$ and $|a_2|^2$; hence, these quantities eventually will acquire the values dictated by the Boltzmann distribution, as shown in Eq. (2.236). But $a_1 a_2^*$ and $a_1^* a_2$ are interference terms strongly influenced by phase fluctuations. Their average over time will vanish and the density matrix will revert to the diagonal form shown in Eq. (2.238).

We conclude that the coherence between $|\psi_1\rangle$ and $|\psi_2\rangle$ in the superposition state is associated with a fixed phase relation between the coefficients a_1 and a_2 that allows for nonvanishing off-diagonal density matrix elements. The fluctuations in the thermal case cause such matrix elements to vanish. In other words, the superposition state is coherent because $|\psi_1\rangle$ and $|\psi_2\rangle$ can interfere, while the thermal case is incoherent due to the impossibility for interference between the two states.

We now shall rewrite the matrix elements shown in Eqs. (2.220) (2.221), and (2.222) under the assumption that the system was initially in thermal equilibrium:

$$\langle k|\rho^{(0)}(t)|l\rangle = \langle k|\rho_0|l\rangle = \rho_{kk}^0\,\delta_{kl}, \tag{2.243}$$

$$\langle k|\rho_a^{(1)}(t)|l\rangle = -\frac{i}{\hbar}e^{-i\omega_{kl}t}\int_{t_0}^{t} dt_1\, V_{kl}(t_1)e^{i\omega_{kl}t_1}\rho_{ll}^0, \tag{2.244a}$$

$$\langle k|\rho_b^{(1)}(t)|l\rangle = \frac{i}{\hbar}e^{-i\omega_{kl}t}\int_{t_0}^{t} dt_1\,\rho_{kk}^0 V_{kl}(t_1)e^{i\omega_{kl}t_1}, \tag{2.244b}$$

$$\langle k|\rho_a^{(2)}(t)|l\rangle = \left(-\frac{i}{\hbar}\right)^2\sum_r e^{-i\omega_{kl}t}\int_{t_0}^{t} dt_1\int_{t_0}^{t_1} dt_2\, V_{kr}(t_1)e^{i\omega_{kr}t_1}V_{rl}(t_2)e^{i\omega_{rl}t_2}\rho_{ll}^0, \tag{2.245a}$$

$$\langle k|\rho_b^{(2)}(t)|l\rangle = \left(\frac{1}{\hbar}\right)^2\sum_r e^{-i\omega_{kl}t}\int_{t_0}^{t} dt_1\int_{t_0}^{t_1} dt_2\, V_{kr}(t_1)e^{i\omega_{kr}t_1}\rho_{rr}^0 V_{rl}(t_2)e^{i\omega_{rl}t_2}, \tag{2.245b}$$

$$\langle k|\rho_c^{(2)}(t)|l\rangle = \left(\frac{1}{\hbar}\right)^2 \sum_r e^{-i\omega_{kl}t} \int_{t_0}^{t} dt_1 \int_{t_0}^{t_1} dt_2 V_{kr}(t_2) e^{i\omega_{kr}t_2} \rho_{rr}^0 V_{rl}(t_1) e^{i\omega_{rl}t_1}, \tag{2.245c}$$

$$\langle k|\rho_d^{(2)}(t)|l\rangle = \left(-\frac{i}{\hbar}\right)^2 \sum_r e^{-i\omega_{kl}t} \int_{t_0}^{t} dt_1 \int_{t_0}^{t_1} dt_2 \rho_{kk}^0 V_{kr}(t_2) e^{i\omega_{kr}t_2} V_{rl}(t_1) e^{i\omega_{rl}t_1}. \tag{2.245d}$$

Under conditions of thermal equilibrium, correlation functions of quantum mechanical operators obey certain theorems, a few of which will be derived now [5]. Let $A_{\mathrm{H}}(t)$ and $B_{\mathrm{H}}(t)$ be two (noncommutative) operators in the Heisenberg representation and let

$$G_{BA}(\tau) \equiv \langle B_{\mathrm{H}}(t) A_{\mathrm{H}}(t+\tau)\rangle = \mathrm{Tr}\{\rho_0 B_{\mathrm{H}}(t) A_{\mathrm{H}}(t+\tau)\} = \langle B A_{\mathrm{H}}(\tau)\rangle \tag{2.246}$$

with $A \equiv A(0)$ and $B \equiv B(0)$. In the last expression, the origin of time in the correlation function has been shifted—a permissible procedure in view of the stationary property of an ensemble at equilibrium. The trace is readily evaluated in a basis $\{|k\rangle\}$ in which ρ_0 is diagonal. Thus,

$$\begin{aligned} G_{BA}(\tau) &= \sum_k \langle k|\rho_0 B_{\mathrm{H}}(t) A_{\mathrm{H}}(t+\tau)|k\rangle = \sum_k \langle k|\rho_0 B A_{\mathrm{H}}(\tau)|k\rangle \\ &= \sum_{kl} \langle k|\rho_0|k\rangle\langle k|B|l\rangle\langle l|A_{\mathrm{H}}(\tau)|k\rangle. \end{aligned} \tag{2.247}$$

After transforming to the Schrödinger representation and adopting the notation

$$\langle k|\rho_0|k\rangle = \rho_{kk}^0, \qquad \langle k|B|l\rangle = B_{kl}, \qquad \langle l|A|k\rangle = A_{lk}, \tag{2.248}$$

we have

$$\langle B A_{\mathrm{H}}(\tau)\rangle = \sum_{kl} \rho_{kk}^0 B_{kl} A_{lk} e^{i\omega_{lk}\tau}. \tag{2.249}$$

Following the same procedure,

$$\langle A_{\mathrm{H}}(\tau) B\rangle = \sum_{kl} \rho_{kk}^0 A_{kl} B_{lk} e^{i\omega_{kl}\tau}. \tag{2.250}$$

Then, if A and B are Hermitian operators, we have the theorem

$$\langle A_{\mathrm{H}}(\tau) B\rangle = \langle B A_{\mathrm{H}}(\tau)\rangle^*. \tag{2.251}$$

One also may write

$$\begin{aligned} \int_0^\infty \langle B A_{\mathrm{H}}(\tau)\rangle e^{-i\omega\tau}\, d\tau &= \int_0^\infty \langle B_{\mathrm{H}}(-\tau) A\rangle e^{-i\omega\tau}\, d\tau \\ &= \int_{-\infty}^0 \langle B_{\mathrm{H}}(\tau) A\rangle e^{i\omega\tau}\, d\tau, \end{aligned} \tag{2.252}$$

in which the last integral arises from a replacement of τ by $-\tau$. Similarly,

$$\int_{-\infty}^{0} \langle BA_{\mathrm{H}}(\tau)\rangle e^{-i\omega\tau}\,d\tau = \int_{-\infty}^{0} \langle B_{\mathrm{H}}(-\tau)A\rangle e^{-i\omega\tau}\,d\tau$$

$$= \int_{0}^{\infty} \langle B_{\mathrm{H}}(\tau)A\rangle e^{i\omega\tau}\,d\tau. \tag{2.253}$$

Combining Eqs. (2.252) and (2.253), we obtain the relation

$$\int_{-\infty}^{\infty} \langle BA_{\mathrm{H}}(\tau)\rangle e^{-i\omega\tau}\,d\tau = \int_{-\infty}^{\infty} \langle B_{\mathrm{H}}(\tau)A\rangle e^{i\omega\tau}\,d\tau. \tag{2.254}$$

Referring to Eq. (2.249),

$$\int_{-\infty}^{\infty} \langle BA_{\mathrm{H}}(\tau)\rangle e^{i\omega\tau}\,d\tau = \sum_{kl} \rho_{kk}^{0} B_{kl} A_{lk} \int_{-\infty}^{\infty} e^{i(\omega_{lk}+\omega)\tau}\,d\tau. \tag{2.255}$$

Similarly, from Eq. (2.250),

$$\int_{-\infty}^{\infty} \langle A_{\mathrm{H}}(\tau)B\rangle e^{i\omega\tau}\,d\tau = \sum_{kl} \rho_{kk}^{0} A_{kl} B_{lk} \int_{-\infty}^{\infty} e^{i(\omega_{kl}+\omega)\tau}\,d\tau$$

$$= \sum_{kl} \rho_{ll}^{0} A_{lk} B_{kl} \int_{-\infty}^{\infty} e^{i(\omega_{lk}+\omega)\tau}\,d\tau, \tag{2.256}$$

where, to obtain the last expression, the summation indices k and l have been interchanged. But

$$\frac{\rho_{kk}^{0}}{\rho_{ll}^{0}} = \frac{e^{-\beta E_k}}{e^{-\beta E_l}} = e^{\beta\hbar\omega_{lk}} \tag{2.257}$$

and

$$\int_{-\infty}^{\infty} e^{i(\omega_{lk}+\omega)\tau}\,d\tau = 2\pi\,\delta(\omega_{lk}+\omega). \tag{2.258}$$

Therefore, upon comparing Eq. (2.255) with Eq. (2.256), we have

$$\int_{-\infty}^{\infty} \langle BA_{\mathrm{H}}(\tau)\rangle e^{i\omega\tau}\,d\tau = e^{-\beta\hbar\omega} \int_{-\infty}^{\infty} \langle A_{\mathrm{H}}(\tau)B\rangle e^{i\omega\tau}\,d\tau \tag{2.259}$$

Other relations follow directly from Eq. (2.259):

$$\int_{-\infty}^{\infty} \langle [B, A_{\mathrm{H}}(\tau)]\rangle e^{i\omega\tau}\,d\tau = (1 - e^{\beta\hbar\omega}) \int_{-\infty}^{\infty} \langle BA_{\mathrm{H}}(\tau)\rangle e^{i\omega\tau}\,d\tau, \tag{2.260}$$

$$\int_{-\infty}^{\infty} \langle BA_{\mathrm{H}}(\tau) + A_{\mathrm{H}}(\tau)B\rangle e^{i\omega\tau}\,d\tau = (1 + e^{\beta\hbar\omega}) \int_{-\infty}^{\infty} \langle BA_{\mathrm{H}}(\tau)\rangle e^{i\omega\tau}\,d\tau, \tag{2.261}$$

$$\int_{-\infty}^{\infty} \langle [B, A_{\rm H}(\tau)] \rangle e^{i\omega\tau}\, d\tau = \frac{1 - e^{\beta\hbar\omega}}{1 + e^{\beta\hbar\omega}} \int_{-\infty}^{\infty} \langle B A_{\rm H}(\tau) + A_{\rm H}(\tau) B \rangle e^{i\omega\tau}\, d\tau$$
$$= -\tanh\left(\frac{\beta\hbar\omega}{2}\right)$$
$$\times \int_{-\infty}^{\infty} \langle B A_{\rm H}(\tau) + A_{\rm H}(\tau) B \rangle e^{i\omega\tau}\, d\tau. \quad (2.262)$$

We shall return to these equations in connection with the fluctuation-dissipation theorem (Section 7.2).

When the Hamiltonian is of the form $\mathscr{H} = \mathscr{H}_0 + V$, all the expressions in this section that have been written with respect to the Heisenberg representation may be rewritten in the interaction representation with the understanding that eigenfunctions and eigenvalues refer to $\mathscr{H}_0$.

2.9 Feynman Diagrams

In Section 2.6 it was shown that the density operator may be expanded to various orders by successive iterations of an integral equation, which, in turn, was derived from the Liouville equation. With increasing order, the expressions soon become unwieldy as one may judge from the second-order expression of Eq. (2.187). Considerable assistance in keeping track of the numerous terms is achieved by means of a diagrammatic technique, analogous to the Feynman diagrams in solid state and elementary particle physics [6–9]. By means of such diagrams, it ultimately will be possible to derive a shorthand method for the computation of nonlinear susceptibilities (see Section 7.3). The present section is the first step toward this objective.

Let us recall that the basic definition (Eq. (2.1)) of the density operator contains a product of a ket $|\psi\rangle$ and a bra $\langle\psi|$. We then may invoke the expansions of the evolution operator and its complex conjugate to evaluate bras and kets such as $\langle\psi^{(n)}(t)|$ and $|\psi^{(n)}(t)\rangle$ in accordance with Eqs. (2.145) and (2.140). Thus, in zero order,

$$\rho^{(0)}(t) = \sum_i |\psi_i^{(0)}(t)\rangle P_i \langle\psi_i^{(0)}(t)|.$$

Employing the expressions shown in Eqs. (2.137a) and (2.142a),

$$|\psi_i^{(0)}(t)\rangle = U^{(0)}(t, t_0)|\psi_i(t_0)\rangle = U_0(t, t_0)|\psi_i(t_0)\rangle$$
$$\langle\psi_i^{(0)}(t)| = \langle\psi_i(t_0)|U^{(0)\dagger}(t, t_0) = \langle\psi_i(t_0)|U_0^\dagger(t, t_0) \quad (2.263)$$

and writing

$$\rho(t_0) = \sum_i |\psi_i(t_0)\rangle P_i \langle\psi_i(t_0)|, \quad (2.264)$$

we have

$$\rho^{(0)}(t) = U_0(t, t_0)\rho(t_0)U_0^\dagger(t, t_0) \tag{2.265}$$

as in Eq. (2.180).

In first order there are two terms,

$$\rho^{(1)}(t) = \rho_a^{(1)}(t) + \rho_b^{(1)}(t), \tag{2.266}$$

where

$$\begin{aligned}\rho_a^{(1)}(t) &= \sum_i |\psi_i^{(1)}(t)\rangle P_i \langle\psi_i^{(1)}(t)| \\ &= \sum_i U^{(1)}(t, t_0)|\psi_i(t_0)\rangle P_i \langle\psi_i(t_0)|U^{(0)\dagger}(t, t_0) \\ &= U^{(1)}(t, t_0)\rho(t_0)U^{(0)\dagger}(t, t_0) \\ &= -\frac{i}{\hbar}\int_{t_0}^{t} dt_1\, U_0(t, t_1)V(t_1)U_0(t_1, t_0)\rho(t_0)U_0^\dagger(t, t_0),\end{aligned} \tag{2.267a}$$

and

$$\begin{aligned}\rho_b^{(1)}(t) &= \sum_i |\psi_i^{(0)}(t)\rangle P_i \langle\psi_i^{(1)}(t)| \\ &= \sum_i U^{(0)}(t, t_0)|\psi_i(t_0)\rangle P_i \langle\psi_i(t_0)|U^{(1)\dagger}(t, t_0) \\ &= \frac{i}{\hbar}\int_{t_0}^{t} dt_1'\, U_0(t, t_0)\rho(t_0)U_0^\dagger(t_1', t_0)V(t_1')U_0^\dagger(t, t_1').\end{aligned} \tag{2.267b}$$

Evidently, Eqs. (2.267a) and (2.267b) are identical with Eqs. (2.185a) and (2.185b), respectively, since the replacement of t_1 by t_1' for the variable of integration in Eq. (2.267b) is of no consequence.

In second order, we write

$$\rho^{(2)}(t) = \sum_i |\psi_i^{(2)}(t)\rangle P_i \langle\psi_i^{(0)}(t)| + \sum_i |\psi_i^{(0)}(t)\rangle P_i \langle\psi_i^{(2)}(t)| + \sum_i |\psi_i^{(1)}(t)\rangle P_i \langle\psi_i^{(1)}(t)|, \tag{2.268a}$$

$$\begin{aligned}&= U^{(2)}(t, t^0)\rho(t_0)U^{(0)\dagger}(t, t_0) + U^{(0)}(t, t_0)\rho(t_0)U^{(2)\dagger}(t, t_0) \\ &\quad + U^{(1)}(t, t_0)\rho(t_0)U^{(1)\dagger}(t, t_0),\end{aligned} \tag{2.268b}$$

where

$$\begin{aligned}U^{(2)}(t, t_0)\rho(t_0)U^{(0)\dagger}(t, t_0) &= \left(-\frac{i}{\hbar}\right)^2 \int_{t_0}^{t} dt_1 \int_{t_0}^{t_1} dt_2\, U_0(t, t_1) \\ &\quad \times V(t_1)U_0(t_1, t_2)V(t_2)U_0(t_2, t_0)\rho(t_0)U_0^\dagger(t, t_0),\end{aligned} \tag{2.269a}$$

$$U^{(0)}(t,t_0)\rho(t_0)U^{(2)\dagger}(t,t_0) = \left(\frac{i}{\hbar}\right)^2 \int_{t_0}^{t} dt'_1 \int_{t_0}^{t'_1} dt'_2\, U_0(t,t_0)$$
$$\times\ \rho(t_0)U_0^\dagger(t'_2,t_0)V(t'_2)U_0^\dagger(t'_1,t'_2)V(t'_1)U_0^\dagger(t,t'_1). \qquad (2.269b)$$

Comparison with Eq. (2.187) indicates that Eqs. (2.269a) and (2.269b) correspond to $\rho_a^{(2)}(t)$ and $\rho_d^{(2)}(t)$, respectively. The evaluation of the third term in Eq. (2.268b) requires that we take into account the two possible orders $t > t_1 > t'_1$ and $t > t'_1 > t_1$, as mentioned at the end of Section 2.5.

$t > t_1 > t'_1$:

$$U^{(1)}(t,t_0)\rho(t_0)U_0^{(1)\dagger}(t,t_0) = \left(-\frac{i}{\hbar}\right)\left(\frac{i}{\hbar}\right)\int_{t_0}^{t} dt_1 \int_{t_0}^{t_1} dt'_1\, U_0(t,t_1)$$
$$\times\ V(t_1)U_0(t_1,t_0)\rho(t_0)U_0^\dagger(t'_1,t_0)V(t'_1)U_0^\dagger(t,t'_1) \qquad (2.270)$$

identical to Eq. (2.187b) for $\rho_b^{(2)}(t)$.

$t > t'_1 > t_1$:

$$U^{(1)}(t,t_0)\rho(t_0)U^{(1)\dagger}(t,t_0) = \left(-\frac{i}{\hbar}\right)\left(\frac{i}{\hbar}\right)\int_{t_0}^{t} dt'_1 \int_{t_0}^{t'_1} dt_1\, U_0(t,t_1)$$
$$\times\ V(t_1)U_0(t_1,t_0)\rho(t_0)U_0^\dagger(t'_1,t_0)V(t'_1)U_0^\dagger(t,t'_1) \qquad (2.271)$$

identical with Eq. (2.187c) for $\rho_c^{(2)}(t)$.

The expressions derived in this section appear to be trivial since they duplicate results already derived in Section 2.6 and merely introduce a slightly different notation. They do provide more insight into the separate roles of the bras and kets, however, and more importantly, they form the basis for the construction of diagrams to represent the various orders of the density operator. Each diagram consists of two directed line segments, one on each side of a vertical time axis with time increasing upward (Fig. 2.1). The left side represents the time-development of a ket and the right side the time-

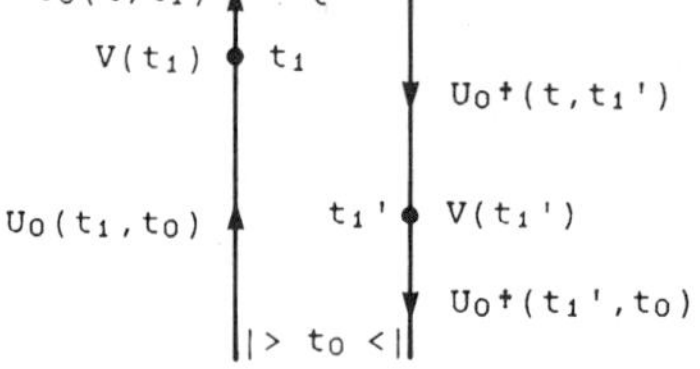

FIGURE 2.1 Feynman diagram for $\rho_b^{(2)}(t)$. The diagram consists of two directed line segments, one on each side of a vertical time axis with time increasing upward. The left side represents the time-development of a ket (|⟩) and the right side that of a bra (⟨|). Primed and unprimed time variables are associated with the bra and ket, respectively; the labelling is described in the text. The time sequence shown is $t > t_1 > t'_1 > t_0$.

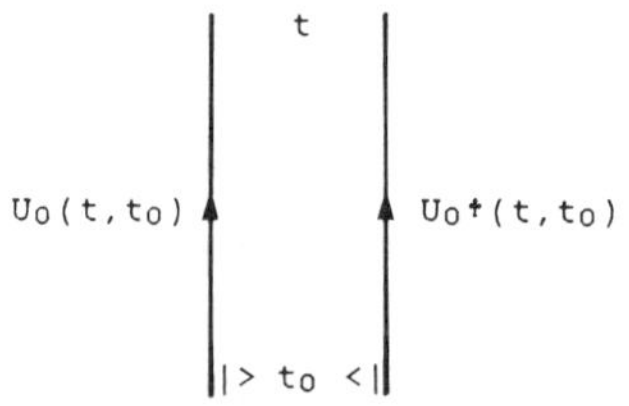

FIGURE 2.2 Diagram for $\rho^0(t) = U_0(t, t_0)\rho(t_0)U_0^\dagger(t, t_0)$.

development of a bra. Intermediate times between t_0 and t (also known as interaction or collision points on the diagram) are arranged in proper sequence, primed time variables on the right (or bra) line segment and unprimed time variables on the left (or ket) line segment.

Additional features of the labelling are best understood by referring simultaneously to Eq. (2.270) and to Fig. 2.1. Starting with the integrand in Eq. (2.270), $U_0(t, t_1)$ corresponds to the portion of the left line segment between t and t_1; $V(t_1)$ corresponds to the interaction point t_1; $U_0(t_1, t_0)$ corresponds to the portion of the left line segment between t_1 and t_0; $\rho(t_0)$ corresponds to the point t_0; $U_0^\dagger(t_1', t_0)$ corresponds to the portion of the right line segment between t_0 and t_1'; $V(t_1')$ corresponds to the interaction point t_1'; and $U_0^\dagger(t, t_1')$ corresponds to the portion of the right line segment between t_1' and t. On the basis of this pattern, we have a diagrammatic representation of the integrand associated with $\rho_b^{(2)}(t)$. Diagrams for $\rho^{(0)}(t)$, $\rho^{(1)}(t)$, and $\rho^{(2)}(t)$ are shown in Figs. 2.2, 2.3, and 2.4, respectively.

It is in connection with matrix elements of the density operator, however, that the diagrams have their greatest utility. In the notation adopted here, the matrix elements evaluated in Section 2.8, for systems initially in thermal

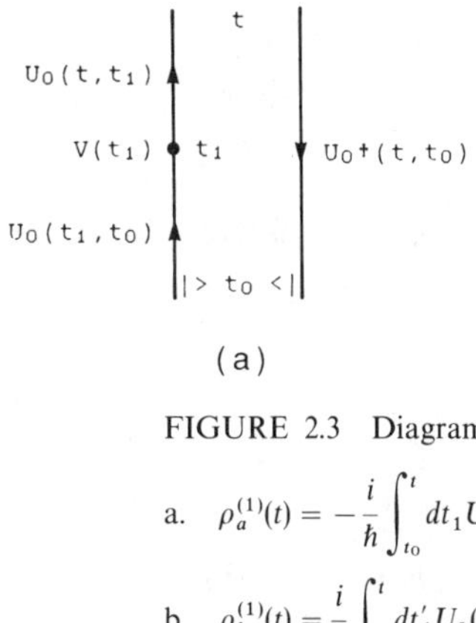

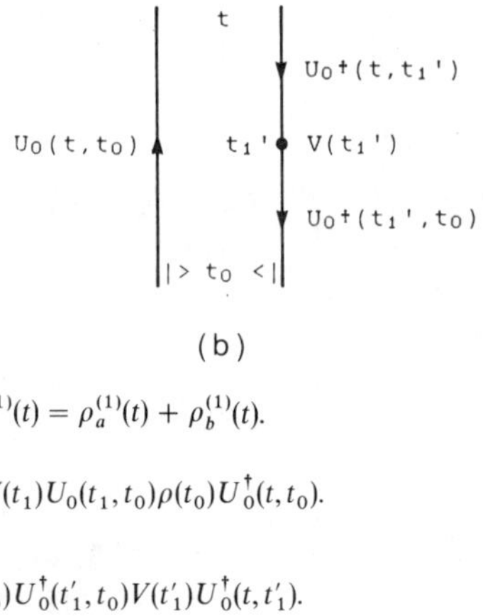

FIGURE 2.3 Diagrams for $\rho^{(1)}(t) = \rho_a^{(1)}(t) + \rho_b^{(1)}(t)$.

a. $$\rho_a^{(1)}(t) = -\frac{i}{\hbar}\int_{t_0}^{t} dt_1 U_0(t, t_1)V(t_1)U_0(t_1, t_0)\rho(t_0)U_0^\dagger(t, t_0).$$

b. $$\rho_b^{(1)}(t) = \frac{i}{\hbar}\int_{t_0}^{t} dt_1' U_0(t, t_0)\rho(t_0)U_0^\dagger(t_1', t_0)V(t_1')U_0^\dagger(t, t_1').$$

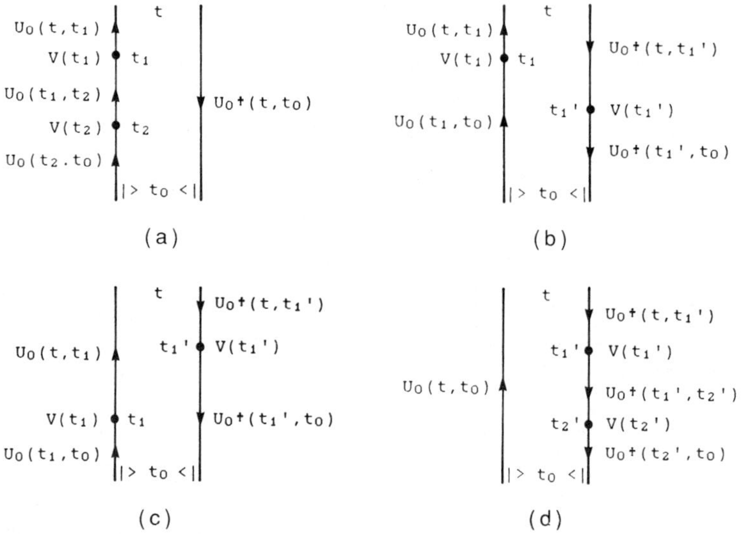

FIGURE 2.4 Diagrams for (a) $\rho_a^{(2)}(t)$ (Eq. (2.269a)), (b) $\rho_b^{(2)}(t)$ (Eq. (2.270)), (c) $\rho_c^{(2)}(t)$ (Eq. (2.271)), (d) $\rho_d^{(2)}(t)$ (Eq. 2.269b)).

equilibrium, are

$$\langle k|\rho^{(0)}(t)|l\rangle = \rho_{kk}^0 \delta_{kl} \tag{2.272}$$

$$\langle k|\rho_a^{(1)}(t)|l\rangle = -\frac{i}{\hbar} e^{-i\omega_{kl}t} \int_{t_0}^{t} dt_1\, V_{kl}(t_1) e^{i\omega_{kl}t_1} \rho_{ll}^0 \tag{2.273a}$$

$$\langle k|\rho_b^{(1)}(t)|l\rangle = \frac{i}{\hbar} e^{-i\omega_{kl}t} \int_{t_0}^{t} dt'_t \rho_{kk}^0 V_{kl}(t'_1) e^{i\omega_{kl}t'_1} \tag{2.273b}$$

$$\langle k|\rho_a^{(2)}(t)|l\rangle = \left(-\frac{i}{\hbar}\right)^2 \sum_r e^{-i\omega_{kl}t} \int_{t_0}^{t} dt_1 \int_{t_0}^{t_1} dt_2\, V_{kr}(t_1) e^{i\omega_{kr}t_1} V_{rl}(t_2) e^{i\omega_{rl}t_2} \rho_{ll}^0 \tag{2.274a}$$

$$\langle k|\rho_b^{(2)}(t)|l\rangle = \left(-\frac{i}{\hbar}\right)\left(\frac{i}{\hbar}\right) \sum_r e^{-i\omega_{kl}t} \int_{t_0}^{t} dt_1 \int_{t_0}^{dt_1} dt'_1\, V_{kr}(t_1) e^{i\omega_{kr}t_1} \rho_{rr}^0 V_{rl}(t'_1) e^{i\omega_{rl}t'_1} \tag{2.274b}$$

$$\langle k|\rho_c^{(2)}(t)|l\rangle = \left(-\frac{i}{\hbar}\right)\left(\frac{i}{\hbar}\right) \sum_r e^{-i\omega_{kl}t} \int_{t_0}^{t} dt'_1 \int_{t_0}^{t'_1} dt_1\, V_{kr}(t_1) e^{i\omega_{kr}t_1} \rho_{rr}^0 V_{rl}(t'_1) e^{i\omega_{rl}t'_1} \tag{2.274c}$$

$$\langle k|\rho_d^{(2)}(t)|l\rangle = \left(\frac{i}{\hbar}\right)^2 \sum_r e^{-i\omega_{kl}t} \int_{t_0}^{t} dt'_1 \int_{t_0}^{t'_1} dt'_2 \rho_{kk}^0 V_{kr}(t'_2) e^{i\omega_{kr}t'_2} V_{rl}(t'_1) e^{i\omega_{rl}t'_1}. \tag{2.274d}$$

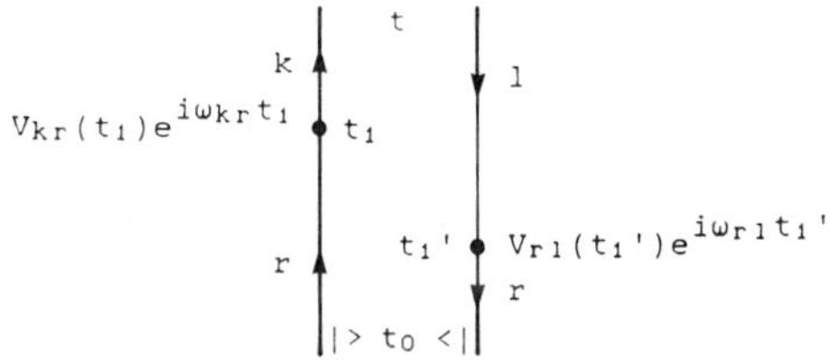

FIGURE 2.5 Feynman diagram for the matrix element $\langle k|\rho_b^{(2)}(t)|l\rangle$. The segment between t_1 and t is labeled k; the segment between t and t_1' is labeled l; the segments between t_1' and t_0 and between t_0 and t_1 are labeled r. With each value of the time we associate factors according to the scheme

$$t: e^{-i\omega_{kl}t}, \qquad t_1: V_{kr}(t_1)e^{i\omega_{kr}t_1},$$
$$t_1': V_{rl}(t_1')e^{i\omega_{rl}t_1'}, \qquad t_0: \rho_{rr}^0.$$

The integration is performed over the interaction points t_1 and t_1'.

The construction of diagrams for matrix elements of the density operator is illustrated in Fig. 2.5 for $\langle k|\rho_b^{(2)}(t)|l\rangle$. We begin again with a two-sided Feynman diagram labelled by the times $t > t_1 > t_1' > t_0$. The segment between t_1 and t is labelled k; the segment between t and t_1' is labelled l; the segments between t_1' and t_0 and between t_0 and t_1 are labelled r. With each value of the time, we associate a factor as follows:

$$\begin{aligned} &t: e^{-i\omega_{kl}t}, \qquad &&t_1: V_{kr}(t_1)e^{i\omega_{kr}t_1}, \\ &t_1': V_{rl}(t_1')e^{i\omega_{rl}t_1'}, \qquad &&t_0: \rho_{rr}^0. \end{aligned} \tag{2.275}$$

The integration is performed over the interaction points, first over t_1' (from t_0 to t_1), then over t_1 (from t_0 to t). Since the integrand no longer contains any operators, the order of the various factors is arbitrary. The summation is over the intermediate states $|r\rangle$; a factor of $i/\hbar$ is included for the interaction point on the right (t_1') and a factor of $-i/\hbar$ for the interaction point on the left (t_1).

Diagrams representing matrix elements of the density operator through the second order are shown in Figs. 2.6, 2.7, and 2.8.

These rules are readily extended to higher order, though it is rarely necessary to go beyond the third order. In the latter case, there are 2^3 diagrams

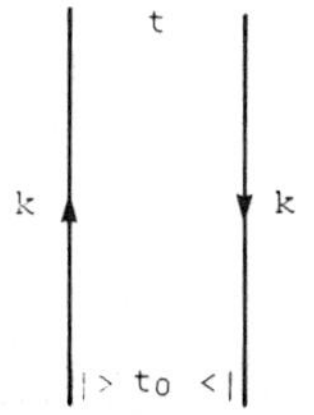

FIGURE 2.6 Diagram for $\langle k|\rho^0(t)|l\rangle = \rho_{kk}^0\delta_{kl}$.

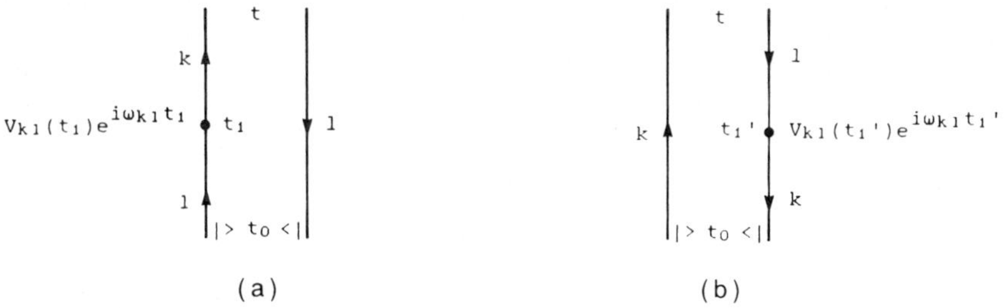

FIGURE 2.7 Diagrams for

$$\langle k|\rho^{(1)}(t)|l\rangle = \langle k|\rho_a^{(1)}(t)|l\rangle + \langle k|\rho_b^{(1)}(t)|l\rangle.$$

a. $\langle k|\rho_a^{(1)}(t)|l\rangle = -\frac{i}{\hbar}e^{-i\omega_{kl}t}\int_{t_0}^{t} dt_1 V_{kl}(t_1)e^{i\omega_{kl}t_1}\rho_{ll}^0.$

b. $\langle k|\rho_b^{(1)}(t)|l\rangle = \frac{i}{\hbar}e^{-i\omega_{kl}t}\int_{t_0}^{t} dt'_1 \rho_{kk}^0 V_{kl}(t'_1)e^{i\omega_{kl}t'_1}.$

corresponding to

$$t > t_1 > t_2 > t_3, t > t_1 > t_2 > t'_1, t > t_1 > t'_1 > t_2, t > t'_1 > t_1 > t_2,$$

$$t > t_1 > t'_1 > t'_2, t > t'_1 > t_1 > t'_2, t > t'_1 > t'_2 > t_1, t > t'_1 > t'_2 > t'_3.$$

An example of a third-order diagram for $t > t_1 > t_2 > t'_1$ is shown in Fig. 2.9.

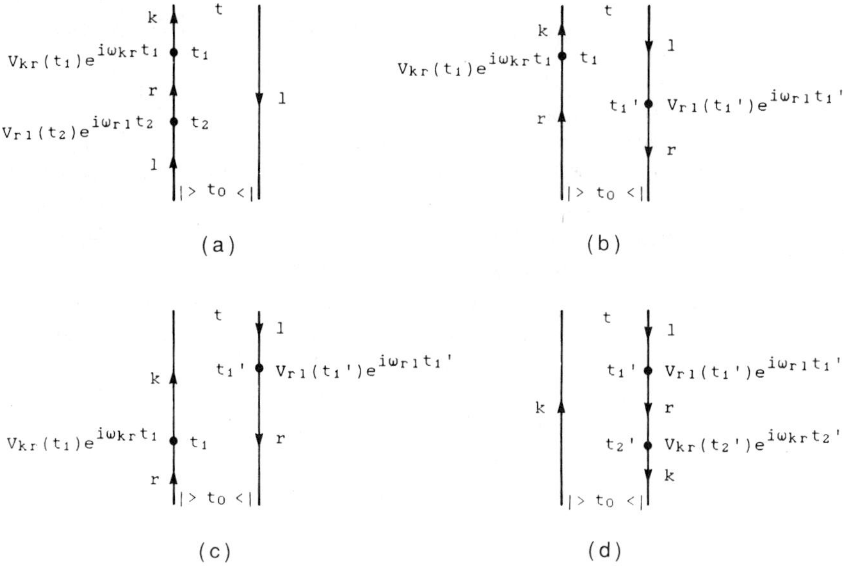

FIGURE 2.8 Diagrams for (a) $\langle k|\rho_a^{(2)}(t)|l\rangle$ (Eq. (2.274a)), (b) $\langle k|\rho_b^{(2)}(t)|l\rangle$ (Eq. (2.274b)), (c) $\langle k|\rho_c^{(2)}(t)|l\rangle$ (Eq. (2.274c)), (d) $\langle k|\rho_d^{(2)}(t)|l\rangle$ (Eq. (2.274d)).

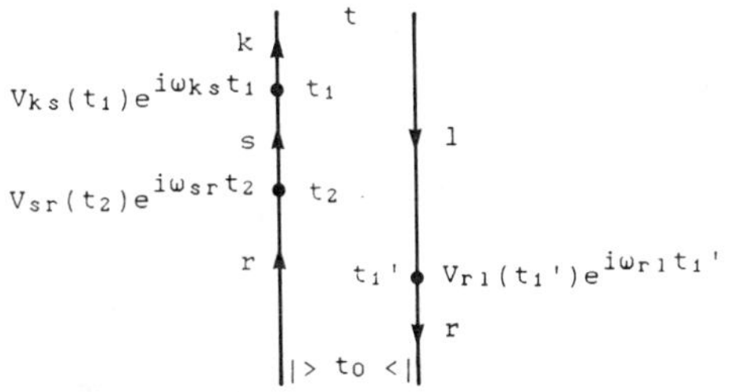

FIGURE 2.9 One of the eight Feynman diagrams that comprise the total matrix element $\langle k|\rho^{(3)}(t)|l\rangle$. For the diagram shown $t > t_1 > t_2 > t_1' > t_0$, and the contribution to the total matrix element is

$$\langle k|\rho_j^{(3)}(t)|l\rangle = \left(-\frac{i}{\hbar}\right)^2\left(\frac{i}{\hbar}\right)\sum_{rs} e^{-i\omega_{kl}t}\int_{t_0}^{t} dt_1 \int_{t_0}^{t_1} dt_2 \int_{t_0}^{t_2} dt_1' V_{ks}(t_1)e^{i\omega_{ks}t_1}V_{sr}(t_2)e^{i\omega_{sr}t_2}\rho_{rr}^0 V_{rl}(t_1')e^{i\omega_{rl}t_1'}.$$

2.10 Green's Functions, Time-Development Operator

In Section 2.3 we introduced the time-development operator $U(t, t_0)$, which transforms a wave function $|\psi(t_0)\rangle$ at the time t_0 to a wave function $|\psi(t)\rangle$ at the later time t, as the system develops in accordance with the time-dependent Schrödinger equation. Additional insight is obtained by relating $U(t, t_0)$ to certain types of propagators that have a broad range of application.

A mathematical digression is in order [10]. Let

$$\zeta(x^+) = -i \lim_{\varepsilon\to 0}\int_0^\infty e^{(ix-\varepsilon)t}\,dt \qquad (\varepsilon > 0) \tag{2.276a}$$

$$= -\lim_{\varepsilon\to 0}\left[\frac{e^{(ix-\varepsilon)t}}{x+i\varepsilon}\right]_0^\infty \tag{2.276b}$$

$$= \lim_{\varepsilon\to 0}\left[\frac{1}{x+i\varepsilon} - \lim_{t\to\infty}\frac{e^{(ix-\varepsilon)t}}{x+i\varepsilon}\right], \tag{2.276c}$$

and $\zeta(x^+) = 0$ for $t < 0$. The limits may be taken in the order $t \to \infty, \varepsilon \to 0$ or in reverse order. In the first case, since ε is a real positive number, $\exp(ix - \varepsilon)$ $t \to 0$ as $t \to \infty$,

$$\zeta(x^+) = \lim_{\varepsilon\to 0}\frac{1}{x+i\varepsilon} \tag{2.277a}$$

$$= \lim_{\varepsilon\to 0}\left[\frac{x}{x^2+\varepsilon^2} - \frac{i\varepsilon}{x^2+\varepsilon^2}\right], \tag{2.277b}$$

in which the first term is the principal value of $1/x$ and the second term is proportional to $\delta(x)$:

$$\lim_{\varepsilon\to 0}\frac{x}{x^2+\varepsilon^2} = P\left(\frac{1}{x}\right) \equiv \begin{cases} \dfrac{1}{x} & x \neq 0, \\ 0 & x = 0, \end{cases} \tag{2.278}$$

$$\lim_{\varepsilon\to 0}\frac{\varepsilon}{x^2+\varepsilon^2} = \pi\,\delta(x) = \begin{cases} 0 & x \neq 0 \\ \infty & x = 0. \end{cases} \tag{2.279}$$

Hence,

$$\zeta(x^+) = P\left(\frac{1}{x}\right) - i\pi\,\delta(x). \tag{2.280}$$

When the limits are taken in the order $\varepsilon \to 0$, $t \to \infty$, we obtain from Eq. (2.276a)

$$\zeta(x^+) = \lim_{t\to\infty}\frac{1-e^{ixt}}{x}, \qquad (t \geq 0), \tag{2.281}$$

whose real and imaginary parts must correspond to Eq. (2.280), i.e.,

$$P\left(\frac{1}{x}\right) = \lim_{t\to\infty}\frac{1-\cos xt}{x} = \lim_{t\to\infty}\int_0^t \sin(xt')\,dt', \tag{2.282a}$$

$$\pi\delta(x) = \lim_{t\to\infty}\frac{\sin xt}{x} = \lim_{t\to\infty}\int_0^t \cos(xt')\,dt'. \tag{2.282b}$$

Combining the two integral forms,

$$\zeta(x^+) = -i\lim_{t\to\infty}\int_0^t e^{ixt'}\,dt' = -i\int_0^\infty e^{ixt}\,dt, \qquad (t \geq 0). \tag{2.283}$$

Thus,

$$\int_0^\infty e^{ixt}\,dt = iP\left(\frac{1}{x}\right) + \pi\,\delta(x). \tag{2.284}$$

A similar development is applicable to $\zeta(x^-)$, defined by

$$\zeta(x^-) = i\lim_{\varepsilon\to 0}\int_{-\infty}^0 e^{(ix+\varepsilon)t}\,dt \qquad (\varepsilon > 0) \tag{2.285}$$

with $\zeta(x^-) = 0$ for $t > 0$. Performing the integration and taking the limits in the order $t \to -\infty$, $\varepsilon \to 0$, we find

$$\zeta(x^-) = \lim_{\varepsilon\to 0}\frac{1}{x - i\varepsilon} = P\left(\frac{1}{x}\right) + i\pi\,\delta(x) = i\int_0^\infty e^{-ixt}\,dt. \tag{2.286}$$

Consequently,

$$\int_0^\infty e^{\pm ixt}\,dt = \pm iP\left(\frac{i}{x}\right) + \pi\,\delta(x), \tag{2.287}$$

and

$$\zeta(x^-) - \zeta(x^+) = 2\pi i\,\delta(x). \tag{2.288}$$

For quantum mechanical applications [11, 12], let

$$G(x^\pm) = \lim_{\varepsilon\to 0}\frac{1}{x - \mathscr{H} \pm i\varepsilon}. \tag{2.289}$$

By analogy with Eqs. (2.280) and (2.286), we have

$$G(x^\pm) = P\left(\frac{1}{x - \mathscr{H}}\right) \mp i\pi\,\delta(x - \mathscr{H}), \tag{2.290}$$

$$G(x^-) - G(x^+) = 2\pi i\,\delta(x - \mathscr{H}). \tag{2.291}$$

Thus

$$\begin{aligned}\frac{1}{2\pi i}\int_{-\infty}^{\infty} e^{-ixt/\hbar}[G(x^-) - G(x^+)]\,dx &= \int_{-\infty}^{\infty} e^{-ixt/\hbar}\,\delta(x - \mathscr{H})\,dx \\ &= e^{-i\mathscr{H}t/\hbar} = U(t),\end{aligned} \tag{2.292}$$

where $\mathscr{H}$ is independent of time. Since $G(x^+) = 0$ for $t < 0$ and $G(x^-) = 0$ for $t > 0$, it is seen that

$$U(t<0) = \frac{1}{2\pi i}\int_{-\infty}^{\infty} e^{-ixt/\hbar}G(x^-)\,dx, \tag{2.293a}$$

$$U(t>0) = -\frac{1}{2\pi i}\int_{-\infty}^{\infty} e^{-ixt/\hbar}G(x^+)\,dx. \tag{2.293b}$$

Now consider the contour integral

$$\frac{1}{2\pi i}\int_{C_1+C_2} e^{-izt/\hbar}G(z)\,dz, \tag{2.294}$$

in which z is a complex variable and

$$G(z) \equiv \frac{1}{z - \mathscr{H}}. \tag{2.295}$$

$G(z)$ is known variously as the *resolvent operator* or *Green's function* or *propagator*. The contours are straight lines parallel to the real axis with C_1 below and C_2 above the real axis (Fig. 2.10). Since the Hamiltonian has real

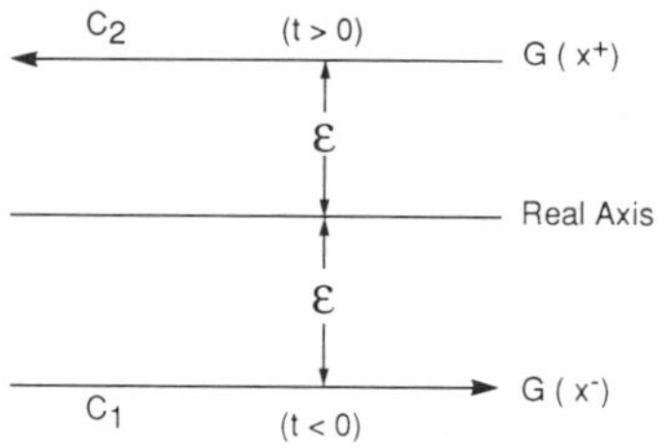

FIGURE 2.10 Contour for the integral representation of the time-development operator $U(t)$.

eigenvalues, all the singularities of $G(z)$ lie on the real axis and the contour $C_1 + C_2$ encloses them. By Cauchy's integral theorem,

$$\frac{1}{2\pi i}\int_{C_1+C_2} e^{-izt/\hbar}G(z)\,dz = \frac{1}{2\pi i}\int_{C_1+C_2} \frac{e^{-izt/\hbar}}{z-\mathscr{H}}\,dz$$
$$= e^{-i\mathscr{H}t/\hbar} = U(t). \tag{2.296}$$

For the two paths C_1 and C_2 considered separately,

$$\int_{C_1} e^{-izt/\hbar}G(z)\,dz = \int_{-\infty}^{\infty} e^{-i(x-i\varepsilon)t/\hbar}G(x^-)\,dx \tag{2.297a}$$

$$\int_{C_2} e^{-izt/\hbar}G(z)\,dz = \int_{\infty}^{-\infty} e^{-i(x+i\varepsilon)t/\hbar}G(x^+)\,dx. \tag{2.297b}$$

In the limit $\varepsilon \to 0$,

$$U(t<0) = \frac{1}{2\pi i}\int_{-\infty}^{\infty} e^{-ixt/\hbar}G(x^-)\,dx = \frac{1}{2\pi i}\int_{C_1} e^{-izt/\hbar}G(z)\,dz, \tag{2.298a}$$

$$U(t>0) = \frac{-1}{2\pi i}\int_{-\infty}^{\infty} e^{-ixt/\hbar}G(x^+)\,dx = \frac{1}{2\pi i}\int_{C_2} e^{-izt/\hbar}G(z)\,dz. \tag{2.298b}$$

Equations (2.298a) and (2.298b) provide the connection between the time-development operator and Green's function.

Let us now explore several additional properties of Green's functions. Readers not interested in the mathematical details may skip to Eq. (2.336).

1. For any two operator A and B, there exists the identity

$$\frac{1}{B} - \frac{1}{A} = \frac{1}{A}(A-B)\frac{1}{B}. \tag{2.299}$$

Then, upon setting

$$G_0 = \frac{1}{z-\mathscr{H}_0} = \frac{1}{A}, \qquad G = \frac{1}{z-\mathscr{H}} = \frac{1}{B}, \qquad \mathscr{H} = \mathscr{H}_0 + V, \tag{2.300}$$

one obtains the series

$$G = G_0 + G_0 V G = G_0 + G_0 V G_0 + G_0 V G_0 V G$$
$$= G_0 \left[1 + \sum_{n=1}^{\infty} (V G_0)^n \right]. \tag{2.301}$$

2. If $|\psi_k\rangle, |\psi_l\rangle, \ldots,$ are eigenstates of $\mathscr{H}_0$ with eigenvalues $\mathrm{E}_k, \mathrm{E}_l, \ldots,$ we have

$$(z - \mathscr{H}_0)|\psi_k\rangle = (z - \mathrm{E}_k)|\psi_k\rangle \tag{2.302a}$$

or, after premultiplication by G_0,

$$G_0|\psi_k\rangle \equiv \frac{1}{z - \mathscr{H}_0}|\psi_k\rangle = \frac{1}{z - \mathrm{E}_k}|\psi_k\rangle. \tag{2.302b}$$

Similarly,

$$\langle\psi_k|G_0 \equiv \langle\psi_k|\frac{1}{z - \mathscr{H}_0} = \frac{1}{z - \mathrm{E}_k}\langle\psi_k|. \tag{2.302c}$$

3. The matrix element of G_0 is

$$\langle\psi_l|G_0|\psi_k\rangle \equiv \langle\psi_l|\frac{1}{z - \mathscr{H}_0}|\psi_k\rangle = \frac{1}{z - \mathrm{E}_k}\delta_{kl}, \tag{2.303}$$

and the matrix elements of the first- and second-order terms in the expansion (2.301) are

$$\langle\psi_l|G^{(1)}|\psi_k\rangle \equiv \langle\psi_l|G_0 V G_0|\psi_k\rangle = \langle\psi_l|\frac{1}{z - \mathscr{H}_0} V \frac{1}{z - \mathscr{H}_0}|\psi_k\rangle$$
$$= \frac{1}{z - \mathrm{E}_l}\langle\psi_l|V|\psi_k\rangle\frac{1}{z - \mathrm{E}_k}, \tag{2.304}$$

$$\langle\psi_l|G^{(2)}|\psi_k\rangle \equiv \langle\psi_l|G_0 V G_0 V G_0|\psi_k\rangle$$
$$= \frac{1}{z - \mathrm{E}_l}\langle\psi_l|V G_0 V|\psi_k\rangle\frac{1}{z - \mathrm{E}_k}. \tag{2.305a}$$

Under the assumption that eigenstates of $\mathscr{H}_0$ form a complete set,

$$\langle\psi_l|G^{(2)}|\psi_k\rangle = \frac{1}{z - \mathrm{E}_l}\sum_m \langle\psi_l|V G_0|\psi_m\rangle\langle\psi_m|V|\psi_k\rangle\frac{1}{z - \mathrm{E}_k}$$
$$= \frac{1}{z - \mathrm{E}_l}\left[\sum_m \frac{\langle\psi_l|V|\psi_m\rangle\langle\psi_m|V|\psi_k\rangle}{z - \mathrm{E}_m}\right]\frac{1}{z - \mathrm{E}_k}. \tag{2.305b}$$

Matrix elements of higher-order terms are evaluated in similar fashion.

4. Again, assuming that the eigenstates of $\mathscr{H}_0$ form a complete set, we define *projection* operators P_k and Q_k:

$$P_k = |\psi_k\rangle\langle\psi_k|, \tag{2.306a}$$

$$Q_k = 1 - P_k = 1 - |\psi_k\rangle\langle\psi_k| = \sum_n |\psi_n\rangle\langle\psi_n| - |\psi_k\rangle\langle\psi_k|$$
$$= \sum_{n \neq k} |\psi_n\rangle\langle\psi_n|. \tag{2.306b}$$

It is easily verified that the projection operators satisfy

$$\sum_k P_k = 1, \qquad \sum_k Q_k = 0, \tag{2.307a}$$

$$P_k^2 = P_k, \qquad Q_k^2 = Q_k, \tag{2.307b}$$

$$P_k G_0 = G_0 P_k = \frac{1}{z - E_k} P_k, \qquad P_k Q_k = 0, \tag{2.308a}$$

$$Q_k G_0 = G_0 Q_k = \sum_{n \neq k} \frac{|\psi_n\rangle\langle\psi_n|}{z - E_n} \tag{2.308b}$$

$$P_k \mathscr{H}_0 = \mathscr{H}_0 P_k = E_k P_k, \qquad Q_k \mathscr{H}_0 = \mathscr{H}_0 Q_k, \tag{2.308c}$$

$$P_k|\psi_n\rangle = |\psi_k\rangle \delta_{kn}, \qquad P_k P_n = P_k \delta_{kn}. \tag{2.308d}$$

Let us now set $z = E_k$; we then may write

$$G_0 Q_k V|\psi_k\rangle = \frac{1}{E_k - \mathscr{H}_0} \sum_{n \neq k} |\psi_n\rangle\langle\psi_n|V|\psi_k\rangle$$
$$= \sum_{n \neq k} \langle\psi_n|V|\psi_k\rangle \frac{1}{E_k - \mathscr{H}_0}|\psi_n\rangle = \sum_{n \neq k} \frac{\langle\psi_n|V|\psi_k\rangle}{E_k - E_n}|\psi_n\rangle, \tag{2.309}$$

which is simply the first-order correction to $|\psi_k\rangle$ due to the perturbation V. That is,

$$G_0 Q_k V|\psi_k\rangle = |\psi_k^{(1)}\rangle. \tag{2.310}$$

Continuing in this fashion,

$$|\psi_k^{(2)}\rangle = G_0 V Q_k V G_0 Q_k V|\psi_k\rangle. \tag{2.311}$$

This pattern is extended readily to higher order.

5. For an arbitrary operator A, let

$$A' \equiv P_k A P_k. \tag{2.312}$$

Then

$$G_0' \equiv \mathrm{P}_k G_0 \mathrm{P}_k = G_0 \mathrm{P}_k^2 = G_0 \mathrm{P}_k = \mathrm{P}_k G_0. \tag{2.313}$$

To evaluate G' we refer to the expansion (2.301) and the properties of the projection operators shown in Eqs. (2.306) and (2.307):

$$G' \equiv \mathrm{P}_k G \mathrm{P}_k = G_0' + \mathrm{P}_k G_0 V G \mathrm{P}_k = G_0' + G_0 \mathrm{P}_k V G \mathrm{P}_k, \tag{2.314}$$

$$V'G' \equiv \mathrm{P}_k V \mathrm{P}_k G \mathrm{P}_k = \mathrm{P}_k V(1 - \mathrm{Q}_k) G \mathrm{P}_k. \tag{2.315}$$

Combining Eqs. (2.314) and (2.315),

$$G' = G_0' + G_0 V'G' + G_0 \mathrm{P}_k V \mathrm{Q}_k G \mathrm{P}_k. \tag{2.316}$$

The product $\mathrm{Q}_k G \mathrm{P}_k$ may be put in another form. Since

$$(z - \mathscr{H})G = (z - \mathscr{H}_0 - V)G = (z - \mathscr{H}_0 - V)(\mathrm{P}_k + \mathrm{Q}_k)G = 1, \tag{3.317a}$$

we have

$$\mathrm{Q}_k(z - \mathscr{H}_0 - V)(\mathrm{P}_k + \mathrm{Q}_k)G\mathrm{P}_k = \mathrm{Q}_k\mathrm{P}_k = 0, \tag{2.317b}$$

as well as

$$\mathrm{Q}_k(z - \mathscr{H}_0)\mathrm{P}_k G \mathrm{P}_k = (z - \mathscr{H}_0)\mathrm{Q}_k \mathrm{P}_k G \mathrm{P}_k = 0. \tag{2.317c}$$

Then,

$$\begin{aligned} \mathrm{Q}_k(z - \mathscr{H}_0 - V)(\mathrm{P}_k + \mathrm{Q}_k)G\mathrm{P}_k &= -\mathrm{Q}_k V \mathrm{P}_k G \mathrm{P}_k + \mathrm{Q}_k(z - \mathscr{H}_0 - V)\mathrm{Q}_k G \mathrm{P}_k \\ &= -\mathrm{Q}_k V \mathrm{P}_k G' + (z - \mathscr{H}_0 - \mathrm{Q}_k V \mathrm{Q}_k)\mathrm{Q}_k G \mathrm{P}_k \\ &= \mathrm{Q}_k \mathrm{P}_k = 0 \end{aligned} \tag{2.318a}$$

or

$$\mathrm{Q}_k G \mathrm{P}_k = \frac{1}{z - \mathscr{H}_0 - \mathrm{Q}_k V \mathrm{Q}_k} \mathrm{Q}_k V \mathrm{P}_k G' = \frac{\mathrm{Q}_k}{z - \mathscr{H}_0 - \mathrm{Q}_k V \mathrm{Q}_k} V \mathrm{P}_k G', \tag{2.318b}$$

the last expression arising from the fact that Q_k commutes with $(z - \mathscr{H}_0 - \mathrm{Q}_k V \mathrm{Q}_k)$. This equation, when inserted into Eq. (2.316), yields

$$G' = G_0' + G_0 V'G' + G_0 \mathrm{P}_k V \frac{\mathrm{Q}_k}{z - \mathscr{H}_0 - \mathrm{Q}_k V \mathrm{Q}_k} V \mathrm{P}_k G', \tag{2.319}$$

and

$$(z - \mathscr{H}_0)G' = \mathrm{P}_k + V'G' + \mathrm{P}_k V \frac{\mathrm{Q}_k}{z - \mathscr{H}_0 - \mathrm{Q}_k V \mathrm{Q}_k} V \mathrm{P}_k G'. \tag{2.320}$$

Thus,

$$(z - \mathscr{H}_0 - R)G' = \mathrm{P}_k \tag{2.321}$$

where

$$R = V' + \mathrm{P}_k V \frac{\mathrm{Q}_k}{z - \mathscr{H}_0 - \mathrm{Q}_k V Q_k} V \mathrm{P}_k. \tag{2.322}$$

We note that $(z - \mathscr{H}_0 - R)G' = (z - \mathscr{H}_0 - R)G'\mathrm{P}_k$; hence, Eq. (2.321) may be satisfied by

$$(z - \mathscr{H}_0 - R)G' = 1 \tag{2.323}$$

since then $(z - \mathscr{H}_0 - R)G'\mathrm{P}_k = \mathrm{P}_k$. Thus,

$$G' = \frac{1}{z - \mathscr{H}_0 - R}. \tag{2.324}$$

Following the procedure leading to Eq. (2.301), the expansion of G' is

$$G' = G_0 + G_0 R G_0 + G_0 R G_0 R G_0 + \cdots. \tag{2.325}$$

6. In view of the relation $\mathrm{P}_k^2 = \mathrm{P}_k$, Eq. (2.324) is equivalent to

$$(z - \mathscr{H}_0 - R)G' = (z - \mathscr{H}_0 - R)\mathrm{P}_k G' = 1. \tag{2.326}$$

The diagonal matrix element then becomes

$$\begin{aligned} \langle\psi_k|(z - \mathscr{H}_0 - R)G'|\psi_k\rangle &= \langle\psi_k|z - \mathscr{H}_0 - R|\psi_k\rangle\langle\psi_k|G'|\psi_k\rangle \\ &= [z - \mathrm{E}_k - \langle\psi_k|R|\psi_k\rangle]\langle\psi_k|G'|\psi_k\rangle \\ &= 1 \end{aligned} \tag{2.327a}$$

or

$$\langle\psi_k|G'|\psi_k\rangle = \langle\psi_k|G|\psi_k\rangle = \frac{1}{z - \mathrm{E}_k - \langle\psi_k|R|\psi_k\rangle}. \tag{2.327b}$$

Matrix elements of R may be approximated by noting that

$$\frac{1}{z - \mathscr{H}_0 - \mathrm{Q}_k V \mathrm{Q}_k} = G_0 + G_0 \mathrm{Q}_k V \mathrm{Q}_k G_0 + G_0 \mathrm{Q}_k V \mathrm{Q}_k G_0 \mathrm{Q}_k V \mathrm{Q}_k G_0 + \cdots \tag{2.328}$$

so that the series expansion of R, defined by Eq. (2.322), is

$$\begin{aligned} R &= V' + \mathrm{P}_k V \mathrm{Q}_k G_0 V \mathrm{P}_k + \mathrm{P}_k V \mathrm{Q}_k G_0 V \mathrm{Q}_k G_0 V \mathrm{P}_k + \cdots \\ &= V' + \mathrm{P}_k V \left[\frac{\mathrm{Q}_k}{z - \mathscr{H}_0} + \frac{\mathrm{Q}_k}{z - \mathscr{H}_0} V \frac{\mathrm{Q}_k}{z - \mathscr{H}_0} + \cdots \right] V \mathrm{P}_k. \end{aligned} \tag{2.329}$$

When the series is terminated after the first two terms and assuming $\langle\psi_k|V|\psi_k\rangle = 0$, the diagonal matrix element of R is

$$\begin{aligned}\langle\psi_k|R|\psi_k\rangle &= \langle\psi_k|V'|\psi_k\rangle + \langle\psi_k|\mathrm{P}_k V \mathrm{Q}_k G_0 V \mathrm{P}_k|\psi_k\rangle \\ &= \langle\psi_k|V|\psi_k\rangle + \langle\psi_k|V \mathrm{Q}_k G_0 V|\psi_k\rangle \\ &= \langle\psi_k|V|\psi_k\rangle + \langle\psi_k|V \sum_{n\neq k} \frac{|\psi_n\rangle\langle\psi_n|}{z - \mathrm{E}_n} V|\psi_k\rangle \\ &= \langle\psi_k|V|\psi_k\rangle + \sum_{n\neq k} \frac{|\langle\psi_k|V|\psi_n\rangle^2|}{z - \mathrm{E}_n}. \end{aligned} \tag{2.330}$$

We may employ this formalism to compute the diagonal matrix element of the time-development operator $U(t)$. Since the main contributions to the sum will come from values of z near E_n, we write, as in Eqs. (2.280) and (2.286),

$$\frac{1}{z - \mathrm{E}_n} \simeq \lim_{\varepsilon\to 0} \frac{1}{x - \mathrm{E}_n \pm i\varepsilon} \equiv F(x^\pm) = \mathrm{P}\left(\frac{1}{x - \mathrm{E}^n}\right) \mp i\pi\,\delta(x - \mathrm{E}^n). \tag{2.331}$$

Then, for perturbations such that $\langle\psi_k|V|\psi_k\rangle = 0$,

$$\langle\psi_k|R(x^\pm)|\psi_k\rangle = \frac{1}{2}(\Delta\mathrm{E} \mp i\hbar\Gamma) \tag{2.332}$$

where

$$\frac{\Delta\mathrm{E}}{2} = \sum_{n\neq k} |\langle\psi_k|V|\psi_n\rangle|^2 \mathrm{P}\left(\frac{1}{x - \mathrm{E}_n}\right), \tag{2.333a}$$

$$\frac{\hbar\Gamma}{2} \equiv \pi \sum_{n\neq k} |\langle\psi_k|V|\psi_n\rangle|^2\, \delta(x - \mathrm{E}_n). \tag{2.333b}$$

Let us now concentrate on $R(x^+)$, which is associated with $U(t)$ for $t > 0$. Inserting

$$\langle\psi_k|R(x^+)|\psi_k\rangle = \frac{1}{2}(\Delta\mathrm{E} - i\hbar\Gamma) \tag{2.334}$$

into Eq. (2.327b), we have

$$\langle\psi_k|G|\psi_k\rangle = \frac{1}{z - \mathrm{E}_k - \frac{1}{2}(\Delta\mathrm{E} - i\hbar\Gamma)}. \tag{2.335}$$

Then, referring to Eq. (2.296),

$$\begin{aligned}\langle\psi_k|U(t)|\psi_k\rangle &= \frac{1}{2\pi i}\int_{\mathrm{C}_1+\mathrm{C}_2} e^{-izt/\hbar}\langle\psi_k|G|\psi_k\rangle\, dz \\ &= \frac{1}{2\pi i}\int_{\mathrm{C}_1+\mathrm{C}_2} e^{-izt/\hbar}\frac{dz}{z - \mathrm{E}_k - \frac{1}{2}(\Delta\mathrm{E} - i\hbar\Gamma)} \\ &= e^{-i(\mathrm{E}_k + 1/2\Delta\mathrm{E})t/\hbar} e^{-1/2\Gamma t}, \end{aligned} \tag{2.336}$$

and

$$|\langle\psi_k|U(t)|\psi_k\rangle|^2 = e^{-i\Gamma t}. \tag{2.337}$$

According to Eq. (2.336), the energy E_k of an eigenstate $|\psi_k\rangle$ of $\mathscr{H}_0$ is modified by the addition of the complex energy $(\Delta E - i\hbar\Gamma)/2$ as a consequence of the perturbation V. The imaginary part $\hbar\Gamma/2$ is responsible for the exponential decay as shown in Eq. (2.337); the real part $\Delta E/2$ represents a shift in energy. In the context of an atom and a radiation field, E_k is an eigenvalue of the zero-order Hamiltonian, that is, in the absence of an interaction between the atom and the field. In that case,

$$\langle\psi_k|U(t)|\psi_k\rangle = e^{-iE_k t/\hbar}, \tag{2.338}$$

consistent with the basic definition (2.292).

An important application of this development appears in the discussion of spontaneous emission from the standpoint of vacuum fluctuations (Section 6.9).

2.11 Reduced Density Matrices

A physical system often may be regarded as a composite of several distinguishable components and it may happen that one wishes to follow the behavior of only one component. It then would be highly desirable to formulate the problem in such a manner that only the variables belonging to the component of interest enter into the computation while the effect of the unwanted components is relegated to an appropriate set of parameters. A situation of this type occurs in the interaction of a system with a reservoir (Chapter VI) as, for example, when atoms are placed in a cavity that can support a range of modes. If our interest is focused on the system, we want to know how the reservoir affects the system without entering into the cumbersome problem of solving for the variables of the reservoir. The reduced density matrix [2, 3] is designed to accomplish this objective and we shall now introduce some of its properties.

Consider a system AB consisting of two components A and B. We assume, at first, that A and B do not interact and are not correlated in any way. All operators associated with A commute with all operators in B and if A and B are in states $|\psi_a\rangle$ and $|\psi_b\rangle$, respectively, the state of the combined (or total) system is $|\psi_{ab}\rangle = |\psi_a\psi_b\rangle$. In a mathematical sense, the components A and B correspond to two independent spaces and the combined system corresponds to the (direct) product space. The assumption that the wave functions $|\psi_a\rangle$, $|\psi_b\rangle$, and $|\psi_{ab}\rangle$ are known implies that the three systems A, B, and AB are all in pure states with density operators

$$\rho(A) = |\psi_a\rangle\langle\psi_a|, \qquad \rho(B) = |\psi_b\rangle\langle\psi_b|, \tag{2.339}$$

$$\rho(AB) = |\psi_{ab}\rangle\langle\psi_{ab}|. \tag{2.340}$$

It now will be shown that a knowledge of $\rho(AB)$ alone is sufficient for the determination of $\rho(A)$ and $\rho(B)$. For this purpose, consider an operator $O(A)$ that acts only on wave functions in A; then

$$\begin{aligned} \langle\psi_{ab}|O(A)|\psi_{a'b'}\rangle &= \langle\psi_a\psi_b|O(A)|\psi_{a'}\psi_{b'}\rangle \\ &= \langle\psi_a|O(A)|\psi_{a'}\rangle\,\delta_{bb'}. \end{aligned} \tag{2.341}$$

For the ensemble average $\langle O(A)\rangle$, we have

$$\begin{aligned} \langle O(A) &= \mathrm{Tr}\{O(A)\rho(AB)\} \\ &= \sum_{ab}\langle\psi_{ab}|O(A)\rho(AB)|\psi_{ab}\rangle \\ &= \sum_{aba'b'}\langle\psi_{ab}|O(A)|\psi_{a'b'}\rangle\langle\psi_{a'b'}|\rho(AB)|\psi_{ab}\rangle \\ &= \sum_{aa'}\langle\psi_a|O(A)|\psi_{a'}\rangle\sum_{b}\langle\psi_{a'b}|\rho(AB)|\psi_{ab}\rangle \end{aligned} \tag{2.342}$$

in which Eq. (2.341) has been used to obtain the last expression. With the definition

$$\mathrm{Tr}_B\,\rho(AB) \equiv \sum_b\langle\psi_b|\rho(AB)|\psi_b\rangle, \tag{2.343}$$

the sum over b in Eq. (2.342) becomes

$$\begin{aligned} \sum_b\langle\psi_{a'b}|\rho(AB)|\psi_{ab}\rangle &= \left\langle\psi_{a'}\left|\sum_b\langle\psi_b|\rho(AB)|\psi_b\rangle\right|\psi_a\right\rangle \\ &= \langle\psi_{a'}|\mathrm{Tr}_B\,\rho(AB)|\psi_a\rangle, \end{aligned} \tag{2.344}$$

and the ensemble average $\langle O(A)\rangle$ now may be written

$$\begin{aligned} \langle O(A)\rangle &= \sum_{aa'}\langle\psi_a|O(A)|\psi_{a'}\rangle\langle\psi_{a'}|\mathrm{Tr}_B\,\rho(AB)|\psi_a\rangle \\ &= \sum_a\langle\psi_a|O(A)\,\mathrm{Tr}_B\,\rho(AB)|\psi_a\rangle. \end{aligned} \tag{2.345}$$

But this is precisely of the form

$$\langle O(A)\rangle = \sum_a\langle\psi_a|O(A)\rho(A)|\psi_a>. \tag{2.346}$$

Hence, we have established the relation

$$\rho(A) = \mathrm{Tr}_B\,\rho(AB). \tag{2.347}$$

Clearly, similar considerations may be applied to an operator $O(B)$ associated with the component B. Hence, it may be concluded that

$$\rho(A) = \mathrm{Tr}_B\, \rho(AB), \qquad \rho(B) = \mathrm{Tr}_A\, \rho(AB), \tag{2.348}$$

which defines the *reduced* density operators $\rho(A)$ and $\rho(B)$.

An additional conclusion may be drawn from Eq. (2.340). Since the components are uncorrelated,

$$\begin{aligned}\rho(AB) &= |\psi_{ab}\rangle\langle\psi_{ab}| \\ &= (|\psi_a\rangle\langle\psi_a|)(|\psi_b\rangle\langle\psi_b|) = \rho(A)\rho(B),\end{aligned} \tag{2.349}$$

which is to be interpreted as a *direct* product since $\rho(A)$ and $\rho(B)$ operate in different spaces.

A simple illustration of these ideas is provided by the spin states of two uncorrelated electrons. Let electrons 1 and 2 correspond to components A and B, and let $|\alpha(1)\rangle$, $|\beta(2)\rangle$, and $|\alpha(1)\beta(2)\rangle$ correspond to $|\psi_a\rangle$, $|\psi_b\rangle$, and $|\psi_{ab}\rangle$, respectively. Since the states are specified, each electron, as well as the combined two-electron system, are all in pure states and the density operators are

$$\rho(1) = |\alpha(1)\rangle\langle\alpha(1)|, \qquad \rho(2) = |\beta(2)\rangle\langle\beta(2)|, \tag{2.350}$$

$$\rho(1,2) = |\alpha(1)\beta(2)\rangle\langle\alpha(1)\beta(2)|. \tag{2.351}$$

To construct the density matrices, we use $\{|\alpha(1)\rangle, |\beta(1)\rangle\}$, $\{|\alpha(2)\rangle, |\beta(2)\rangle\}$, and $\{|\alpha(1)\alpha(2)\rangle, |\alpha(1)\beta(2)\rangle, |\beta(1)\alpha(2)\rangle, |\beta(1)\beta(2)\rangle\}$ as the bases for $\rho(1)$, $\rho(2)$, and $\rho(1,2)$, respectively. These are shown in Fig. 2.11. It is observed that the direct product $\rho(1)\rho(2)$ is obtained by multiplying each element of the two-dimensional matrix $\rho(1)$ by each element of the two-dimensional matrix $\rho(2)$; the result is the four-dimensional matrix $\rho(1,2)$, thereby verifying Eq. (2.349).

To verify Eq. (2.348) for this case, we note that in accordance with Eq. (2.343),

$$\begin{aligned}\mathrm{Tr}_2\, \rho(1,2) &= \langle\alpha(2)|\rho(1,2)|\alpha(2)\rangle + \langle\beta(2)|\rho(1,2)|\beta(2)\rangle \\ &= \langle\alpha(2)|\alpha(1)\beta(2)\rangle\langle\alpha(1)\beta(2)|\alpha(2)\rangle \\ &\quad + \langle\beta(2)|\alpha(1)\beta(2)\rangle\langle\alpha(1)\beta(2)|\beta(2)\rangle.\end{aligned} \tag{2.352}$$

As a consequence of the orthonormality of the spin states, the first term is zero and the second term is simply $|\alpha(1)\rangle\langle\alpha(1)|$ or $\rho(1)$.

We now turn to the case where the components of a composite system are correlated in some fashion such as a coupling scheme or an interaction. In that event, if $|\psi_a\rangle$ and $|\psi_b\rangle$ are the states of the components A and B in the absence of any correlation, the state $|\psi_{ab}\rangle$ of the combined, correlated system is no

ρ(1)	α(1)	β(1)
α(1)	1	0
β(1)	0	0

(a)

ρ(2)	α(2)	β(2)
α(2)	0	0
β(2)	0	1

(b)

ρ(1,2)	α(1) α(2)	α(1) β(2)	β(1) α(2)	β(1) β(2)
α(1) α(2)	0	0	0	0
α(1) β(2)	0	1	0	0
β(1) α(2)	0	0	0	0
β(1) β(2)	0	0	0	0

(c)

FIGURE 2.11 Density matrices for two uncorrelated electrons. $|\alpha(1)\rangle$ and $|\beta(2)\rangle$ are the states of electrons 1 and 2, respectively. (a) $\rho(1)$ is the density matrix of electron 1; (b) $\rho(2)$ is the density matrix of electron 2; (c) $\rho(1,2)$ is the density matrix of the combined system. Electrons 1 and 2, as well as the combined system, are in pure states. $\rho(1) = \mathrm{Tr}_2\,\rho(1,2)$, $\rho(2) = \mathrm{Tr}_1\,\rho(1,2)$ and $\rho(1,2)$ is the direct product of $\rho(1)$ and $\rho(2)$.

longer simply a product of $|\psi_a\rangle$ and $|\psi_b\rangle$ but, instead, is a sum of terms

$$|\psi_{ab}\rangle = \sum_{ab} c_{ab}|\psi_a\psi_b\rangle, \tag{2.353}$$

in which the c_{ab} are constants. Rather than proceed with the general formalism, which becomes quite cumbersome, it will be simpler and more transparent to illustrate the results with a two-electron system in a triplet state, one of whose wave functions is

$$|\psi(1,2)\rangle = \frac{1}{\sqrt{2}}[|\alpha(1)\beta(2)\rangle + |\beta(1)\alpha(2)\rangle]. \tag{2.354}$$

In contrast to the previous case, we now have a state in which the electrons are correlated so that, despite our knowledge of the wave function for the composite system, the wave functions for the individual electrons cannot be constructed. In other words, the composite system is in a pure state whereas the individual electrons are in mixed states.

For the composite system, the density operator is

$$\rho(1,2) = |\psi(1,2)\rangle\langle\psi(1,2)|, \tag{2.355}$$

and its matrix representation in the basis

$$\begin{aligned} |\alpha(1)\alpha(2)\rangle &\equiv a_1, \qquad |\alpha(1)\beta(2)\rangle \equiv a_2, \\ |\beta(1)\alpha(2)\rangle &\equiv a_3, \qquad |\beta(1)\beta(2)\rangle \equiv a_4, \end{aligned} \tag{2.356}$$

is shown in Fig. 2.12a. As in the previous case, we consider an operator $O(1)$ whose ensemble average is

$$\begin{aligned} \langle O(1)\rangle &= \mathrm{Tr}\{O(1)\rho(1,2)\} \\ &= \langle a_1|O(1)\rho(1,2)|a_1\rangle + \langle a_2|O(1)\rho(1,2)|a_2\rangle \\ &\quad + \langle a_3|O(1)\rho(1,2)|a_3\rangle + \langle a_4|O(1)\rho(1,2)|a_4\rangle. \end{aligned} \tag{2.357}$$

Each of the four terms may be expanded into products as, for example,

$$\begin{aligned} \langle a_1|O(1)\rho(1,2)|a_1\rangle &= \langle a_1|O(1)|a_1\rangle\langle a_1|\rho(1,2)|a_1\rangle \\ &\quad + \langle a_1|O(1)|a_2\rangle\langle a_2|\rho(1,2)|a_1\rangle \\ &\quad + \langle a_1|O(1)|a_3\rangle\langle a_3|\rho(1,2)|a_1\rangle \\ &\quad + \langle a_1|O(1)|a_4\rangle\langle a_4|\rho(1,2)|a_1\rangle. \end{aligned} \tag{2.358}$$

ρ(1,2)	α(1) α(2)	α(1) β(2)	β(1) α(2)	β(1) β(2)
α(1) α(2)	0	0	0	0
α(1) β(2)	0	1/2	1/2	0
β(1) α(2)	0	1/2	1/2	0
β(1) β(2)	0	0	0	0

(a)

ρ(1)	α(1)	β(1)
α(1)	1/2	0
β(1)	0	1/2

(b)

ρ(2)	α(2)	β(2)
α(2)	1/2	0
β(2)	0	1/2

(c)

FIGURE 2.12 Density matrices for two electrons correlated by the wave function

$$|\psi(1,2)\rangle = \frac{1}{\sqrt{2}}[|\alpha(1)\beta(2)\rangle + |\beta(1)\rangle\alpha(2)\rangle].$$

(a) $\rho(1,2)$ is the density matrix of the combined system; (b) $\rho(1)$ is the density matrix of electron 1; (c) $\rho(2)$ is the density matrix of electron 2. The combined system is in a pure state but electrons 1 and 2 are in mixed states. $\rho(1) = \mathrm{Tr}_2\,\rho(1,2)$, $\rho(2) = \mathrm{Tr}_1\,\rho(1,2)$, but $\rho(1,2)$ is *not* the direct product of $\rho(1)$ and $\rho(2)$.

Consulting the matrix elements shown in Fig. 2.12a, it is found that

$$\begin{aligned}\langle O(1)\rangle &= \langle\alpha(1)\beta(2)|O(1)|\alpha(1)\beta(2)\rangle\langle\alpha(1)\beta(2)|\rho(1,2)|\alpha(1)\beta(2)\rangle \\ &\quad + \langle\beta(1)\alpha(2)|O(1)|\beta(1)\alpha(2)\rangle\langle\beta(1)\alpha(2)|\rho(1,2)|\beta(1)\alpha(2)\rangle \\ &= \frac{1}{2}[\langle\alpha(1)|O(1)|\alpha(1)\rangle + \langle\beta(1)|O(1)|\beta(1)\rangle].\end{aligned} \tag{2.359}$$

We now shall arrive at the same result via another route. From the general definition of Eq. (2.343)

$$\mathrm{Tr}_2\,\rho(1,2) = \langle\alpha(2)|\rho(1,2)|\alpha(2)\rangle + \langle\beta(2)|\rho(1,2)|\beta(2)\rangle. \tag{2.360}$$

Referring to Eqs. (2.354) and (2.355), it is found that

$$\begin{aligned}\langle\alpha(1)|\,\mathrm{Tr}_2\,\rho(1,2)|\alpha(1)\rangle &= \langle\beta(1)|\,\mathrm{Tr}_2\,\rho(1,2)|\beta(1)\rangle = \frac{1}{2}, \\ \langle\alpha(1)|\,\mathrm{Tr}_2\,\rho(1,2)|\beta(1)\rangle &= \langle\beta(1)|\,\mathrm{Tr}_2\,\rho(1,2)|\alpha(1)\rangle = 0.\end{aligned} \tag{2.361}$$

These matrix elements are shown in Fig. 2.12b and are labelled $\rho(1)$ in anticipation of the interpretation of $\rho(1)$ as the density operator associated with electron 1. To verify this feature, it is noted that

$$\begin{aligned}\mathrm{Tr}\{O(1)\rho(1)\} &= \langle\alpha(1)|O(1)\rho(1)|\alpha(1)\rangle + \langle\beta(1)|O(1)\rho(1)|\beta(1)\rangle, \\ &= \langle\alpha(1)|O(1)|\alpha(1)\rangle\langle\alpha(1)|\rho(1)|\alpha(1)\rangle \\ &\quad + \langle\alpha(1)|O(1)|\beta(1)\rangle\langle\beta(1)|\rho(1)|\alpha(1)\rangle \\ &\quad + \langle\beta(1)|O(1)|\alpha(1)\rangle\langle\alpha(1)|\rho(1)|\beta(1)\rangle \\ &\quad + \langle\beta(1)|\rho(1)|\beta(1)\rangle\langle\beta(1)|O(1)|\beta(1)\rangle, \\ &= \frac{1}{2}[\langle\alpha(1)|O(1)|\alpha(1)\rangle + \langle\beta(1)|O(1)|\beta(1)\rangle]\end{aligned} \tag{2.362}$$

which is identical with Eq. (2.359) for the ensemble average $\langle O(1)\rangle$. The treatment of $\rho(2)$ is similar (Fig. 2.12c); we conclude that

$$\rho(1) = \mathrm{Tr}_2\,\rho(1,2), \qquad \rho(2) = \mathrm{Tr}_1\,\rho(1,2). \tag{2.363}$$

More generally, one finds

$$\rho(A) = \mathrm{Tr}_B\,\rho(AB), \qquad \rho(B) = \mathrm{Tr}_A\,\rho(AB), \tag{2.364}$$

as in Eq. (2.348). Thus, the derivation of the reduced density operators in the correlated case is exactly the same as in the uncorrelated case.

Distinct differences exist, however, between the two cases. Referring again to Fig. 2.12, it is seen that

$$\mathrm{Tr}\,\rho^2(1) = \mathrm{Tr}\,\rho^2(2) = \frac{1}{2}, \tag{2.365}$$

$$\mathrm{Tr}\,\rho^2(1,2) = 1, \tag{2.366}$$

which verifies the fact that electrons 1 and 2 are in mixed states but the composite system is in a pure state, whereas in the uncorrelated case all three were in pure states. Also, the direct product $\rho(1)\rho(2)$ is a diagonal matrix totally different from $\rho(1,2)$ so that, in general, for the correlated case

$$\rho(A)\rho(B) \neq \rho(AB) \tag{2.367}$$

in contrast to Eq. (2.349).

One of the important applications of the reduced density matrix formalism occurs in perturbation theory when the Hamiltonian has the form

$$\mathscr{H}_{AB} = \mathscr{H}_A + \mathscr{H}_B + V \tag{2.368}$$

where V represents an interaction between two physical systems A and B. If V is turned on at a time $t = t_0$, then for $t \leq t_0$ the two systems are uncorrelated; $\mathscr{H}_A$ and $\mathscr{H}_B$, separately, generate eigenstates $|\psi_a\rangle$ and $|\psi_b\rangle$ whose products are eigenstates of $\mathscr{H}_{AB}$. The components A and B and the composite system AB are all in pure states and ρ_{AB} is the direct product of ρ_A and ρ_B, as in the example discussed at the beginning of this section. But for $t > t_0$, that is, after the onset of the interaction, AB is in a pure state, whereas A and B are in mixed states since there are no wave functions for A and B in the presence of the interaction. In the latter case, all the quantum mechanical information concerning the components is contained in the reduced density operators.

We also may obtain the time-development of the reduced density operator. Starting with $\rho_{AB}^{(t)}$ for the composite system, the equation of motion is

$$i\hbar\dot{\rho}_{AB}(t) = [\mathscr{H}_{AB}, \rho_{AB}(t)], \tag{2.369}$$

and, for the reduced density operator,

$$\begin{aligned} i\hbar\dot{\rho}_A(t) &= i\hbar\,\mathrm{Tr}_B\,\dot{\rho}_{AB}(t) \\ &= \mathrm{Tr}_B\,[(\mathscr{H}_A + \mathscr{H}_B + V), \rho_{AB}(t)]. \end{aligned} \tag{2.370}$$

But

$$\mathrm{Tr}_B\,[\mathscr{H}_A, \rho_{AB}(t)] = [\mathscr{H}_A, \mathrm{Tr}_B\,\rho_{AB}(t)] = [\mathscr{H}_A, \rho_A(t)], \tag{2.371}$$

$$\begin{aligned} \mathrm{Tr}_B\,[\mathscr{H}_B, \rho_{AB}(t)] &= \sum_B [\langle\psi_B|\mathscr{H}_B\rho_{AB}(t)|\psi_B\rangle - \langle\psi_B|\rho_{AB}(t)\mathscr{H}_B|\psi_B\rangle] \\ &= \sum_B [\mathrm{E}_B\langle\psi_B|\rho_{AB}(t)|\psi_B\rangle - \langle\psi_B|\rho_{AB}(t)|\psi_B\rangle\mathrm{E}_B] = 0. \end{aligned} \tag{2.372}$$

Therefore,

$$i\hbar\dot{\rho}_A(t) = [\mathscr{H}_A, \rho_A] + \mathrm{Tr}_B[V, \rho_{AB}(t)], \tag{2.373}$$

and by the same procedure for component B,

$$i\hbar\dot{\rho}_B(t) = [\mathscr{H}_B, \rho_B] + \mathrm{Tr}_A[V, \rho_{AB}(t)]. \tag{2.374}$$

For the perturbation calculations, it is often more convenient to employ the interaction representation where

$$i\hbar\frac{d}{dt}\tilde{\rho}_{AB}(t) = [\tilde{V}(t), \tilde{\rho}_{AB}(t)], \tag{2.375}$$

as in Eq. (2.164) or

$$i\hbar\frac{d}{dt}\tilde{\rho}_{AB}(t) = [\tilde{V}(t), \rho_{AB}(t_0)] - \frac{i}{\hbar}\int_{t_0}^{t} dt'[\tilde{V}(t), [\tilde{V}(t'), \tilde{\rho}_{AB}(t')]], \tag{2.376}$$

as in Eq. (2.168). Then

$$i\hbar\frac{d}{dt}\tilde{\rho}_A(t) = \mathrm{Tr}_B[\tilde{V}(t), \tilde{\rho}_{AB}(t)] \tag{2.377}$$

or, in integral form,

$$i\hbar\frac{d}{dt}\tilde{\rho}_A(t) = \mathrm{Tr}_B[\tilde{V}(t), \rho_{AB}(t_0)] - \frac{i}{\hbar}\int_{t_0}^{t} dt'\,\mathrm{Tr}_B[\tilde{V}(t), [\tilde{V}(t'), \tilde{\rho}_{AB}(t')]]. \tag{2.378}$$

We note that

$$\rho_{AB}(t_0) = \rho_A(t_0)\rho_B(t_0) \tag{2.379}$$

but not for $t > t_0$. In the iterated form of Eq. (2.378),

$$i\hbar\frac{d}{dt}\tilde{\rho}_A(t) = \mathrm{Tr}_B[\tilde{V}(t), \rho_A(t_0)\rho_B(t_0)]$$

$$- \frac{i}{\hbar}\int_{t_0}^{t} dt'\,\mathrm{Tr}_B[\tilde{V}(t), [\tilde{V}(t'), \rho_A(t_0)\rho_B(t_0)]] + \cdots. \tag{2.380}$$

References

[1] K. Blum, *Density Matrix Theory and Applications.* Plenum Press, New York, 1981.

[2] W. H. Louisell, *Quantum Statistical Properties of Radiation.* J. Wiley, New York, 1973.

[3] M. Weissbluth, *Atoms and Molecules.* Academic Press, New York, 1978.

[4] F. T. Arecchi, E. Courtens, R. Gilmore, and H. Thomas, *Phys. Rev.* A **6**, 2211(1972).

[5] H. J. Zeiger and G. W. Pratt, *Magnetic Interactions in Solids.* Clarendon Press, Oxford, 1973.

[6] S. Y. Yee, T. K. Gustafson, S. A. J. Druet, and J.-P. E. Taran, *Optics Comm.* **23**, 1(1977).

[7] S. Y. Yee and T. K. Gustafson, *Phys. Rev.* A **18**, 1597(1978).
[8] J. P. Uyemura, *IEEE J. Quant. Electr.* **QE-16**, 472(1980).
[9] J. Fujimoto and T. K. Yee, *IEEE J. Quant. Electr.* **QE-19**, 861(1983).
[10] W. Heitler, *The Quantum Theory of Radiation.* Oxford Press, Oxford, 1954.
[11] A. Messiah, *Quantum Mechanics.* North-Holland, Amsterdam, 1962.
[12] C. Cohen-Tannoudji, *In Cargese Lectures in Physics.* **2**, 347(1968), (M. Levy, ed.), Gordon and Breach, New York.

III Magnetic Two-Level System

Magnetic resonance methods are employed extensively in the study of matter containing unpaired nuclear and/or electronic spins. The material under investigation is subjected to two magnetic fields—one constant in time and space and the other rotating (or oscillating) in a plane perpendicular to the constant field. With a proper choice of parameters, it is possible to established resonance conditions from which a great deal of physical and chemical information can be extracted [1–4]. It turns out that much of the formalism describing the magnetic resonance of a spin-1/2 system is readily adaptable to the description of a two-level atomic or molecular system interacting with a radiation field. Some of the terminology employed in the optical region has been borrowed from magnetic resonance; numerous optical phenomena are described on the basis of their magnetic analogues.

3.1 Classical Motion of a Magnetic Moment

The classical and quantum mechanical descriptions of the interaction of a magnetic moment with a magnetic field have many features in common. One important advantage of the former, however, is that it has easily visualizable geometrical attributes that are of considerable assistance in the interpretation and design of experiments, both in the magnetic and optical domains.

An object with a magnetic moment **M** and angular momentum **L**, when subjected to a magnetic induction **B**, experiences a torque

$$\mathbf{T} = \frac{d\mathbf{L}}{dt} = \mathbf{M} \times \mathbf{B}. \tag{3.1}$$

For negatively charged particles (e.g., electrons), the magnetic moment is proportional to the angular momentum and oppositely directed:

$$\mathbf{M} = -\gamma \mathbf{L}, \tag{3.2}$$

where γ is a positive constant known as the *gyromagnetic ratio.* Equation then becomes

$$\dot{\mathbf{M}} = -\gamma \mathbf{M} \times \mathbf{B}. \tag{3.3}$$

To interpret this equation it is noted that

$$\mathbf{M} \cdot \dot{\mathbf{M}} = \frac{1}{2}\frac{d}{dt} M^2 = -\gamma \mathbf{M} \cdot \mathbf{M} \times \mathbf{B} = 0, \tag{3.4}$$

indicating that M^2 is constant in time or that $|\mathbf{M}|$—the magnitude of the magnetic moment—remains constant. Also, for constant **B**,

$$\mathbf{B} \cdot \dot{\mathbf{M}} = \frac{d}{dt}(\mathbf{B} \cdot \mathbf{M}) = -\gamma \mathbf{B} \cdot \mathbf{M} \times \mathbf{B} = 0 \tag{3.5}$$

from which it is concluded that the angle between **B** and **M** does not change with time. Hence, Eq. (3.3) describes the precession of a magnetic moment **M**, at a fixed cone angle, about the direction of the field (Fig. 3.1).

Since the direction of the field is arbitrary, its orientation may be chosen to be along the z direction:

$$\mathbf{B} = B_0 \hat{\mathbf{k}} = \mathbf{B}_0. \tag{3.6}$$

Then, at the *Larmor frequency* defined by $\omega_0 \equiv \gamma B_0$, the components of Eq. (3.3) are

$$\dot{M}_x = -\omega_0 M_y, \qquad \dot{M}_y = \omega_0 M M_x, \qquad \dot{M}_z = 0, \tag{3.7}$$

and the solutions with the initial conditions

$$\mathbf{M}(0) = M_x(0)\hat{\mathbf{i}} + M_z(0)\hat{\mathbf{k}} \tag{3.8}$$

are readily seen to be

$$M_x(t) = M_x(0)\cos\omega_0 t, \qquad M_y(t) = M_x(0)\sin\omega_0 t, \qquad M_z(t) = M_z(0). \tag{3.9}$$

Thus, at $t = 0$ the magnetic moment lies in the xz plane and as time progresses, it executes a precessional motion at the frequency ω_0 in the direction of a right-

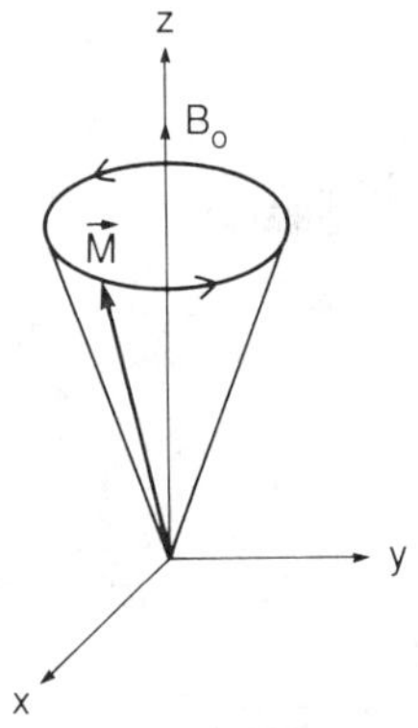

FIGURE 3.1 The precession of a magnetic moment **M** about a magnetic field oriented in the positive z direction. For the motion shown it was assumed that $\mathbf{M} = -\gamma\mathbf{L}$ where **L** is the angular momentum and γ, the gyromagnetic ratio, is a positive constant. The precessional (Larmor) frequency is $\omega_0 = \gamma B_0$.

handed screw advancing in the direction of the field (positive z direction). When the magnetic moment is projected on the xy plane, the rotation is seen to be counterclockwise (Fig. 3.1).

It is instructive to view the motion from the standpoint of an observer attached to a rotating coordinate system. Let the magnetic moment vector **M** be transformed according to

$$\mathbf{M}' = R\mathbf{M} \tag{3.10}$$

where R is the rotation matrix

$$R = \begin{pmatrix} \cos\omega t & \sin\omega t & 0 \\ -\sin\omega t & \cos\omega t & 0 \\ 0 & 0 & 1 \end{pmatrix}. \tag{3.11}$$

The vector $\mathbf{M}'$ is the magnetic moment vector referred to a coordinate system rotating about the z axis at a frequency ω and in the same sense as the precessional motion. The components of $\mathbf{M}'$ are

$$M_{x'} = M_x \cos\omega t + M_y \sin\omega t, \qquad M_{y'} = -M_x \sin\omega t + M_y \cos\omega t,$$
$$M_{z'} = M_z. \tag{3.12}$$

Equations (3.7) now may be transformed to the rotating coordinates

$$\dot{M}_{x'} = -\Delta M_{y'}, \qquad \dot{M}_{y'} = \Delta M_{x'}, \qquad \dot{M}_{z'} = 0, \tag{3.13}$$

where $\Delta = \omega_0 - \omega$. To simplify the notation we shall omit the primes in equations that refer to a rotating coordinate system and instead, such equations will be indicated by (R). Thus, Eq. (3.13) will be written

$$(R) \qquad \dot{M}_x = -\Delta M_y, \qquad \dot{M}_y = \Delta M_x, \qquad \dot{M}_z = 0, \tag{3.14}$$

or, upon combining the three equations into a single vector equation,

$$(R) \qquad \dot{\mathbf{M}} = -\gamma \mathbf{M} \times \mathbf{B}_{\text{eff}}, \tag{3.15}$$

where

$$\mathbf{B}_{\text{eff}} = \left(B_0 - \frac{\omega}{\gamma}\right)\hat{\mathbf{k}} = \frac{\Delta}{\gamma}\hat{\mathbf{k}}. \tag{3.16}$$

At resonance ($\Delta = 0$), Eq. (3.14) reduce to

$$(R) \qquad \dot{M}_x = \dot{M}_y = \dot{M}_z = 0. \tag{3.17}$$

This means that to an observer in a coordinate system rotating at the frequency ω_0 (about the z axis), that is, in synchronism with the precession of the magnetic moment, the latter appears to be a stationary vector. Moreover, when $\omega = \omega_0$, $\mathbf{B}_{\text{eff}} = 0$; this is consistent with the rotating observer's conclusion that since the magnetic moment is stationary, there is no torque acting on it, hence, there can be no magnetic field. When $\omega < \omega_0$, $\mathbf{B}_{\text{eff}} < \mathbf{B}_0$ and when $\omega > \omega_0$, $\mathbf{B}_{\text{eff}}$ reverses its direction. Also, when $\omega \neq \omega_0$, the apparent motion of the magnetic moment in the rotating frame is a precession about $\mathbf{B}_{\text{eff}}$.

The results obtained thus far may be generalized by invoking a rule for the transformation of a vector and its time derivative from a stationary coordinate system to one that is rotating. The rule is to perform the replacement

$$\mathbf{M} \to \mathbf{M}', \qquad \dot{\mathbf{M}} \to \dot{\mathbf{M}}' + \boldsymbol{\omega} \times \mathbf{M}'. \tag{3.18}$$

Applying this rule to Eq. (3.3), one obtains, in the present notation,

$$(R) \qquad \dot{\mathbf{M}} = -\gamma \mathbf{M} \times \mathbf{B}_{\text{eff}}, \tag{3.19}$$

where

$$\mathbf{B}_{\text{eff}} = \mathbf{B} - \frac{\boldsymbol{\omega}}{\gamma}. \tag{3.20}$$

Comparing Eq. (3.3) with Eq. (3.19), it is seen that one may transform the equation of motion of a magnetic moment in a magnetic field from the stationary (or laboratory) frame to a rotating frame simply by replacing $\mathbf{B}$ with $\mathbf{B}_{\text{eff}}$.

To observe magnetic resonance, it is necessary to augment the constant field in the z direction by a time-varying field of the form

$$\mathbf{b}_1(t) = b_1(\hat{\mathbf{i}} \cos \omega t + \hat{\mathbf{j}} \sin \omega t). \tag{3.21}$$

It will be noted that $\mathbf{b}_1(t)$ rotates in a plane perpendicular to the constant field and in the same sense as the precessing magnetic moment. The rotating field exerts an additional torque that changes the angle between $\mathbf{M}$ and $\mathbf{B}_0$. We now have for the equation of motion

$$\dot{\mathbf{M}} = -\gamma \mathbf{M} \times \mathbf{B} = -\gamma \mathbf{M} \times [B_0 \hat{\mathbf{k}} + b_1(\hat{\mathbf{i}} \cos \omega t + \hat{\mathbf{j}} \sin \omega t)], \quad (3.22)$$

or, in component form,

$$\dot{M}_x = -\omega_0 M_y + \gamma b_1 M_z \sin \omega t, \qquad \dot{M}_y = \omega_0 M_x - \gamma b_1 M_z \cos \omega t,$$
$$\dot{M}_z = -\gamma b_1 (M_x \sin \omega t - M_y \cos \omega t). \quad (3.23)$$

The solutions at resonance ($\omega = \omega_0$) are obtained easily. Assuming $\mathbf{M}(0) = M_z(0) \equiv M_0$,

$$M_x(t) = M_0 \sin \omega_0 t \sin \gamma b_1 t,$$
$$M_y(t) = -M_0 \cos \omega_0 t \sin \gamma b_1 t, \qquad (\omega = \omega_0) \quad (3.24)$$
$$M_z(t) = M_0 \cos \gamma b_1 t.$$

M_x and M_y oscillate rapidly with a frequency ω_0 modulated by a frequency γb_1 while M_z oscillates at a frequency γb_1. In a time $t = \pi/\gamma b_1$, the magnetic moment vector, which initially pointed in the positive z direction, has been completely over flipped to point in the negative z direction.

As was done previously, we transform Eq. (3.22) to a rotating frame, specifically, to a frame that rotates in synchronism with $\mathbf{b}_1(t)$ at the frequency ω. Clearly, in such a frame, $\mathbf{b}_1$ will appear to be constant in time and if the stationary and rotating frames coincide at $t = 0$, the magnetic induction in the rotating frame will be

$$(R) \qquad \mathbf{B}_{\text{eff}} = \left(B_0 - \frac{\omega}{\gamma}\right)\hat{\mathbf{k}} + b_1 \hat{\mathbf{i}}. \quad (3.25)$$

From Eq. (3.15) the equation of motion is

$$(R) \qquad \dot{\mathbf{M}} = -\gamma \mathbf{M} \times \mathbf{B}_{\text{eff}} = -\gamma \mathbf{M} \times \left[\left(B_0 - \frac{\omega}{\gamma}\right)\hat{\mathbf{k}} + b_1 \hat{\mathbf{i}}\right], \quad (3.26)$$

or, in component form, with $\Delta = \gamma B_0 - \omega = \omega_0 - \omega$,

$$(R) \qquad \dot{M}_x = -\Delta M_y, \qquad \dot{M}_y = \Delta M_x - \gamma b_1 M_z, \qquad \dot{M}_z = \gamma b_1 M_y. \quad (3.27)$$

At resonance, with the initial condition

$$\mathbf{M}(0) = M_x(0)\hat{\mathbf{i}} + M_y(0)\hat{\mathbf{j}} + M_z(0)\hat{\mathbf{k}}, \quad (3.28)$$

the solutions to Eq. (3.27) are

$$(R) \qquad M_x(t) = M_x(0), \qquad M_y(t) = -M_z(0) \sin \gamma b_1 t + M_y(0) \cos \gamma b_1 t,$$
$$M_z(t) = M_y(0) \sin \gamma b_1 t + M_z(0) \cos \gamma b_1 t, \quad (3.29)$$

which reduce to

$$(R) \qquad M_x(t) = 0, \qquad M_y(t) = -M_0 \sin \gamma b_1 t, \qquad M_z(t) = M_0 \cos \gamma b_1 t, \quad (3.30)$$

when $\mathbf{M}(0) = M_z(0) \equiv M_0$. Comparing Eq. (3.30) with Eq. (3.24) it is seen that $M_z(t)$ has the same form in both stationary and rotating coordinates. Hence, the observer in the rotating system also sees the magnetic moment reversing its direction in a time $t = \pi/\gamma b_1$. To the latter observer, however, the magnetic moment appears to rotate about the x' axis (i.e., the x axis in the rotating frame) at a frequency γb_1 (Fig. 3.2). In other words, the magnetic moment $\mathbf{M}$ precesses about the vector $\mathbf{B}_{\text{eff}} = b_1 \mathbf{i}'$ which, in the rotating frame, is independent of time and is oriented along the x' axis. More generally, for frequencies ω not necessarily equal to ω_0, Eq. (3.25) implies that, in the rotating system, $\mathbf{M}$ precesses about $\mathbf{B}_{\text{eff}}$ (Fig. 3.3) at the *Rabi* frequency [5] given by

$$\Omega = \gamma |\mathbf{B}_{\text{eff}}| = \sqrt{(\gamma B_0 - \omega)^2 + (\gamma b_1)^2} = \sqrt{\Delta^2 + (\gamma b_1)^2}. \qquad (3.31)$$

If the detuning Δ is a function of time, as for example when one is tuning through resonance or in cases where pulses at various frequencies are employed, the orientation of the vector $\mathbf{B}_{\text{eff}}$ becomes a function of time. Under such circumstances, the magnetization vector may or may not be able to follow closely the changes in direction of $\mathbf{B}_{\text{eff}}$. If it does, the condition is known as *adiabatic following*. This will occur if in the time for a small change in the direction of $\mathbf{B}_{\text{eff}}$, the vector $\mathbf{M}$ has precessed through many revolutions.

With the initial condition of Eq. (3.28) and $\Delta \neq 0$, the solution to the set of Eqs. (3.27) yields the components of the magnetization vector in the rotating

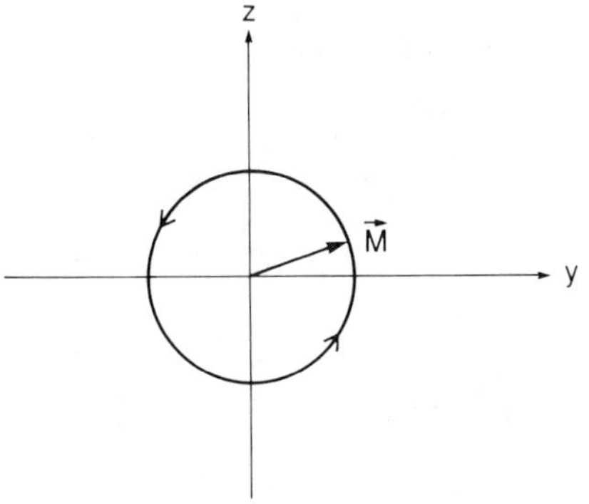

FIGURE 3.2 In the rotating frame, the magnetic moment $\mathbf{M}$, whose initial value is $M_z(0)$, rotates about the x axis with a period of $2\pi/\gamma b_1$ where $\mathbf{b}_1(t) = b_1(\mathbf{i} \cos \omega t + \mathbf{j} \sin \omega t)$ and $\omega = \omega_0$ is the Larmor precession frequency.

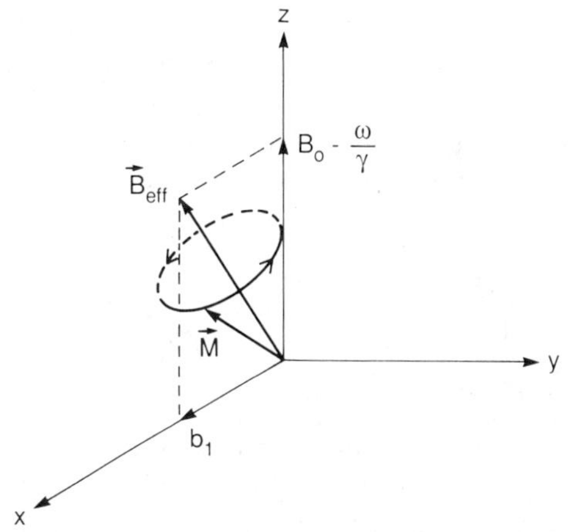

FIGURE 3.3 The precession of the magnetic moment **M**, when $\omega \neq \omega_0$, about $\mathbf{B}_{\text{eff}} = (B_0 - \omega/\gamma)\mathbf{k} + b_1\mathbf{i}$ in the rotating frame.

frame:

$$M_x(t) = M_x(0)\frac{(\gamma b_1)^2 + \Delta^2 \cos \Omega t}{\Omega^2} - M_y(0)\frac{\Delta}{\Omega}\sin \Omega t - M_z(0)\frac{\Delta \gamma b_1}{\Omega^2}(\cos \Omega t - 1),$$

(R) $$M_y(t) = M_x(0)\frac{\Delta}{\Omega}\sin \Omega t + M_y(0)\cos \Omega t - M_z(0)\frac{\gamma b_1}{\Omega}\sin \Omega t, \qquad (3.32)$$

$$M_z(t) = -M_x(0)\frac{\Delta \gamma b_1}{\Omega^2}(\cos \Omega t - 1) + M_y(0)\frac{\gamma b_1}{\Omega}\sin \Omega t + M_z(0)\left[1 + \frac{(\gamma b_1)^2}{\Omega^2}(\cos \Omega t - 1)\right].$$

These equations reduce to Eq. (3.29) at resonance ($\Delta = 0$). For $M_x(0) = M_y(0) = 0$, $M_z(0) \equiv M_0$,

$$M_x(t) = -M_0\frac{\Delta \gamma b_1}{\Omega^2}(\cos \Omega t - 1), \qquad M_y(t) = -M_0\frac{\gamma b_1}{\Omega}\sin \Omega t,$$

(R) $$M_z(t) = M_0\left[1 + \frac{(\gamma b_1)^2}{\Omega^2}(\cos \Omega t - 1)\right]. \qquad (3.33)$$

If the counterclockwise rotating field (Eq. (3.21)) is replaced by a field that rotates in the clockwise sense, the term containing $(B_0 - \omega/\gamma)$ in Eq. (3.25) would be replaced by one with $(B_0 + \omega/\gamma)$. In the counterclockwise rotating frame, at a frequency $\omega = \gamma B_0$, the counterclockwise field rotates in synchronism with the precession of the magnetic moment and therefore exerts a constant torque on the latter. But the clockwise field reverses itself at a frequency 2ω; its effect is then very small and usually may be ignored. In that case,

$$\begin{aligned}\mathbf{B}_1 &= B_1\hat{\mathbf{i}}\cos\omega t = 2b_1\hat{\mathbf{i}}\cos\omega t \\ &= b_1(\hat{\mathbf{i}}\cos\omega t + \hat{\mathbf{j}}\sin\omega t) + b_1(\hat{\mathbf{i}}\cos\omega t - \hat{\mathbf{j}}\sin\omega t) \\ &\simeq b_1(\hat{\mathbf{i}}\cos\omega t + \hat{\mathbf{j}}\sin\omega t).\end{aligned} \tag{3.34}$$

The neglect of the clockwise term in the case of the oscillating field is known as the *rotating wave approximation* (*RWA*). It should be noted, however, that in magnetic fields of high intensity, the clockwise term produces a small shift in the resonance frequency known as the *Bloch-Siegert shift* [6]:

$$\delta\omega = \frac{(\gamma b_1)^2}{4\omega_0}. \tag{3.35}$$

3.2 Hamiltonian

In the previous section, the motion of a magnetic moment in an external magnetic field was described in classical terms. We now turn to the quantum mechanical formulation—specifically, the case of an electronic spin-1/2 system.

The interaction of an electronic magnetic moment with an external magnetic field is governed by the Hamiltonian

$$\mathscr{H} = -\mathbf{m}\cdot\mathbf{B}, \tag{3.36}$$

in which the magnetic moment operator $\mathbf{m}$ is related to the Pauli spin operator $\boldsymbol{\sigma}$ by

$$\mathbf{m} = -\frac{e\hbar}{2m_e}\boldsymbol{\sigma} = -\mu_B\boldsymbol{\sigma}. \tag{3.37}$$

The minus sign appears as a consequence of the negative charge of the electron; e is the (absolute) charge (1.602×10^{-19} C) and m_e is the mass (9.109×10^{-31} kg) of the electron; μ_B is the *Bohr magneton* (9.274×10^{-24} J/T). One also may write

$$\mathbf{m} = -\frac{1}{2}\gamma\hbar\boldsymbol{\sigma}, \tag{3.38}$$

in which γ, the gyromagnetic ratio, is related to the Bohr magneton by

$$\gamma = \frac{2\mu_B}{\hbar} = 1.761 \times 10^{11}\ T^{-1}s^{-1} \tag{3.39}$$

The macroscopic magnetization **M** for a system of N identical spin-1/2 particles per unit volume is $N\langle \mathbf{m} \rangle$ where $\langle \mathbf{m} \rangle$ is the ensemble average of the magnetic moment operator. With **m** as in Eq. (3.38),

$$\mathbf{M} = N\langle \mathbf{m} \rangle = -\frac{N\gamma\hbar}{2}\langle \boldsymbol{\sigma} \rangle, \tag{3.40}$$

and

$$\mathcal{H} = \frac{1}{2}\gamma\hbar\boldsymbol{\sigma} \cdot \mathbf{B}. \tag{3.41}$$

If $\mathbf{B} = B_0\mathbf{k}$, the Hamiltonian

$$\mathcal{H} = \frac{1}{2}\gamma\hbar B_0\sigma_z \equiv \mathcal{H}_0 \tag{3.42}$$

is diagonal in the basis consisting of the spin states $|\alpha\rangle$ and $|\beta\rangle$ with eigenvalues $\frac{1}{2}\gamma\hbar B_0$ and $-\frac{1}{2}\gamma\hbar B_0$, respectively (Fig. 3.4). Thus, the energy difference between the two spin states is

$$\Delta E = E_\alpha - E_\beta = \gamma\hbar B_0 = \hbar\omega_0, \tag{3.43}$$

where ω_0, which is equal to γB_0, is identical to the frequency of precession of the magnetic moment in the classical formulation. At thermal equilibrium, the populations in the two states are determined by the Boltzmann distribution.

Now suppose that in addition to the constant field in the z direction we apply a rotating magnetic field in the xy plane. Then

$$\mathbf{B} = B_0\hat{\mathbf{k}} + \mathbf{b}_1 = B_0\hat{\mathbf{k}} + b_1(\hat{\mathbf{i}}\cos\omega t + \hat{\mathbf{j}}\sin\omega t), \tag{3.44}$$

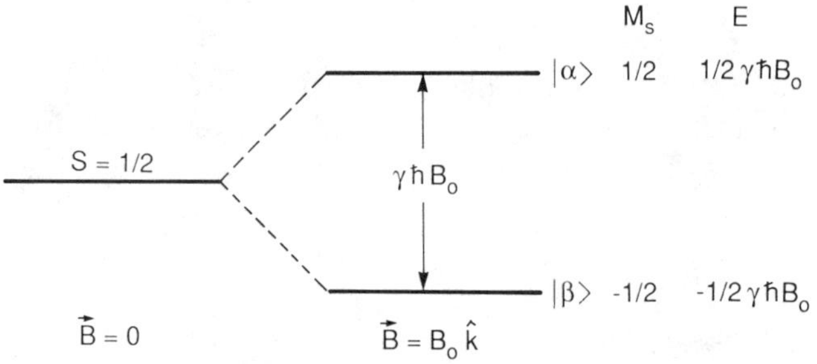

FIGURE 3.4 Energy level diagram for an electronic spin-1/2 system. When $\mathbf{B} = 0$ the two spin states are degenerate; when $\mathbf{B} = B_0\mathbf{k}$ the two spin states are split apart with a separation of $\gamma\hbar B_0$.

and the Hamiltonian becomes

$$\mathscr{H} = \frac{\gamma\hbar}{2}[B_0\sigma_z + b_1(\sigma_x \cos\omega t + \sigma_y \sin\omega t)] \tag{3.45}$$

$$= \frac{\gamma\hbar}{2}[B_0\sigma_z + b_1(\sigma^+ e^{-i\omega t} + \sigma^- e^{i\omega t})], \tag{3.46}$$

where σ^+ and σ^- are defined by Eq. (2.39). Another useful form of the Hamiltonian is derivable from the expression

$$f(\phi) = e^{i\phi\sigma_z/2}\sigma_x e^{-i\phi\sigma_z/2}. \tag{3.47}$$

With the help of the commutation properties (Eqs. (2.28)) for the Pauli spin operators, we find

$$\frac{df}{d\phi} = -e^{i\phi\sigma_z/2}\sigma_y e^{-i\phi\sigma_z/2}, \tag{3.48}$$

$$\frac{d^2f}{d\phi} = -f(\phi). \tag{3.49}$$

If the solution to Eq. (3.49) is written

$$f(\phi) = A\cos\phi + B\sin\phi, \tag{3.50}$$

the constants are

$$f(0) = A = \sigma_x, \qquad \left(\frac{df}{d\phi}\right)_{\phi=0} = B = -\sigma_y, \tag{3.51}$$

and

$$f(\phi) = \sigma_x\cos\phi - \sigma_y\sin\phi. \tag{3.52}$$

With $\phi = -\omega t$, the Hamiltonian (Eq. (3.45)) acquires the form

$$\mathscr{H} = \frac{\gamma\hbar}{2}(B_0\sigma_z + b_1 e^{-i\omega t\sigma_z/2}\sigma_x e^{i\omega t\sigma_z/2}). \tag{3.53}$$

Let us now transform to rotating coordinates as was done for the classical case. In the context of quantum mechanics, this is accomplished by means of a unitary operator. If ψ satisfies the time-dependent Schrödinger equation

$$i\hbar\frac{\partial\psi}{\partial t} = \mathscr{H}\psi, \tag{3.54}$$

the unitary transformation

$$\psi' = O\psi \tag{3.55}$$

gives

$$i\hbar\frac{\partial\psi}{\partial t} = i\hbar\left[O^{-1}\frac{\partial\psi'}{\partial t} + \frac{\partial O^{-1}}{\partial t}\psi'\right], \tag{3.56}$$

$$\mathscr{H}\psi = \mathscr{H}O^{-1}\psi'. \tag{3.57}$$

In view of Eq. (3.54)

$$i\hbar\left[O^{-1}\frac{\partial\psi'}{\partial t} + \frac{\partial O^{-1}}{\partial t}\psi'\right] = \mathscr{H}O^{-1}\psi' \tag{3.58}$$

or

$$i\hbar\frac{\partial\psi'}{\partial t} = \left[O\mathscr{H}O^{-1} - i\hbar O\frac{\partial O^{-1}}{\partial t}\right]\psi' \equiv \mathscr{H}'\psi'. \tag{3.59}$$

Since $\mathscr{H}$ is Hermitian and O is unitary, there are two equivalent forms for $\mathscr{H}'$:

$$\mathscr{H}' = O\mathscr{H}O^{-1} - i\hbar O\frac{\partial O^{-1}}{\partial t} = O\mathscr{H}O^{\dagger} + i\hbar\frac{\partial O}{\partial t}O^{\dagger}. \tag{3.60}$$

We now tranform the Hamiltonian (Eq. (3.53)) with the unitary operator

$$O = e^{i\omega t\sigma_z/2}. \tag{3.61}$$

Then

$$O\mathscr{H}O^{\dagger} = \frac{\gamma\hbar}{2}(B_0\sigma_z + b_1\sigma_x), \qquad i\hbar\frac{\partial O}{\partial t}O^{\dagger} = -\frac{1}{2}\hbar\omega\sigma_z, \tag{3.62}$$

$$\mathscr{H}' = \frac{\gamma\hbar}{2}\left[\left(B_0 - \frac{\omega}{\gamma}\right)\sigma_z + b_1\sigma_x\right] = \frac{\gamma\hbar}{2}\boldsymbol{\sigma}\cdot\mathbf{B}_{\text{eff}}, \tag{3.63}$$

where

$$\mathbf{B}_{\text{eff}} = \left(B_0 - \frac{\omega}{\gamma}\right)\hat{\mathbf{k}} + b_1\hat{\mathbf{i}}. \tag{3.64}$$

Thus, by virtue of the unitary transformation with the operator (Eq. (3.61)), the transformed Hamiltonian $\mathscr{H}'$ is independent of time. We note that Eq. (3.64) is identical with Eq. (3.60); hence, the effect of the unitary operator (Eq. (3.61)) in the quantum mechanical context is analogous to the transformation to rotating coordinates in the classical case. Either classically or quantum mechanically, the transformation is implemented by the replacement of $\mathbf{B}$ with $\mathbf{B}_{\text{eff}}$.

For future reference we list the transformation of the Pauli operators by the unitary operator $\exp(i\phi\sigma_z/2)$:

$$
\begin{aligned}
e^{i\phi\sigma_z/2}\sigma_x e^{-i\phi\sigma_z/2} &= \sigma_x \cos\phi - \sigma_y \sin\phi = \sigma^+ e^{i\phi} + \sigma^- e^{-i\phi}, \\
e^{i\phi\sigma_z/2}\sigma_y e^{-i\phi\sigma_z/2} &= \sigma_x \sin\phi + \sigma_y \cos\phi = \sigma^+ e^{-i\phi} + \sigma^- e^{i\phi}, \\
e^{i\phi\sigma_z/2}\sigma_z e^{-i\phi\sigma_z/2} &= \sigma_z, \\
e^{i\phi\sigma_z/2}\sigma^+ e^{-i\phi\sigma_z/2} &= \sigma^+ e^{i\phi}, \\
e^{i\phi\sigma_z/2}\sigma^- e^{-i\phi\sigma_z/2} &= \sigma^- e^{-i\phi}, \\
e^{i\phi\sigma_z/2}\sigma^+\sigma^- e^{-i\phi\sigma_z/2} &= \sigma^+\sigma^-, \\
e^{i\phi\sigma_z/2}\sigma^-\sigma^+ e^{-i\phi\sigma_z/2} &= \sigma^-\sigma^+.
\end{aligned}
\tag{3.65}
$$

3.3 Transition Probability, Rabi Formula

In a constant magnetic field, the eigenstates of the Hamiltonian (Eq. (3.42)) are the stationary spin states $|\alpha\rangle$ and $|\beta\rangle$ with an energy separation $E_\alpha - E_\beta = \gamma\hbar B_0 = \hbar\omega_0$. Upon addition of the time-dependent (rf) transverse rotating field, as in Eq. (3.21), the states $|\alpha\rangle$ and $|\beta\rangle$ no longer remain stationary. We now calculate the effect of the rotating field on the temporal behavior of the spin states.

The general form of the normalized wave function for a magnetic two-level system in a field of the form of Eq. (3.44) consists of a linear superposition of the two spin states, as in Eq. (2.36). Since we now are concerned with the time-dependent of the wave function, the coefficients will be functions of the item. Thus,

$$|\psi(t)\rangle = a(t)|\alpha\rangle + b(t)|\beta\rangle, \tag{3.66}$$

with the normalization condition

$$|a(t)|^2 + |b(t)|^2 = 1. \tag{3.67}$$

As noted previously (Section 2.2), the wave function $|\psi(t)\rangle$ is a coherent superposition of $|\alpha\rangle$ and $|\beta\rangle$ since the density matrix Eq. (2.37) contains nonvanishing off-diagonal elements. We shall consider transitions $|\beta\rangle \to |\alpha\rangle$ assuming that initially $|\beta\rangle$ is occupied and $|\alpha\rangle$ is vacant; i.e.,

$$a(0) = 0, \qquad b(0) = 1. \tag{3.68}$$

As time progresses $a(t)$ and $b(t)$ depart from their initial values due to the effect of the rotating field.

The simplifications resulting from a transformation to a rotating coordinate system already have been noted. We therefore transform $\psi(t)$ by means of the unitary operator in Eq. (3.61):

$$\psi(t) = O^{-1}\psi'(t) = e^{-i\omega t\sigma_z/2}\psi'(t). \tag{3.69}$$

Since $\mathscr{H}'$ (Eq. (3.63)) is independent of time, it is permissible to express the time-development of $\psi'(t)$ by

$$\psi'(t) = e^{-i\mathscr{H}'t/\hbar}\psi'(0) = e^{-i\mathscr{H}'t/\hbar}\psi(0) = e^{-i\mathscr{H}'t/\hbar}|\beta\rangle, \tag{3.70}$$

in which the initial conditions of Eq. (3.68) have been inserted. Combining Eqs. (3.66), (3.69), and (3.70),

$$\psi(t) = a(t)|\alpha\rangle + b(t)|\beta\rangle = e^{-i\omega t\sigma_z/2}e^{-i\mathscr{H}'t/\hbar}|\beta\rangle, \tag{3.71}$$

or, in view of the orthonormality of the spin functions,

$$a(t) = \langle\alpha|e^{-i\omega t\sigma_z/2}e^{-i\mathscr{H}'t/\hbar}|\beta\rangle. \tag{3.72}$$

To evaluate the matrix element, it is convenient to define

$$\boldsymbol{\Omega} = (\gamma B_0 - \omega)\hat{\mathbf{k}} + \gamma b_1\hat{\mathbf{i}} = \gamma\mathbf{B}_{\text{eff}} = \Delta\hat{\mathbf{k}} + \gamma b_1\hat{\mathbf{i}}, \tag{3.73a}$$

$$\Omega = \sqrt{(\gamma B_0 - \omega)^2 + (\gamma b_1)^2}, \tag{3.73b}$$

$$\sin\theta = \frac{\gamma b_1}{\Omega}, \qquad \cos\theta = \frac{\gamma B_0 - \omega}{\Omega} = \frac{\omega_0 - \omega}{\Omega} = \frac{\Delta}{\Omega}. \tag{3.74}$$

The Hamiltonian $\mathscr{H}'$ (Eq. (3.63)) now may be written

$$\mathscr{H}' = \frac{\gamma\hbar}{2}\left[\left(B_0 - \frac{\omega}{\gamma}\right)\sigma_z + b_1\sigma_x\right] = \frac{\hbar\Omega}{2}(\cos\theta\sigma_z + \sin\theta\sigma_x), \tag{3.75}$$

which then yields

$$e^{-i\mathscr{H}'t/\hbar} = e^{-i\Omega(\cos\theta\sigma_z + \sin\theta\sigma_x)t/2}$$

$$= \cos\left[\frac{\Omega}{2}(\cos\theta\sigma_z + \sin\theta\sigma_x)t\right] - i\sin\left[\frac{\Omega}{2}(\cos\theta\sigma_z + \sin\theta\sigma_x)t\right]. \tag{3.76}$$

Writing

$$\cos\theta\sigma_z + \sin\theta\sigma_x = \begin{pmatrix}\cos\theta & \sin\theta\\ \sin\theta & -\cos\theta\end{pmatrix}, \tag{3.77}$$

it is readily verified that even powers of the matrix are equal to the unit matrix I and that odd powers simply reproduce the original matrix. Therefore,

$$\cos\left[\frac{\Omega}{2}(\cos\theta\sigma_z + \sin\theta\sigma_x)t\right] = \cos\frac{\Omega t}{2} I, \tag{3.78}$$

$$\sin\left[\frac{\Omega}{2}(\cos\theta\sigma_z + \sin\theta\sigma_x)t\right] = (\cos\theta\sigma_z + \sin\theta\sigma_x)\sin\frac{\Omega t}{2}, \tag{3.79}$$

$$e^{-i\mathscr{H}'t/\hbar} = \cos\frac{\Omega t}{2} I - i(\cos\theta\sigma_z + \sin\theta\sigma_x)\sin\frac{\Omega t}{2}. \tag{3.80}$$

The matrix element in Eq. (3.72) now becomes

$$\begin{aligned} a(t) = \langle\alpha|e^{-i\omega t\sigma_z/2}e^{-i\mathscr{H}'t/2}|\beta\rangle &= \cos\frac{\Omega t}{2}\langle\alpha|e^{-i\omega t\sigma_z/2}|\beta\rangle \\ &-i\sin\frac{\Omega t}{2}\cos\theta\langle\alpha|e^{-i\omega t\sigma_z/2}\sigma_z|\beta\rangle \\ &-i\sin\frac{\Omega t}{2}\sin\theta\langle\alpha|e^{-i\omega t\sigma_z/2}\sigma_z|\beta\rangle. \end{aligned} \tag{3.81}$$

Since $\sigma_z|\beta\rangle = -|\beta\rangle$, $\sigma_x|\beta\rangle = |\alpha\rangle$, and $\langle\alpha\,|\,\beta\rangle = 0$, only the last of the three terms on the right side of the equation is nonvanishing; hence,

$$a(t) = -i\sin\frac{\Omega t}{2}\sin\theta\langle\alpha|\mathrm{e}^{-i\omega t\sigma_z/2}|\alpha\rangle = -i\sin\frac{\Omega t}{2}\sin\theta e^{-i\omega t/2} \tag{3.82}$$

and the probability of finding the system in the state $|\alpha\rangle$ is given by

$$\begin{aligned} |a(t)|^2 &= \sin^2\left(\frac{\Omega t}{2}\right)\sin^2\theta = \left(\frac{\gamma b_1}{\Omega}\right)^2\sin^2\left(\frac{\Omega t}{2}\right) \\ &= \frac{(\gamma b_1)^2}{\Delta^2 + (\gamma b_1)^2}\sin^2\frac{t}{2}\sqrt{\Delta^2 + (\gamma b_1)^2}. \end{aligned} \tag{3.83}$$

This is known as the *Rabi formula* [5] and Ω, according to Eq. (3.73b), is the Rabi frequency, also known as the *nutational* or *flopping frequency.*

The Rabi formula indicates that the population in the upper state oscillates in time (Fig. 3.5). At resonance ($\Delta = 0$), $\gamma b_1 \equiv \Omega_0$; the Rabi formula then reduces to

$$|a(t)|^2 = \sin^2\frac{\gamma b_1 t}{2} = \sin^2\frac{\Omega_0 t}{2}, \tag{3.84}$$

in which case the oscillation is between zero and one, with the higher value reached in a time $t = \pi/\gamma b_1 = \pi/\Omega_0$. This means that if at $t = 0$ the system is in

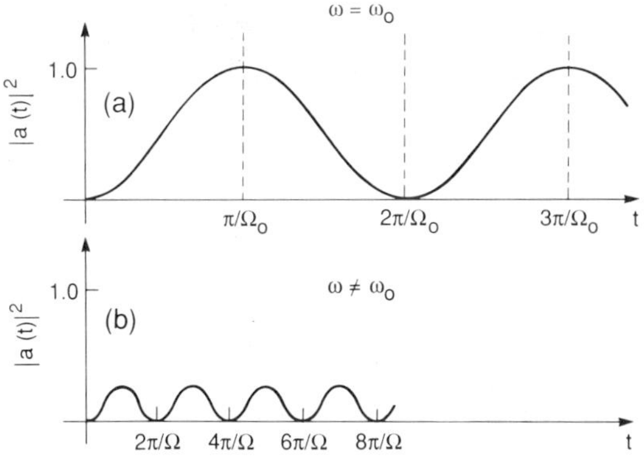

FIGURE 3.5 When the initial population of a two-level system is entirely in the lower state, the population in the upper state oscillates in time (Rabi oscillation). (a) $\Delta = \omega_0 - \omega = 0$, the oscillation is between 0 and 1; (b) $\Delta \neq 0$, the oscillation is between 0 and a value less than 1.

the state $|\beta\rangle$, then after a time $\pi'\Omega_0$ elapses, the system occupies the state $|\alpha\rangle$. The larger the value of b_1—the amplitude of the rotating field—the shorter is the time required for a spin reversal. In the terminology of magnetic resonance, π/Ω_0 is known as a π-pulse.

Evidently, there is a strong resemblance between the classical and quantum mechanical descriptions. This is seen most clearly at resonance. In the former it was shown, in connection with Eq. (3.30), that in the rotating frame the magnetic moment reverses direction in a time $\pi/\gamma b_1$ which is precisely the time required for the transition $|\beta\rangle \rightarrow |\alpha\rangle$ as deduced from the Rabi formula.

In view of the normalization condition of Eq. (3.67) the population in $|\beta\rangle$ also must oscillate in time (Fig. 3.6). Thus, at resonance, when the initial conditions are those given in Eq. (3.68),

$$|b(t)|^2 = \cos^2 \frac{\gamma b_1 t}{2} = \cos^2 \frac{\Omega_0 t}{2}. \tag{3.85}$$

The two relations in Eqs. (3.84) and (3.85) imply that, on the average, the states $|\alpha\rangle$ and $|\beta\rangle$ are equally populated. When $\omega \neq \omega_0$, however, $|a(t)|^2 < 1$; the average population in the lower state $|\beta\rangle$ is then higher than in the upper state $|\alpha\rangle$ (Figs. 3.5 and 3.6).

In the experimental realization of magnetic resonance it is preferable to employ oscillating rather than rotating magnetic fields. But the oscillating field is simply a superposition of two counter-rotating fields of which only one is effective within the rotating wave approximation. We then may replace b_1 in the Rabi formula by $B_1/2$ where B_1 is the amplitude of the oscillating field;

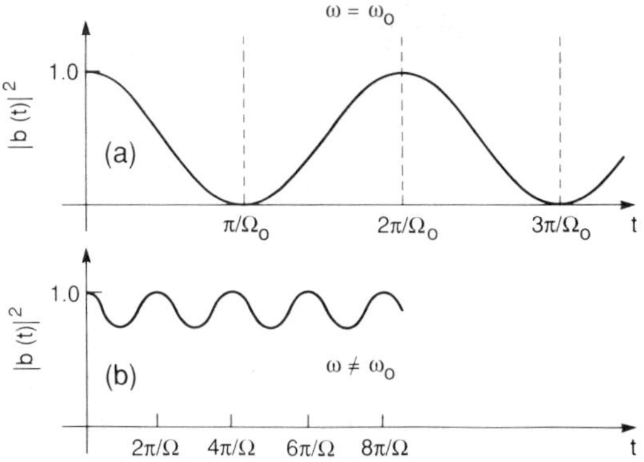

FIGURE 3.6 Oscillation of the population in the lower state of a two-level system under the same initial condition as in Fig. 3.5. (a) $\Delta = \omega_0 - \omega = 0$, the oscillation is between 1 and 0 (i.e., in quadrature with the oscillation in the upper state); (b) $\Delta \neq 0$, the oscillation is between a maximum of 1 and a value greater than 0.

thus,

$$|a(t)|^2 = \frac{\left(\frac{\gamma B_1}{2}\right)^2}{\Delta^2 + \left(\frac{\gamma B_1}{2}\right)^2} \sin^2 \frac{t}{2} \sqrt{\Delta^2 + \left(\frac{\gamma B_1}{2}\right)^2}. \tag{3.86}$$

The Rabi formula predicts the probability for transitions between two spin states induced by a rotating magnetic field. As will be discussed more fully at a later stage, excited states also may revert to the ground state by spontaneous emission which does not require the presence of a rotating field. This feature is not taken into account in the Rabi formula. Spontaneous emission decreases rapidly with decreasing separation between the two states, however, and is negligible for the small separations encountered in magnetic transitions even in experiments employing high fields.

3.4 Equations of Motion

The fundamental equation for the temporal behavior of the density operator is given by Eq. (2.151). Specializing to a two-level system, we derived the set of differential equations (2.192) which now may be applied to the spin-1/2 system. Since $E_e - E_g = \hbar\omega_0$ by definition, and for the spin case

$$E_\alpha - E_\beta = \hbar\omega_0, \tag{3.87}$$

we have the correspondence $|g\rangle \rightarrow |\beta\rangle$ and $|e\rangle \rightarrow |\alpha\rangle$. We now write the Hamiltonian (Eq. (3.45) or (3.46) in the form

$$\mathscr{H} = \mathscr{H}_0 + V \tag{3.88}$$

where

$$\mathscr{H}_0 = \frac{\gamma\hbar}{2} B_0 \sigma_z = \frac{1}{2}\hbar\omega_0\sigma_z, \tag{3.89}$$

$$V = \frac{\gamma\hbar}{2} b_1(\sigma_x \cos\omega t + \sigma_y \sin\omega t) = \frac{\gamma\hbar}{2} b_1(\sigma^+ e^{-i\omega t} + \sigma^- e^{i\omega t}). \tag{3.90}$$

The matrix elements of V are readily evaluated:

$$\begin{aligned} V_{\alpha\alpha} &= V_{\beta\beta} = 0, \\ V_{\alpha\beta} &= \frac{\gamma\hbar}{2} b_1 e^{-i\omega t} = V^*_{\beta\alpha}. \end{aligned} \tag{3.91}$$

With these expressions and the density matrices (2.35) and (2.41) the differential equations (2.192) yield the following:

$$\begin{aligned} \dot{\rho}_{\alpha\alpha} &= \frac{d}{dt}\langle\sigma^+\sigma^-\rangle = \frac{i\gamma b_1}{2}[e^{i\omega t}\langle\sigma^-\rangle - e^{-i\omega t}\langle\sigma^+\rangle] \\ &= \frac{i\gamma b_1}{2}[e^{i\omega t}\rho_{\alpha\beta} - e^{-i\omega t}\rho_{\beta\alpha}] = -\dot{\rho}_{\beta\beta} = -\frac{d}{dt}\langle\sigma^-\sigma^+\rangle, \end{aligned} \tag{3.92}$$

$$\begin{aligned} \dot{\rho}_{\alpha\beta} &= \frac{d}{dt}\langle\sigma^-\rangle = -i\gamma B_0\langle\sigma^-\rangle + i\gamma b_1 e^{-i\omega t}\left(\langle\sigma^+\sigma^-\rangle - \frac{1}{2}\right) \\ &= -i\omega_0\langle\sigma^-\rangle + \frac{i\gamma b_1}{2} e^{-i\omega t}\langle\sigma_z\rangle \\ &= -i\omega_0\rho_{\alpha\beta} + \frac{i\beta b_1}{2} e^{-i\omega t}(\rho_{\alpha\alpha} - \rho_{\beta\beta}), \end{aligned} \tag{3.93}$$

$$\begin{aligned} \dot{\rho}_{\beta\alpha} &= \frac{d}{dt}\langle\sigma^+\rangle = i\gamma\beta_0\langle\sigma^+\rangle - i\gamma b_1 e^{i\omega t}\left(\langle\sigma^+\sigma^-\rangle - \frac{1}{2}\right) \\ &= i\omega_0\langle\sigma^+\rangle - \frac{i\gamma b_1}{2} e^{i\omega t}\langle\sigma_z\rangle \\ &= i\omega_0\rho_{\beta\alpha} - \frac{i\gamma b_1}{2} e^{i\omega t}(\rho_{\alpha\alpha} - \rho_{\beta\beta}). \end{aligned} \tag{3.94}$$

From these equations we may construct an equivalent set,

$$\begin{aligned}
\frac{d}{dt}\langle\sigma_x\rangle &= \frac{d}{dt}(\langle\sigma^+\rangle + \langle\sigma^-\rangle) = -\omega_0\langle\sigma_y\rangle + \gamma b_1\langle\sigma_z\rangle \sin\omega t, \\
\frac{d}{dt}\langle\sigma_y\rangle &= i\frac{d}{dt}(\langle\sigma^-\rangle - \langle\sigma^+\rangle) = \omega_0\langle\sigma_x\rangle - \gamma b_1\langle\sigma_z\rangle \cos\omega t, \qquad (3.95)\\
\frac{d}{dt}\langle\sigma_z\rangle &= 2\frac{d}{dt}\langle\sigma^+\sigma^-\rangle = \gamma b_1(\langle\sigma_y\rangle\cos\omega t - \langle\sigma_x\rangle\sin\omega t),
\end{aligned}$$

which can be written more compactly as a single vector equation

$$\frac{d}{dt}\langle\boldsymbol{\sigma}\rangle = -\gamma\langle\boldsymbol{\sigma}\rangle \times \mathbf{B} \qquad (3.96)$$

where

$$\langle\boldsymbol{\sigma}\rangle = \langle\sigma_x\rangle\hat{\mathbf{i}} + \langle\sigma_y\rangle\hat{\mathbf{j}} + \langle\sigma_z\rangle\hat{\mathbf{k}}, \qquad (3.97)$$

$$\mathbf{B} = B_0\hat{\mathbf{k}} + b_1(\hat{\mathbf{i}}\cos\omega t + \hat{\mathbf{j}}\sin\omega t). \qquad (3.98)$$

A comparison of Eq. (3.22), for the time rate of change of the magnetization, with Eq. (3.96) for the spin system indicates that the two equations are identical. Clearly, this is not an unexpected result since the magnetization is proportional to $\langle\boldsymbol{\sigma}\rangle$ as in Eq. (3.40). The solutions to Eq. (3.95), at resonance, are therefore of the same form as Eq. (3.24), namely

$$\begin{aligned}
\langle\sigma_x\rangle &= A\sin\omega_0 t \sin\gamma b_1 t, \\
\langle\sigma_y\rangle &= -A\cos\omega_0 t \sin\gamma b_1 t, \qquad (3.99)\\
\langle\sigma_z\rangle &= A\cos\gamma b_1 t,
\end{aligned}$$

where, according to Eq. (3.40), $A = -2M/N\gamma\hbar$, and M is the magnitude of the vector $\mathbf{M}$.

It has been noted previously that the expression for the time rate of change of an operator in the Heisenberg representation (Eq. (2.102), has the same form as the time rate of change of the expectation value of the operator (Eq. (2.103)). Therefore, the equations of motion of the Pauli operators, in the Heisenberg representation, are identical in form with the equations of motion for the expectation values. Thus, for example, the Heisenberg operator $(\sigma_{\mathrm{H}})_x$ is defined by

$$(\sigma_{\mathrm{H}})_x = e^{i\mathscr{H}t/\hbar}\sigma_x e^{-i\mathscr{H}t/\hbar} \qquad (3.100)$$

where σ_x is in the Schrödinger representation. We then may write

$$\frac{d}{dt}(\sigma_{\mathrm{H}})_x = -\omega_0(\sigma_{\mathrm{H}})_y + \gamma b_1(\sigma_{\mathrm{H}})_z \sin\omega t, \qquad (3.101)$$

by analogy with the first equation in (3.95).

It will be recalled that the Hamiltonian, consisting of the two terms (3.89) and (3.90) was derived with reference to a stationary (or laboratory) coordinate system (Section 3.2). We now wish to carry out a similar development in a rotating coordinate system in which the Hamiltonian $\mathcal{H}'$ is given by Eq. (3.63). Writing

$$\mathcal{H}' = \mathcal{H}'_0 + V', \tag{3.102}$$

with

$$\mathcal{H}'_0 = \frac{\gamma\hbar}{2}\left(B_0 - \frac{\omega}{\gamma}\right)\sigma_z = \frac{1}{2}\hbar\,\Delta\sigma_z \tag{3.103}$$

$$V' = \frac{\gamma\hbar}{2} b_1 \sigma_x, \tag{3.104}$$

the matrix elements of V' are

$$V'_{\alpha\alpha} = V'_{\beta\beta} = 0 \qquad V'_{\alpha\beta} = V'_{\beta\alpha} = \frac{\gamma\hbar}{\varrho} b_1. \tag{3.105}$$

The matrix elements of the density operator now may be evaluated by means of the fundamental equation

$$i\hbar\dot{\rho} = [\mathcal{H}', \rho] \tag{3.106}$$

or, more simply, by substitution in Eq. (2.192) with

$$V_{ge} = V_{eg} = V'_{\alpha\beta} = V'_{\beta\alpha} = \frac{\gamma\hbar}{2} b_1. \tag{3.107}$$

The resulting equations are

$$\dot{\rho}_{\alpha\alpha} = \frac{d}{dt}\langle\sigma^+\sigma^-\rangle = \frac{i\gamma b_1}{2}[\langle\sigma^-\rangle - \langle\sigma^+\rangle]$$

$$= \frac{i\gamma b_1}{2}(\rho_{\alpha\beta} - \rho_{\beta\alpha}) = -\dot{\rho}_{\beta\beta} = -\frac{d}{dt}\langle\sigma^-\sigma^+\rangle, \tag{3.108}$$

(R)

$$\dot{\rho}_{\alpha\beta} = \frac{d}{dt}\langle\sigma^-\rangle = -i\,\Delta\langle\sigma^-\rangle + \frac{i\gamma b_1}{2}\langle\sigma_z\rangle$$

$$= -i\,\Delta\rho_{\alpha\beta} + \frac{i\gamma b_1}{2}(\rho_{\alpha\alpha} - \rho_{\beta\beta}). \tag{3.109}$$

$$\dot{\rho}_{\beta\alpha} = \frac{d}{dt}\langle\sigma^+\rangle = i\,\Delta\langle\sigma^+\rangle - \frac{i\gamma b_1}{2}\langle\sigma_z\rangle$$

$$= i\,\Delta\rho_{\beta\alpha} - \frac{i\gamma b_1}{2}(\rho_{\alpha\alpha} - \rho_{\beta\beta}), \tag{3.110}$$

or

$$\frac{d}{dt}\langle\sigma_x\rangle = \frac{d}{dt}(\langle\sigma^+\rangle + \langle\sigma^-\rangle) = -\Delta\langle\sigma_y\rangle,$$

(R) $$\frac{d}{dt}\langle\sigma_y\rangle = i\frac{d}{dt}(\langle\sigma^-\rangle - \langle\sigma^+\rangle) = \Delta\langle\sigma_x\rangle - \gamma b_1\langle\sigma_z\rangle, \tag{3.111}$$

$$\frac{d}{dt}\langle\sigma_z\rangle = 2\frac{d}{dt}\langle\sigma^+\sigma^-\rangle = \gamma b_1\langle\sigma_y\rangle.$$

The equivalent vector equation is

(R) $$\frac{d}{dt}\langle\boldsymbol{\sigma}\rangle = -\gamma\langle\boldsymbol{\sigma}\rangle \times \mathbf{B}_{\text{eff}} \tag{3.112}$$

where

$$\langle\boldsymbol{\sigma}\rangle = \langle\sigma_x\rangle\hat{\mathbf{i}} + \langle\sigma_y\rangle\hat{\mathbf{j}} + \langle\sigma_z>\hat{\mathbf{k}}, \tag{3.113}$$

$$\mathbf{B}_{\text{eff}} = \left(B_0 - \frac{\omega}{\gamma}\right)\hat{\mathbf{k}} + b_1\hat{\mathbf{i}}. \tag{3.114}$$

It should be observed that $\gamma\mathbf{B}_{\text{eff}}$ is just the vector $\boldsymbol{\Omega}$ (3.73a) whose magnitude is the Rabi frequency.

Another form of the equations of motion is obtained by defining

(R) $$u_1 = \rho_{\alpha\beta} + \rho_{\beta\alpha'} \qquad u_2 = i(\rho_{\alpha\beta} - \rho_{\beta\alpha}), \qquad u_3 = \rho_{\alpha\alpha} - \rho_{\beta\beta'} \tag{3.115a}$$

$$u_1^2 + u_2^2 + u_3^2 = 1. \tag{3.115b}$$

The last equation is a consequence of $\det\rho = 0$, as shown in Eq. (2.38), and $\operatorname{Tr}\rho = 1$. Differentiating Eq. (3.115a) with respect to time and inserting the results into Eqs. (3.108)–(3.110), one gets the set of equations

$$\dot{u}_1 = \dot{\rho}_{\alpha\beta} + \dot{\rho}_{\beta\alpha} = -\Delta u_2,$$

(R) $$\dot{u}_2 = i(\dot{\rho}_{\alpha\beta} - \dot{\rho}_{\beta\alpha}) = \Delta u_1 - \gamma b_1 u_3, \tag{3.116}$$

$$\dot{u}_3 = \dot{\rho}_{\alpha\alpha} - \dot{\rho}_{\beta\beta} = \gamma b_1 u_2,$$

or

(R) $$\dot{\boldsymbol{\beta}} = -\boldsymbol{\beta} \times \boldsymbol{\Omega} \tag{13.117}$$

where

$$\boldsymbol{\beta} = u_1\hat{\mathbf{i}} + u_2\hat{\mathbf{j}} + u_3\hat{\mathbf{k}}, \tag{3.118}$$

$$\boldsymbol{\Omega} = \gamma\mathbf{B}_{\text{eff}} = (\gamma B_0 - \omega)\hat{\mathbf{k}} + \gamma b_1\hat{\mathbf{i}} = \Omega_0\hat{\mathbf{i}} + \Delta\hat{\mathbf{k}}, \tag{3.119}$$

$$\beta^2 = u_1^2 + u_2^2 + u_3^2 = 1. \tag{3.120}$$

The vector $\boldsymbol{\beta}$ is known as the *Bloch vector* and $\boldsymbol{\Omega}$ the *torque* vector. Equations (3.116) constitute a special case of the *Bloch* equations when relaxation processes are not significant.

We note that the equations governing the time rate of change of $\mathbf{M}$ (Eq. (3.26)), $\langle \boldsymbol{\sigma} \rangle$ (Eq. 3.112)), and $\boldsymbol{\beta}$ (Eq. 3.117)) in the rotating coordinate system all have the same form. The solution to the Bloch equations therefore will be of the same form as Eq. (3.22) with the replacement of $M_x(0)$, $M_y(0)$, and $M_z(0)$ by $u_1(0)$, $u_2(0)$, and $u_3(0)$, respectively.

A simple example will illustrate the connection between the Bloch vector $\boldsymbol{\beta}$ and the occupation of the spin states. For the wave function (3.66), the density matrix is shown in Eq. (2.37). If, initially, the state $|\alpha\rangle$ is occupied and $|\beta\rangle$ is vacant, $\rho_{\alpha\alpha} = 1$ and all other density matrix elements as zero. Then, according to Eq. (3.115a), $u_1(0) = u_2(0) = 0$, $u_3(0) = 1$, or $\boldsymbol{\beta} = \hat{\mathbf{k}}$. Conversely, for $|\beta\rangle$ occupied and $|\alpha\rangle$ vacant, $\rho_{\beta\beta} = 1$, $u_3(0) = -1$, and $\boldsymbol{\beta} = -\hat{\mathbf{k}}$.

3.5 Bloch Equations

The temporal behavior of the magnetization vector in a rotating frame has been described by Eq. (3.26) and (3.27). These equations are valid as far as they go; however, they do not take into account the fact that in real physical systems unavoidable dissipative processes always are present. Such effects will be examined more closely in a subsequent chapter (Chapter VI); but for the present, it is sufficient to adopt a phenomenological approach by simply adding a dissipative (or relaxation) term to each equation in Eq. (3.27):

$$(R) \qquad \begin{aligned} \dot{M}_x &= \Delta M_y - \frac{M_x}{T_2}, \\ \dot{M}_y &= \Delta M_x - \gamma b_1 M_z - \frac{M_y}{T_2}, \\ \dot{M}_z &= \gamma b_1 M_y - \frac{M_z - M_z^0}{T_1}. \end{aligned} \qquad (3.121)$$

These are the Bloch equations [1] in which T_1 is the *longitudinal* and T_2 the *transverse* relaxation time, and M_z^0 is the value of M_z at thermal equilibrium.

To understand the significance of T_1 and T_2, let us remove the rotating field. The resulting equations in the stationary frame are given then by Eq. (3.7), augmented by the relaxation terms. Thus,

$$\dot{M}_x = -\omega_0 M_y - \frac{M_x}{T_2}, \qquad \dot{M}_y = \omega_0 M_x - \frac{M_y}{T_2}, \qquad \dot{M}_z = -\frac{M_z - M_z^0}{T_1}. \qquad (3.122)$$

Assuming the initial condition of Eq. (3.8), the solutions are

$$M_x(t) = M_x(0)e^{-t/T_2}\cos\omega_0 t, \qquad M_y(t) = M_x(0)e^{-t/T_2}\sin\omega_0 t, \quad (3.123)$$
$$M_z(t) = [M_z(0) - M_z^0]e^{-t/T_1} + M_z^0.$$

Thus, in the absence of a time-varying field, the rate of return of M_z from its initial value $M_z(0)$ to its value M_z^0 at the thermal equilibrium is governed by T_1 while T_2 is the characteristic time for the transverse components to decay to zero. That differences in behavior between the longitudinal component M_z and the transverse components M_x and M_y are to be expected also may be seen as arising from the fact that the static magnetic field provides only a single preferred direction in space while transverse directions remain undefined.

The microscopic origin of the relaxation times T_1 and T_2 is to be found in the specific interactions of the spin system with its surroundings, often called the *lattice* or *heat bath* or *reservoir*. Speaking generally, such interactions produce random, fluctuating perturbations as, for example, when atoms or molecules in a gas suffer collisions with one another. If the random perturbations contain B_x and B_y components near the resonant frequency ω_0, they will have the same effect on the spin system as the rotating field $\mathbf{b}_1(t)$; that is, they will induce transitions between the spin states, resulting in an alteration in the spin populations. We then may expect, from general thermodynamic considerations, that the spin populations will change in the direction of thermal equilibrium. Since M_z is proportional to $\langle\sigma_z\rangle$ and the latter, according to Eq. (2.38)–(2.213), is related to the diagonal elements of the density matrix by

$$\langle\sigma_z\rangle = \rho_{\alpha\alpha} - \rho_{\beta\beta}, \quad (3.124)$$

M_z will revert to thermal equilibrium as has been shown in Eq. (3.123). The time constant for this process is T_1.

In a similar vein, if the random perturbations contain low-frequency B_z components, the energies of the spin states also must fluctuate. Interpreted classically, this means that the spins do not all precess at the same frequency; eventually, they get sufficiently out of phase to cause the transverse components of the magnetization to vanish. A mechanism of this type is but one of several that contribute to the relaxation associated with T_2. We shall return to this topic in the context of the optical Bloch equations (Chapter VI).

The inclusion of the rotating field has no significant effect on the interpretation of T_1 and T_2. But it is necessary to recognize that the rotating field tends to equalize the populations in the two spin states, as we have indicated in connection with the Rabi formula. Therefore, in the presence of the rotating field, the energy-exchange interactions that are responsible for T_1 cannot impose thermal equilibrium upon the system. The equilibrium that is achieved by the two competing processes is one in which the upper state has a greater population than it would have under thermal equilibrium.

General analytic solutions to the Bloch equations (3.121) are not available. They may be solved for special cases, however, and among these are the steady state solutions that arise when the observation time is long compared with any relaxation time. Then

$$\dot{M}_x = \dot{M}_y = \dot{M}_z = 0, \tag{3.125}$$

and the Bloch equations reduce to three algebraic equations whose solutions are

$$(R) \qquad M_x = M_z^0 \frac{\gamma b_1 T_2^2 \Delta}{D}, \qquad M_y = -M_z^0 \frac{\gamma b_1 T_2}{D},$$

$$M_z = M_z^0 \frac{1 + T_2^2 \Delta^2}{D}, \tag{3.126}$$

where

$$D = 1 + T_2^2 \Delta^2 + \gamma^2 b_1^2 T_1 T_2. \tag{3.127}$$

The transformation to the laboratory frame is accomplished by means of the relation

$$\mathbf{M} = R^{-1}\mathbf{M}' \tag{3.128}$$

where R is the matrix in Eq. (3.11). Equations (3.126) then transform to

$$M_x = M_z^0 \frac{\gamma b_1 T_2^2 \Delta \cos \omega t + \gamma b_1 T_2 \sin \omega t}{D},$$

$$M_y = M_z^0 \frac{\gamma b_1 T_1^2 \Delta \sin \omega t - \gamma b_1 T_2 \cos \omega t}{D}, \tag{3.129}$$

$$M_z = M_z^0 \frac{1 + T_2^2 \Delta^2}{D}.$$

The steady state solutions for M_x and M_y in the laboratory frame may be written in another form by defining

$$\chi'(\omega) = M_z^0 \gamma \frac{T_2^2 \Delta}{D}, \qquad \chi''(\omega) = M_z^0 \gamma \frac{T_2}{D}. \tag{3.130}$$

Then

$$M_x = b_1[\chi'(\omega) \cos \omega t + \chi''(\omega) \sin \omega t], \tag{3.131}$$

$$M_y = b_1[\chi'(\omega) \sin \omega t - \chi''(\omega) \cos \omega t].$$

When b_1 is sufficiently small that

$$\gamma^2 b_1^2 T_1 T_2 \ll 1, \tag{3.132}$$

the condition is said to be *unsaturated*. In this condition $\chi'(\omega)$ and $\chi''(\omega)$ reduce to

$$\chi'(\omega) = M_z^0 \gamma \frac{T_2^2 \Delta}{1 + T_2^2 \Delta^2}, \tag{3.133}$$

$$\chi''(\omega) = M_z^0 \gamma \frac{T_2}{1 + T_2^2 \Delta^2}.$$

It is seen (Fig. 3.7) that $\chi'(\omega)$ corresponds to dispersion and $\chi''(\omega)$ corresponds to absorption. This suggests that $\chi'(\omega)$ and $\chi''(\omega)$ may be regarded as the real and imaginary components of a complex susceptibility $\chi(\omega)$, i.e.,

$$\chi(\omega) = \chi'(\omega) + i\chi''(\omega). \tag{3.134}$$

It may be noted that if $\chi''(\omega)$ is written in the form

$$\chi''(\omega) = M_z^0 \gamma \pi g(\omega) \tag{3.135}$$

with

$$g(\omega) = \frac{1}{\pi} \frac{T_2}{1 + T_2^2 \Delta^2}, \tag{3.136}$$

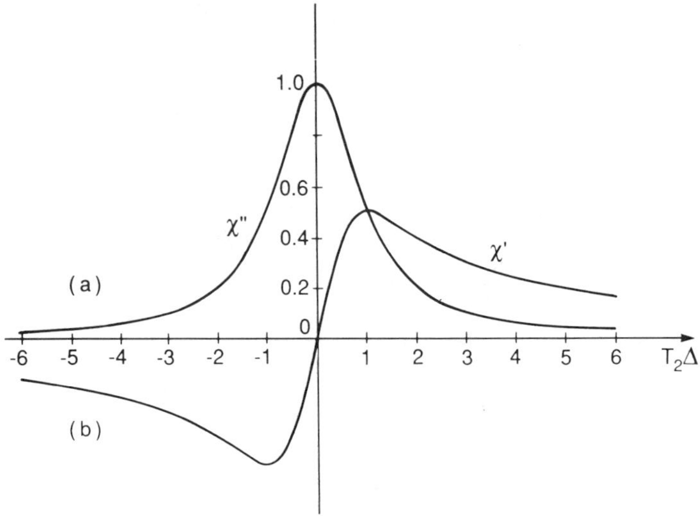

FIGURE 3.7 The real and imaginary parts of the susceptibility at low saturation. (a) the imaginary part χ'' corresponds to absorption and (b) the real part χ' to dispersion.

the function $g(\omega)$ is a normalized Lorentzian (Fig. 3.7b) with a maximum at the resonance condition $\omega = \omega_0$ and a full width at half maximum equal to $2/T_2$.

The magnetization at thermal equilibrium may be computed from the excess of β spins over α spins when the populations in the two states correspond to a Boltzmann distribution. Assuming $kT \gg \hbar\omega_0 = \gamma\hbar B_0$, which is satisfied at room temperature even for large fields,

$$\frac{N_\alpha}{N_\beta} = e^{-\gamma\hbar B_0/kT} \simeq (1 - \gamma\hbar B_0/kT), \tag{3.137}$$

$$\Delta N \equiv N_\beta - N_\alpha = \frac{\gamma\hbar B_0}{kT} N_\beta \simeq \frac{\gamma\hbar B_0}{2kT} N. \tag{3.138}$$

But since

$$N = N_\alpha + N_\beta, \qquad N_\alpha \simeq N_\beta \simeq \frac{N}{2}, \tag{3.139}$$

we have

$$M_z^0 = \frac{\gamma\hbar}{2}\Delta N = \frac{N\gamma^2\hbar^2 B_0}{4kT}. \tag{3.140}$$

In this section we concentrated on the equation of motion for the magnetization; but the Bloch vector $\boldsymbol{\beta}$ (Eq. (3.188) obeys the same type of equation. Hence, the general Bloch equations may be written

$$\begin{aligned} \dot{u}_1 &= -\Delta u_2 - \frac{u_1}{T_2}, \\ \dot{u}_2 &= \Delta u_1 - \gamma b_1 u_3 - \frac{u_2}{T_2}, \\ \dot{u}_3 &= \gamma b_1 u_2 - \frac{u_3 - u_3^0}{T_1}, \end{aligned} \tag{3.141}$$

in which u_3^0 is the value of u_3 at thermal equilibrium.

Let us return to Eqs. (3.24) which describe the temporal behavior of the magnetization vector at resonance. Assuming the system had been prepared initially so that

$$M_x(0) = M_y(0) = 0, \qquad M_z(0) = M_0 \tag{3.142}$$

(that is, the magnetization vector $\mathbf{M}$ initally points in the positive z direction (Fig. 3.8a)), let us apply the rotating magnetic field $\mathbf{b}_1$ (which could be a circularly polarized component of an oscillating electromagnetic field). After a time $t = \pi/2\gamma b_1$ ($\pi/2$ pulse) $\mathbf{M}$ rotates into the $-y$ direction (Figs. 3.2 and

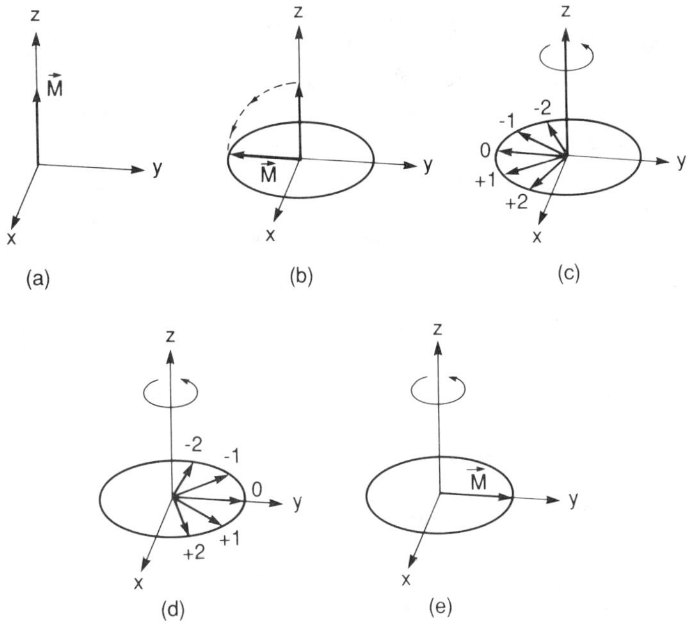

FIGURE 3.8 Stages in the development of a spin echo. (a) At $t = 0$ the magnetization vector **M** is oriented in the positive z direction; (b) after the application of a $\pi/2$ pulse the vector **M** has been rotated into the negative y direction; (c) as a consequence of various inhomogeneities, the individual magnetic moments **m**, where $\mathbf{M} = N\langle\mathbf{m}\rangle$, acquire different precessional frequencies and are no longer in phase with one another; (d) the application of a π pulse rotates each magnetic moment about the x axis so that the ones with the higher frequenceis are now behind the ones with the lower frequencies; (e) since the magnetic moments continue to precess in the same sense the faster magnetic moments catch up with the slower ones and eventually the system rephases.

3.8b). Let the rotating field be removed at this stage; **M** then begins to precess (in a positive sense) in the xy plane about the constant magnetic field B_0 along the z axis. The components of the magnetization vector, therefore, acquire the values

$$M_x(t) = M_0 \sin \omega_0 t, \qquad M_y(t) = -M_0 \cos \omega_0 t, \qquad M_z(t) = 0. \quad (3.143)$$

Such motion results in the emission of electromagnetic radiation, which is the *free induction signal* (Fig. 3.9). This condition cannot continue indefinitely, however. Apart from radiation losses, which will ultimately damp out the motion, various inhomogeneities and other random interactions will produce phase differences among the precessing magnetic moments that make up the macroscopic magnetization. As an example, suppose the magnetic field is not uniform throughout the sample. Each individual magnetic moment **m**, which contributes to the macroscopic magnetization $\mathbf{M} = N\langle\mathbf{m}\rangle$, then will have

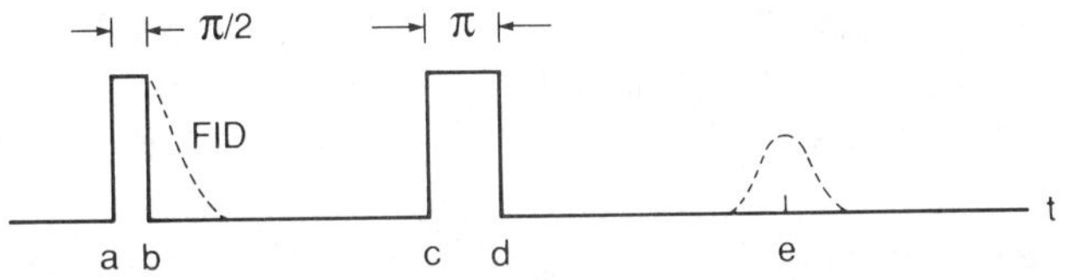

FIGURE 3.9 Pulse sequence to produce a spin echo. The labels *a* through *e* correspond to Fig. 3.8a through *e*. The free induction decay (FID) arises from the precession of the magnetization **M** in the static magnetic field. Owing to the inhomogeneities that randomize the phases of the individual magnetic moments **m**, the magnetization **M** decays and so does the free induction signal. After the application of a π pulse, the magnetic moments begin to rephase resulting in the reconstitution of the magnetization whose precession produces the echo signal at *e*.

its own precessional frequency and, in time, the phases among the various magnetic moments will be randomized (Fig. 3.8c). The macroscopic magnetization and the emitted radiation then vanish. The rate at which the free induction signal decays to zero (FID in Fig. 3.9) provides a measure of the transverse relaxation time T_2.

One might suppose that free induction decay is an irreversible process so that after the phases of the magnetic moments have been randomized there is no returning to a coherent state. It turns out that this is not so and that a coherent state can indeed be reconstructed. This is accomplished by the application of a π pulse (i.e., a pulse of duration $\pi/\gamma b_1$) at the time τ (measured from $t = 0$) that is greater than the duration of the free induction signal (Fig. 3.9). The π pulse, applied at the same frequency as the initial $\pi/2$ pulse, has the effect of rotating the direction of each magnetic moment through an angle π about the x axis. That is, a magnetic moment lying in the xy plane returns to the xy plane after the π pulse with its x coordinate unchanged but with its y coordinate inverted (Fig. 3.8d). At the end of the π pulse, each magnetic moment continues to precess about the z axis in the same (positive) sense. The same mechanisms that caused one magnetic moment to precess faster and therefore get ahead of a slower one are operative once again. But owing to the inversion, the faster magnetic moment now is located behind the slower one. After the elapse of an additional time τ, the faster moment will catch up with the slower one. Thus, at $t = 2\tau$ all the individual magnetic moments are back in phase and have merged again to reconstruct the macroscopic magnetization (Fig. 3.8e) which, as it continues to precess, again emits electromagnetic radiation. In an obvious terminology the latter is known as the *echo* signal (Fig. 3.9). Since all the dipoles are back in phase, the radiation is coherent. In time, the echo signal decays and if it is so desired, the process may be repeated.

We could just as well have presented all of the above in terms of spins rather than magnetic moments since the two are linearly related (and antiparallel, according to Eq. (3.37)). For this reason, the echo signals are more commonly called *spin echoes* [7]. The pulse sequence described above is by no means unique; many other combinations of pulses have been devised for specific experiments.

Similar phenomena, known as *photon echoes* have been observed in the optical region (Section 6.5).

References

[1] F. Bloch, *Phys. Rev.* **70,** 460(1946).
[2] A. Abragam, *Principles of Nuclear Magnetism.* Oxford Press, Oxford, 1961.
[3] A. Carrington and A. D. McLachlan, *Introduction to Magnetic Resonance.* Harper and Row, New York, 1967.
[4] C. P. Slichter, *Principles of Magnetic Resonance.* Harper and Row, New York, 1963.
[5] I. I. Rabi, *Phys. Rev.* **51**, 652(1937).
[6] F. Bloch and A. J. F. Siegert, *Phys. Rev.* **57**, 522(1940).
[7] E. L. Hahn, *Phys. Rev.* **80**, 580(1950).

IV The Radiation Field

We now begin the main theme of this book: the interaction of a radiation field with matter. As a first step, most of the present chapter is devoted to the properties of the radiation field alone, with emphasis on its quantized form and the statistics of the various states arising from the quantization

4.1 Polarization, Density Matrices, Angular Momentum

Consider a plane wave propagating in the z direction with the electric vector along the x axis:

$$\mathbf{E}(z, t) = \frac{1}{2}[E_x e^{i(k_z z - \omega t)} + E_x^* e^{-i(k_z z - \omega t)}]\hat{\boldsymbol{\varepsilon}}_x = E_x(z, t)\hat{\boldsymbol{\varepsilon}}_x. \tag{4.1}$$

$E_x(z, t)$ is the real field, E_x is the complex amplitude, and $\hat{\boldsymbol{\varepsilon}}_x$ is the unit polarization vector. If we take

$$E_x = E_{x0} e^{i\phi}, \tag{4.2}$$

the real field becomes

$$E_x(z, t) = E_{x0} \cos(k_z z - \omega t + \phi). \tag{4.3}$$

Let us now combine two fields, one polarized along x and the other along y and differing in phase by the angle ϕ. Choosing an arbitrary value of z, say $z = 0$,

$$\mathbf{E}(t) = E_x(t)\hat{\boldsymbol{\varepsilon}}_x + E_y(t)\hat{\boldsymbol{\varepsilon}}_y = \hat{\boldsymbol{\varepsilon}}_x E_{x0} \cos \omega t + \hat{\boldsymbol{\varepsilon}}_y E_{y0} \cos(\omega t - \phi). \tag{4.4}$$

We recall that $\hat{\boldsymbol{\varepsilon}}_x$, $\hat{\boldsymbol{\varepsilon}}_y$, and the unit vector along the direction of propagation define a right-handed, orthogonal coordinate system. Special cases are of interest.

1. $\phi = 0$,

$$\begin{aligned} \mathbf{E}(t) &= (\hat{\boldsymbol{\varepsilon}}_x E_{x0} + \hat{\boldsymbol{\varepsilon}}_y E_{y0}) \cos \omega t \\ &= E_0(\hat{\boldsymbol{\varepsilon}}_x \cos \theta + \hat{\boldsymbol{\varepsilon}}_y \sin \theta) \cos \omega t = \hat{\boldsymbol{\varepsilon}} E_0 \cos \omega t \end{aligned} \tag{4.5}$$

where

$$E_0^2 = E_{x0}^2 + E_{y0}^2, \qquad \tan \theta = \frac{E_{y0}}{E_{x0}}, \tag{4.6}$$

$$\hat{\boldsymbol{\varepsilon}} = \hat{\boldsymbol{\varepsilon}}_x \cos \theta + \hat{\boldsymbol{\varepsilon}}_y \sin \theta. \tag{4.7}$$

The resultant field is plane (linearly) polarized with the plane of polarization inclined at an angle θ to the x axis. The two orthogonal fields oscillate in phase. When $\phi = \pi$, the resultant fields also is polarized linearly but the plane of polarization is inclined at $-\theta$ to the x axis.

2. $\phi = \dfrac{\pi}{2}$, $\qquad E_{x0} = E_{y0} = E_0$,

$$\begin{aligned} \mathbf{E}(t) &= E_0(\hat{\boldsymbol{\varepsilon}}_x \cos \omega t + \hat{\boldsymbol{\varepsilon}}_y \sin \omega t) \\ &= E_0 \operatorname{Re}[(\hat{\boldsymbol{\varepsilon}}_x + i\hat{\boldsymbol{\varepsilon}}_y)e^{-i\omega t}] = E_0 \operatorname{Re}[\hat{\boldsymbol{\varepsilon}} e^{-i\omega t}] \end{aligned} \tag{4.8}$$

where

$$\hat{\boldsymbol{\varepsilon}} = \hat{\boldsymbol{\varepsilon}}_x + i\hat{\boldsymbol{\varepsilon}}_y. \tag{4.9}$$

As time progresses, this field rotates in the *counterclockwise* (mathematically, positive) sense as seen by an observer facing the oncoming beam. Such a field is said to be *left circularly polarized* according to standard optical convention. (The reader should be cautioned, however, that this convention is not universal.) Changing ϕ to $-\pi/2$ produces a *right circularly polarized* field with the polarization vector

$$\hat{\boldsymbol{\varepsilon}} = \hat{\boldsymbol{\varepsilon}}_x - i\hat{\boldsymbol{\varepsilon}}_y. \tag{4.10}$$

3. $\phi = \dfrac{\pi}{2}, \qquad E_{x0} \neq E_{y0},$

$$\begin{aligned}\mathbf{E}(t) &= \hat{\boldsymbol{\varepsilon}}_x E_{x0} \cos \omega t + \hat{\boldsymbol{\varepsilon}}_y E_{y0} \sin \omega t \\ &= E_0 \operatorname{Re}[(\hat{\boldsymbol{\varepsilon}}_x \cos\theta + i\hat{\boldsymbol{\varepsilon}}_y \sin\theta)e^{-i\omega t}] = E_0 \operatorname{Re}[\hat{\boldsymbol{\varepsilon}} e^{-i\omega t}]\end{aligned} \tag{4.11}$$

where E_0 and θ are defined by Eq. (4.6) and

$$\hat{\boldsymbol{\varepsilon}} = \hat{\boldsymbol{\varepsilon}}_x \cos\theta + i\hat{\boldsymbol{\varepsilon}}_y \sin\theta. \tag{4.12}$$

This is an elliptically polarized field with a counterclockwise sense of rotation, as in Case 2. The major and minor axes of the elllipse are along the x and y axes. Here, too, changing ϕ to $-\pi/2$ reverses the sense of rotation.

4. $\phi \neq \pm\dfrac{\pi}{2}, \qquad E_{x0} \neq E_{y0},$

$$\begin{aligned}\mathbf{E}(t) &= \hat{\boldsymbol{\varepsilon}}_x E_{x0} \cos \omega t + \hat{\boldsymbol{\varepsilon}}_y E_{y0} \cos(\omega t - \phi) \\ &= E_0 \operatorname{Re}[(\hat{\boldsymbol{\varepsilon}}_x \cos\theta + \hat{\boldsymbol{\varepsilon}}_y \sin\theta e^{i\phi})e^{-i\omega t}] = E_0 \operatorname{Re}[\hat{\boldsymbol{\varepsilon}} e^{-i\omega t}]\end{aligned} \tag{4.13}$$

with the same definitions for E_0 and θ and

$$\hat{\boldsymbol{\varepsilon}} = \hat{\boldsymbol{\varepsilon}}_x \cos\theta + \hat{\boldsymbol{\varepsilon}}_y \sin\theta e^{i\phi}. \tag{4.14}$$

This is the general expression for the polarization of a transverse electromagnetic wave; all possible polarizations are obtained by assigning appropriate values to θ and ϕ.

Not all fields are polarized; partially polarized and unpolarized fields also exist. It therefore becomes necessary to devise a measure for the degree of polarization; the polarization density operator

$$\rho = |\hat{\boldsymbol{\varepsilon}}\rangle\langle\hat{\boldsymbol{\varepsilon}}| \tag{4.15}$$

may be used for this purpose. With $\hat{\boldsymbol{\varepsilon}}_x$ and $\hat{\boldsymbol{\varepsilon}}_y$ as the basis set, and $\hat{\boldsymbol{\varepsilon}}$ given by Eq. (4.14), we construct the polarization density matrix

$$\rho = \begin{pmatrix} \langle\hat{\boldsymbol{\varepsilon}}_x|\rho|\hat{\boldsymbol{\varepsilon}}_x\rangle & \langle\hat{\boldsymbol{\varepsilon}}_x|\rho|\hat{\boldsymbol{\varepsilon}}_y\rangle \\ \langle\hat{\boldsymbol{\varepsilon}}_y|\rho|\hat{\boldsymbol{\varepsilon}}_x\rangle & \langle\hat{\boldsymbol{\varepsilon}}_y|\rho|\hat{\boldsymbol{\varepsilon}}_y\rangle \end{pmatrix} \tag{4.16a}$$

$$= \begin{pmatrix} \cos^2\theta & \sin\theta\cos\theta e^{-i\phi} \\ \sin\theta\cos\theta e^{i\phi} & \sin^2\theta \end{pmatrix}. \tag{4.166}$$

Now let

$$\cos^2\theta = \frac{1}{2}(1 + \xi_3), \qquad \sin^2\theta = \frac{1}{2}(1 - \xi_3), \qquad \sin\theta\cos\theta e^{\pm i\phi} = \frac{1}{2}(\xi_1 + i\xi_2), \tag{4.17}$$

or

$$\begin{aligned}\xi_1 &= 2\sin\theta\cos\theta\cos\phi,\\ \xi_2 &= 2\sin\theta\cos\theta\sin\phi,\\ \xi_3 &= \cos^2\theta - \sin^2\theta,\end{aligned} \tag{4.18}$$

in which ξ_1, ξ_2, and ξ_3 are known as the *Stokes parameters*. The polarization density matrix then acquires a form identical with Eq. (2.35) for the spin-1/2 system:

$$\begin{aligned}\rho &= \frac{1}{2}\begin{pmatrix} 1+\xi_3 & \xi_1 - i\xi_2 \\ \xi_1 + i\xi_2 & 1-\xi_3 \end{pmatrix}\\ &= \frac{1}{2}[I + \xi_1\sigma_x + \xi_2\sigma_y + \xi_3\sigma_z]\end{aligned} \tag{4.19}$$

where I is the 2×2 unit matrix and σ_x, σ_y, and σ_z are the Pauli spin matrices. It is further observed that

$$\xi_1^2 + \xi_2^2 + \xi_3^2 = 1, \tag{4.20}$$

$$\mathrm{Tr}\,\rho^2 = \frac{1}{2}(1 + \xi_1^2 + \xi_2^2 + \xi_3^2) = 1. \tag{4.21}$$

Equations (4.20) and (4.21) have been derived for polarized fields, whether plane, circular, or elliptical. In view of the general relation, however,

$$\mathrm{Tr}\,\rho^2 \le 1. \tag{4.22}$$

The density matrix approach can be extended to include the partially polarized and unpolarized cases as well. To this end let

$$P = \sqrt{\xi_1^2 + \xi_2^2 + \xi_3^2}, \qquad 0 \le P \le 1, \tag{4.23}$$

be the definition of the degree of polarization. Thus, when $P = 1$ the field is completely polarized, and since $\mathrm{Tr}\,\rho^2 = 1$, the field is said to be in a pure state of polarization. At the other extreme, when

$$\xi_1 = \xi_2 = \xi_3 = 0 \tag{4.24}$$

or $P = 0$, the field is completely unpolarized. The density matrix then reduces to

$$\rho = \frac{1}{2}\begin{pmatrix} 1 & 0 \\ 0 & 1 \end{pmatrix}, \qquad \mathrm{Tr}\,\rho^2 = \frac{1}{2}, \tag{4.25}$$

which indicates that the unpolarized field is in a mixed state. An important example in this category is the light emitted by a thermal source with all directions of polarization equally probable. A partially polarized field is characterized by an intermediate value of P,

$$0 < \mathrm{P} < 1. \tag{4.26}$$

The rotational properties of an electromagnetic field are intimately related to the polarization [1, 2]. For a vector field, such as the electromagnetic field, the rotational properties are characterized by an orbital angular momentum operator $\mathbf{L}$ and a spin angular momentum operator $\mathbf{S}$. The components of $\mathbf{L}$ are

$$L_x = -i\left(y\frac{a}{\partial z} - z\frac{\partial}{\partial y}\right), \qquad L_y = -i\left(z\frac{\partial}{\partial x} - x\frac{\partial}{\partial z}\right),$$
$$L_z = -i\left(x\frac{\partial}{\partial y} - y\frac{\partial}{\partial x}\right), \tag{4.27}$$

with the commutation rules

$$[L_x, L_y] = iL_z, \qquad [L_y, L_z] = iL_x, \qquad [L_z, L_x] = iL_y, \tag{4.28}$$

or $\mathbf{L} \times \mathbf{L} = i\mathbf{L}$. The spherical harmonics $Y_{lm}(\theta, \phi)$ are eigenstates of L^2 and L_z:

$$L^2 Y_{lm}(\theta, \phi) = l(l+1) Y_{lm}(\theta, \phi),$$
$$L_z Y_{lm}(\theta, \phi) = m Y_{lm}(\theta, \phi),$$
$$l = 0, 1, 2, \ldots, \qquad m = -l, -l+1, \ldots, l.$$

The values $l = 1, 2, 3, \ldots$ correspond to dipole, quadrupole, octupole, ... terms in the mutipole expansion of the field; no term corresponds to $l = 0$ due to the assumption of transversality. A further classification arises from parity considerations. Thus, for $l = 1$ we have both *electric* and *magnetic* dipole terms designated by E1 and M1, respectively. For E1 the electric field $\mathbf{E}(\mathbf{r})$ has even parity and the magnetic induction $\mathbf{B}(\mathbf{r})$, odd parity; for M1 the parities are reversed. By convention, the parity of an electromagnetic field is defined as the parity of $\mathbf{B}(\mathbf{r})$; hence, electric dipole radiation (E1) has odd parity and magnetic dipole radiation (M1), even parity. More generally, and El multipole has odd (even) parity for l odd (even); and Ml multipole has odd (even) parity for l even (odd).

Turning now to the spin properties of a vector field, the components of $\mathbf{S}$ are

$$S_x = -i\begin{pmatrix} 0 & 0 & 0 \\ 0 & 0 & 1 \\ 0 & -1 & 0 \end{pmatrix}, \qquad S_y = -i\begin{pmatrix} 0 & 0 & -1 \\ 0 & 0 & 0 \\ 1 & 0 & 0 \end{pmatrix},$$
$$S_z = -i\begin{pmatrix} 0 & 1 & 0 \\ -1 & 0 & 0 \\ 0 & 0 & 0 \end{pmatrix}, \tag{4.30}$$

with the commutation properties embodied in $\mathbf{S} \times \mathbf{S} = i\mathbf{S}$. The eigenvalue equations for S^2 and S_z are

$$S^2\hat{\boldsymbol{\varepsilon}}_q = S(S+1)\hat{\boldsymbol{\varepsilon}}_q, \qquad S_z\hat{\boldsymbol{\varepsilon}}_q = q\hat{\boldsymbol{\varepsilon}}, \tag{4.31}$$

where

$$S = 1, \qquad q = 1, 0, -1, \tag{4.32}$$

and

$$\hat{\boldsymbol{\varepsilon}}_{+1} = -\frac{1}{\sqrt{2}}\begin{pmatrix}1\\ i\\ 0\end{pmatrix}, \qquad \hat{\boldsymbol{\varepsilon}}_0 = \begin{pmatrix}0\\ 0\\ 1\end{pmatrix}, \qquad \hat{\boldsymbol{\varepsilon}}_{-1} = -\frac{1}{\sqrt{2}}\begin{pmatrix}1\\ -i\\ 0\end{pmatrix}. \tag{4.33}$$

If one writes the cartesian unit vectors in the form

$$\hat{\boldsymbol{\varepsilon}}_x = \begin{pmatrix}1\\ 0\\ 0\end{pmatrix}, \qquad \hat{\boldsymbol{\varepsilon}}_y = \begin{pmatrix}0\\ 1\\ 0\end{pmatrix}, \qquad \hat{\boldsymbol{\varepsilon}}_z = \begin{pmatrix}0\\ 0\\ 1\end{pmatrix}, \tag{4.34}$$

it is verified readily that

$$\hat{\boldsymbol{\varepsilon}}_{+1} = -\frac{1}{\sqrt{2}}(\hat{\boldsymbol{\varepsilon}}_x + i\hat{\boldsymbol{\varepsilon}}_y), \qquad \hat{\boldsymbol{\varepsilon}}_{-1} = \frac{1}{\sqrt{2}}(\hat{\boldsymbol{\varepsilon}}_x - i\hat{\boldsymbol{\varepsilon}}_y), \qquad \hat{\boldsymbol{\varepsilon}}_0 = \hat{\boldsymbol{\varepsilon}}_z. \tag{4.35}$$

We note that, apart from an inconsequential sign (which is associated with the adoption of the Condon-Shortley phase convention for spherical harmonics) and normalization constant, the eigenvectors $\hat{\boldsymbol{\varepsilon}}_{+1}$ and $\hat{\boldsymbol{\varepsilon}}_{-1}$ correspond to the circular polarization vectors in Eqs. (4.9) and (4.10), respectively. Thus, we have a dual interpretation for $\hat{\boldsymbol{\varepsilon}}_{\pm 1}$ as circular polarization vectors and as simultaneous eigenvectors of S^2 and S_z for a spin-1 field such as the electromagnetic field. For transverse fields, $\hat{\boldsymbol{\varepsilon}}_0 = \hat{\boldsymbol{\varepsilon}}_z = 0$. Fields with polarizations $\hat{\boldsymbol{\varepsilon}}_{+1}$ (left circular) or $\hat{\boldsymbol{\varepsilon}}_{-1}$ (right circular) are said to have positive or negative helicity.

4.2 Classical Hamiltonian

The transition from classical to quantum mechanics is facilitated by the construction of an appropriate Hamiltonian function. This is equally true for fields as it is for particles. In this section we derive the classical Hamiltonian for a transverse electromagnetic field in which **E** and **B** are perpendicular to one another and both are perpendicular to the direction of propagation.

For the purpose at hand, it is mathematically advantageous to express the transverse radiation fields as expansions in terms of normal modes. This may be accomplished by assuming that the fields are enclosed in a cavity; normal modes then arise as a consequence of the boundary conditions on the walls of the cavity. If the dimensions of the cavity are much larger than the longest wavelengths, the results of physical measurements will be independent of the size and shape of the cavity, which then may be assumed to be in the shape of a cube of side L. Furthermore, to confine the field inside the cavity, the walls will be assumed to be perfectly reflecting.

Although there are several sets of equally satisfactory boundary conditions, the most convenient ones for our purpose are periodic conditions at each face of the cube. In the x direction, for example, the cavity modes—assumed to be plane transverse wave—must satisfy

$$e^{ik_x x} = e^{ik_x(x+L)} \tag{4.36}$$

or

$$k_x = (2\pi/L)N_x, \qquad N_x = 0, \pm 1, \pm 2, \ldots, \tag{4.37}$$

similarly for the y and z directions. Hence, the general propagation vector $\mathbf{k}$, with absolute magnitude k, may be written

$$\mathbf{k} = k_x\hat{\mathbf{i}}_1 + k_y\hat{\mathbf{i}}_2 + k_z\hat{\mathbf{i}}_3 = \frac{2\pi}{L}(N_x\hat{\mathbf{i}}_1 + N_y\hat{\mathbf{i}}_2 + N_z\hat{\mathbf{i}}_3). \tag{4.38}$$

The normalized (or unit) propagation vector $\hat{\mathbf{k}}$ and the frequency ω_k are related by

$$\mathbf{k} = k\hat{\mathbf{k}}, \qquad k = \frac{\omega_k}{c}, \tag{4.39}$$

where c is the velocity of electromagnetic waves in free space.

Each set of three integers N_x, N_y, and N_z defines a propagation mode and an arbitrary configuration of the field is expressible as a linear superposition of these modes. Although the number of modes is infinite, it is a denumerable (countable) infinity that is mathematically simpler than the continuous infinity of modes existing in an unbounded region. In an interval defined by ΔN_x, ΔN_y, and ΔN_z, the total number of modes is

$$\Delta N = \Delta N_x \Delta N_y \Delta N_z = \left(\frac{V}{8\pi^3}\right)\Delta k_x \Delta k_y \Delta k_z \tag{4.40}$$

in which $V = L^3$. But under the assumption that the wavelengths are small compared with the dimensions of the cavity, the waves will be densely packed in k space. We then may approximate ΔN by

$$dN = \left(\frac{V}{8\pi^3}\right) dk_x\, dk_y\, dk_z \tag{4.41}$$

or, in terms of spherical coordinates in k space,

$$dN = \left(\frac{V}{8\pi^3}\right)k^2 \sin\theta \, dk \, d\theta \, d\phi$$

$$= \left(\frac{V}{8\pi^3}\right)k^2 \, dk \, d\Omega = \left(\frac{V}{8\pi^3 c^3}\right)\omega_k^2 \, d\omega_k \, d\Omega, \tag{4.42}$$

where $d\Omega$ is an element of solid angle in the direction of propagation. Integration of Eq. (4.42) over all directions gives the number of modes between ω_k and $\omega_k + d\omega_k$:

$$\frac{V\omega_k^2 \, d\omega_k}{8\pi^3 c^3}\int d\Omega = \frac{V\omega_k^2 \, d\omega_k}{2\pi^2 c^3} \equiv dN(\omega_k). \tag{4.43}$$

Each mode has two independent polarizations; when these are counted as separate modes the expressions for dN are doubled.

The radiation field under consideration consists of an infinite number of transverse plane waves propagating in all possible directions in a loss-free cubical cavity. The general vector potential is then a superposition of plane waves

$$\mathbf{A}(\mathbf{r}, t) = \sum_{\mathbf{k}\lambda} \hat{\boldsymbol{\varepsilon}}_{\mathbf{k}\lambda}[A_{\mathbf{k}\lambda}e^{i(\mathbf{k}\cdot\mathbf{r} - \omega_k t)} + A_{\mathbf{k}\lambda}^{*}e^{-i(\mathbf{k}\cdot\mathbf{r} - \omega_k t)}], \tag{4.44}$$

and since the two terms in the square brackets are complex conjugates, $\mathbf{A}(\mathbf{r}, t)$ is real. $A_{\mathbf{k}\lambda}$ is the complex amplitude of the vector potential associated with a single mode; $\hat{\boldsymbol{\varepsilon}}_{\mathbf{k}\lambda}$ is a (real) polarization vector. The sum over the vector $\mathbf{k}$ is a shorthand notation for a sum over the triad N_x, N_y, N_z and the sum over λ is a sum over the two independent polarizations that must satisfy the orthogonality conditions

$$\hat{\boldsymbol{\varepsilon}}_{\mathbf{k}\lambda} \cdot \hat{\boldsymbol{\varepsilon}}_{\mathbf{k}\lambda'} = \delta_{\lambda\lambda'}, \qquad \hat{\boldsymbol{\varepsilon}}_{\mathbf{k}\lambda} \cdot \hat{\mathbf{k}} = 0, \qquad \lambda, \lambda' = 1, 2. \tag{4.45}$$

The fields $\mathbf{E}$ and $\mathbf{B}$ now may be obtained directly from the vector potential. In SI units,

$$\mathbf{E}(\mathbf{r}, t) = -\frac{\partial}{\partial t}\mathbf{A}(\mathbf{r}, t)$$

$$= i\sum_{\mathbf{k}\lambda} \omega_k \hat{\boldsymbol{\varepsilon}}_{\mathbf{k}\lambda}[A_{\mathbf{k}\lambda}e^{i(\mathbf{k}\cdot\mathbf{r} - \omega_k t)} - A_{\mathbf{k}\lambda}^{*}e^{-i(\mathbf{k}\cdot\mathbf{r} - \omega_k t)}], \tag{4.46}$$

$$\mathbf{B}(\mathbf{r}, t) = \nabla \times \mathbf{A}(\mathbf{r}, t)$$

$$= \frac{i}{c}\sum_{\mathbf{k}\lambda} \omega_k(\hat{\mathbf{k}} \times \hat{\boldsymbol{\varepsilon}}_{\mathbf{k}\lambda})[A_{\mathbf{k}\lambda}e^{i(\mathbf{k}\cdot\mathbf{r} - \omega_k t)} - A_{\mathbf{k}\lambda}^{*}e^{-i(\mathbf{k}\cdot\mathbf{r} - \omega_k t)}]. \tag{4.47}$$

To derive the Hamiltonian for the electromagnetic field in the cavity, it is necessary to express the field energy W in terms of canonical variables. As a first step, we evaluate

$$W = \frac{1}{2}\int_V (\varepsilon_0 \mathbf{E}\cdot\mathbf{E} + \mu_0 \mathbf{H}\cdot\mathbf{H})\,dV, \qquad \mathbf{H} = \frac{\mathbf{B}}{\mu_0}, \tag{4.48}$$

in terms of the amplitudes of the vector potential using Eqs. (4.46) and (4.47). Noting that the periodic boundary condition of Eq. (4.36) implies that

$$\int_0^L e^{\pm i k_x x}\,dx = \begin{cases} 0, & k_x \neq 0, \\ L, & k_x = 0, \end{cases} \tag{4.49}$$

or, in three dimensions,

$$\int_V e^{\pm i(\mathbf{k}-\mathbf{k}')\cdot\mathbf{r}}\,dV = \delta_{\mathbf{k}\mathbf{k}'}V, \tag{4.50}$$

it is found that

$$\begin{aligned}\int_V \varepsilon_0 \mathbf{E}\cdot\mathbf{E}\,dV &= 2\varepsilon_0 V \sum_{\mathbf{k}\lambda} \omega_k^2 A_{\mathbf{k}\lambda}(t) A^*_{\mathbf{k}\lambda}(t) \\ &\quad - \varepsilon_0 V \sum_{\mathbf{k}\lambda\lambda'} \omega_k^2 \hat{\varepsilon}_{\mathbf{k}\lambda}\cdot\hat{\varepsilon}_{-\mathbf{k}\lambda'}(A_{\mathbf{k}\lambda}(t)A_{-\mathbf{k}\lambda'}(t) + A^*_{\mathbf{k}\lambda}(t)A^*_{-\mathbf{k}\lambda}(t)),\end{aligned} \tag{4.51}$$

where

$$A_{\mathbf{k}\lambda}(t) = A_{\mathbf{k}\lambda} e^{-i\omega_k t}. \tag{4.52}$$

To compute the magnetic contribution to the field energy, the vector identity

$$(\mathbf{A}\times\mathbf{B})\cdot(\mathbf{C}\times\mathbf{D}) = (\mathbf{A}\cdot\mathbf{C})(\mathbf{B}\cdot\mathbf{D}) - (\mathbf{A}\cdot\mathbf{D})(\mathbf{B}\cdot\mathbf{C}) \tag{4.53}$$

yields

$$(\hat{\mathbf{k}}\times\hat{\varepsilon}_{\mathbf{k}\lambda})\cdot(\hat{\mathbf{k}}\times\hat{\varepsilon}_{\mathbf{k}\lambda'}) = \delta_{\lambda\lambda'}, \tag{4.54}$$

$$(\hat{\mathbf{k}}\times\hat{\varepsilon}_{\mathbf{k}\lambda})\cdot(-\hat{\mathbf{k}}\times\hat{\varepsilon}_{-\mathbf{k}\lambda'}) = -\hat{\varepsilon}_{\mathbf{k}\lambda}\cdot\hat{\varepsilon}_{-\mathbf{k}\lambda'}. \tag{4.55}$$

Then, with $\varepsilon_0\mu_0 c^2 = 1$,

$$\begin{aligned}\int_V \mu_0 \mathbf{H}\cdot\mathbf{H}\,dV &= 2\varepsilon_0 V \sum_{\mathbf{k}\lambda} \omega_k^2 A_{\mathbf{k}\lambda}(t) A^*_{\mathbf{k}\lambda}(t) \\ &\quad + \varepsilon_0 V \sum_{\mathbf{k}\lambda\lambda'} \omega_k^2 \hat{\varepsilon}_{\mathbf{k}\lambda}\cdot\hat{\varepsilon}_{-\mathbf{k}\lambda'}(A_{\mathbf{k}\lambda}(t)A_{-\mathbf{k}\lambda'}(t) + A^*_{\mathbf{k}\lambda}(t)A^*_{-\mathbf{k}\lambda'}(t)).\end{aligned} \tag{4.56}$$

Combining Eqs. (4.48), (4.51), and (4.56), the field energy acquires the simple form

$$W = 2\varepsilon_0 V \sum_{\mathbf{k}\lambda} \omega_k^2 A_{\mathbf{k}\lambda}(t) A^*_{\mathbf{k}\lambda}(t) = 2\varepsilon_0 V \sum_{\mathbf{k}\lambda} \omega_k^2 A_{\mathbf{k}\lambda} A^*_{\mathbf{k}\lambda}. \tag{4.57}$$

We now introduce a new set of variables

$$Q_{\mathbf{k}\lambda} = \sqrt{\varepsilon_0 V}\,[A_{\mathbf{k}\lambda} + A^*_{\mathbf{k}\lambda}], \qquad P_{\mathbf{k}\lambda} = -i\omega_k\sqrt{\varepsilon_0 V}\,[A_{\mathbf{k}\lambda} - A^*_{\mathbf{k}\lambda}], \tag{4.58a}$$

$$A_{\mathbf{k}\lambda} = \frac{1}{2\omega_k\sqrt{\varepsilon_0 V}}[\omega_k Q_{\mathbf{k}\lambda} + iP_{\mathbf{k}\lambda}], \tag{4.58b}$$

$$A^*_{\mathbf{k}\lambda} = \frac{1}{2\omega_k\sqrt{\varepsilon_0 V}}[\omega_k Q_{\mathbf{k}\lambda} - iP_{\mathbf{k}\lambda}],$$

which transform the field energy into

$$W = \frac{1}{2}\sum_{\mathbf{k}\lambda}(P^2_{\mathbf{k}\lambda} + \omega_k^2 Q^2_{\mathbf{k}\lambda}). \tag{4.59}$$

With the time-dependence of $A_{\mathbf{k}\lambda}(t)$ as in Eq. (4.52),

$$\dot{Q}_{\mathbf{k}\lambda} = P_{\mathbf{k}\lambda}, \qquad \dot{P}_{\mathbf{k}\lambda} = -\omega_k^2 Q_{\mathbf{k}\lambda}, \tag{4.60}$$

and

$$\frac{\partial W}{\partial Q_{\mathbf{k}\lambda}} = \omega_k^2 Q_{\mathbf{k}\lambda} = -\dot{P}_{\mathbf{k}\lambda}, \qquad \frac{\partial W}{\partial P_{\mathbf{k}\lambda}} = P_{\mathbf{k}\lambda} = \dot{Q}_{\mathbf{k}\lambda}. \tag{4.61}$$

Thus, the field energy W, when written in the form of Eq. (4.59), satisfies the classical Hamilton Eqs. (4.61); hence, $Q_{\mathbf{k}\lambda}$ and $P_{\mathbf{k}\lambda}$ are canonical variables and Eq. (4.59) is the Hamiltonian function for the electromagnetic field, that is,

$$\mathscr{H} = \frac{1}{2}\sum_{\mathbf{k}\lambda}(P^2_{\mathbf{k}\lambda} + \omega^2 Q^1_{\mathbf{k}\lambda}). \tag{4.62}$$

The expression for $\mathscr{H}$ is identical with the Hamiltonian for an assembly of simple harmonic oscillators. Each mode of the radiation field therefore is formally equivalent to a single harmonic oscillator, and it is this feature that facilitates the transition from the classical electromagnetic field to the quantized field.

For future reference, it is worth noting that a plane wave

$$\mathbf{A}(\mathbf{r}, t) = \hat{\boldsymbol{\varepsilon}}_{\mathbf{k}} A_{\mathbf{k}} e^{i(\mathbf{k}\cdot\mathbf{r} - \omega_k t)}, \tag{4.63}$$

with its polarization vector perpendicular to the direction of propagation satisfies

$$\nabla \cdot \mathbf{A}(\mathbf{r}, t) = i\mathbf{k} \cdot \hat{\boldsymbol{\varepsilon}}_{\mathbf{k}} A_{\mathbf{k}} e^{i(\mathbf{k}\cdot\mathbf{r} - \omega_k t)} = 0. \tag{4.64}$$

Therefore, for the vector potential (Eq. (4.44)), $\nabla \cdot \mathbf{A} = 0$. An alternative statement is that the condition $\nabla \cdot \mathbf{A} = 0$, known as the *Columb gauge*, ensures the transversality of the fields.

The description of the electromagnetic field in terms of running plane waves, discussed in this section, is probably the simplest type of normal mode expansion for most applications. Nevertheless, the reader should be aware that other types of modes may be employed.

4.3 Harmonic Oscillator, Boson Operators

In the previous section, it was shown that the Hamiltonian of a radiation field is formally identical to the Hamiltonian of an assembly of simple harmonic oscillators. Since our aim is to quantize the radiation field, the present section summarizes the essential properties of the quantized simple harmonic oscillator.

A one-dimensional oscillator of mass m, frequency ω, displacement q, and momentum p is described by the Hamiltonian

$$\mathscr{H} = \frac{p^2}{2m} + \frac{m}{2}\omega^2 q^2. \tag{4.65}$$

With a change of variables to

$$P^2 = \frac{p^2}{m}, \qquad Q^2 = mq^2, \tag{4.66}$$

the Hamiltonian becomes

$$\mathscr{H} = \frac{1}{2}(P^2 + \omega^2 Q^2). \tag{4.67}$$

To obtain the quantized form, it is necessary, according to the fundamental axioms of quantum mechanics, to reinterpret P and Q as Hermitian operators obeying the commutation rule

$$[Q, P] = i\hbar \tag{4.68}$$

which is equivalent to the replacement of P by $-i\hbar\partial/\partial Q$. With the definitions

$$a = \frac{1}{\sqrt{2\hbar\omega}}(\omega Q + iP), \qquad a^\dagger = \frac{1}{\sqrt{2\hbar\omega}}(\omega Q - iP), \tag{4.69}$$

$$Q = \sqrt{\frac{\hbar}{2\omega}}(a^\dagger + a), \qquad P = i\sqrt{\frac{\hbar\omega}{2}}(a^\dagger - a), \tag{4.70}$$

one arrives at the commutation rule

$$[a, a^\dagger] = 1, \tag{4.71}$$

characteristic of particles that obey Bose-Einstein statistics, or *bosons* for short. In terms of a and $a^\dagger$, the Hamiltonian becomes

$$\mathscr{H} = \tfrac{1}{2}\hbar\omega(a^\dagger a + aa^\dagger) = \hbar\omega(a^\dagger a + \tfrac{1}{2}) = \hbar\omega(N + \tfrac{1}{2}) \tag{4.72}$$

where

$$N \equiv a^\dagger a. \tag{4.73}$$

The non-Hermitian operators a and $a^\dagger$ are known as *annihilation* and *creation* operators, respectively. The Hermitian operator N is called the *number* operator; its real eigenvalues are designated by n and the corresponding eigenstates by $|n\rangle$. Thus,

$$N|n\rangle = n|n\rangle. \tag{4.74}$$

For future reference, we list a number of properties of the quantized harmonic oscillator:

1. n is a positive integer: $0, 1, 2, \ldots$.

2. The eigenstates $|n\rangle$, known as *occupation number* states or *Fock* states, satisfy

$$\mathscr{H}|n\rangle = \hbar\omega(n + \tfrac{1}{2})|n\rangle, \qquad \langle n|n'\rangle = \delta'_{nn} \tag{4.75}$$

and may be represented by column matrices:

$$|0\rangle = \begin{pmatrix} 1 \\ 0 \\ 0 \\ 0 \\ \vdots \end{pmatrix}, \qquad |1\rangle = \begin{pmatrix} 0 \\ 1 \\ 0 \\ 0 \\ \vdots \end{pmatrix}, \qquad |2\rangle = \begin{pmatrix} 0 \\ 0 \\ 1 \\ 0 \\ \vdots \end{pmatrix}, \qquad \text{etc.} \tag{4.76}$$

3. The energy of the oscillator, from Eq. (4.72), is

$$E = \hbar\omega\langle n|N + \tfrac{1}{2}|n\rangle = \hbar\omega(n + \tfrac{1}{2}). \tag{4.77}$$

In the lowest energy state, $n = 0$ and $E = \frac{1}{2}\hbar\omega$; this is known as the zero point energy.

4. The annihilation and creation operators are so called because of the properties

$$\begin{aligned} &a|n\rangle = \sqrt{n}|n-1\rangle, \qquad a^\dagger|n\rangle = \sqrt{n+1}|n+1\rangle, \\ &a|0\rangle = 0, \qquad a|1\rangle = |0\rangle, \qquad a^\dagger|0\rangle = |1\rangle, \end{aligned} \tag{4.78a}$$

$$\begin{aligned} &\langle n|a = \sqrt{n+1} < n+1|, \qquad \langle n|a^\dagger = \sqrt{n} < n-1|, \\ &\langle 0|a = \langle 1|, \qquad \langle 1|a^\dagger = \langle 0|, \qquad \langle 0|a^\dagger = 0. \end{aligned} \tag{4.78b}$$

The distinction between the number 0 and the lowest state of the harmonic oscillator described by the ket $|0\rangle$ or bra $\langle 0|$) should be noted.

5. Any state $|n\rangle$ may be generated from the state $|0\rangle$ by repeated application of the creation operator $a^\dagger$:

$$|n\rangle = \frac{1}{\sqrt{n!}}(a^\dagger)^n|0\rangle. \tag{4.79}$$

6. The commutation rule (Eq. (4.71)) leads to

$$aN^n = (N+1)^n a, \qquad a^\dagger N^n = (N-1)^n a^\dagger, \tag{4.80a}$$

which imply the relations

$$[a, \mathscr{H}] = \hbar\omega a, \qquad [a^\dagger, \mathscr{H}] = -\hbar\omega a t. \tag{4.80b}$$

and

$$a e^{i\omega t N} = e^{i\omega t(N+1)} a, \qquad a^\dagger e^{i\omega t N} = e^{i\omega t(N-1)} a^\dagger. \tag{4.80c}$$

7. Several important matrix elements are the following:

$$\begin{aligned} \langle n-1|a|n\rangle &= n^{1/2}, \\ \langle n+1|a^\dagger|n\rangle &= (n+1)^{1/2}, \\ \langle n|a^\dagger a|n\rangle &= \langle n|N|n\rangle = n, \\ \langle n|aa^\dagger|n\rangle &= \langle n|N+1|n\rangle = n+1, \\ \langle n|\mathscr{H}|n\rangle &= \hbar\omega(n+\tfrac{1}{2}), \\ \langle n|a|n\rangle &= \langle n|a^\dagger|n\rangle = 0, \\ \langle n|aa|n\rangle &= \langle n|a^\dagger a^\dagger|n\rangle = 0. \end{aligned} \tag{4.81}$$

8. In the Heisenberg representation, a and $a^\dagger$ must satisfy Eq. (2.107). With $\mathscr{H} = \mathscr{H}_H = \hbar\omega(a_H^\dagger a_H + 1/2)$,

$$\begin{aligned} \dot{a}_{\mathrm{H}}(t) &= -\frac{i}{\hbar}[a_{\mathrm{H}}(t), \mathscr{H}] = -i\omega a_{\mathrm{H}}(t), \\ \dot{a}_{\mathrm{H}}^\dagger(t) &= -\frac{i}{\hbar}[a_{\mathrm{H}}^\dagger(t), \mathscr{H}] = i\omega a_{\mathrm{H}}^\dagger(t), \end{aligned} \tag{4.82}$$

or

$$a_{\rm H}(t) = a_{\rm H}(0)e^{-i\omega t} = ae^{-i\omega t},$$
$$a_{\rm H}^\dagger(t) = a_{\rm H}^\dagger(0)e^{i\omega t} = a^\dagger e^{i\omega t}. \quad (4.83)$$

These equations also be obtained on the basis of the following theorem [3]. Let

$$F(x) = e^{xa^\dagger a}ae^{-xa^\dagger a} \quad (4.84a)$$

$$\frac{dF}{dx} = e^{xa^\dagger a}(a^\dagger aa - aa^\dagger a)e^{-xa^\dagger a}. \quad (4.84b)$$

With the help of the commutation rule (Eq. (4.71)),

$$\frac{dF}{dx} = -e^{xa^\dagger a}ae^{-xa^\dagger a} = -F(x) \quad (4.85)$$

so that

$$F(x) = F(0)e^{-x} \quad (4.86a)$$

or

$$e^{xa^\dagger a}ae^{-xa^\dagger a} = ae^{-x}. \quad (4.86b)$$

We now may apply this theorem to transform a and $a^\dagger$ to the Heisenberg representation. According to Eq. (2.105), with $t_0 = 0$,

$$a_{\rm H}(t) = e^{i\mathscr{H}t/\hbar}ae^{-i\mathscr{H}t/\hbar}. \quad (4.87)$$

Substituting Eq. (4.72) for the Hamiltonian and referring to Eq. (4.86) with $x = i\omega t$,

$$a_{\rm H}(t) = e^{i\omega ta^\dagger a}ae^{-i\omega ta^\dagger a} = ae^{-i\omega t}. \quad (4.88)$$

The conversion of $a^\dagger$ follows in the same fashion.

9. In the interaction representation, with $\mathscr{H} = \mathscr{H}_0 + V$ and $\mathscr{H}_0 = \hbar\omega(a^\dagger a + 1/2)$,

$$\tilde{a}(t) = ae^{-i\omega t}, \qquad \tilde{a}^\dagger(t) = a^\dagger e^{i\omega t}. \quad (4.89)$$

10. There are several definitions of boson operator products. Consider, for example, the product $A = a^\dagger aa^\dagger$. By use of the commutation rule $[a, a^\dagger] = 1$, we may write A in either of the two forms

$$A_n = a^\dagger a^\dagger a + a^\dagger, \qquad A_a = aa^\dagger a^\dagger - a^\dagger. \quad (4.90)$$

In A_n all creation operators are placed to the left of annihilation operators, while in A_a the operators are reversed in position—creation operators to the right of annihilation operators. A_n is said to be in *normal order* and A_a in *antinormal order*. Although the operator products are written in different forms, they are nevertheless equal, i.e.,

$$A = A_n = A_a. \tag{4.91}$$

Operators in normal order, such as A_n, are often written $:A:$. To generalize these statements, any function $f(a, a^\dagger)$ that can be expressed as a power series in a and $a^\dagger$ may be converted by means of the commutation rule into normal order

$$:f(a, a^\dagger): \equiv f_n(a, a^\dagger) = \sum_{ij} \alpha_{ij} (a^\dagger)^i a^j \tag{492a}$$

or antinormal order

$$f_a(a, a^\dagger) = \sum_{ij} \beta_{ij} a^i (a^\dagger)^j, \tag{4.92b}$$

where α_{ij} and β_{ij} are numerical coefficients. Thus, if

$$f(a, a^\dagger) = (a + a^\dagger)^2 = a^2 + aa^\dagger + a^\dagger a + a^{\dagger 2}, \tag{4.93a}$$

the normal order is

$$:f(a, a^\dagger): = a^2 + 2a^\dagger a + a^{\dagger 2} + 1, \tag{4.93b}$$

and the antinormal order is

$$f_a(a, a^\dagger) = a^2 + 2aa^\dagger + a^{\dagger 2} - 1. \tag{4.93c}$$

4.4 Quantized Fields, Photon-Number States

Now that we have the quantized form of the harmonic oscillator as described in the previous section, the process of quantizing the radiation field [1–4] becomes a matter of simply reinterpreting the canonical variables $Q_{\mathbf{k}\lambda}$ and $P_{\mathbf{k}\lambda}$ in Eq. (4.62) as operators that satisfy commutation relations similar to Eq. (4.68). We therefore postulate

$$\begin{aligned} [Q_{\mathbf{k}\lambda}, P_{\mathbf{k}'\lambda'}] &= i\hbar\, \delta_{\mathbf{k}\mathbf{k}'} \delta_{\lambda\lambda'}, \\ [Q_{\mathbf{k}\lambda}, Q_{\mathbf{k}'\lambda'}] &= [P_{\mathbf{k}\lambda}, P_{\mathbf{k}'\lambda'}] = 0. \end{aligned} \tag{4.94}$$

The annihilation and creation operators, defined by

$$a_{\mathbf{k}\lambda} = \frac{1}{\sqrt{2\hbar\omega_k}} (\omega_k Q_{\mathbf{k}\lambda} + iP_{\mathbf{k}\lambda}), \qquad a_{\mathbf{k}\lambda}^\dagger = \frac{1}{\sqrt{2\hbar\omega_k}} (\omega_k Q_{\mathbf{k}\lambda} - iP_{\mathbf{k}\lambda}), \tag{4.95a}$$

$$Q_{\mathbf{k}\lambda} = \sqrt{\frac{\hbar}{2\omega_k}} (a_{\mathbf{k}\lambda}^\dagger + a_{\mathbf{k}\lambda}), \qquad P_{\mathbf{k}\lambda} = i\sqrt{\frac{\hbar\omega_k}{2}} (a_{\mathbf{k}\lambda}^\dagger - a_{\mathbf{k}\lambda}), \tag{4.95b}$$

then satisfy the boson commutation rules

$$[a_{\mathbf{k}\lambda}, a^\dagger_{\mathbf{k}'\lambda'}] = \delta_{\mathbf{k}\mathbf{k}'}\,\delta_{\lambda\lambda'},$$
$$[a_{\mathbf{k}\lambda}, a_{\mathbf{k}'\lambda'}] = [a^\dagger_{\mathbf{k}\lambda}, a^\dagger_{\mathbf{k}'\lambda'}] = 0. \tag{4.96}$$

As in the case of the harmonic oscillator, the Hamiltonian for the radiation field can be written in terms of the annihilation and creation operators

$$\mathscr{H} = \sum_{\mathbf{k}\lambda} \mathscr{H}_{\mathbf{k}\lambda} = \sum_{\mathbf{k}\lambda} \hbar\omega_k\left(a^\dagger_{\mathbf{k}\lambda}a_{\mathbf{k}\lambda} + \frac{1}{2}\right) = \sum_{\mathbf{k}\lambda} \hbar\omega_k\left(N_{\mathbf{k}\lambda} + \frac{1}{2}\right) \tag{4.97}$$

in which

$$N_{\mathbf{k}\lambda} = a^\dagger_{\mathbf{k}\lambda}a_{\mathbf{k}\lambda} \tag{4.98}$$

is the *photon number* or *occupation number* operator for the $\mathbf{k}\lambda$ mode of the radiation field. The eigenvalue equations

$$N_{\mathbf{k}\lambda}|n_{\mathbf{k}\lambda}\rangle = n_{\mathbf{k}\lambda}|n_{\mathbf{k}\lambda}\rangle, \tag{4.99a}$$

$$\mathscr{H}_{\mathbf{k}\lambda}|n_{\mathbf{k}\lambda}\rangle = E_{k\lambda}|n_{\mathbf{k}\lambda}\rangle = \hbar\omega_k(n_{\mathbf{k}\lambda} + \tfrac{1}{2}), \tag{4.99b}$$

with

$$\langle n_{k\lambda}|n_{k'\lambda'}\rangle = \delta_{kk'}\,\delta_{\lambda\lambda'} \tag{4.100}$$

define the *photon-number* states $|n_{k\lambda}\rangle$ whose eigenvalues $n_{k\lambda}$ are positive integers, 0, 1, 2,.... By analogy with Eq. (4.98), we have

$$a_{\mathbf{k}\lambda}|n_{\mathbf{k}\lambda}\rangle = \sqrt{n_{\mathbf{k}\lambda}}\,|n_{\mathbf{k}\lambda} - 1\rangle, \qquad a_{\mathbf{k}\lambda}|0\rangle = 0,$$
$$a^\dagger_{\mathbf{k}\lambda}|n_{\mathbf{k}\lambda}\rangle = \sqrt{n_{\mathbf{k}\lambda} + 1}\,|n_{\mathbf{k}\lambda} + 1\rangle. \tag{4.101}$$

Also, as in Eq. (4.80b),

$$[a_{\mathbf{k}\lambda}, \mathscr{H}] = \hbar\omega_k a_{\mathbf{k}\lambda}, \qquad [a^\dagger_{\mathbf{k}\lambda}, \mathscr{H}] = -\hbar\omega_k a^\dagger_{\mathbf{k}\lambda}. \tag{4.102}$$

The transition to quantum mechanics by means of the preceding formalism lends itself to the following interpretation: $a_{\mathbf{k}\lambda}$ and $a^\dagger_{\mathbf{k}\lambda}$ are annihilation and creation operators, respectively, for a *photon* with propagation vector $\mathbf{k}$, polarization $\hat{\varepsilon}_{\mathbf{k}\lambda}$, momentum $\hbar\mathbf{k}$, frequency ω_k, and energy $\hbar\omega_k$. The frequency ω_k depends on the magnitude k but is independent of the direction of propagation and the polarization. In the quantized field, $n_{k\lambda}$ gives the number of photons characterized by $\mathbf{k}$ and λ.

It should be recognized that a pure photon-number state is an abstraction not unlike that of an ideal classical monochromatic electromagnetic wave. Neither can be realized in practice; nevertheless, the concepts are very useful and help us to understand the behavior of real systems.

The operators $a_{\mathbf{k}\lambda}$ and $a^\dagger_{\mathbf{k}\lambda}$ are in the Schrödinger representation. To convert them to the Heisenberg and interaction representations, one simply may follow the pattern established by Eqs. (4.83) and (4.89).

$$\{a_{\mathbf{k}\lambda}(t)\}_{\mathrm{H}} = a_{\mathbf{k}\lambda}e^{-i\omega_k t}, \qquad \{a^{\dagger}_{\mathbf{k}\lambda}(t)\}_{\mathrm{H}} = a^{\dagger}_{\mathbf{k}\lambda}e^{i\omega_k t}. \tag{4.103}$$

$$\tilde{a}_{\mathbf{k}\lambda}(t) = a_{\mathbf{k}\lambda}e^{-i\omega_k t}, \qquad \tilde{a}^{\dagger}_{\mathbf{k}\lambda}(t) = a^{\dagger}_{\mathbf{k}\lambda}e^{i\omega_k t}. \tag{4.104}$$

A complete specification of the quantized radiation field consists of an enumeration of the photon numbers $n_{\mathbf{k}\lambda}$. Since each mode is independent, and the total Hamiltonian $\mathscr{H}$ is the sum of all the partial Hamiltonians $\mathscr{H}_{\mathbf{k}\lambda}$, the product of all the eigenstates $|n_{\mathbf{k}\lambda}\rangle$ is an eigenstate of $\mathscr{H}$. A multimode photon-number state of the field therefore is written

$$|n_{\mathbf{k}_1\lambda_1}\rangle|n_{\mathbf{k}_2\lambda_2}\rangle\cdots|n_{\mathbf{k}_j\lambda_j}\rangle\cdots \equiv |n_{\mathbf{k}_1\lambda_1}, n_{\mathbf{k}_2\lambda_2},\ldots,n_{\mathbf{k}_j\lambda_j},\ldots\rangle. \tag{4.105}$$

In a less cumbersome notation, with

$$n_{\mathbf{k}_j\lambda_j} \equiv n_j, \tag{4.106}$$

the multimode photon-number state is written

$$|n_1, n_2,\ldots,n_j,\ldots\rangle = |\{n_i\}\rangle = \prod_i |n_i\rangle. \tag{4.107}$$

A state of this type is an eigenstate of the field Hamiltonian (Eq. (4.97))

$$\mathscr{H}|\{n_i\}\rangle = E|\{n_i\}\rangle, \qquad E = \sum_i \hbar\omega_i\left(n_i + \frac{1}{2}\right), \tag{4.108}$$

and is subject to the orthogonality condition

$$\langle n_1, n_2,\ldots,n_j,\ldots|n'_1, n'_2,\ldots,n'_j,\ldots\rangle = \delta_{n_1n'_1}\delta_{n_2n'_2}\cdots\delta_{n_jn'_j}\cdots. \tag{4.109}$$

With the definitions

$$a_j \equiv a_{\mathbf{j}_j\lambda'_j} \qquad a^{\dagger}_j \equiv a^{\dagger}_{\mathbf{k}_j\lambda'_j} \tag{4.110}$$

the relations analogous to Eq. (4.78a) are

$$\begin{aligned} a_j|n_1, n_2,\ldots,n_j,\ldots\rangle &= n_j^{1/2}|n_1, n_2,\ldots,n_j - 1,\ldots\rangle, \\ a_j|n_1, n_2,\ldots,0_j,\ldots\rangle &= 0 \\ a^{\dagger}_j|n_1, n_2,\ldots,n_j,\ldots\rangle &= (n_j + 1)^{1/2}|n_1, n_2,\ldots,n_j + 1,\ldots\rangle. \end{aligned} \tag{4.111}$$

When a particular $n_{\mathbf{k}\lambda}$ is zero, the mode $\mathbf{k}\lambda$ is in its lowest state and when $n_{\mathbf{k}\lambda} = 0$ for all $\mathbf{k}$, the field as a whole is said to be in the ground or *vacuum* state, designated by

$$|0\rangle \equiv |\mathrm{vac}\rangle \equiv |0_1, 0_2,\ldots,0_j,\ldots\rangle. \tag{4.112}$$

For this state,

$$a_j|0\rangle = 0 \tag{4.113}$$

for all j, and by repeated use of the creation operator, it is possible to generate an arbitrary state of the field,

$$|\{n_i\}\rangle = \prod_i \frac{(a_i^\dagger)^{n_i}}{\sqrt{n_i!}}|0\rangle. \tag{4.114}$$

In the course of the transition from classical to quantum mechanics, embodied in the commutation rules (Eq. (4.94) or (4.96)), the classical vector potential (Eq. (4.44)) becomes a quantum mechanical operator. This is a result of both the connection between $A_{\mathbf{k}\lambda}$, $A_{\mathbf{k}\lambda}^*$ and $Q_{\mathbf{k}\lambda}$, $P_{\mathbf{k}\lambda}$, as in Eq. (4.58) and the relation of the latter quantities to the annihilation and creation operators $a_{\mathbf{k}\lambda}$, $a_{\mathbf{k}\lambda}^\dagger$ given by Eq. (4.95). We now may express the classical $\to$ quantum mechanical transition as follows:

$$\begin{aligned} A_{\mathbf{k}\lambda}e^{i(\mathbf{k}\cdot\mathbf{r}-\omega_k t)} &\to \sqrt{\frac{\hbar}{2\varepsilon_0\omega_k V}}\, a_{\mathbf{k}\lambda}e^{i(\mathbf{k}\cdot\mathbf{r}-\omega_k t)} \\ &= \sqrt{\frac{\hbar}{2\varepsilon_0\omega_k V}}\,(a_{\mathbf{k}\lambda})_{\mathrm{H}}e^{i\mathbf{k}\cdot\mathbf{r}}, \end{aligned} \tag{4.115}$$

where the subscript H denotes that the operator is in the Heisenberg representation. A similar expression is written for the complex conjugate term; hence, for the complete vector potential, the transition from the classical to the quantum mechanical form is given by

$$\mathbf{A}(\mathbf{r}, t) \to \mathbf{A}_{\mathrm{H}}(\mathbf{r}, t) = \sum_{\mathbf{k}\lambda} \sqrt{\frac{\hbar}{2\varepsilon_0\omega_k V}}\,\hat{\boldsymbol{\varepsilon}}_{\mathbf{k}\lambda}[(a_{\mathbf{k}\lambda})_{\mathrm{H}}e^{i\mathbf{k}\cdot\mathbf{r}} + (a_{\mathbf{k}\lambda}^\dagger)_{\mathrm{H}}e^{-i\mathbf{k}\cdot\mathbf{r}}]. \tag{4.116}$$

Both sides of the quantum mechanical vector potential now may be transformed to the Schrödinger representation, giving

$$\mathbf{A}(r) = \sum_{\mathbf{k}\lambda} \sqrt{\frac{\hbar}{2\varepsilon_0\omega_k V}}\,\hat{\boldsymbol{\varepsilon}}_{\mathbf{k}\lambda}[a_{\mathbf{k}\lambda}e^{i\mathbf{k}\cdot\mathbf{r}} + a_{\mathbf{k}\lambda}^\dagger e^{-i\mathbf{k}\cdot\mathbf{r}}]. \tag{4.117}$$

The quantized forms of **E** and **B** follow directly from the basic relations $\mathbf{E} = -\partial\mathbf{A}/\partial t$ and $\mathbf{B} = \nabla \times \mathbf{A}$. To obtain the electric field **E**, we revert to the Heisenberg representation and the corresponding equation of motion:

$$\mathbf{E}_{\mathrm{H}}(\mathbf{r}, t) = -\frac{\partial \mathbf{A}_{\mathrm{H}}(\mathbf{r}, t)}{\partial t} = \frac{i}{\hbar}[\mathbf{A}_{\mathrm{H}}(\mathbf{r}, t), \mathscr{H}]. \tag{4.118}$$

With $\mathbf{A}_{\mathrm{H}}(\mathbf{r}, t)$ provided by Eq. (4.116), the commutator is evaluated by referring to Eq. (4.82). The result is

$$\mathbf{E}_{\mathrm{H}}(\mathbf{r}, t) = i\sum_{k\lambda} \sqrt{\frac{\hbar\omega_k}{2\varepsilon_0 V}}\,\hat{\boldsymbol{\varepsilon}}_{\mathbf{k}\lambda}[(a_{\mathbf{k}\lambda})_{\mathrm{H}}e^{i\mathbf{k}\cdot\mathbf{r}} - (a_{\mathbf{k}\lambda}^\dagger)_{\mathrm{H}}e^{-i\mathbf{k}\cdot\mathbf{r}}], \tag{4.119a}$$

and the corresponding electric field in the Schrödinger representation is

$$\mathbf{E}(\mathbf{r}) = i\sum_{\mathbf{k}\lambda}\sqrt{\frac{\hbar\omega_k}{2\varepsilon_0 V}}\,\hat{\boldsymbol{\varepsilon}}_{\mathbf{k}\lambda}[a_{\mathbf{k}\lambda}e^{i\mathbf{k}\cdot\mathbf{r}} - a^{\dagger}_{\mathbf{k}\lambda}e^{-i\mathbf{k}\cdot\mathbf{r}}]. \tag{4.119b}$$

The **B** field is derived directly from Eq. (4.117):

$$\mathbf{B}(\mathbf{r}) = \nabla \times \mathbf{A}(\mathbf{r})$$

$$= \frac{i}{c}\sum_{\mathbf{k}\lambda}(\hat{\mathbf{k}} \times \hat{\boldsymbol{\varepsilon}}_{\mathbf{k}\lambda})\sqrt{\frac{\hbar\omega_k}{2\varepsilon_0 V}}\,[a_{\mathbf{k}\lambda}e^{i\mathbf{k}\cdot\mathbf{r}} - a^{\dagger}_{\mathbf{k}\lambda}e^{-i\mathbf{k}\cdot\mathbf{r}}]. \tag{4.120}$$

It is often convenient to write **E**(**r**) as the sum of two terms:

$$\mathbf{E}^{(+)}(\mathbf{r}) = i\sum_{\mathbf{k}\lambda}\sqrt{\frac{\hbar\omega_k}{2\varepsilon_0 V}}\,\hat{\boldsymbol{\varepsilon}}_{\mathbf{k}\lambda}a_{\mathbf{k}\lambda}e^{i\mathbf{k}\cdot\mathbf{r}}, \tag{4.121a}$$

$$\mathbf{E}^{(-)}(\mathbf{r}) = [\mathbf{E}^{(+)}(\mathbf{r})]^{\dagger}. \tag{4.121b}$$

Similarly, for **B**(**r**) we define

$$\mathbf{B}^{(+)}(\mathbf{r}) = \frac{i}{c}\sum_{\mathbf{k}\lambda}(\hat{\mathbf{k}} \times \hat{\boldsymbol{\varepsilon}}_{\mathbf{k}\lambda})\sqrt{\frac{\hbar\omega_k}{2\varepsilon_0 V}}\,a_{\mathbf{k}\lambda}e^{i\mathbf{k}\cdot\mathbf{r}}, \tag{4.122a}$$

$$\mathbf{B}^{(-)}(\mathbf{r}) = [\mathbf{B}^{(+)}(\mathbf{r})]^{\dagger}. \tag{4.122b}$$

(In these definitions, the superscript $(+)$ is associated with the annihilation operator a and the superscript $(-)$ is associated with $a^{\dagger}$. This convention is not universal: some authors relate $(+)$ with $a^{\dagger}$ and $(-)$ with a.)

In Section 4.1 we called attention to the rotational properties of the electromagnetic field. These features are carried along in the process of quantization. In the context of the photon picture, we say that the photon has an intrinsic spin $S = 1$ with $M_S = \pm 1$; the value $M_S = 0$ is missing on account of the transversality condition. For a given mode, the z axis may be chosen to lie in the direction of propagation, that is, in the direction of the vector **k**. If the z axis also serves as the axis of quantization, the two possible values of M_S indicate that the spin vector has only two possible orientations—parallel or antiparallel to the photon momentum $\hbar\mathbf{k}$. In the parallel case, the photon is said to have positive *helicity* $(+1)$ or left circular polarization represented by the complex unit polarization vector

$$\hat{\boldsymbol{\varepsilon}}_{\mathbf{k},+1} = -\frac{1}{\sqrt{2}}(\hat{\boldsymbol{\varepsilon}}_{\mathbf{k}1} + i\hat{\boldsymbol{\varepsilon}}_{\mathbf{k}2}). \tag{4.123a}$$

For the antiparallel case, the photon has negative helicity (-1) or right circular polarization given by the vector

$$\hat{\boldsymbol{\varepsilon}}_{\mathbf{k},-1} = \frac{1}{\sqrt{2}}(\hat{\boldsymbol{\varepsilon}}_{\mathbf{k}1} - i\hat{\boldsymbol{\varepsilon}}_{\mathbf{k}2}). \tag{4.123b}$$

The two polarization vectors given above are merely generalizations of the corresponding two vectors in Eq. (4.35).

The photon picture that emerges from the quantization of the electromagnetic field contains a number of singularities that make the theory somewhat less than totally consistent internally. The expression for the energy in Eq. (4.108), for example, contains a zero point energy of $1/2\hbar\omega$ for each mode, even when the photon number n is zero. Since there are an infinite number of modes, the total energy is infinite for every state $|n_1, n_2, \ldots, n_j, \ldots\rangle$, including the vacuum state in Eq. (4.112), which has no photons at all.

The expectation value of the electric field and its square also give rise to unexpected results when viewed from a classical standpoint. Thus, for a single mode with photon number n and frequency ω, the expectation value of $\mathbf{E}(\mathbf{r})$ is

$$\langle n|\mathbf{E}(\mathbf{r})|n\rangle = \langle n|\mathbf{E}^{(+)}(\mathbf{r})|n\rangle + \langle n|\mathbf{E}^{(-)}(\mathbf{r})|n\rangle. \tag{4.124}$$

The two terms on the right are proportional to $\langle n|a|n\rangle$ and $\langle n|a^\dagger|n\rangle$, respectively, both of which are zero as one easily may verify on the basis of Eq. (4.101) or (4.81). Hence,

$$\langle n|\mathbf{E}(\mathbf{r})|n\rangle = 0 \tag{4.125}$$

for all values of the photon number, no matter how large. This result holds for all modes, which means, then, that the expectation value of the electric field in any many-photon state is zero. On the other hand, the expectation value of $E^2(\mathbf{r})$ is proportional to $\langle n|a^\dagger a + aa^\dagger|n\rangle$, which leads to the conclusion that for a single mode

$$\langle n|E^2(\mathbf{r})|n\rangle = \frac{\hbar\omega}{\varepsilon_0 V}\left(n + \frac{1}{2}\right). \tag{4.126}$$

For this case, as for the energy, the expectation value is nonvanishing even when $n = 0$, with the result that for the total field, consisting of an infinite number of modes, the expectation value of $E^2(\mathbf{r})$ is infinite for all many-photon states including the vacuum state.

The same comments apply to the mean square fluctuation (variance), which is defined by

$$\begin{aligned}(\Delta E)^2 &= \langle n|E^2(\mathbf{r})|n\rangle - (\langle n|E(\mathbf{r})|n\rangle)^2 \\ &= \langle n|E^2(\mathbf{r})|n\rangle = \frac{\hbar\omega}{\varepsilon_0 V}\left(n + \frac{1}{2}\right).\end{aligned} \tag{4.127}$$

These and other shortcomings notwithstanding, the quantum theory of radiation is extremely successful; in the low energy regime particularly, which will be our main concern, the singularities are of little consequence and can be ignored in most cases.

4.5 Coherent States

Thus far, the quantum mechanical description of the radiation field has been based on the photon-number formalism. For some theoretical discussions, an alternative basis is preferable. The radiation from a well-stabilized laser oscillator operating in a single mode, for example, is not in a pure photon-number state. To a good approximation, such radiation is in a *coherent* state [5–8] whose properties we now will develop. We retain the same general picture of a radiation field confined within a cubical cavity and described as a superposition of quantized normal modes. Since the latter are independent, the discussion may be simplified by focusing attention on a single mode.

A coherent state $|\alpha\rangle$ is defined as a normalized eigenstate of the annihilation operator,

$$a|\alpha\rangle = \alpha|\alpha\rangle \tag{4.128}$$

with

$$\langle\alpha|\alpha\rangle = 1. \tag{4.129}$$

Here $|\alpha\rangle$ is the eigenstate and α is the eigenvalue that may be complex since the annihilation operator is non-Hermitian. Note that we have not imposed an orthogonality condition on two coherent states, $|\alpha\rangle$ and $|\beta\rangle$. The initial objective is to express $|\alpha\rangle$ in terms of the photon-number states $|n\rangle$, that is, to evaluate $\langle n|\alpha\rangle$ in the expansion

$$|\alpha\rangle = \sum_n |n\rangle\langle n|\alpha\rangle. \tag{4.130}$$

Since $a|n\rangle = n^{1/2}|n-1\rangle$ and $a|0\rangle = 0$,

$$\begin{aligned} a|\alpha\rangle &= \sum_{n=0}^{\infty} \langle n|\alpha\rangle a|n\rangle = \sum_{n=1}^{\infty} \langle n|\alpha\rangle n^{1/2}|n-1\rangle \\ &= \sum_{n=0}^{\infty} \langle n+1|\alpha\rangle (n+1)^{1/2}|n\rangle. \end{aligned} \tag{4.131}$$

Also, from Eq. (4.130),

$$a|\alpha\rangle = \alpha|\alpha\rangle = \sum_{n=0}^{\infty} \langle n|\alpha\rangle \alpha|n\rangle. \tag{4.132}$$

Equating the two expressions for $a|\alpha\rangle$ yields

$$\sum_n \langle n+1|\alpha\rangle (n+1)^{1/2}|n\rangle = \sum_n \langle n|\alpha\rangle a|n\rangle, \tag{4.133}$$

and upon premultiplying both sides by the photon-number state $\langle m|$ and invoking the orthogonality property $\langle m|n\rangle = \delta_{mn}$, we obtain the recursion relation

$$\langle n+1|\alpha\rangle = \frac{\langle n|\alpha\rangle}{\sqrt{n+1}}\alpha. \tag{4.134}$$

All the coefficients $\langle n|\alpha\rangle$ may be expressed now in terms of the single coefficient $c_0 \equiv \langle 0|\alpha\rangle$:

$$\langle n|\alpha\rangle = \frac{\alpha^n}{\sqrt{n!}}c_0. \tag{4.135}$$

Substituting in Eq. (4.130),

$$|\alpha\rangle = c_0 \sum_{n=0} \frac{\alpha^n}{\sqrt{n!}}|n\rangle. \tag{4.136}$$

To evaluate c_0 we use the normalization condition $\langle\alpha|\alpha\rangle = 1$ and the orthogonality property of the photon-number states. Thus,

$$\langle\alpha|\alpha\rangle = |c_0|^2 \sum_{m,n} \frac{(\alpha^*)^m\alpha^n}{\sqrt{m!n!}}\langle m|n\rangle = |c_0|^2 \sum_n \frac{(|\alpha|^2)^n}{n!}$$

$$= |c_0|^2 e^{|\alpha|^2} = 1 \tag{4.137}$$

or

$$c_0 = e^{-1/2|\alpha|^2}. \tag{4.138}$$

Therefore, the expansion of the coherent state $|\alpha\rangle$ in terms of photon-number states is given by

$$|\alpha\rangle = e^{-1/2|\alpha|^2} \sum_n \frac{\alpha^n}{\sqrt{n!}}|n\rangle = e^{-1/2|\alpha|^2 + \alpha a^\dagger}|0\rangle, \tag{4.139}$$

where, in the last expression of this equation, we used the relation

$$|n\rangle = \frac{(a^\dagger)^n}{\sqrt{n!}}|0\rangle. \tag{4.140}$$

The relation $a|\alpha\rangle = \alpha|\alpha\rangle$ implies the conjugate relations

$$\langle\alpha|a^\dagger = \alpha^*\langle\alpha|, \tag{4.141}$$

in which

$$\langle\alpha| = e^{-1/2|\alpha|^2} \sum_n \frac{(\alpha^*)^n}{\sqrt{n!}}\langle n| = \langle 0|e^{-1/2|\alpha|^2 + \alpha^* a}. \tag{4.142}$$

From the basic definition of a coherent state as an eigenfunction of the annihilation operator, one obtains the relations

$$\langle\alpha|N|\alpha\rangle = \langle\alpha|a^\dagger a|\alpha\rangle = |\alpha|^2 \equiv \langle n\rangle_c, \tag{4.143a}$$

$$\langle\alpha|aa^\dagger|\alpha\rangle = \langle\alpha|N+1|\alpha\rangle = |\alpha|^2 + 1, \tag{4.143b}$$

$$\langle\alpha|N^2|\alpha\rangle = |\alpha|^4 + |\alpha|^2 \equiv \langle n^2\rangle_c \tag{4.143c}$$

$$\langle\alpha|(a^\dagger)^i a^j|\alpha\rangle = (\alpha^*)^i\alpha^j, \tag{4.143d}$$

in which $\langle\alpha|N|\alpha\rangle \equiv \langle n\rangle_c$ is the average number of photons in the coherent state $|\alpha\rangle$.

The expressions for $|\alpha\rangle$ and $\langle\alpha|$ given by Eqs. (4.139) and (4.142) enable us to compute the matrix elements

$$\langle n|\alpha\rangle = e^{-1/2|\alpha|^2}\frac{\alpha^n}{\sqrt{n!}}, \qquad \langle\alpha|n\rangle = e^{-1/2|\alpha|^2}\frac{(\alpha^*)^n}{\sqrt{n!}}, \tag{4.144}$$

$$|\langle n|\alpha\rangle|^2 = \frac{e^{-|\alpha|^2}|\alpha|^{2n}}{n!} = \frac{e^{-\langle n\rangle_c}\langle n\rangle_c^n}{n!} \tag{4.145}$$

The quantity $|\langle n|\alpha\rangle|^2$ is interpretable as the probability of finding n photons in a coherent state $|\alpha\rangle$; therefore, according to the last expression, the probability has a Poissonian distribution.

The coherent states are normalized, as we have seen, but they are not orthogonal; in fact,

$$\begin{aligned}\langle\alpha|\beta\rangle &= e^{-1/2|\alpha|^2}e^{-1/2|\beta|^2}\sum_{m,n}\frac{(\alpha^*)^m\beta^n}{\sqrt{m!n!}}\langle m|n\rangle \\ &= e^{-1/2(|\alpha|^2+|\beta|^2)}\sum_n\frac{(\alpha^*\beta)^n}{\sqrt{n!}} = e^{-1/2(|\alpha|^2+|\beta|^2)}e^{\alpha^*\beta}\end{aligned} \tag{4.146}$$

or

$$|\langle\alpha|\beta\rangle^2 = e^{-|\alpha-\beta|^2}. \tag{4.147}$$

Coherent states, although they lack the property of orthogonality, nevertheless satisfy a closure relation of a special type. We write

$$\frac{1}{\pi}\int|\alpha\rangle\langle\alpha|\,d^2\alpha = \frac{1}{\pi}\sum_{m,n}\frac{|m\rangle\langle n|}{\sqrt{m!n!}}\int e^{-|\alpha|^2}\alpha^m(\alpha^*)^n\,d^2\alpha, \tag{4.148}$$

where $d^2\alpha$ is an element of area in the complex plane, i.e.,

$$d^2\alpha = d(\mathrm{Re}\,\alpha)\,d(\mathrm{Im}\,\alpha), \tag{4.149a}$$

or, in polar coordinates,

$$\alpha = re^{i\theta}, \qquad d^2\alpha = r\,dr\,de. \tag{4.149b}$$

Then

$$\int e^{-|\alpha|^2}\alpha^m(\alpha^*)^n\,d^2\alpha = \int_0^\infty dre^{-r^2}r^{m+n+1}\int_0^{2\pi} e^{i(m-n)\theta}\,d\theta$$

$$= 2\pi\int_0^\infty dre^{-r^2}r^{m+n+1}\,\delta_{mn} = 2\pi\int_0^\infty dre^{-r^2}r^{2n+1}. \tag{4.150}$$

With a change of variable to $\xi = r^2$, the right side transforms to

$$\pi\int_0^\infty d\xi e^{-\xi}\xi^n = \pi n!. \tag{4.151}$$

Using this relation to evaluate Eq. (4.148) and remembering that $m = n$ as required by the θ integral, we obtain the closure relation for coherent states:

$$\frac{1}{\pi}\int |\alpha\rangle\langle\alpha|\,d^2\alpha = \sum_n |n\rangle\langle n| = 1. \tag{4.152}$$

An arbitrary state $|f\rangle$ then may be written

$$|f\rangle = \frac{1}{\pi}\int |\alpha\rangle\langle\alpha|f\rangle\,d^2\alpha. \tag{4.153}$$

As an example of Eq. (4.153) we may obtain an expression for a photon-number state $|n\rangle$ in a coherent state basis, that is, the inverse of Eq. (4.139). Thus,

$$|n\rangle = \frac{1}{\pi}\int |\alpha\rangle\langle\alpha|n\rangle\,d^2\alpha$$

$$= \frac{1}{\pi}\int d^2\alpha e^{-1/2|\alpha|^2}\frac{(\alpha^*)^n}{\sqrt{n!}}|\alpha\rangle, \tag{4.154}$$

where $\langle\alpha|n\rangle$ was obtained from Eq. (4.144). As another example, let $|f\rangle$ be a coherent state $|\beta\rangle$; then, owing to the nonorthogonality of the coherent states, one obtains the rather unusual result of a coherent state expressed as an integral over other coherent states.

$$|\beta\rangle = \frac{1}{\pi}\int |\alpha\rangle\langle\alpha|\beta\rangle\,d^2\alpha = \frac{1}{\pi}\int |\alpha\rangle e^{-1/2(|\alpha|^2+|\beta|^2)+\alpha^*\beta}\,d^2\alpha, \tag{4.155}$$

in which Eq. (4.146) for $\langle\alpha|\beta\rangle$ has been inserted in the integral. With the normalization condition and Eq. (4.147), we also have

$$\begin{aligned}\langle\beta|\beta\rangle &= \frac{1}{\pi}\int|\langle\alpha|\beta\rangle|\,d^2\alpha \\ &= \frac{1}{\pi}\int e^{-|\alpha-\beta|^2}\,d^2\alpha = 1.\end{aligned} \tag{4.156}$$

An arbitrary operator O also may be expressed in a coherent state basis. Two applications of the closure relation in Eq. (4.152) yields

$$\begin{aligned}O &= \frac{1}{\pi}\int O|\alpha\rangle\langle\alpha|\,d^2\alpha \\ &= \frac{1}{\pi^2}\int d^2\beta\int d^2\alpha|\beta\rangle\langle\beta|O|\alpha\rangle\langle\alpha|.\end{aligned} \tag{4.157}$$

But

$$\begin{aligned}\langle\beta|O|\alpha\rangle &= \sum_{m,n}\langle\beta|m\rangle\langle m|O|n\rangle\langle n|\alpha\rangle \\ &= e^{-1/2(|\alpha|^2+|\beta|^2)}\sum_{m,n}\langle m|O|n\rangle\frac{(\beta^*)^m\alpha^n}{\sqrt{m!n!}},\end{aligned} \tag{4.158}$$

in which, again, we have referred to Eq. (4.144) for $\langle\beta|m\rangle$ and $\langle n|\alpha\rangle$. It is convenient to define

$$R(\alpha,\beta^*) = \langle\beta|O|\alpha\rangle e^{1/2(|\alpha|^2+|\beta|^2)} = \sum_{m,n}\langle m|O|n\rangle\frac{(\beta^*)^m\alpha^n}{\sqrt{m!n!}} \tag{4.159}$$

so that the arbitrary operator O may be written in the form

$$O = \frac{1}{\pi^2}\int d^2\beta\int d^2\alpha R(\alpha,\beta^*)e^{-1/2(|\alpha|^2+|\beta|^2)}|\beta\rangle\langle\alpha|. \tag{4.160}$$

It also is observed that Eq. (4.152) enables us to write

$$\begin{aligned}\mathrm{Tr}\,O = \sum_n\langle n|O|n\rangle &= \frac{1}{\pi}\int d^2\alpha\sum_n\langle n|O|\alpha\rangle\langle\alpha|n\rangle \\ &= \frac{1}{\pi}\int d^2\alpha\sum_n\langle\alpha|n\rangle\langle n|O|\alpha\rangle = \frac{1}{\pi}\int d^2\alpha\langle\alpha|O|\alpha\rangle.\end{aligned} \tag{4.161}$$

A coherent state will evolve in time according to the general relation

$$|\alpha(t)\rangle = e^{-i\mathscr{H}t/\hbar}|\alpha\rangle \tag{4.162}$$

where $|\alpha\rangle \equiv |\alpha(O)\rangle$ and $\mathscr{H} = \hbar\omega(N + 1/2)$. Using the expansion (4.139) for the coherent state in terms of the photon-number states,

$$|\alpha(t)\rangle = \sum_n e^{-1/2|\alpha|^2} \frac{\alpha^n}{\sqrt{n!}} e^{-i\omega(N+1/2)t}|n\rangle$$
$$= e^{-i\omega t/2} \sum_n \frac{(\alpha e^{-i\omega t})}{\sqrt{n!}} e^{-1/2|\alpha|^2}|n\rangle = e^{-i\omega t/2}|\alpha e^{-i\omega t}\rangle, \qquad (4.163)$$

from which it may be concluded that a coherent state remains coherent as time progresses.

According to Eq. (4.125), we found that the expectation value of the electric field in a photon-number state vanishes. In a coherent state, however, the expectation value of the field in a single mode is

$$\langle\alpha|\mathbf{E}(\mathbf{r})|\alpha\rangle = i\sqrt{\frac{\hbar\omega}{2\varepsilon_0 V}}\hat{\varepsilon}(\langle\alpha|a|\alpha\rangle e^{i\mathbf{k}\cdot\mathbf{r}} - \langle\alpha|a^\dagger|\alpha\rangle e^{-i\mathbf{k}\cdot\mathbf{r}})$$
$$= i\sqrt{\frac{\hbar\omega}{2\varepsilon_0 V}}\hat{\varepsilon}(\alpha e^{i\mathbf{k}\cdot\mathbf{r}} - \alpha^* e^{-i\mathbf{k}\cdot\mathbf{r}}), \qquad (4.164)$$

which is not automatically zero. Indeed, $\langle\alpha|\mathbf{E}(\mathbf{r})|\alpha\rangle$ represents a classical traveling plane wave whose phase is that of α and whose amplitude is proportional to the absolute value of α which, according to Eq. (4.143a), is the square root of the average number of photons. Neither does the expectation

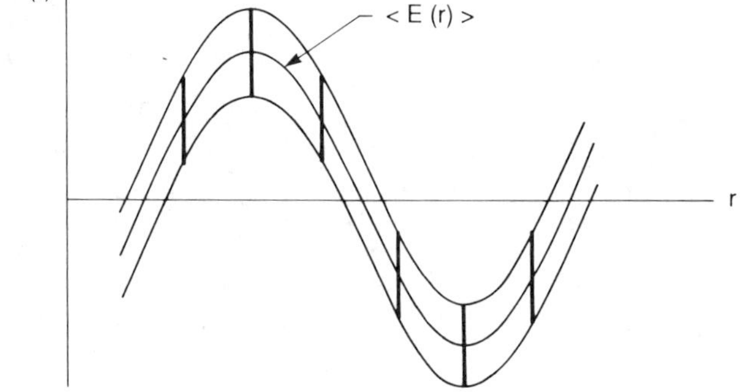

FIGURE 4.1 The middle curve is the expectation value $\langle E(r)\rangle = \langle\alpha|E(r)|\alpha\rangle$ of a (scalar) electric field in a coherent state $|\alpha\rangle$. The vertical bars indicate the root mean square fluctuation ΔE where $(\Delta E)^2 = \langle\alpha|E^2(r)|\alpha\rangle - (\langle\alpha|E(r)|\alpha\rangle)^2$. The rms fluctuation ΔE is independent of r and α.

value of $E^2(\mathbf{r})$ vanish; thus, when $\mathbf{k} \cdot \mathbf{r} \ll 1$, for example,

$$\langle\alpha|E^2(\mathbf{r})|\alpha\rangle = \frac{\hbar\omega}{2\varepsilon_0 V}(1 - (\alpha - \alpha^*)^2), \tag{4.165}$$

and the mean square fluctuation for this case is

$$(\Delta E)^2 = \langle\alpha|E^2(\mathbf{r})|\alpha\rangle - (\langle\alpha|\mathbf{E}(\mathbf{r})|\alpha\rangle)^2 = \frac{\hbar\omega}{2\varepsilon_0 V}. \tag{4.166}$$

This result (Fig. 4.1), which is finite and independent of α, contrasts sharply with the mean square fluctuation in the photon-number basis (Eq. (4.127)), which tends to infinity with increasing photon number n. Note, also, that $(\Delta E)^2$ for a coherent state (Eq. (4.166)) is equal to $(\Delta E)^2$ for a photon-number state (Eq. (4.127)) only when $n = 0$. That is, the quantum fluctuations in a coherent state are precisely the zero-point fluctuations of the vacuum.

When the electric field is written in the Heisenberg representation in order to include the time-dependence, it is seen that there is a close correspondence between a single mode coherent state and a classical plane wave. The multi-mode coherent state

$$|\alpha_1, \alpha_2, \ldots, \alpha_j, \ldots\rangle \equiv |\{\alpha_i\}\rangle = \prod_i |\alpha_i\rangle \tag{4.167}$$

corresponds to a superposition of plane waves.

4.6 Displacement Operator and Characteristic Functions

This section is devoted to the derivation of a number of results which reappear in connection with squeezed states (Section 4.9) and with photon statistics (Section 5.13).

Coherent states may be discussed in terms of a *translation* or *displacement* operator $D(\alpha)$ [8] defined by

$$D(\alpha) = e^{\alpha a^\dagger - \alpha^* a}. \tag{4.168}$$

Although $D(\alpha)$ cannot be written as a simple product of $e^{\alpha a^\dagger}$ and $e^{-\alpha^* a}$ a useful theorem exists known as the *disentangling theorem*, which is a special case of the Campbell-Baker-Hausdorff formula [3]. The theorem states that it A and B are two noncommuting operators that satisfy the conditions

$$[A, [A, B]] = [B, [A, B]] = 0, \tag{4.169}$$

then

$$e^{A+B} = e^A e^B e^{-1/2[A,B]} = e^B e A e^{1/2[A,B]}. \tag{4.170}$$

Letting $A = \alpha a^\dagger$, $B = -\alpha^* a$, and noting that $[a, a^\dagger] = 1$, we have

$$[A, B] = [\alpha a^\dagger, -\alpha^* a] = |\alpha|^2,$$
$$[A, [A, B]] = [\alpha a^\dagger, |\alpha|^2] = 0, \tag{4.171}$$
$$[B, [A, B]] = [-\alpha^* a, |\alpha|^2] = 0.$$

It now becomes possible to express $D(\alpha)$ in the form

$$D(\alpha) = e^{\alpha a^\dagger - \alpha^* a} = e^{-1/2|\alpha|^2} e^{\alpha a^\dagger} e^{-\alpha^* a} \tag{4.172a}$$
$$= e^{1/2|\alpha|^2} e^{-\alpha^* a} e^{\alpha a^\dagger}. \tag{4.172b}$$

Note that in Eq. (4.172a), the product of the exponentials is in normal order since the creation operators will appear to the left of annihilation operators when the exponentials are written as power series; similarly, Eq. (4.172b) is in antinormal order. The Hermitian conjugate of Eq. (4.172a) is

$$D^\dagger(\alpha) = e^{-1/2|\alpha|^2} e^{-\alpha a^\dagger} e^{\alpha^* a}, \tag{4.173}$$

and when this expression is premultiplied by $D(\alpha)$ from Eq. (4.172b), we have

$$D(\alpha) D^\dagger(\alpha) = 1. \tag{4.174}$$

Therefore, $D(\alpha)$ is a unitary operator:

$$D^\dagger(\alpha) = D^{-1}(\alpha) = e^{-1/2|\alpha|^2} e^{-\alpha a^\dagger} e^{\alpha^* a} = e^{1/2|\alpha|^2} e^{\alpha^* a} e^{-\alpha a^\dagger}$$
$$= e^{-\alpha a^\dagger + \alpha^* a} = D(-\alpha). \tag{4.175}$$

Another general operator theorem, derivable from Eq. (4.170), states that if A and B are two noncommuting operators,

$$e^A B e^{-A} = B + [A, B] + \frac{1}{2!}[A, [A, B]] + \frac{1}{3!}[A, [A, [A, B]]] + \cdots. \tag{4.176}$$

In the present context, let

$$A = -\alpha a^\dagger + \alpha^* a, \qquad B = a. \tag{4.177}$$

Then,

$$e^A = e^{-\alpha a^\dagger + \alpha^* a} = D^{-1}(\alpha),$$
$$e^{-A} = e^{\alpha a^\dagger - \alpha^* a} = D(\alpha), \tag{4.178}$$
$$[A, B] = -\alpha[a^\dagger, a] = \alpha,$$

and all higher order commutators vanish. Thus,

$$D^{-1}(\alpha) a D(\alpha) = a + \alpha, \qquad D^{-1}(\alpha) a^\dagger D(\alpha) = a^\dagger + \alpha^*, \tag{4.179a}$$

or

$$D(\alpha)aD^{-1}(\alpha) = a - \alpha, \qquad D(\alpha)a^{\dagger}D^{-1}(\alpha) = a^{\dagger} - \alpha^*. \tag{4.179b}$$

The reason for calling $D(\alpha)$ a translation or displacement operator now becomes apparent. When the annihilation and creation operators a and $a^{\dagger}$ are subjected to a unitary transformation by $D(\alpha)$, as in Eq. (4.179), a and $a^{\dagger}$ are augmented by α and α^* respectively.

The coherent state has been defined by the basic relation $a|\alpha\rangle = \alpha|\alpha\rangle$. Then, in view of Eq. (4.179), we may write

$$\begin{aligned} D^{-1}(\alpha)aD(\alpha)D^{-1}(\alpha)|\alpha\rangle &= D^{-1}(\alpha)a|\alpha\rangle \\ &= \alpha D^{-1}(\alpha)|\alpha\rangle = (a + \alpha)D^{-1}(\alpha)|\alpha\rangle \end{aligned} \tag{4.180}$$

which means that

$$aD^{-1}(\alpha)|\alpha\rangle = 0. \tag{4.181}$$

Since $a|0\rangle = 0$, it is possible to make the identification

$$D^{-1}(\alpha)|\alpha\rangle = |0\rangle \qquad \text{or} \qquad |\alpha\rangle = D(\alpha)|0\rangle. \tag{4.182}$$

In other words, a coherent state $|\alpha\rangle$ can be generated from the vacuum state $|0\rangle$ by means of the operator $D(\alpha)$ or, alternatively, a coherent state $|\alpha\rangle$ may be transformed to the vacuum state by $D^{-1}(\alpha)$. Indeed, the various properties of coherent states, derived in the previous section starting with Eq. (4.128), may be derived just as easily from Eq. (4.182) which then serves as the definition of a coherent state.

An alternative verification of Eq. (4.181) may be obtained by referring to the power series expansion of an exponential operator. Thus,

$$e^{-\alpha^* a}|0\rangle = \sum_n \frac{(-\alpha^* a)^n}{n!}|0\rangle = !0\rangle, \tag{4.183}$$

since only the term with $n = 0$ is nonvanishing, and

$$e^{\alpha a^{\dagger}}e^{-\alpha^* a}|0\rangle = e^{\alpha a^{\dagger}}|0\rangle = \sum_n \frac{(\alpha a^{\dagger})^n}{n!}|0\rangle = \sum_n \frac{\alpha^n}{\sqrt{n!}}|n\rangle. \tag{4.184}$$

Therefore, from Eqs. (4.172), (4.184), and (4.180),

$$\begin{aligned} D(\alpha)|0\rangle &= e^{-1/2|\alpha|2}e^{\alpha a^{\dagger}}e^{-\alpha^* a}|0\rangle = e^{-1/2|\alpha|2}\sum_n \frac{(\alpha a^{\dagger})^n}{n!}|0\rangle \\ &= e^{-1/2|\alpha|2}\sum_n \frac{\alpha^n}{\sqrt{n!}}|n\rangle = |\alpha\rangle. \end{aligned} \tag{4.185}$$

as in Eq. (4.182).

Characteristic functions are defined [8] as follows:

$$\chi(\lambda) = \mathrm{Tr}\{\rho D(\lambda)\} = \mathrm{Tr}\{\rho e^{\lambda a^\dagger - \lambda^* a}\}, \tag{4.186a}$$

$$\chi_n(\lambda) = \mathrm{Tr}\{\rho e^{\lambda a^\dagger} e^{-\lambda^* a}\}, \tag{4.186b}$$

$$\chi_a(\lambda) = \mathrm{Tr}\{\rho e^{-\lambda^* a} e^{\lambda a^\dagger}\}, \tag{4.186c}$$

where $\chi_n(\lambda)$ is in normal order and $\chi_a(\lambda)$ is in antinormal order; λ is a complex variable. The three functions are related by the disentangling theorem (4.170). Thus, based on Eq. (4.172), we have

$$\chi(\lambda) = \chi_a(\lambda) e^{1/2|\lambda|^2} = \chi_n(\lambda) e^{-1/2|\lambda|^2}. \tag{4.187}$$

The characteristic functions have many interesting properties. It is found, for example, that

$$\left.\frac{\partial}{\partial\lambda}\frac{\partial}{\partial(-\lambda^*)}\chi_n(\lambda)\right|_{\lambda=0} = \left.\mathrm{Tr}\{\rho a^\dagger e^{\lambda a^\dagger} e^{-\lambda^* a} a\}\right|_{\lambda=0}$$
$$= \mathrm{Tr}\{\rho a^\dagger a\} = \langle a^\dagger a\rangle. \tag{4.188}$$

This is generalized easily to

$$\left.\frac{\partial^n}{\partial\lambda^n}\frac{\partial^m}{\partial(-\lambda^*)^m}\chi_n(\lambda)\right|_{\lambda=0} = \mathrm{Tr}\{\rho(a^\dagger)^n a^m\}. \tag{4.189a}$$

In similar fashion,

$$\left.\frac{\partial^n}{\partial\lambda^n}\frac{\partial^m}{\partial(-\lambda^*)^m}\chi_a(\lambda)\right|_{\lambda=0} = \mathrm{Tr}\{\rho a^m (a^\dagger)^n. \tag{4.189b}$$

By virtue of the theorem (4.161), the characteristic functions are expressible in the form of integrals. In particular,

$$\chi_a(\lambda) = \mathrm{Tr}\{\rho e^{-\lambda^* a} e^{\lambda a^\dagger}\} = \mathrm{Tr}\{e^{\lambda a^\dagger}\rho e^{-\lambda^* a}\}$$
$$= \frac{1}{\pi}\int d^2\alpha \langle\alpha|e^{\lambda a^\dagger}\rho e^{-\lambda^* a}|\alpha\rangle = \frac{1}{\pi}\int d^2\alpha\langle\alpha|\rho|\alpha\rangle e^{\lambda\alpha^* - \lambda^*\alpha}. \tag{4.190}$$

The last integral is but a thinly disguised expression for the two-dimensional Fourier transform of $\langle\alpha|\rho|\alpha\rangle$. To appreciate this fact, let the complex quantities α and λ be written

$$\alpha = \frac{(q + ip)}{\sqrt{2},} \qquad \lambda = \frac{(x + ik)}{\sqrt{2}.} \tag{4.191}$$

Then,

$$\langle\alpha|\rho|\alpha\rangle = \langle q, p|\rho|q, p\rangle, \qquad \lambda\alpha^* - \lambda^*\alpha = i(kq - xp),$$

$$d^2\alpha = d(\mathrm{Re}\,\alpha)\,d(\mathrm{Im}\,\alpha) = \frac{1}{2}dq\,dp, \tag{4.192}$$

and

$$\chi_a(x, k) = \frac{1}{2\pi} \int \langle q, p|\rho|q, p\rangle e^{i(kq - xp)}\, dq\, dp. \tag{4.193}$$

The inverse Fourier transform is

$$\langle q, p|\rho|q, p\rangle = \frac{1}{2\pi} \int \chi_a(x, k) e^{i(xp - kq)}\, dx\, dk, \tag{4.194}$$

which reverts to

$$\langle\alpha|\rho|\alpha\rangle = \frac{1}{\pi} \int \chi_a(\lambda) e^{\lambda^*\alpha - \lambda\alpha^*}\, d\lambda. \tag{4.195}$$

We now shall develop a number of theorems pertaining to coherent states. Referring to Eq. (4.139) and (4.142),

$$\begin{aligned} |\alpha\rangle\langle\alpha| &= e^{(\alpha a^\dagger - 1/2|\alpha|^2)}|0\rangle\langle 0|e^{(\alpha^* a - 1/2|\alpha|^2)} \\ &= e^{-\alpha^*\alpha} e^{\alpha a^\dagger}|0\rangle\langle 0|e^{\alpha^* a}. \end{aligned} \tag{4.196}$$

Hence,

$$\frac{\partial}{\partial\alpha}|\alpha\rangle\langle\alpha| = (a^\dagger - \alpha^*)|\alpha\rangle\langle\alpha| \tag{4.197a}$$

or

$$a^\dagger|\alpha\rangle\langle\alpha| = \left(\frac{\partial}{\partial\alpha} + \alpha^*\right)|\alpha\rangle\langle\alpha|. \tag{4.197b}$$

The Hermitian adjoint yields

$$|\alpha\rangle\langle\alpha|a = \left(\frac{\partial}{\partial\alpha^*} + \alpha\right)|\alpha\rangle\langle\alpha|. \tag{4.198}$$

Note that if $\alpha = x + iy$ and $\alpha^* = x - iy$,

$$\frac{\partial}{\partial\alpha} = \frac{1}{2}\left(\frac{\partial}{\partial x} - i\frac{\partial}{\partial y}\right), \qquad \frac{\partial}{\partial\alpha^*} = \frac{1}{2}\left(\frac{\partial}{\partial x} + i\frac{\partial}{\partial y}\right). \tag{4.199}$$

since

$$a|\alpha\rangle\langle\alpha| = \alpha|\alpha\rangle\langle\alpha|, \qquad |\alpha\rangle\langle\alpha|a^\dagger = \alpha^*|\alpha\rangle\langle\alpha|, \tag{4.200}$$

we have

$$a^\dagger a|\alpha\rangle\langle\alpha| = \left(\alpha\frac{\partial}{\partial\alpha} + |\alpha|^2\right)|\alpha\rangle\langle\alpha|. \tag{4.201}$$

Similarly,

$$|\alpha\rangle\langle\alpha|a^\dagger a = \left(a^*\frac{\partial}{\partial\alpha^*} + |\alpha|^2\right)|\alpha\rangle\langle\alpha|. \tag{4.202}$$

These relations may be extended to arbitrary products of a and $a^\dagger$; for example,

$$|\alpha\rangle\langle\alpha|a^\dagger aa = \alpha\left(\frac{\partial}{\partial\alpha^*} + \alpha\right)|\alpha\rangle\langle\alpha|a = \alpha\left(\frac{\partial}{\partial\alpha^*} + \alpha\right)^2|\alpha\rangle\langle\alpha|, \tag{4.203a}$$

$$\begin{aligned}|\alpha\rangle\langle\alpha|aa^\dagger a &= \left(\frac{\partial}{\partial\alpha^*} + \alpha\right)|\alpha\rangle\langle\alpha|a^\dagger a \\ &= \left(\frac{\partial}{\partial\alpha^*} + \alpha\right)\alpha^*\left(\frac{\partial}{\partial\alpha^*} + \alpha\right)|\alpha\rangle\langle\alpha|.\end{aligned} \tag{4.203b}$$

Thus, if $F(a, a^\dagger)$ is a function of a and $a^\dagger$ expressible as a power series in the operators, and $|\psi\rangle$ is an arbitrary wave function, we have the *classical-quantum* correspondence

$$\langle\alpha|F(a, a^\dagger)|\psi\rangle = F\left(\frac{\partial}{\partial\alpha^*} + \alpha, \alpha^*\right)\langle\alpha|\psi\rangle \tag{4.204}$$

whereby the annihilation operator a is replaced by $(\partial/\partial\alpha^*) + \alpha$, and the creation operator $a^\dagger$ is replaced by α^*. What Eq. (4.204) reveals is that matrix elements of the quantum mechanical operators a and $a^\dagger$ may be computed by c number operations, i.e., operations that do not involve quantum mechanical operators.

From the relation

$$a|\alpha\rangle\langle\alpha|a^\dagger = |\alpha|^2|\alpha\rangle\langle\alpha| \tag{4.205}$$

one obtains

$$a^\dagger a|\alpha\rangle\langle\alpha| - 2a|\alpha\rangle\langle\alpha|a^\dagger + |\alpha\rangle\langle\alpha|a^\dagger a = \left(\alpha\frac{\partial}{\partial\alpha} + \alpha^*\frac{\partial}{\partial\alpha^*}\right)|\alpha\rangle\langle\alpha|. \tag{4.206}$$

Several additional relations that may be verified by similar methods are the following:

$$aa^\dagger|\alpha\rangle\langle\alpha| = (a^\dagger a + 1)|\alpha\rangle\langle\alpha| = \left(\alpha\frac{\partial}{\partial\alpha} + |\alpha|^2 + 1\right)|\alpha\rangle\langle\alpha|, \tag{4.207a}$$

$$|\alpha\rangle\langle\alpha|aa^\dagger = |\alpha\rangle\langle\alpha|(a^\dagger a + 1) = \left(\alpha^\dagger\frac{\partial}{\partial\alpha^*} + |\alpha|^2 + 1\right)|\alpha\rangle\langle\alpha|, \tag{4.207b}$$

$$a^\dagger|\alpha\rangle\langle\alpha|a = \left(\frac{\partial}{\partial\alpha} + \alpha^*\right)|\alpha\rangle\langle\alpha|a = \left(\frac{\partial}{\partial\alpha} + \alpha^*\right)\left(\frac{\partial}{\partial\alpha^*} + \alpha\right)|\alpha\rangle\langle\alpha|$$

$$= \left(\frac{\partial^2}{\partial\alpha\partial\alpha^*} + \alpha^*\frac{\partial}{\partial\alpha^*} + \alpha\frac{\partial}{\partial\alpha} + |\alpha|^2 + 1\right)|\alpha\rangle\langle\alpha|. \qquad (4.207c)$$

From these we obtain

$$aa^\dagger|\alpha\rangle\langle\alpha| - 2a^\dagger|\alpha\rangle\langle\alpha|a + |\alpha\rangle\langle\alpha|aa^\dagger$$

$$= -\left(2\frac{\partial^2}{\partial\alpha\partial\alpha^*} + \alpha\frac{\partial}{\partial\alpha} + \alpha^*\frac{\partial}{\partial\alpha^*}\right)|\alpha\rangle\langle\alpha|. \qquad (4.208)$$

4.7 Statistical Properties of Photon-Number States

The statistical properties of the radiation field are most conveniently treated by means of the density matrix formalism. In the photon number basis that contains the closure property

$$\sum_n |n\rangle\langle n| = 1, \qquad (4.209)$$

the density operator may be written

$$\rho = \sum_{mn} |m\rangle\langle m|\rho|n\rangle\langle n|. \qquad (4.210)$$

When the radiation field is under thermal equilibrium with the walls of the cavity, ρ becomes the thermal density operator ρ_0 whose general form is given by Eq. (2.226). For a single mode, with $\mathscr{H} = N + 1/2 \equiv a^\dagger a + 1/2$,

$$\frac{1}{Z}\langle m|e^{-\beta\mathscr{H}}|n\rangle = \frac{1}{Z}\langle m|e^{-\beta\hbar\omega(N+1/2)}|n\rangle = \frac{1}{Z}e^{-\beta\hbar\omega(n+1/2)}\,\delta_{mn}, \qquad (4.211)$$

$$Z = \mathrm{Tr}\{e^{-\beta\mathscr{H}}\} = \sum_n \langle n|e^{-\beta\hbar\omega(N+1/2)}|n\rangle = e^{-\beta\hbar\omega/2}\sum_n e^{-\beta n\hbar\omega}$$

$$= \frac{e^{-\beta\hbar\omega/2}}{1 - e^{-\beta\hbar\omega}}. \qquad (4.212)$$

We have

$$\langle m|\rho_0|n\rangle = (1 - e^{-\beta\hbar\omega})e^{-\beta n\hbar\omega}\,\delta_{mn}, \qquad (4.213)$$

and

$$\rho_0 = (1 - e^{-\beta\hbar\omega})\sum_n e^{-\beta n\hbar\omega}|n\rangle\langle n|. \qquad (4.214)$$

The diagonal matrix element $\langle n|\rho_0|n\rangle$ gives the probability of finding n photons at the frequency ω when the electromagnetic field is in thermal equilibrium at the temperature T.

Having obtained an expression for the thermal density operator of an electromagnetic field in a cavity, we now may calculate various average properties. The first important quantity is the average number of photons:

$$\begin{aligned}\langle n\rangle &= \mathrm{Tr}\{\rho_0 N\} = (1 - e^{-\beta\hbar\omega})\,\mathrm{Tr}\left\{\sum_n e^{-n\beta\hbar\omega}|n\rangle\langle n|N\right\}\\ &= (1 - e^{-\beta\hbar\omega})\sum_{m,n} e^{-n\beta\hbar\omega}\langle m|n\rangle\langle n|N|m\rangle\\ &= (1 - e^{-\beta\hbar\omega})\sum_n n e^{-n\beta\hbar\omega}.\end{aligned} \tag{4.215}$$

Upon setting $x = \beta\hbar\omega$,

$$\begin{aligned}\sum_n n e^{-n\beta\hbar\omega} &= \sum_n n e^{-nx} = -\frac{\partial}{\partial x}\sum_n e^{-nx} = -\frac{\partial}{\partial x}\frac{1}{1 - e^{-x}}\\ &= \frac{e^{-x}}{(1 - e^{-x})^2} = \frac{e^{-\beta\hbar\omega}}{(1 - e^{\beta\hbar\omega})^2}.\end{aligned} \tag{4.216}$$

Thus,

$$\langle n\rangle = \frac{e^{-\beta\hbar\omega}}{1 - e^{-\beta\hbar\omega}} = \frac{1}{e^{\beta\hbar\omega} - 1} = \begin{cases} 1/\beta\hbar\omega = kT/\hbar\omega, & (kT \gg \hbar\omega) \quad (4.217a)\\ e^{-\beta\hbar\omega} = e^{-\hbar\omega/kT}, & (kT \ll \hbar\omega). \quad (4.217b)\end{cases}$$

At room temperature, the average number of photons in the optical region of the spectrum is extremely small ($\sim 10^{-40}$) but grows rapidly for longer wavelengths ($\langle n\rangle \simeq 1$ for $\lambda = 10 - 100\ \mu$m). For microwaves, $\langle n\rangle \gg 1$; to obtain $\langle n\rangle \simeq 1$ in the visible region, a temperature of $\sim 10^4$ K is required. Note also that $\langle n\rangle \to 0$ when $T \to 0$, independent of frequency.

Correlation functions under thermal equilibrium were described in Section 2.8. For the creation and annihilation operators, we have

$$\langle a_\mathrm{H}^\dagger(\tau)a\rangle = e^{i\omega\tau}\langle a^\dagger a\rangle = e^{i\omega\tau}\langle n\rangle, \tag{4.218a}$$

$$\langle a_\mathrm{H}(\tau)a^\dagger\rangle = e^{-i\omega\tau}\langle aa^\dagger\rangle = e^{-i\omega\tau}[\langle a^\dagger a\rangle + 1] = e^{-i\omega\tau}[\langle n\rangle + 1], \tag{4.218b}$$

$$\langle a^\dagger a_\mathrm{H}(\tau)\rangle = e^{-i\omega\tau}\langle a^\dagger a\rangle = e^{-i\omega\tau}\langle n\rangle, \tag{4.218c}$$

$$\langle a a_\mathrm{H}^\dagger(\tau)\rangle = e^{i\omega\tau}\langle aa^\dagger\rangle = e^{i\omega\tau}[\langle n\rangle + 1], \tag{4.218d}$$

$$\langle a_\mathrm{H}^\dagger(\tau)a^\dagger\rangle = \langle a_\mathrm{H}(\tau)a\rangle = 0, \tag{4.218e}$$

where the conversion of $a_\mathrm{H}^\dagger$ and a_H in the Heisenberg representation to $a^\dagger$ and a in the Schrödinger representation was made according to Eq. (4.83) and $\langle n\rangle$ is given by Eq. (4.217). We note that a and $a^\dagger$ belong to the same radiation mode; all correlation functions vanish when a and $a^\dagger$ belong to different modes.

The expression for the average number of photons may be used to derive a temperature-independent form for the density operator. Combining Eqs. (4.214) and (4.217),

$$\rho_0 = \frac{1}{1 + \langle n \rangle} \sum_n \left(\frac{\langle n \rangle}{1 + \langle n \rangle} \right)^n |n\rangle\langle n|, \tag{4.219a}$$

$$\langle n|\rho_0|n\rangle = \frac{1}{1 + \langle n \rangle} \left(\frac{\langle n \rangle}{1 + \langle n \rangle} \right)^n. \tag{4.219b}$$

The diagonal matrix element, when written in this form, gives the probability of finding n photons in a mode whose average number of photons is $\langle n \rangle$. Although these expressions were derived specifically for the case of thermal equilibrium, they are, nevertheless, also applicable to the general case of a random (chaotic) photon distribution, which may have been produced by a nonthermal randomizing process.

For thermal photon distributions, the mean square fluctuation (variance) is given by

$$(\Delta n)^2 = \langle n^2 \rangle - \langle n \rangle^2 \tag{4.220}$$

where

$$\begin{aligned}\langle n^2 \rangle &= \mathrm{Tr}\{\rho_0 N^2\} \\ &= (1 - e^{-\beta\hbar\omega})\,\mathrm{Tr}\left\{\sum_n e^{-n\beta\hbar\omega}|n\rangle\langle n|N^2\right\}.\end{aligned} \tag{4.221}$$

But

$$\begin{aligned}\mathrm{Tr}\left\{\sum_n e^{-n\beta\hbar\omega}|n\rangle\langle n|N^2\right\} &= \sum_{m,n} e^{-n\beta\hbar\omega}\langle m|n\rangle\langle n|N^2|m\rangle \\ &= \sum_n e^{-n\beta\hbar\omega}\langle n|N^2|n\rangle = \sum_n n^2 e^{-n\beta\hbar\omega}.\end{aligned} \tag{4.222}$$

Again, setting $x = \beta\hbar\omega$ and writing

$$\sum_n n^2 e^{-nx} = \frac{\partial^2}{\partial x^2} \sum_n e^{-nx} = \frac{\partial^2}{\partial x^2} \frac{1}{1 - e^{-x}}, \tag{4.223}$$

one finds, with the expression for $\langle n \rangle$ given by Eq. (4.217),

$$\langle n^2 \rangle = \frac{e^{\beta\hbar\omega} + 1}{(e^{\beta\hbar\omega} - 1)^2} = \langle n \rangle + 2\langle n \rangle^2, \tag{4.224}$$

$$(\Delta n)^2 = e^{\beta\hbar\omega}\langle n \rangle^2 = \langle n \rangle(1 + \langle n \rangle). \tag{4.225}$$

$$\eta \equiv \frac{\Delta n}{\langle n \rangle} = \frac{[\langle n \rangle(1 + \langle n \rangle)]^{1/2}}{\langle n \rangle} = \begin{cases} 1, & \langle n \rangle \gg 1 \\ \langle n \rangle^{-1/2}, & \langle n \rangle \ll 1. \end{cases} \tag{4.226}$$

The quantity η is known as the *relative uncertainty.* Under thermal equilibrium, η approaches infinity as $\langle n\rangle \to 0$. It is the uncertainty (or fluctuations) in this regime that is associated with quantum noise.

The *Planck radiation law* follows directly from the average photon number. All that is required is to multiply the average energy $\hbar\omega\langle n\rangle$ by the number of modes per unit interval in ω in a unit volume, taking into account the two independent directions of polarization. From Eqs. (4.43) and (4.217) we have

$$U(\omega) = 2\hbar\omega\langle n\rangle \frac{\omega^2}{2\pi^2 c^3} = \frac{\hbar\omega^3}{\pi^2 c^3} \frac{1}{e^{\beta\hbar\omega} - 1}. \tag{4.227}$$

$U(\omega)$ is the average energy per unit volume per unit interval in ω, for a distribution of photons at thermal equilibrium; it is also known as the "black body distribution." It is of interest to observe that, despite the presence of the zero-point energy in the Hamiltonian (Eq. (4.97)), $U(\omega)$ is finite at all temperature and vanishes when $T \to 0$. If the unit interval in ω is replaced by unit intervals in $\nu(=\omega/2\pi)$ or $\lambda(=c/\nu)$, the relations among the distributions are

$$U(\omega)\, d\omega = U(\nu)\, d\nu = U(\lambda)\, d\lambda \tag{4.228}$$

or

$$U(\omega) = 2\pi U(\nu) = \frac{(2\pi)^3 c}{\omega^3} U(\lambda). \tag{4.229a}$$

One then derives

$$U(\nu) = \frac{8\pi h\nu^3}{c^3} \frac{1}{e^{\beta h\nu} - 1}, \tag{4.229b}$$

$$U(\lambda) = \frac{8\pi hc}{\lambda^5} \frac{1}{e^{\beta hc/\lambda} - 1}, \tag{4.229c}$$

where $U(\nu)$ is the average energy per unit volume per unit interval in ν, $U(\lambda)$ is the average energy per unit volume per unit interval in λ, and $h = 2\pi\hbar$. A plot of $U(\lambda)$ for $T = 2000$, 3000, and 4000 K is shown in Fig. 4.2.

In the high-temperature limit, with $\langle n\rangle$ as in Eq. (4.217a), we obtain the *Rayleigh limit*

$$U(\omega) = \frac{\omega^2}{\pi^2 c^3 \beta}, \qquad \left(\frac{1}{\beta} = kT \gg \hbar\omega\right), \tag{4.230}$$

whereas in the low-temperature limit, with $\langle n\rangle$ given by Eq. (4.217b),

$$U(\omega) = \frac{\hbar\omega^3}{\pi^2 c^3} e^{-\beta\hbar\omega}. \tag{4.231}$$

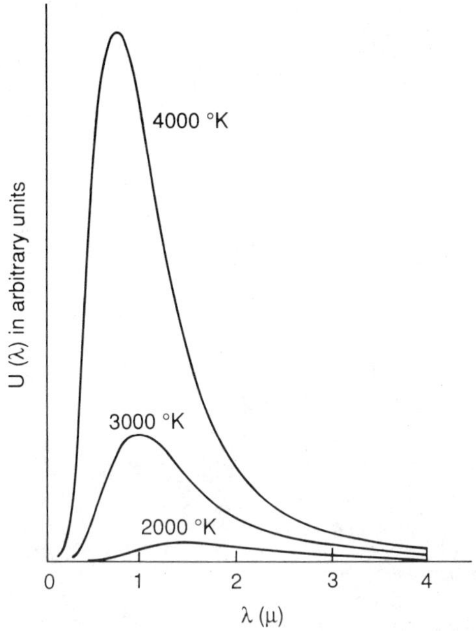

FIGURE 4.2 Curves of black-body radiation at $T = 2000°\text{K}$, $3000°\text{K}$ and $4000°\text{K}$.

The average energy per unit volume is obtained by integrating $U(\omega)$ over all frequencies:

$$U = \int_0^\infty U(\omega)\,d\omega = \frac{\hbar}{\pi^2 c^3}\int_0^\infty \frac{\omega^3}{e^{\beta\hbar\omega} - 1}\,d\omega$$

$$= \frac{\pi^2}{15\beta^4 c^3 \hbar^3} = 7.56 \times 10^{-16}\,\text{K}^4, \qquad (\text{J/m}^3). \tag{4.232}$$

The average number of photons per unit volume N_V is

$$N_V = \int_0^\infty \frac{U(\omega)}{\hbar\omega}\,d\omega = \int_0^\infty \langle n\rangle \frac{\omega^2}{\pi^2 c^3}\,d\omega = \frac{1}{\pi^2 c^3}\int_0^\infty \frac{\omega^2}{e^{\beta\hbar\omega}}\,d\omega$$

$$= 2.028 \times 10^7\,\text{K}^3, \qquad (\text{photons/m}^3). \tag{4.233}$$

The ratio U/N_V is the average energy per photon

$$E = \frac{U}{N_V} = 3.73 \times 10^{-23}\,\text{K}, \qquad (\text{J}). \tag{4.234}$$

We also may compute the average number of photons associated with a beam produced by a thermal source. Since the general relation between energy density and intensity is $I = cU$, we have, from Eq. (4.227),

$$\langle n\rangle = \frac{\pi^2 c^3}{\hbar\omega^3} U(\omega) = \frac{\pi^2 c^2}{\hbar\omega^3} I(\omega) = 8.41 \times 10^{51} \frac{I(\omega)}{\omega^3}. \tag{4.235}$$

Here, $I(\omega)$ is the intensity (Jm^{-2}) per unit interval in ω.

4.8 Statistical Properties of Coherent States

The expression for a general operator in a coherent state basis (Eq. (4.160)) may be applied to the density operator

$$\begin{aligned} \rho &= \frac{1}{\pi^2} \int d^2\beta \int d^2\alpha |\beta\rangle\langle\beta|\rho|\alpha\rangle\langle\alpha| \\ &= \frac{1}{\pi^2} \int d^2\beta \int d^2\alpha R(\alpha, \beta^*) e^{-1/2(|\alpha|^2 + |\beta|^2)} |\beta\rangle\langle\alpha|, \end{aligned} \tag{4.236}$$

where, according to Eq. (4.159),

$$R(\alpha, \beta^*) = \sum_{m,n} \langle m|\rho|n\rangle \frac{(\beta^*)^m \alpha^n}{\sqrt{m!n!}}, \tag{4.237}$$

and

$$\langle\beta|\rho|\alpha\rangle = e^{-1/2(|\alpha|^2 + |\beta|^2)} R(\alpha, \beta^*). \tag{4.238}$$

Furthermore, from Eq. (4.161) we have

$$\mathrm{Tr}\,\rho = \frac{1}{\pi} \int d^2\alpha \langle\alpha|\rho|\alpha\rangle = \frac{1}{\pi} \int R(\alpha, \alpha^*) e^{-|\alpha|^2} d^2\alpha = 1. \tag{4.239}$$

Under thermal equilibrium, $R(\alpha, \beta^*)$ becomes

$$R(\alpha, \beta^*) = \sum_{n,n} \langle m|\rho_0|n\rangle \frac{(\beta^*)^m \alpha^n}{\sqrt{m!n!}}. \tag{4.240}$$

Using Eqs. (4.213) and (4.217), the special forms of $R(\alpha, \beta^*)$ under thermal equilibrium are

$$R(\alpha, \beta^*) = (1 - e^{-\beta\hbar\omega}) \sum_n \frac{(\beta^*\alpha)^n}{n!} e^{-n\beta\hbar\omega} = \frac{1}{1 + \langle n\rangle} \sum_n \left(\frac{\langle n\rangle}{1 + \langle n\rangle}\right)^n \frac{(\beta^*\alpha)^n}{n!} \tag{4.241a}$$

$$= \frac{1}{1 + \langle n\rangle} \exp\left(\frac{\langle n\rangle}{1 + \langle n\rangle} \beta^*\alpha\right). \tag{4.241b}$$

These expressions may be inserted into Eq. (4.238) to obtain the matrix element of the thermal density operator in a coherent state basis. In particular, the diagonal matrix element is

$$\langle\alpha|\rho_0|\alpha\rangle = (1 - e^{-\beta\hbar\omega})e^{-|\alpha|^2}\sum_n \frac{|\alpha|^{2n}}{n!}e^{-n\beta\hbar\omega}$$

$$= \frac{1}{1+\langle n\rangle}\exp\left(-\frac{|\alpha|^2}{1+\langle n\rangle}\right). \tag{4.242}$$

The expression for the density operator given by Eq. (4.236) is the general form of the density operator in a coherent state basis. It may be simplified, under certain conditions, by defining a new quantity $P(\alpha)$ such that

$$\rho = \int d^2\alpha P(\alpha)|\alpha\rangle\langle\alpha|. \tag{4.243}$$

This is known as the *P-representation* or the *diagonal representation* of the density operator [7, 8]. Since the condition $\mathrm{Tr}\,\rho = 1$ must be satisfied, we have

$$\mathrm{Tr}\,\rho = \sum_n \langle n|\rho|n\rangle = \int d^2\alpha P(\alpha)\sum_n \langle n|\alpha\rangle\langle\alpha|n\rangle = 1. \tag{4.244}$$

But

$$\sum_n \langle n|\alpha\rangle\langle\alpha|n\rangle = \sum_n \langle\alpha|n\rangle\langle n|\alpha\rangle = \langle\alpha|\alpha\rangle = 1; \tag{4.245}$$

hence, we arrive at the normalization property of the P-representation

$$\int P(\alpha)\, d^2\alpha = 1. \tag{4.246}$$

For a general density operator ρ written in terms of $P(\alpha)$, as in Eq. (4.243), the diagonal element is

$$\langle\beta|\rho|\beta\rangle = \int d^2\alpha P(\alpha)\langle\beta|\alpha\rangle\langle\alpha|\beta\rangle = \int d^2\alpha P(\alpha)e^{-|\alpha-\beta|^2}, \tag{4.247}$$

in which the last expression is based on Eq. (4.147).

These properties suggest that $P(\alpha)$ behaves like an ordinary probability density, which is often the case; but $P(\alpha)$ also may become highly singular or negative. Hence, it cannot be regarded as a legitimate probability density under all circumstances. As an example, a pure coherent state consistency between $\rho = |\alpha\rangle\langle\alpha|$ and Eq. (4.243) requires that

$$P(\alpha) = \delta^{(2)}(\alpha - \alpha') = \delta[\mathrm{Re}(\alpha - \alpha')]\,\delta[\mathrm{Im}(\alpha - \alpha')]. \tag{4.248}$$

In terms of $P(\alpha)$, the ensemble average of an operator O is

$$\langle O\rangle = \mathrm{Tr}\{\rho O\} = \sum_n \langle n| \int d^2\alpha P(\alpha)|\alpha\rangle\langle\alpha|O|n\rangle$$

$$= \int d^2\alpha P(\alpha)\sum_n \langle n|\alpha\rangle\langle\alpha|O|n\rangle = \int d^2\alpha P(\alpha)\langle\alpha|O|\alpha\rangle. \tag{4.249}$$

Thus, the expectation value of products of a and $a^\dagger$ in normal order may be written

$$\langle (a^\dagger)^i a^j\rangle = \mathrm{Tr}\{\rho(a^\dagger)^i a^j\} = \int d^2\alpha P(\alpha)\langle\alpha|(a^\dagger)^i a^j|\alpha\rangle = \int d^2\alpha P(\alpha)(\alpha^*)^i\alpha^j. \tag{4.250}$$

Here, too, as in Eq. (4.204), we have an example of a quantum mechanical average evaluated by a c-number procedure. More generally,

$$\langle :O(a, a^\dagger):\rangle = \int P(\alpha)O(\alpha, \alpha^*)\, d^2\alpha, \tag{4.251}$$

that is, the expectation value of an operator in normal order is expressible as a c-number average.

We also may obtain an expression for $P(\alpha)$ when the normally ordered operator is $\chi_n(\lambda)$ as defined by Eq. (4.186b). Again, we refer to Eq. (4.249b) to write

$$\chi_n(\lambda) = \mathrm{Tr}\{\rho e^{\lambda a^\dagger} e^{-\lambda^* a}\} = \int P(\alpha)\langle\alpha|e^{\lambda a^\dagger}e^{-\lambda^* a}|\alpha\rangle\, d^2\alpha = \int P(\alpha)e^{\lambda\alpha^* - \lambda^*\alpha}\, d^2\alpha, \tag{4.252}$$

whose inverse Fourier transform, as described in Section 4.6, yields

$$P(\alpha) = \frac{1}{\pi^2}\int e^{\lambda^*\alpha - \lambda\alpha^*}\chi_n(\lambda)\, d^2\lambda. \tag{4.253}$$

Let us now derive the P-representation of the density operator at thermal equilibrium [7, 8]. We begin by computing the characteristic function $\chi_a(\lambda)$ in accordance with Eq. (4.190) where the diagonal matrix element of the thermal density operator is given by Eq. (4.242):

$$\chi_a(\lambda) = \frac{1}{\pi}\int d^2\alpha\langle\alpha|\rho_0|\alpha\rangle e^{\lambda\alpha^* - \lambda^*\alpha}$$

$$= \frac{1}{\pi(1 + \langle n\rangle)}\int d^2\alpha \exp\left(-\frac{|\alpha|^2}{1 + \langle n\rangle}\right)e^{\lambda\alpha^* - \lambda^*\alpha}. \tag{4.254}$$

In terms of the real and imaginary components defined in Eq. (4.191), we have

$$\chi_a(x, k) = \frac{1}{2\pi(1 + \langle n \rangle)} \int \exp\left(-\frac{q^2 + p^2}{2(1 + \langle n \rangle)} \right) e^{i(kq - xp)}\, dq\, dp. \quad (4.255)$$

To facilitate the evaluation of the integral, note that

$$\int \exp(-at^2) e^{\pm i\omega t}\, dt = \sqrt{\frac{\pi}{a}} \exp\left(\frac{-\omega^2}{4a} \right), \quad (4.256)$$

which then leads to the simple result

$$\chi_a(\lambda) = \exp[-(1 + \langle N \rangle)|\lambda|^2] = \chi_n(\lambda) \exp[-|\lambda|^2]. \quad (4.257)$$

The second equality makes reference to Eq. (4.187). Since $P(\alpha)$ is related to $\chi_n(\lambda)$ by Eq. (4.263), we have

$$P(\alpha) = \frac{1}{\pi^2} \int \exp[-\langle n \rangle |\lambda|^2] e^{\lambda^* \alpha - \lambda \alpha^*}\, d^2\lambda = \frac{1}{\pi \langle n \rangle} \exp\left[\frac{-|\alpha|^2}{\langle n \rangle} \right] \quad (4.258)$$

which again, has been evaluated with the help of Eq. (4.255). We note that if $\langle n \rangle \gg 1$, $P(\alpha)$ merges with $\langle \alpha | \rho_0 | \alpha \rangle$ in Eq. (4.242). It is seen that $P(\alpha)$ for thermal or, more generally, chaotic radiation is a Gaussian function; it therefore may be interpreted as a probability distribution. For the multimode case,

$$P(\{\alpha_k\}) = \prod_k \frac{1}{\pi \langle n_k \rangle} \exp\left[-\frac{|\alpha_k|^2}{\langle n_k \rangle} \right]. \quad (4.259)$$

The insertion of Eq. (4.258) into Eq. (4.243) yields the P-representation of the thermal density operator

$$\rho_0 = \frac{1}{\pi \langle n \rangle} \int d^2\alpha \exp\left[-\frac{|\alpha|^2}{\langle n \rangle} \right] |\alpha\rangle\langle\alpha| \quad (4.260)$$

for a single mode.

Let us now compare some of the statistical properties of coherent states with the corresponding properties of photon-number states. In the latter, the number of photons n is precisely determined. Thus, with $N = a^\dagger a$,

$$\langle n \rangle_n = \langle n | N | n \rangle = n, \quad (4.261a)$$

$$\langle n^2 \rangle_n = \langle n | N^2 | n \rangle = n^2. \quad (4.261b)$$

We then obtain for the mean square fluctuation and the relative uncertainty

$$(\Delta n)_n^2 = \langle n^2 \rangle_n - \langle n \rangle_n^2 = 0, \quad (4.262)$$

$$\eta = \frac{(\Delta n)_n}{\langle n \rangle_n} = 0. \quad (4.263)$$

In the coherent state basis, however,

$$\langle n\rangle_c = \langle\alpha|N|\alpha\rangle = |\alpha|^2, \qquad \langle n^2\rangle_c = \langle\alpha|N^2|\alpha\rangle = |\alpha|^4 + |\alpha|^2, \quad (4.264)$$

$$(\Delta n)_c^2 = \langle n^2\rangle_c - \langle n\rangle_c^2 = |\alpha|^2 = \langle n\rangle_c, \quad (4.265)$$

consistent with Poissonian statistics. Also,

$$\eta_c = \frac{(\Delta n)_c}{\langle n\rangle_c} = |\alpha|^{-1} = \langle n\rangle_c^{-1/2}. \quad (4.266)$$

We see, then, that in a coherent state the number of photons is not precise since, according to Eq. (4.265), the mean square fluctuation is equal to the average number of photons. But the relative uncertainty, η_c, approaches zero with increasing mean photon number $\langle n\rangle_c$.

Another important comparison between photon-number and coherent states is obtained when one calculates the product $\Delta Q\,\Delta P$ for each case. Referring to Eq. (4.69),

$$\begin{aligned} \langle n|Q|n\rangle &= \langle n|P|n\rangle = 0, \\ \langle n|Q^2|n\rangle &= \frac{\hbar}{2\omega}\langle n|(a^\dagger a)^2|n\rangle = \frac{\hbar}{2\omega}(2n+1), \\ \langle n|P^2|n\rangle &= -\frac{\hbar\omega}{2}\langle n|(a^\dagger - a)^2|n\rangle = \frac{\hbar\omega}{2}(2n+1). \end{aligned} \quad (4.267)$$

Thus,

$$(\Delta Q)_n^2 \equiv \langle n|Q^2|n\rangle - \langle n|Q|n\rangle^2 = \frac{\hbar}{2\omega}(2n+1), \quad (4.268)$$

$$\begin{aligned} (\Delta P)_n^2 &\equiv \langle n|P^2|n\rangle - \langle n|P|n\rangle^2 = \frac{\hbar\omega}{2}(2n+1), \\ (\Delta Q\,\Delta P)_n &= \frac{\hbar}{2}(2n+1) \geq \frac{\hbar}{2}. \end{aligned} \quad (4.269)$$

The minimum uncertainty, i.e., $\Delta Q\,\Delta P = \hbar/2$, is achieved only in the vacuum state ($n = 0$). When the calculation is repeated for a coherent state we obtain, with the help of the relations in Eq. (4.143),

$$\begin{aligned} \langle\alpha|Q|\alpha\rangle &= \sqrt{\frac{\hbar}{2\omega}}(\alpha^* + \alpha), \qquad \langle\alpha|P|\alpha\rangle = i\sqrt{\frac{\hbar\omega}{2}}(\alpha^* - \alpha), \\ \langle\alpha|Q^2|\alpha\rangle &= \frac{\hbar}{2\omega}(\alpha^{*2} + 2|\alpha|^2 + \alpha^2 + 1), \\ \langle\alpha|P^2|\alpha\rangle &= \frac{\hbar\omega}{2}(\alpha^{*2} - 2|\alpha|^2 + \alpha^2 - 1). \end{aligned} \quad (4.270)$$

Then

$$(\Delta Q)_c^2 \equiv \langle\alpha|Q^2|\alpha\rangle - \langle\alpha|Q|\alpha\rangle^2 = \frac{\hbar}{2\omega},$$

$$(\Delta P)_c^2 \equiv \langle\alpha|P^2|\alpha\rangle - \langle\alpha|P|\alpha\rangle^2 = \frac{\hbar\omega}{2}, \tag{4.271}$$

$$(\Delta Q\,\Delta P)_c = \frac{\hbar}{2}. \tag{4.272}$$

The coherent state is therefore a *minimum uncertainty state*—a feature shared with the vacuum state in the photon-number basis.

4.9 Squeezed States

Squeezed states are among the discoveries involving laser radiation that have an important bearing on potential applications to low-noise detection systems. We begin by writing the annihilation and creation operators for a single mode in the form

$$a = \frac{1}{\sqrt{2}}(a_1 + ia_2), \qquad a^\dagger = \frac{1}{\sqrt{2}}(a_1 - ia_2),$$

$$a_1 = \frac{1}{\sqrt{2}}(a^\dagger + a), \qquad a_2 = \frac{i}{\sqrt{2}}(a^\dagger - a), \tag{4.273}$$

in which a_1 and a_2, known as the *quadrature components* of the radiation mode, are Hermitian operators. They are related closely to the position and momentum operators of the field since, according to Eq. (4.95),

$$Q = \sqrt{\frac{\hbar}{2\omega}}(a^\dagger a) = \sqrt{\frac{\hbar}{\omega}}\,a_1,$$

$$P = i\sqrt{\frac{\hbar\omega}{2}}(a^\dagger - a) = \sqrt{\hbar\omega}\,a_2. \tag{4.274}$$

The reason for referring to a_1 and a_2 as quadrature components may be seen by writing the electric field operator in terms of a_1 and a_2. For a single mode in the Heisenberg representation, we have, from Eq. (4.119),

$$\mathbf{E}_{\mathrm{H}}(\mathbf{r}, t) = K\hat{\varepsilon}i[ae^{i\theta} - a^\dagger e^{-i\theta}]$$

$$= \frac{K}{\sqrt{2}}\hat{\varepsilon}i[(a_1 + ia_2)e^{i\theta} - (a_1 - ia_2)e^{-i\theta}]$$

$$= -K\hat{\varepsilon}\sqrt{2}(a_2\cos\theta + a_1\sin\theta) \tag{4.275}$$

in which

$$K = \sqrt{\frac{\hbar\omega}{2\varepsilon_0 V}}, \qquad \theta = \mathbf{k}\cdot\mathbf{r} - \omega t. \tag{4.276}$$

Thus, in terms of a_1 and a_2 (or in terms of P and Q), the two components of the radiation mode are 90° out of phase (i.e., in quadrature).

The commutator relation $[a, a^\dagger] = 1$ enables us to derive a commutator relation for a_1 and a_2. Applying the definitions (4.273), we find

$$[a, a^\dagger] = i[a_2, a_1] = 1 \tag{4.277a}$$

or

$$[a_1, a_2] = i. \tag{4.277b}$$

But the general uncertainty principle states that if two observables A and B satisfy the commutator relation $[A, B] = ic$ where c is a numerical constant, then $\Delta A\, \Delta B \geq c/2$. Therefore, in view of Eq. (4.277), we have, in the photon-number basis,

$$\Delta a_1\, \Delta a_2 = \frac{1}{2}(2n+1) \geq \frac{1}{2}. \tag{4.278}$$

The same result may be deduced from the position-momentum uncertainty relation $\Delta Q\, \Delta P \geq \hbar/2$.

The variances of Q and P in a coherent state basis were given given by Eq. (4.271); thus,

$$(\Delta Q)_c^2 = \frac{\hbar}{\omega}(\Delta a_1)_c^2 = \frac{\hbar}{2\omega}, \qquad (\Delta P)_c^2 = \hbar\omega(\Delta a_2)_c^2 = \frac{\hbar\omega}{2}. \tag{4.279}$$

Then

$$(\Delta a_1)_c^2 = (\Delta a_2)_c^2 = \frac{1}{2}, \qquad (\Delta a_1\, \Delta a_2)_c = \frac{1}{2}, \tag{4.280}$$

which also may be obtained directly from Eq. (4.273). We conclude that in a coherent state, which is a minimun uncertainty state as shown by Eq. (4.272), the variances of the two quadrature components are equal. The same property is found for the vacuum state in the photon-number basis, as is evident from Eq. (4.269). *Squeezed states* [9–11] are members of a broader class of minimum uncertainty states in which the variances of the two quadrature components are *not* equal. This implies that the quantum fluctuations in one quadrature component may be reduced at the expense of increased quantum fluctuations in the other quadrature component. Or, in terms of an electromagnetic field consisting of two components that are 90° out of phase, as in Eq. (4.275), squeezing means that the quantum noise in one component is reduced while the other component suffers an increase in quantum noise. The condition for squeezing therefore, may, be written

$$\Delta a_1 \, \Delta a_2 = \frac{1}{2}, \qquad \text{but } (\Delta a_i)^2 < \frac{1}{2} \text{ for } i = 1 \text{ or } 2. \tag{4.281}$$

We note that a photon-number state is not a minimum-uncertainty state, as shown by Eq. (4.269), and therefore cannot be squeezed.

Squeezed states may be generated by means of a *squeeze operator* defined by

$$S(z) = \exp\left[\frac{1}{2}(za^2 - z^*a^{\dagger 2})\right] \equiv e^{iA}, \qquad z = re^{-i\phi}. \tag{4.282}$$

Since the operator

$$A = -\frac{i}{2}(za^2 - z^*a^{\dagger 2}) \tag{4.283}$$

is Hermitian, $S(z)$ must be unitary, i.e.,

$$S^\dagger(z) = S^{-1}(z). \tag{4.284}$$

Let us now perform a unitary transformation on the annihilation operator a by means of the squeeze operator:

$$\begin{aligned} t &\equiv S(z)aS^\dagger(z) = e^{iA}ae^{-iA} \\ &= a + [iA, a] + \frac{1}{2!}[iA, [iA, a]] + \frac{1}{3!}[iA, [iA, [iA, a]]] + \cdots \end{aligned} \tag{4.285}$$

in which the theorem (4.170) has been used. With the commutation property $[a, a^\dagger] = 1$, the result is

$$\begin{aligned} t &= a + z^*a^\dagger + \frac{1}{2!}|z|^2 a + \frac{1}{3!}|z|^2 z^* a^\dagger + \frac{1}{4!}|z|^4 a + \cdots, \\ &= a\left(1 + \frac{1}{2!}r^2 + \frac{1}{4!}r^4 + \cdots\right) + a^\dagger e^{i\phi}\left(r + \frac{1}{3!}r^3 + \frac{1}{5!}r^5 \cdots\right), \end{aligned} \tag{4.286a}$$

or

$$t = a\cos hr + a^\dagger e^{i\phi}\sin hr, \qquad t^\dagger = a^\dagger \cos hr + ae^{-i\phi}\sin hr, \tag{4.286b}$$

which also is known as the *Bogoliubov transformation*. The transformed operators t and $t^\dagger$ then satisfy

$$t = \mu^* a - \nu a^\dagger, \qquad |\mu|^2 - |\nu|^2 = 1, \tag{4.287}$$

and the commutation rule

$$[t, t^\dagger] = 1. \tag{4.288}$$

We now define the c-numbers τ and τ^*:

$$\tau = \alpha \cos hr + \alpha^* e^{i\phi} \sin hr, \qquad \tau^* = \alpha^* \cos hr + \alpha e^{-i\phi} \sin hr, \tag{4.289a}$$

$$\tau t^\dagger - \tau^* t = \alpha a^\dagger - \alpha^* a. \tag{4.289b}$$

The definition of a squeezed state follows a pattern similar to the definition of a coherent state, which was shown to satisfy the relations

$$\begin{aligned} a|\alpha\rangle &= \alpha|\alpha\rangle, \qquad & a|O\rangle &= O, \\ |\alpha\rangle &= D(\alpha)|O\rangle, \qquad & D(\alpha) &= e^{\alpha a^\dagger - \alpha^* a}. \end{aligned} \tag{4.290}$$

We now may define an operator,

$$D(\tau) = e^{\tau t^\dagger - \tau^* t} = e^{\alpha a^\dagger - \alpha^* a} = D(\alpha), \tag{4.291}$$

in which the equality $D(\tau) = D(\alpha)$ follows from Eq. (4.286b). Whereas the coherent state $|\alpha\rangle$ is generated from the vacuum state $|O\rangle$ by the displacement operator $D(\alpha)$, the squeezed state $|\tau\rangle$ is generated from the vacuum by the relation

$$|\tau\rangle = D(\alpha)S(z)|O\rangle. \tag{4.292}$$

The vacuum state with respect to the transformed annihilation operator t, written $|O_t\rangle$, is defined by

$$|O_t\rangle = S(z)|O\rangle, \tag{4.293}$$

since

$$\begin{aligned} t|O_t\rangle &= tS(z)|O\rangle = S(z)aS^\dagger(z)S(z)|O\rangle \\ &= S(z)a|O\rangle = O. \end{aligned} \tag{4.294}$$

Substituting Eqs. (4.293) and (4.291) into Eq. (4.292), it is seen that

$$|\tau\rangle = D(\tau)|O_t\rangle, \qquad D^{-1}(\tau)|\tau\rangle = |O_t\rangle; \tag{4.295}$$

that is, the squeezed state is generated by the displacement operator $D(\tau)$ from the vacuum state with respect to the operator t. This is an exact parallel to the manner in which a coherent state is generated. Furthermore, it may be shown that $|\tau\rangle$ is an eigenstate of t with the complex eigenvalue τ. Thus, noting that $D(\tau)$ is subject to the same type of analysis as that leading to Eq. (4.179), we have

$$D^{-1}(\tau)tD(\tau) = t + \tau, \qquad D^{-1}(\tau)t^\dagger D(\tau) = t^\dagger + \tau^*, \tag{4.296a}$$

or

$$t = D(\tau)(t + \tau)D^{-1}(\tau), \qquad t^\dagger = D(\tau)(t^\dagger + \tau^*)D^{-1}(\tau). \tag{4.296b}$$

Therefore, in view of Eqs. (4.296b) and (4.294),

$$t|\tau\rangle = D(\tau)(t+\tau)D^{-1}(\tau)|\tau\rangle = D(\tau)tD^{-1}(\tau)|\tau\rangle + \tau|\tau\rangle$$
$$= D(\tau)t|O_t\rangle + \tau|\tau\rangle = \tau|\tau\rangle. \tag{4.297}$$

One may regard Eq. (4.297) as an alternative definition of a squeezed state.

It may appear that up to this point nothing has been accomplished other than to duplicate the formalism of the coherent state with a change of notation. Nevertheless, it will be shown shortly that the statistics of the squeezed states differ in an important respect from the statistics of coherent states. The commutation rule (Eq. (4.288)) and the definition (4.297) for the squeezed state enable us to write expressions similar in form to those in Eq. (4.143) as, for example

$$\langle\tau|(t^\dagger)^m t^n|\tau\rangle = (\tau^*)^m\tau^n, \qquad \langle\tau|tt^\dagger|\tau\rangle = |\tau|^2 + 1. \tag{4.298}$$

We now shall transform the quadrature components a_1 and a_2 with $\phi = 0$:

$$t_1 = \frac{1}{\sqrt{2}}(t^\dagger + t) = \frac{1}{\sqrt{2}}S(z)(a^\dagger + a)S^\dagger(z)$$
$$= \frac{1}{\sqrt{2}}(a^\dagger + a)(\cos hr + \sin hr) = a_1 e^r, \tag{4.299a}$$

$$t_2 = \frac{i}{\sqrt{2}}(t^\dagger - t) = \frac{i}{\sqrt{2}}S(z)(a^\dagger - a)S^\dagger(z)$$
$$= \frac{i}{\sqrt{2}}(a^\dagger - a)(\cos hr - \sin hr) = a_2 e^{-r}. \tag{4.299b}$$

These relations, together with Eq. (4.298), make it possible to demonstrate the essential difference between a coherent state and a squeezed state: coherent and squeezed states share the common property of minimum uncertainty but whereas the variances of a_1 and a_2 are equal in a coherent state, they are not equal in a squeezed state. For this purpose, it is necessary to compute several matrix elements. With the understanding that $\phi = 0$,

$$\langle\tau|t_1|\tau\rangle = \frac{1}{\sqrt{2}}\langle\tau|t^\dagger + t|\tau\rangle = \frac{1}{\sqrt{2}}(\tau^* + \tau) \tag{4.300}$$

$$\langle\tau|t_1^2|\tau\rangle = \frac{1}{2}\langle\tau|t^{\dagger 2} + t^\dagger t + tt^\dagger + t^2|\tau\rangle = \frac{1}{2}(\tau^{*2} + 2|\tau|^2 + \tau^2 + 1), \tag{4.301}$$

$$(\Delta t_1)^2 \equiv \langle\tau|t_1^2|\tau\rangle - \langle\tau|t_1|\tau\rangle^2 = \frac{1}{2}. \tag{4.302}$$

In the same way,

$$(\Delta t_2)^2 = \frac{1}{2}. \tag{4.303}$$

From Eq. (4.299), however,

$$\langle\tau|t_1|\tau\rangle = \langle\tau|a_1|\tau\rangle e^r, \qquad \langle\tau|t_1^2|\tau\rangle = \langle\tau|a_1^2|\tau\rangle e^{2r}, \tag{4.304a}$$

$$\langle\tau|t_2|\tau\rangle = \langle\tau|a_2|\tau\rangle e^{-r}, \qquad \langle\tau|t_2^2|\tau\rangle = \langle\tau|a_2^2|\tau\rangle e^{-2r}. \tag{4.304b}$$

Consequently,

$$(\Delta a_1)_\tau^2 = (\Delta t_1)^2 e^{-2r} = \frac{1}{2} e^{-2r}, \tag{4.305a}$$

$$(\Delta a_2)_\tau^2 = (\Delta t_2)^2 e^{2r} = \frac{1}{2} e^{2r}, \tag{4.305b}$$

$$(\Delta a_1 \, \Delta a_2)_\tau = \frac{1}{2}. \tag{4.305c}$$

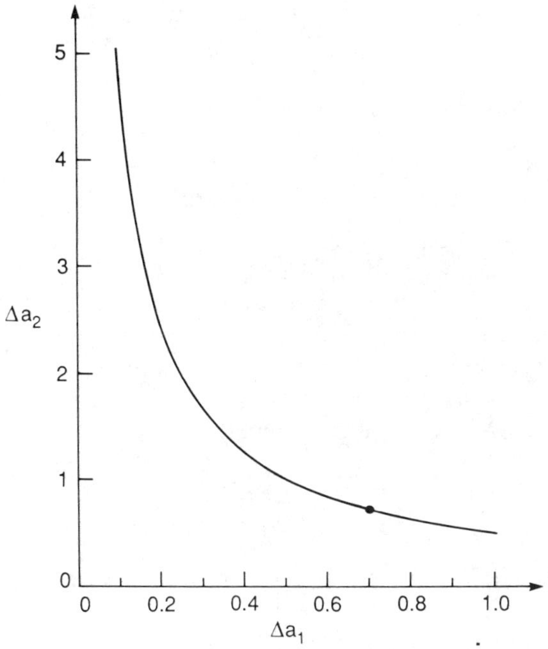

FIGURE 4.3 A plot of Δa_2 *vs* Δa_1 where $\Delta a_1 \, \Delta a_2 = 1/2$. The coherent state is the single point at $\Delta a_1 = \Delta a_2 = (1/2)^{1/2}$. Squeezed states correspond to other points on the curve where $\Delta a_1 \neq \Delta a_2$.

The conclusion, then, is that in a squeezed state the magnitude of the fluctuations or the noise level in one of the two quadrature components can be reduced below the noise level of a coherent state, which, we saw, corresponds to the vacuum limit. Clearly, the noise diminution in one quadrature is achieved at the expense of noise enhancement in the other quadrature. A plot of Eq. (4.305) is shown in Fig. 4.3.

Various nonlinear processes have been proposed for the production of squeezed states. Among them are parametric amplifiers [12], degenerate and nondegenerate four-wave mixers [13–17], resonance fluorescence [18, 19], and interaction of electromagnetic waves with plasmas [20]. Experimental verification employing phase-sensitive detection methods has been demonstrated by four-wave mixing [21–23] and by degenerate parametric down-conversion [24]. Applications of squeezed states to communication [25] and to gravitational radiation detectors [26] have been suggested.

4.10 Gauge Transformations

In classical electromagnetic theory, a gauge transformation on the scalar potential $\phi(\mathbf{r}, t)$ and the vector potential $\mathbf{A}(\mathbf{r}, t)$ is defined by the relations

$$\begin{aligned} \phi'(\mathbf{r}, t) &= \phi(\mathbf{r}, t) - \frac{\partial}{\partial t} f(\mathbf{r}, t), \\ \mathbf{A}'(\mathbf{r}, t) &= \mathbf{A}(\mathbf{r}, t) + \nabla f(\mathbf{r}, t), \end{aligned} \tag{4.306}$$

where $f(\mathbf{r}, t)$ is an arbitrary differentiable scalar function of space and time. Since the fields are related to the potentials by

$$\begin{aligned} \mathbf{E}(\mathbf{r}, t) &= -\nabla\phi(\mathbf{r}, t) - \frac{\partial}{\partial t}\mathbf{A}(\mathbf{r}, t), \\ \mathbf{B}(\mathbf{r}, t) &= \nabla \times \mathbf{A}(\mathbf{r}, t), \end{aligned} \tag{4.307}$$

a gauge transformation of the potentials has no effect on the fields; that is, when ϕ and $\mathbf{A}$ are replaced by ϕ' and $\mathbf{A}'$, the fields are unchanged. Hence, the potentials are not determined uniquely.

It now will be shown that the transformations (4.306) also may arise in a quantum mechanical context. The fundamental Hamiltonian for the interaction of a nonrelativistic electron with an external electromagnetic field [1, 2] is

$$\mathscr{H}(\mathbf{r}, t) = \frac{1}{2m_e}[\mathbf{p} + e\mathbf{A}(\mathbf{r}, t)]^2 - e\phi(\mathbf{r}, t), \tag{4.308}$$

where m_e, e, and p are the mass, charge, and linear momentum, respectively, of the electron; $\phi(\mathbf{r}, t)$ and $\mathbf{A}(\mathbf{r}, t)$ are the scalar and vector potentials, respectively, of the external field. It is recalled (Section 3.2) that the unitary transformation

$$\psi'(\mathbf{r}, t) = O\psi(\mathbf{r}, t) \tag{4.309}$$

transforms the time-dependent Schrödinger equation

$$i\hbar\frac{\partial}{\partial t}\psi(\mathbf{r}, t) = \mathscr{H}\psi(\mathbf{r}, t) \tag{4.310}$$

into

$$i\hbar\frac{\partial}{\partial t}\psi'(\mathbf{r}, t) = \mathscr{H}'\psi'(\mathbf{r}, t), \tag{4.311}$$

where

$$\mathscr{H}' = O\mathscr{H}O^\dagger + i\hbar\frac{\partial O}{\partial t}O^\dagger. \tag{4.312}$$

Now, let

$$O = e^{-ief/\hbar}, \qquad f \equiv f(\mathbf{r}, t). \tag{4.313}$$

With $p = -i\hbar\nabla$,

$$\mathbf{p}O^\dagger = eO^\dagger\nabla f + O^\dagger\mathbf{p}, \qquad (\mathbf{p} + e\mathbf{A})O^\dagger = O^\dagger[\mathbf{p} + e(\mathbf{A} + \nabla f)], \tag{4.314}$$

$$(\mathbf{p} + e\mathbf{A})^2 O^\dagger = (\mathbf{p} + e\mathbf{A})O^\dagger[\mathbf{p} + e(\mathbf{A} + \nabla f)] = O^\dagger[\mathbf{p} + e(\mathbf{A} + \nabla f)]^2, \tag{4.315}$$

we have

$$O(p + eA)^2 O^\dagger = [p + e(A + \nabla f)]^2, \tag{4.316}$$

$$O\phi O^\dagger = \phi, \qquad i\hbar\frac{\partial O}{\partial t}O^\dagger = e\frac{\partial f}{\partial t}. \tag{4.317}$$

The Hamiltonian (Eq. (4.308)) is transformed then into

$$\mathscr{H}' = \frac{1}{2m}[\mathbf{p} + e(\mathbf{A} + \nabla f)]^2 - e\left(\phi - \frac{\partial f}{\partial t}\right) = \frac{1}{2m}[\mathbf{p} + e\mathbf{A}']^2 - e\phi'. \tag{4.318}$$

What has been shown is that the operator O defined by Eq. (4.313) acting through the transformation (4.312), generates a gauge transformation of the type of Eq. (4.306) when the Hamiltonian has the form of Eq. (4.308). This, then, is the quantum mechanical counterpart to the classical gauge transformation. Since O is a unitary operator, its effect on the wave function ψ is to alter the phase but not the absolute value. Consequently, the physical content of quantum mechanics cannot depend on the specific choice of the gauge function $f(\mathbf{r}, t)$.

Let us now consider the case of a electron interacting with an electromagnetic field that has no scalar potential ($\phi = 0$). The field then, is, described completely by the vector potential alone, as in the case of a radiation field. The Hamiltonian (Eq. (4.308)) reduces to

$$\mathscr{H} = \frac{1}{2m}[\mathbf{p} + e\mathbf{A}(r, t)]^2, \tag{4.319}$$

and the transformed Hamiltonian (Eq. (4.318)) reduces to

$$\mathscr{H}' = \frac{1}{2m}[\mathbf{p} + e(\mathbf{A} + \nabla f)]^2 + e\frac{\partial f}{\partial t}. \tag{4.320}$$

We now choose a gauge function in the form

$$f(r, t) = -\mathbf{A}(t)\cdot\mathbf{r}, \tag{4.321}$$

in which the vector potential is independent of spatial coordinates. Since we are concerned mainly with atomic or molecular systems, it is sufficient for **A** (or the fields **E** and **B**) not to vary over a distance of a few Angstroms. In other words, the wavelengths associated with **A** must be long in comparison with atomic dimensions, or,

$$\mathbf{k} \cdot \mathbf{r} \ll 1. \tag{4.322}$$

In the context of a multipole expansion, this condition is recognized as the approximation whereby all multipoles except the leading one—the dipole term—are ignored. Specifically, it is the electric dipole term ($E1$) that is of primary interest.

With the gauge function (Eq. (4.321)) we have

$$\nabla f = -\mathbf{A}(t), \qquad \frac{\partial f}{\partial t} = -r \cdot \frac{\partial \mathbf{A}}{\partial t} = \mathbf{r} \cdot \mathbf{E}(t), \tag{4.323}$$

and

$$\mathscr{H}' = \frac{p^2}{2m} + e\mathbf{r} \cdot \mathbf{E}(t). \tag{4.324}$$

Thus, the effect of the gauge transformation with the gauge function (Eq. (4.321)) has been to eliminate the vector potential from the Hamiltonian.

The Hamiltonian, Eq. (4.319), contains the terms $e(\mathbf{p} \cdot \mathbf{A} + \mathbf{A} \cdot \mathbf{p})$. If $g(\mathbf{r}, t)$ is an arbitrary function of position and time,

$$\mathbf{p} \cdot \mathbf{A}g(\mathbf{r}, t) = \mathbf{A} \cdot [\mathbf{p}g(\mathbf{r}, t)] + g(\mathbf{r}, t)[\mathbf{p} \cdot \mathbf{A}]. \tag{4.325}$$

Since we are interested exclusively in radiation fields obeying the transversality condition, the vector potential is subject to the Coulomb gauge

(Section 4.2) $\nabla \cdot \mathbf{A} = 0$. Therefore,

$$(\mathbf{p} \cdot \mathbf{A}) = -i\hbar\nabla \cdot \mathbf{A} = 0, \tag{4.326}$$

and

$$\mathbf{p} \cdot \mathbf{A} = \mathbf{A} \cdot \mathbf{p}. \tag{4.327}$$

We see, then, that the interaction term in $\mathscr{H}$ is

$$\frac{e}{m}\mathbf{p} \cdot \mathbf{A} + \left(\frac{e}{2m}\right)^2 A^2. \tag{4.328}$$

On the other hand, the interaction term in the gauge-transformed Hamiltonian (Eq. (4.324)) is

$$e\mathbf{r} \cdot \mathbf{E}(t) = -\mathbf{d} \cdot \mathbf{E}(t) \tag{4.329}$$

where $\mathbf{d} = -e\mathbf{r}$ is the electronic dipole moment operator. Thus, within the dipole approximation, the single term (Eq. (4.329)) is equivalent to the two terms (Eq. (4.328)).

One should be aware, however, that although Eq. (4.329) is used widely as the interaction Hamiltonian in quantum optics, more delicate considerations may need to be taken into account [27–29].

4.11 Density Matrix for Interactions with Monochromatic Fields

For later applications to nonlinear phenomena, we now shall evaluate the density matrix elements shown in Eqs. (2.274) and (2.275) for systems initially in thermal equilibrium when $V(t)$ is the dipole interaction operator

$$V(t) = -\mathbf{d} \cdot \mathbf{E}(t) \tag{4.330}$$

as in Eq. (4.329). Following the convention due to Shen [30], the complex classical field (with the spatial dependence omitted) is written in terms of its Fourier components.

$$\mathbf{E}(t) = \sum_i \mathbf{E}(\omega_r)e^{-i\omega_r t}. \tag{4.331}$$

Employing the definitions

$$v_{kl}(\omega_r) \equiv -\mathbf{d}_{kl} \cdot \mathbf{E}(\omega_r), \tag{4.332}$$

and the assumption that d_{kl} are real matrix elements, the matrix element of the interaction operator, Eq. (4.330), for a single monochromatic field of frequency ω, is

$$V_{kl}(t) = v_{kl}(\omega)e^{-i\omega t}, \tag{4.333}$$

and the matrix element (Eq. (2.274a)) becomes

$$\langle k|\rho_a^{(1)}(t)|l\rangle = -\frac{i}{\hbar}e^{-i\omega_{kl}t}\rho_{ll}^0 v_{kl}(\omega)\int_{t_0}^{t} dt_1 e^{i(\omega_{kl}-\omega)t_1}. \tag{4.335}$$

The integral is readily evaluated, bearing in mind that $V(t) = 0$ for $t \le t_0$ as discussed in Section 2.8. Thus,

$$\langle k|\rho_a^{(1)}(t)|l\rangle = -\frac{1}{\hbar}\rho_{ll}^0 \frac{v_{kl}(\omega)e^{-i\omega t}}{(\omega_{kl}-\omega)}. \tag{4.336}$$

Following the same procedure, the matrix element (Eq. (2.274b)) becomes

$$\langle k|\rho_b^{(1)}(t)|l\rangle = \frac{1}{\hbar}\rho_{kk}^0 \frac{v_{kl}(\omega)e^{-i\omega t}}{(\omega_{kl}-\omega)}. \tag{4.337}$$

Combining Eqs. (4.336) and (4.337),

$$\begin{aligned}\langle k|\rho^{(1)}(t)|l\rangle &= \langle k|\rho_a^{(1)}(t)|l\rangle + \langle k|\rho_b^{(1)}(t)|l\rangle \\ &= \frac{1}{\hbar}(\rho_{kk}^0 - \rho_{ll}^0)\frac{v_{kl}(\omega)e^{-i\omega t}}{(\omega_{kl}-\omega)}.\end{aligned} \tag{4.338}$$

When the field contains more than one frequency, it is necessary merely to write an expression like the one above for each frequency; the final matrix element is obtained by summing over all frequencies, i.e.,

$$\langle k|\rho^{(1)}(t)|l\rangle \equiv \rho_{kl}^{(1)}(t) = \frac{1}{\hbar}(\rho_{kk}^0 - \rho_{ll}^0)\sum_r \frac{v_{kl}(\omega_r)e^{-i\omega_r t}}{(\omega_{kl}-\omega_r)}. \tag{4.339}$$

It is clear from this expression that the individual frequencies do not mix with one another—each frequency behaves independently. Because of this property, $\rho^{(1)}(t)$ is said to be associated with linear effects.

Nonlinear effects begin to appear when we investigate $\rho^{(2)}(t)$. For this purpose, let

$$\mathbf{E}(t) = \mathbf{E}(\omega_1)e^{-i\omega_1 t} + \mathbf{E}(\omega_2)e^{-i\omega_2 t}. \tag{4.340}$$

The matrix element of the interaction operator then may be written

$$V_{kl}(t) = v_{kl}(\omega_1)e^{-i\omega_1 t} + v_{kl}(\omega_2)e^{-i\omega_2 t}. \tag{4.341}$$

With this interaction, we can generate a density matrix whose time-dependence is proportional to $\exp[-i(\omega_1 + \omega_2)t]$. Should we wish to generate a density matrix proportional to $\exp[-i(\omega_1 - \omega_2)t]$, the Fourier components

$$\mathbf{E}(t) = \mathbf{E}(\omega_1)e^{-i\omega_1 t} + \mathbf{E}^*(\omega_2)e^{i\omega_2 t}, \tag{4.342}$$

and the interaction matrix element

$$V_{kl}(t) = v_{kl}(\omega_1)e^{-i\omega_1 t} + v_{kl}^*(\omega_2)e^{i\omega_2 t} \tag{4.343}$$

would be required. As a general rule, the time-dependence of the density matrix will be written in the form $\exp[-i(\pm\omega_1 \pm \omega_2 \pm \cdots \pm \omega_n)t]$. To illustrate the above, we compute the terms in $\langle k|\rho^{(2)}(t)|l\rangle$ whose frequency dependence is of the form $\exp[-i(\omega_1 - \omega_2)t]$.

Consider the matrix element (Eq. (2.275a) that contains the product $V_{kr}(t_1)V_{rl}(t_2)$ with the time variables in the order $t > t_1 > t_2 > t_0$. Choosing

$$\begin{aligned} V_{kr}(t_1) &= v_{kr}(\omega_1)e^{-i\omega_1 t_1}, \\ V_{rl}(t_2) &= v_{rl}^*(\omega_2)e^{i\omega_2 t_2}, \end{aligned} \tag{4.344}$$

produces the integral

$$\begin{aligned} \int_{t_0}^{t} dt_1 \int_{t_0}^{t_1} dt_2 e^{i(\omega_{kr}-\omega_1)t_1} e^{i(\omega_{rl}+\omega_2)t_2} &= \int_{t_0}^{t} dt_1 \frac{e^{i(\omega_{kr}-\omega_1)t_1} e^{i(\omega_{rl}+\omega_2)t_1}}{i(\omega_{rl}+\omega_2)} \\ &- \frac{e^{i(\omega_{kl}-\omega_1+\omega_2)t}}{i(\omega_{rl}+\omega_2)i(\omega_{kl}-\omega_1+\omega_2)}, \end{aligned} \tag{4.345}$$

in which $\omega_{kl} = \omega_{kr} + \omega_{rl}$. Hence, the contribution to Eq. (2.275a) from the interaction shown in Eq. (4.344) is

$$\left(\frac{1}{\hbar}\right)^2 \rho_{ll}^0 \sum_r v_{kr}(\omega_1)v_{rl}^*(\omega_2) \frac{e^{-i(\omega_1-\omega_2)t}}{(\omega_{rl}+\omega_2)(\omega_{kl}-\omega_1+\omega_2)}. \tag{4.346}$$

The same procedure yields the matrix elements (Eq. (2.275b), (2.275c), and (2.275d)) for the particular choice shown in Eq. (4.344). The results for the four matrix elements now are written

$t > t_1 > t_2 > t_0$:

$$\begin{aligned} &V_{kr}(t_1)V_{rl}(t_2) \\ &\quad = v_{kr}(\omega_1)e^{-i\omega_1 t_1} v_{rl}^*(\omega_2)e^{i\omega_2 t_2}\langle k|\rho_a^{(2)}[t, v_{kr}(\omega_1), v_{rl}^*(\omega_2)]|l\rangle \\ &\quad = \left(\frac{1}{\hbar}\right)^2 \rho_{ll}^0 \sum_r v_{kr}(\omega_1)v_{rl}^*(\omega_2) \frac{e^{-i(\omega_1-\omega_2)t}}{(\omega_{rl}+\omega_2)(\omega_{kl}-\omega_1+\omega_2)} \end{aligned} \tag{4.347a}$$

$t > t_1 > t_1' > t_0$:

$$\begin{aligned} &V_{kr}(t_1)V_{rl}(t_1') \\ &\quad = v_{kr}(\omega_1)e^{-i\omega_1 t_1} v_{rl}^*(\omega_2)e^{i\omega_2 t_1'}\langle k|\rho_b^{(2)}[t, v_{kr}(\omega_1), v_{rl}^*(\omega_2)]|l\rangle \\ &\quad = -\left(\frac{1}{\hbar}\right)^2 \sum_r \rho_{rr}^0 v_{kr}(\omega_1)v_{rl}^*(\omega_2) \frac{e^{-i(\omega_1-\omega_2)t}}{(\omega_{rl}+\omega_2)(\omega_{kl}-\omega_1+\omega_2)}, \end{aligned} \tag{4.347b}$$

$t > t_1' > t_1 > t_0$:

$$\begin{aligned} &V_{kr}(t_1)V_{rl}(t_1') \\ &\quad = v_{kr}(\omega_1)e^{-i\omega_1 t_1}v_{rl}^*(\omega_2)e^{i\omega_2 t_1'}\langle k|\rho_c^{(2)}[t, v_{kr}(\omega_1), v_{rl}^*(\omega_2)]|l\rangle \\ &\quad = -\left(\frac{1}{\hbar}\right)^2 \sum_r \rho_{rr}^0 v_{kr}(\omega_1)v_{rl}^*(\omega_2)\frac{e^{-i(\omega_1-\omega_2)t}}{(\omega_{kr}-\omega_1)(\omega_{kl}-\omega_1+\omega_2)}, \end{aligned} \tag{4.347c}$$

$t > t_1' > t_2' > t_0$:

$$\begin{aligned} &V_{kr}(t_2')V_{rl}(t_1') \\ &\quad = v_{kr}(\omega_1)e^{-i\omega_1 t_2'}v_{rl}^*(\omega_2)e^{i\omega_2 t_1'}\langle k|\rho_d^{(2)}[t, v_{kr}(\omega_1), v_{rl}^*(\omega_2)] \\ &\quad = \left(\frac{1}{\hbar}\right)^2 \rho_{kk}^0 \sum_r v_{kr}(\omega_1)v_{rl}^*(\omega_2)\frac{e^{-i(\omega_1-\omega_2)t}}{(\omega_{kr}-\omega_1)(\omega_{kl}-\omega_1+\omega_2)}. \end{aligned} \tag{4.347d}$$

The frequency dependence $\exp[-i(\omega_1 - \omega_2)t]$ of the four terms in Eq. (4.347) came about as a result of the choice of Eq. (4.344). There is, however, another choice of interaction terms that leads to the same frequency dependence as we now show. Let

$$V_{kr}(t_1) = v_{kr}^*(\omega_2)e^{i\omega_2 t_1},\ V_{rl}(t_2) = v_{rl}(\omega_1)e^{-i\omega_1 t_2}. \tag{4.348}$$

We now have another set of terms analogous to Eq. (4.347):

$t > t_1 > t_2 > t_0$:

$$\begin{aligned} &V_{kr}(t_1)V_{rl}(t_2) \\ &\quad = v_{kr}^*(\omega_2)e^{i\omega_2 t_1}v_{rl}(\omega_1)e^{-i\omega_1 t_1'}\langle k|\rho_b^{(2)}[t, v_{kr}^*(\omega_2), v_{rl}(\omega_1)]|l\rangle \\ &\quad = \left(\frac{1}{\hbar}\right)^2 \rho_{ll}^0 \sum_r v_{kr}^*(\omega_2)v_{rl}(\omega_1)\frac{e^{-i(\omega_1-\omega_2)t}}{(\omega_{rl}-\omega_1)(\omega_{kl}-\omega_1+\omega_2)}, \end{aligned} \tag{4.349a}$$

$t > t_1 > t_1' > t_0$:

$$\begin{aligned} &V_{kr}(t_1)V_{rl}(t_1') \\ &\quad = v_{kr}^*(\omega_2)e^{i\omega_2 t_1}v_{rl}(\omega_1)e^{-i\omega_1 t_1'}\langle k|\rho_b^{(2)}[t, v_{kr}^*(\omega_2), v_{rl}(\omega_1)]|l\rangle \\ &\quad = -\left(\frac{1}{\hbar}\right)^2 \sum_r \rho_{rr}^0 v_{kr}^*(\omega_2)v_{rl}(\omega_1)\frac{e^{-i(\omega_1-\omega_2)t}}{(\omega_{rl}-\omega_1)(\omega_{kl}-\omega_1+\omega_2)}, \end{aligned} \tag{4.349b}$$

$t > t_1' > t_1 > t_0$:

$$\begin{aligned} &V_{kr}(t_1)V_{rl}(t_1') \\ &\quad = v_{kr}^*(\omega_2)e^{i\omega_2 t_1}v_{rl}(\omega_1)e^{-i\omega_1 t_1'}\langle k|\rho_c^{(2)}[t, v_{kr}^*(\omega_2), v_{rl}(\omega_1)]|l\rangle \\ &\quad = -\left(\frac{1}{\hbar}\right)^2 \sum_r \rho_{rr}^0 v_{kr}^*(\omega_2)v_{rl}(\omega_1)\frac{e^{-i(\omega_1-\omega_2)t}}{(\omega_{kr}+\omega_2)(\omega_{kl}-\omega_1+\omega_2)}, \end{aligned} \tag{4.349c}$$

$t > t_1' > t_2' > t_0$:

$$V_{kr}(t_2')V_{rl}(t_1')$$
$$= v_{kr}^*(\omega_2)e^{i\omega_2 t_2'}v_{rl}(\omega_1)e^{-i\omega_1 t_1'}\langle k|\rho_d^{(2)}[t, v_{kr}^*(\omega_2), v_{rl}(\omega_1)]|l\rangle$$
$$= \left(\frac{1}{\hbar}\right)^2 \rho_{kk}^0 \sum_r v_{kr}^*(\omega_2)v_{rl}(\omega_1)\frac{e^{-i(\omega_1-\omega_2)t}}{(\omega_{kr}+\omega_2)(\omega_{kl}-\omega_1+\omega_2)}. \quad (4.349d)$$

The complete matrix element $\langle k|\rho^{(2)}(t)|l\rangle$ with frequency dependence $\exp[-i(\omega_1-\omega_2)t]$ is the sum of the eight terms contained in Eqs. (4.347) and (4.349). Contributions to the matrix element at other frequencies are calculated in the same way.

Let us now extend the Feynman diagram formalism [31–34] introduced in Section 2.9 to interactions described by Eqs. (4.330) and (4.331). In first order, there are two diagrams as shown in Figs. 2.3 and 2.7, with an interaction point at t_1 (left or ket side) and an interaction point t_1' (right or bra side). In Fig. 4.4 the same diagrams are shown without any labels. At each interaction point

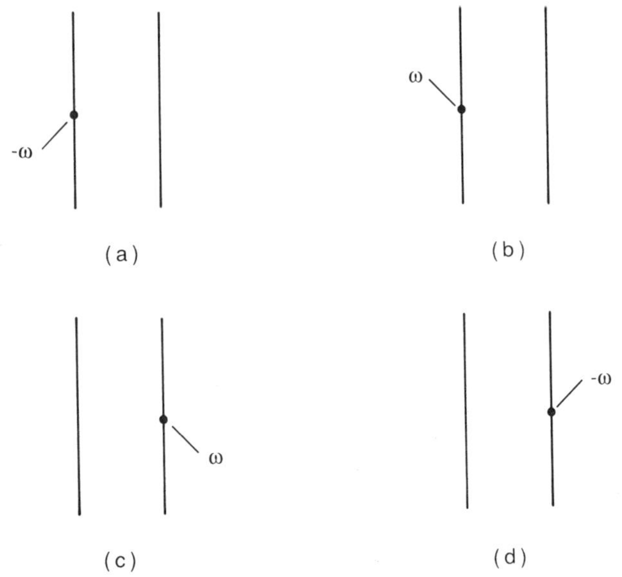

FIGURE 4.4 Conventions employed in Feynman diagrams for interactions with monochromatic fields. At each interaction point a sloping line (/) or (\) is attached. The label $-\omega$ is attached to (/), shown in (a) and (d), to correspond to $v(\omega)\exp[-i\omega t]$ where $v(\omega) = -\mathbf{d}\cdot\mathbf{E}(\omega)$; the label ω is attached to (\), shown in (b) and (c), to correspond to $v^*(\omega)\exp[i\omega t]$ where $v^*(\omega) = -\mathbf{d}\cdot\mathbf{E}^*(\omega)$.

attach a sloping line according to the following scheme:

A line sloping southwest ↔ northwest (/) is labeled $-\omega$ and is associated with $v(\omega)\exp[-i\omega t]$.

A line sloping southwest ↔ northwest (\) is labeled ω and is associated with $v^*(\omega)\exp[i\omega t]$.

Since there are two possible sloping lines at each interaction point, we now have a total of four diagrams of which two are associated with the factor $\exp[-i\omega t]$ and two with $\exp[i\omega t]$. The terms in Eqs. (4.336) and (4.337) are represented in Fig. 4.5; after labeling the line segments, (defined in Section 2.9) the matrix element of $v(\omega)$ is determined by the labels on the line segments above and below the interaction point, taking into account the direction of the arrows. The matrix element of ρ^0 is determined by the labels on the ket and bra line segments at t_0. Proceeding above the interaction point, the labels on the ket and bra line segments determine the resonance denominator $\omega_{kl} - \omega$. Finally, a factor of $-1/\hbar$ is inserted when the interaction point is on the ket side and a factor of $1/\hbar$ for an interaction point on the bra side.

In second order, there are four basic diagrams (Figs. 2.4 and 2.8). On each diagram there are two interaction points (which no longer need to be labelled). Since we have selected a time-dependence in the form $\exp[-i(\omega_1 - \omega_2)t]$, each interaction point can be labeled $-\omega_1$ or ω_2, thereby doubling the number of diagrams to eight (Figs. 4.6–4.9). Following the rules stated previously,

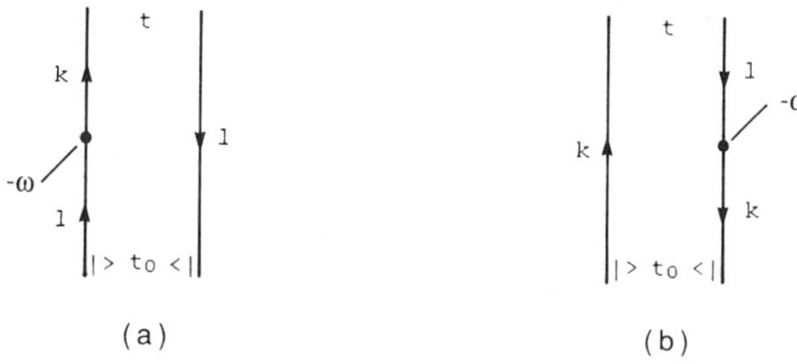

FIGURE 4.5 Diagrams for

$$\langle k|\rho^{(1)}(t)|l\rangle = \langle k|\rho_a^{(1)}(t)|l\rangle + \langle k|\rho_b^{(1)}(t)|l\rangle$$

in which each term contains the factor $\exp[-i\omega t]$.

a. $\langle k|\rho_a^{(1)}(t)|l\rangle = -\dfrac{1}{\hbar}\rho_{ll}^0 \dfrac{v_{kl}(\omega)e^{-i\omega t}}{(\omega_{kl} - \omega)}.$

b. $\langle k|\rho_b^{(1)}(t)|l\rangle = \dfrac{1}{\hbar}\rho_{kk}^0 \dfrac{v_{kl}(\omega)e^{-i\omega t}}{(\omega_{kl} - \omega)}.$

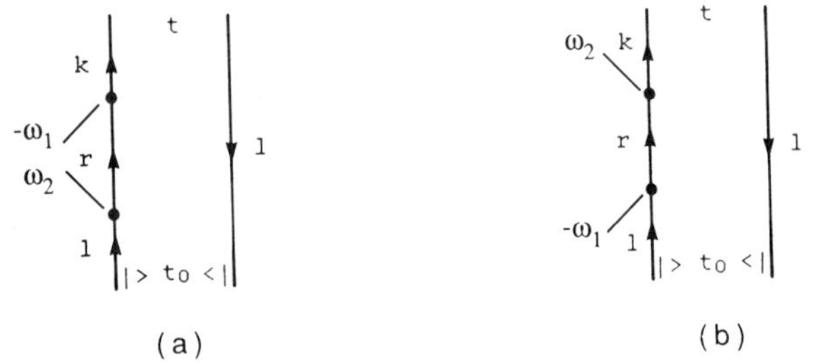

FIGURE 4.6 Diagrams to represent (a) Eq. (4.347a) and (b) Eq. (4.349a).

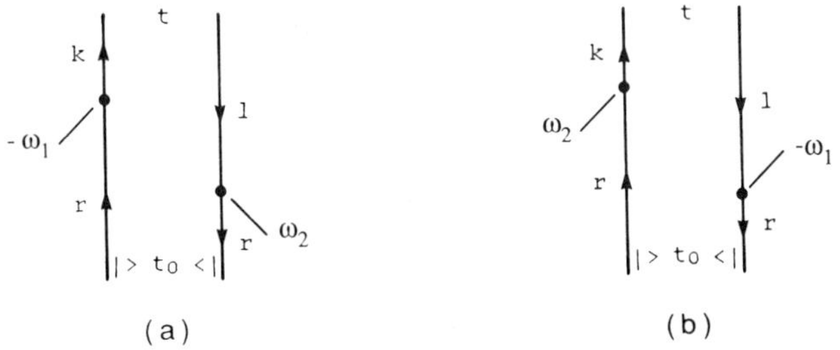

FIGURE 4.7 Diagrams to represent (a) Eq. (4.347b) and (b) Eq. (4.349b).

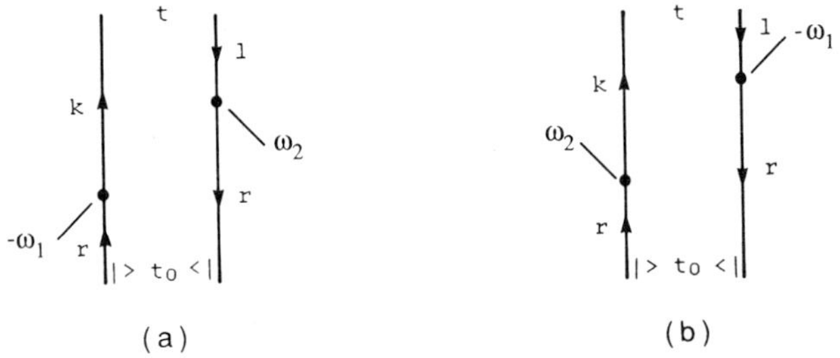

FIGURE 4.8 Diagrams to represent (a) Eq. (4.347c) and (b) Eq. (4.349c).

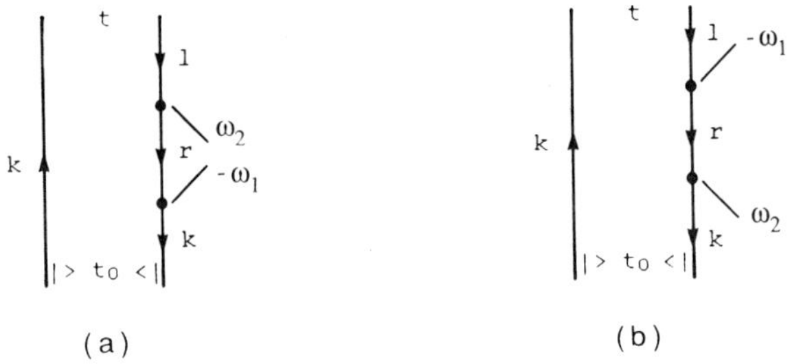

FIGURE 4.9 Diagrams to represent (a) Eq. (4.347d) and (b) Eq. (4.349d).

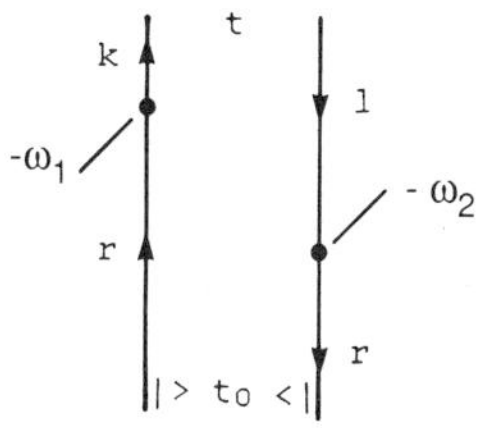

FIGURE 4.10 A diagram that represents a single term among the terms belonging to $\rho_{kl}^{(2)}(t)$ with the factor $\exp[-i(\omega_1 + \omega_2)t]$. The diagram shown corresponds to

$$-\left(\frac{1}{\hbar}\right)^2 \sum_r \rho_{rr}^0 v_{kr}(\omega_1) v_{rl}(\omega_2) \frac{e^{-i(\omega_1+\omega_2)t}}{(\omega_{rl} - \omega_2)(\omega_{kl} - \omega_1 - \omega_2)}.$$

we identify the matrix elements of ρ^0, $v(\omega_1)v^*(\omega_2)$, and the products of $\pm 1/\hbar$. The rule for obtaining the resonance denominator for the first-order case now must be extended to a more general rule illustrated by Fig. 4.6a: after passing the first interaction point (ω_2), the labels on the ket and bra line segments contribute the resonance denominator $\omega_{rl} + \omega$; after passing the second interaction point, the labels on the ket and bra line segments contribute the resonance denominator $\omega_{kl} - \omega_1 + \omega_2$. Each time an interaction point is crossed, the contribution to the resonance denominator contains all positive and negative frequencies below it. The final step is to sum over intermediate states.

An example of a second-order diagram for density matrix elements whose time-dependence is of the form $\exp[-i(\omega_1 + \omega_2)]$ is shown in Fig. 4.10 while Fig. 4.11 illustrates a third-order diagram with the time-dependence

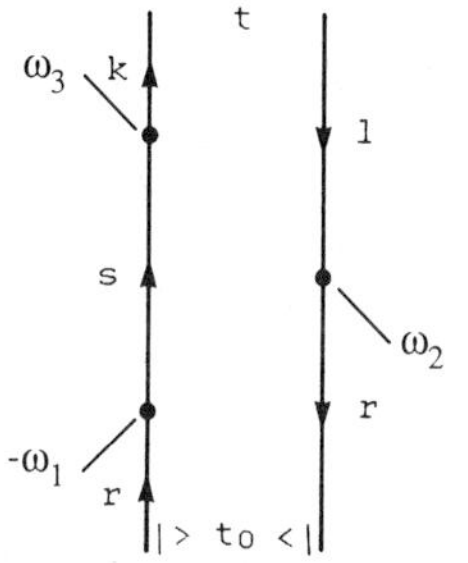

FIGURE 4.11 A diagram that represents a single term among the terms belonging to $\rho_{kl}^{(3)}(t)$ with the factor $\exp[-i(\omega_1 - \omega_2 - \omega_3)t]$. The diagram shown corresponds to

$$\left(-\frac{1}{\hbar}\right)^2 \left(\frac{1}{\hbar}\right) \sum_{rs} \rho_{rr}^0 v_{ks}^*(\omega_3) v_{sr}(\omega_1) v_{rl}^*(\omega_2) \frac{e^{-i(\omega_1-\omega_2-\omega_3)t}}{(\omega_{sr} - \omega_1)(\omega_{sl} - \omega_1 + \omega_2)(\omega_{kl} - \omega_1 + \omega_2 + \omega_3)}.$$

$\exp[-i(\omega_1 - \omega_2 - \omega_3)]$. The corresponding matrix elements are

$$\rho_{kl}^{(2)}(t) = \div\left(\frac{1}{\hbar}\right)^2 e^{-i(\omega_1+\omega_2)t}\sum_r \rho_{rr}^0 \frac{v_{kr}(\omega_1)v_{rl}(\omega_2)}{(\omega_{rl}-\omega_2)(\omega_{kl}-\omega_1-\omega_2)}, \tag{4.350}$$

$$\rho_{kl}^{(3)}(t) = \left(\frac{-1}{\hbar}\right)^2\left(\frac{1}{\hbar}\right)e^{-i(\omega_1-\omega_2-\omega_3)t} \times \sum_{rs}\rho_{rr}^0 \frac{v_{ks}^*(\omega_3)v_{sr}(\omega_1)v_{rl}^*(\omega_2)}{(\omega_{sr}-\omega_1)(\omega_{sl}-\omega_1+\omega_2)(\omega_{kl}-\omega_1+\omega_2+\omega_3)}. \tag{4.351}$$

Finally, it should be noted that the conventions for the construction of diagrams are not universal. Therefore, one must exercise some care in comparing diagrams by various authors.

References

[1] W. Heitler, *The Quantum Theory of Radiation.* Oxford Press, Oxford, 1954.
[2] M. Weissbluth, *Atoms and Molecules.* Academic Press, New York, 1978.
[3] W. H. Louisell, *Quantum Statistical Properties of Radiation.* J. Wiley, New York, 1973.
[4] A. Messiah, *Quantum Mechanics.* North-Holland, Amsterdam, 1962.
[5] R. J. Glauber, *Phys. Rev.* **130**, 2529(1963).
[6] R. J. Glauber, In *Quantum Optics and Electronics.* (C. DeWitt, A. Blandin, and C. Cohen-Tannoudji, eds.), Gordon and Breach, New York, 1965.
[7] R. J. Glauber, In *Quantum Optics.* (R. J. Glauber, ed.), Academic Press, New York, 1969.
[8] R. J. Glauber, In *Laser Handbook.* (F. T. Arecchi and E. O. Schulz-Dobois, eds.), North-Holland, Amsterdam, 1972.
[9] H. P. Yuen, *Phys. Rev.* A**13**, 2226(1976).
[10] D. F. Walls, *Nature*, **306**, 141(1983).
[11] C. M. Caves and B. L. Schumaker, *Phys. Rev.* A**31**, 3068(1985).
[12] M. J. Collett and C. W. Gardiner, *Phys. Rev.* A**30**, 1386(1984).
[13] M. D. Reid and D. F. Walls, *Phys. Rev.* A**31**, 1622(1985).
[14] M. D. Reid and D. F. Walls, *Phys. Rev.* A**33**, 4465(1986).
[15] B. Yurke, *Phys. Rev.* A**29**, 408 (1984).
[16] J. R. Klauder, S. L. McCall, and B. Yurke, *Phys. Rev.* A**33**, 3204(1986).
[17] R. M. Shelby, M. D. Levenson, D. F. Walls, and A. Aspect, *Phys. Rev.* A**33**, 4008(1986).
[18] R. Loudon, *Opt. Commun.* **49**, 24(1984).
[19] R. Loudon, *Opt. Commun.* **49**, 67(1984).
[20] Y. Ben-Aryeh and A. Mann, *Phys. Rev.* A**32**, 552(1985).
[21] R. E. Slusher, L. W. Hollberg, B. Yurke, J. C. Mertz, and J. F. Valey, *Phys. Rev. Lett.* **55**, 2409(1985).
[22] B. L. Schumaker, S. H. Perlmutter, R. M. Shelby, and M. D. Levenson, *Phys. Rev. Lett.* **58**, 357(1987).
[23] R. S. Bondurant, P. Kumar, J. H. Shapiro and M. Maeda, *Phys. Rev.* **A30**, 343(1984).
[24] L-A. Wu, H. J. Kimble, J. L. Hall, and H. Wu, *Phys. Rev. Lett.* **57**, 2520(1986).

[25] H. P. Yuen and J. H. Shapiro, *IEEE Trans. Inf. Theory* **IT-24**, 657(1978).
[26] C. M. Caves, *Phys. Rev.* D**23**, 1693(1981).
[27] E. A. Power and S. Zienau, *Phil. Trans. Roy. Soc.* A**251**, 427(1959).
[28] D. H. Kobe and K-H. Yang, *Phys. Rev.* A**32**, 952(1985).
[29] Y. Aharonov and C. K. Au, *Phys. Rev.* A**20**, 1553(1979).
[30] Y. R. Shen, *The Principles of Nonlinear Optics.* J. Wiley, New York, 1984.
[31] J.P. Uyemura, *IEEE. J. Quant. Electr.* **QE-16**, 472(1980).
[32] J. Fujimoto and T. K. Yee. *IEEE. J. Quant. Electr.* **QE-19**, 861(1983).
[33] S. Y. Yee. T. K. Gustafson, S. A. J. Druet, and J.-P. E. Taran, *Optics Comm.* **23**, 1(1977).
[34] S. Y. Yee and T. K. Gustafson, *Phys. Rev.* A**18**, 1597(1978).

V Absorption, Emission, and Scattering in Weak Fields

The general problem of a physical system interacting with light involves the complex internal structure of the system (atom, molecule, solid, etc.) and the infinite number of modes of the electromagnetic field. The problem is simplified considerably if the light is nearly monochromatic at a frequency ω, and there exist two states $|g\rangle$ and $|e\rangle$ with energy separation $E_e - E_g = \hbar\omega_0$ such that $\omega_0 \simeq \omega$, and there are no other levels with separations close to $\hbar\omega$. We then may adopt a simplified model in which the real system interacting with the radiation field is replaced by a two-level system, $|g\rangle$ and $|e\rangle$. The high degree of monochromaticity achieved with laser sources renders validity to such a model in numerous physical experiments and permits one to focus on the radiation interaction process in the simplest possible form. To avoid a ponderous nomenclature, we shall enlarge the term "atom" to include any medium (atoms, molecules, crystals, etc.) in which radiative interactions are investigated.

Fundamentally, the theory that describes the interactions between an atom and a radiation field treats both the atom and the field as quantum mechanical systems. For many types of experiments, such as those in nonlinear optics where strong, nearly monochromatic fields with small fluctuations are employed, it is quite sufficient to ignore the quantum mechanical features of the radiation field and to treat it classically on the basis of the Maxwell equations. This format, consisting of a quantized atom and a classical field, is known as the *semiclassical theory* [1]. Nevertheless, there are important situations where such simplification is not possible. We therefore must treat both the semiclassical and the fully quantized theory. This chapter deals with processes in lowest order and constitutes the basis for higher order interactions.

5.1 Two-Level Operators

Atomic wave functions, even in the simplest cases, are complicated functions that reflect the various internal interactions, e.g., Coulombic interactions of electrons with nuclei and with each other, spin-orbit interactions, hyperfine interactions, etc. But in the present context, the specific details of atomic states are not relevant. When we specialize to a two-level system, the states $|g\rangle$ (ground state) and $|e\rangle$ (excited state) serve merely as labels to distinguish one state from the other.

By limiting the discussion to a two-level system, one gains another simplification from the fact that a spin-1/2 system, characterized by the Pauli spin operators and the eigenstates $|\alpha\rangle$ and $|\beta\rangle$, serves as a prototype for any system consisting of two states [2]. The formalism of Section 2.2 then, is, adaptable immediately to an atomic two-level system by a simple redefinition of the spin states $|\alpha\rangle$ and $|\beta\rangle$. We write

$$|g\rangle = \begin{pmatrix} 0 \\ 1 \end{pmatrix}, \qquad |e\rangle = \begin{pmatrix} 1 \\ 0 \end{pmatrix}, \tag{5.1}$$

thereby establishing a correspondence between $|g\rangle$ and $|\beta\rangle$ and between $|e\rangle$ and $|\alpha\rangle$. The opposite correspondence ($|g\rangle \leftrightarrow |\alpha\rangle, |e\rangle \leftrightarrow |\beta\rangle$) might have been chosen. To be consistent with the development in Section 3.2, however, where $|\alpha\rangle$ is the excited state and $|\beta\rangle$ is the ground state (Fig. 3.4), requires the choice in Eq. (5.1). It is important to recognize that the transfer of the Pauli scheme to the atomic case is a purely formal procedure and has nothing to do with the intrinsic spin properties of the atomic states represented by $|g\rangle$ and $|e\rangle$.

We now write, in analogy with the spin formalism,

$$\begin{aligned} \sigma^+ &= \frac{1}{2}(\sigma_x + i\sigma_y) = \begin{pmatrix} 0 & 1 \\ 0 & 0 \end{pmatrix} = |e\rangle\langle g|, \\ \sigma^- &= \frac{1}{2}(\sigma_x - i\sigma_y) = \begin{pmatrix} 0 & 0 \\ 1 & 0 \end{pmatrix} = |g\rangle\langle e|, \end{aligned} \tag{5.2}$$

$$\sigma^+\sigma^- = \frac{1}{2}(1 + \sigma_z) = \begin{pmatrix} 1 & 0 \\ 0 & 0 \end{pmatrix} = |e\rangle\langle e|, \tag{5.3}$$

$$\sigma^-\sigma^+ = \frac{1}{2}(1 - \sigma_z) = \begin{pmatrix} 0 & 0 \\ 0 & 1 \end{pmatrix} = |g\rangle\langle g|, \tag{5.4}$$

$$(\sigma^+)^2 = (\sigma^-)^2 = 0,$$

$$\sigma^+\sigma^- + \sigma^-\sigma^+ = \begin{pmatrix} 1 & 0 \\ 0 & 1 \end{pmatrix} = |e\rangle\langle e| + |g\rangle\langle g| = I \tag{5.5}$$

$$\sigma_x = \sigma^+ + \sigma^- = \begin{pmatrix} 0 & 1 \\ 1 & 0 \end{pmatrix} = |e\rangle\langle g| + |g\rangle\langle e|,$$

$$\sigma_y = -i(\sigma^+ - \sigma^-) = \begin{pmatrix} 0 & -i \\ i & 0 \end{pmatrix} = -i[|e\rangle\langle g| - |g\rangle\langle e|], \tag{5.6}$$

$$\sigma_z = \sigma^+\sigma^- - \sigma^-\sigma^+ = \begin{pmatrix} 1 & 0 \\ 0 & -1 \end{pmatrix} = |e\rangle\langle e| - |g\rangle\langle g|.$$

On the basis of these definitions, we obtain the commutator relations

$$\begin{aligned} &[\sigma^+, \sigma_x] = \sigma_z, && [\sigma^-, \sigma_x] = -\sigma_z, \\ &[\sigma^+, \sigma_y] = i\sigma_z, && [\sigma^-, \sigma_y] = i\sigma_z, \\ &[\sigma^+, \sigma_z] = -2\sigma^+, && [\sigma^-, \sigma_z] = 2\sigma^-. \end{aligned} \tag{5.7}$$

The σ operators acting on the states $|g\rangle$ and $|e\rangle$ yield the following

$$\begin{aligned} &\sigma_x|g\rangle = |e\rangle, && \sigma_x|e\rangle = |g\rangle, \\ &\sigma_y|g\rangle = -i|e\rangle, && \sigma_y|e\rangle = i|g\rangle, \\ &\sigma_z|g\rangle = -|g\rangle, && \sigma_z|e\rangle = |e\rangle, \end{aligned} \tag{5.8}$$

$$\begin{aligned} &\sigma^+|g\rangle = |e\rangle, && \sigma^+|e\rangle = 0, && \langle g|\sigma^+ = 0, && \langle e|\sigma^+ = \langle g|, \\ &\sigma^-|g\rangle = 0, && \sigma^-|e\rangle = |g\rangle, && \langle g|\sigma^- = \langle e|, && \langle e|\sigma^- = 0. \end{aligned} \tag{5.9}$$

One often refers to σ^+ and σ^- as *raising* and *lowering* operators, respectively, since σ^+, in effect, produces a transition from the lower to the upper state (Fig. 5.1) while σ^- has the opposite effect.

An alternative procedure may be developed on the basis of the fermion operators c (annihilation) and $c^\dagger$ (creation) that obey the commutation rules

$$c_k c_1 + c_1 c_k = c_k^+ c_1^+ + c_1^+ c_k^+ = 0, \tag{5.10a}$$

$$c_k c_1^\dagger + c_1^\dagger c_k \equiv \{c_k, c_1^\dagger\} = \delta_{k1}.$$

The quantity $\{c_k, c_1^\dagger\}$ is known as an *anticommutator*. The annihilation operator c_k annihilates in electron in the state $|k\rangle$ provided $|k\rangle$ initially contained an electron; if $|k\rangle$ is vacant, $c_k|k\rangle = 0$. Similarly, $c_k^\dagger$ creates an

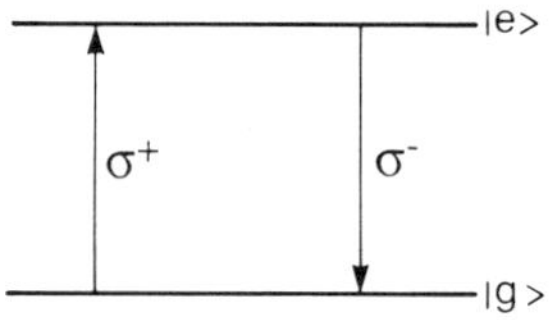

FIGURE 5.1 In a two-level system, the raising operator σ^+ operates on the lower state $|\mathrm{g}\rangle$ to produce a transition to the upper state $|\mathrm{e}\rangle$, i.e., $\sigma^+|\mathrm{g}\rangle = |\mathrm{e}\rangle$. Similarly, the lowering operator σ^- operates on the upper state $|\mathrm{e}\rangle$ to produce a transition to $|\mathrm{g}\rangle$, or $\sigma^-|\mathrm{e}\rangle = |\mathrm{g}\rangle$. The energies of the upper and lower states are E_e and E_g, respectively, and $E_\mathrm{e} - E_\mathrm{g} = \hbar\omega_0$.

electron in $|k\rangle$ provided $|k\rangle$ was initially vacant; otherwise, the effect is zero. A more general form of Eq. (5.10a) includes labels μ and μ' to identify different atoms:

$$c_{k\mu}c_{l\mu'} + c_{l\mu'}c_{k\mu} = c^\dagger_{k\mu}c^\dagger_{l\mu'} + c^\dagger_{l\mu'}c^\dagger_{k\mu} = 0$$
$$c_{k\mu}c^\dagger_{l\mu'} + c^\dagger_{l\mu'}c_{k\mu} = \delta_{kl}\delta_{\mu\mu'}. \tag{5.10b}$$

For the two-level system, the connection between the Pauli operators and the fermion operators is given by

$$\sigma^+ = c^\dagger_\mathrm{e}c_\mathrm{g} \qquad \sigma^- = c^\dagger_\mathrm{g}c_\mathrm{e}. \tag{5.11}$$

When $c^\dagger_\mathrm{e}c_\mathrm{g}$ operates on the occupied state $|g\rangle$, for example, the effect of c_g is to annihilate the electron in $|g\rangle$, thereby producing the vacuum state $|0\rangle$; the operator $c^\dagger_\mathrm{e}$ then creates an electron in $|0\rangle$ to produce the occupied state $|e\rangle$. Thus,

$$c^\dagger_\mathrm{e}c_\mathrm{g}|g\rangle = c^\dagger_\mathrm{e}|0\rangle = |\mathrm{e}\rangle, \qquad c^\dagger_\mathrm{e}c_\mathrm{g}|\mathrm{e}\rangle = 0,$$

which is consistent with $\sigma^+|\mathrm{g}\rangle = |\mathrm{e}\rangle$ and $\sigma^+|\mathrm{e}\rangle = 0$. We list several additional relations between the Pauli and fermion operators:

$$\sigma^+\sigma^- = c^\dagger_\mathrm{e}c_\mathrm{e}, \qquad \sigma^-\sigma^+ = c^\dagger_\mathrm{g}c_\mathrm{g}, \tag{5.13}$$

$$c^\dagger_\mathrm{g}c_\mathrm{g} + c^\dagger_\mathrm{e}c_\mathrm{e} = 1, \qquad c^\dagger_\mathrm{e}c_\mathrm{e} - c^\dagger_\mathrm{g}c_\mathrm{g} = \sigma_z. \tag{5.14}$$

Both the Pauli and fermion schemes are used in the literature; neither one has a distinct advantage over the other.

The Hamiltonian for the atomic two-level system with $E_\mathrm{e} - E_\mathrm{g} = \hbar\omega_0$ may be written in any one of several forms:

$$\mathscr{H}_\mathrm{A} = \frac{1}{2}(E_\mathrm{e} - E_\mathrm{g})\sigma_z = \frac{1}{2}\hbar\omega_0\sigma_z = \frac{1}{2}\hbar\omega_0(|\mathrm{e}\rangle\langle\mathrm{e}| - |\mathrm{g}\rangle\langle\mathrm{g}|)$$
$$= \frac{1}{2}\hbar\omega_0(c^\dagger_\mathrm{e}c_\mathrm{e} - c^\dagger_\mathrm{g}c_\mathrm{g}) \tag{5.15}$$

with

$$\mathscr{H}_{\mathrm{A}}|\mathrm{e}\rangle = \frac{1}{2}\hbar\omega_0|\mathrm{e}\rangle, \qquad \mathscr{H}_{\mathrm{A}}|\mathrm{g}\rangle = -\frac{1}{2}\hbar\omega_0|\mathrm{g}\rangle. \tag{5.16}$$

For this case, the zero of energy lies midway between the two levels. If one writes

$$\begin{aligned}\mathscr{H}'_{\mathrm{A}} &= \frac{1}{2}\hbar\omega_0(\sigma_z + 1) = \hbar\omega_0\sigma^+\sigma^- \\ &= \hbar\omega_0|\mathrm{e}\rangle\langle\mathrm{e}| = \hbar\omega_0 c_{\mathrm{e}}^\dagger c_{\mathrm{e}},\end{aligned} \tag{5.17}$$

the zero of energy is located at the lower level, i.e., $E_{\mathrm{g}} = 0$. Still another form with an undefined zero of energy is written

$$\begin{aligned}\mathscr{H}''_{\mathrm{A}} &= \frac{1}{2}(E_{\mathrm{e}} + E_{\mathrm{g}}) + \frac{1}{2}(E_{\mathrm{e}} - E_{\mathrm{g}})\sigma_z = E_{\mathrm{e}}\sigma^+\sigma^- + E_{\mathrm{g}}\sigma^-\sigma^+ \\ &= E_{\mathrm{g}} + (E_{\mathrm{e}} - E_{\mathrm{g}})\sigma^+\sigma^- = E_{\mathrm{g}}c_{\mathrm{g}}^\dagger c_{\mathrm{g}} + E_{\mathrm{e}}c_{\mathrm{e}}^\dagger c_{\mathrm{e}}.\end{aligned} \tag{5.18}$$

The transformation of a general operator to the Heisenberg representation is defined by Eq. (2.105). For an atomic operator such as σ_x, the transformation (with $t_0 = 0$) acquires the form

$$\begin{aligned}(\sigma_{\mathrm{H}})_x &= e^{i\mathscr{H}_{\mathrm{A}}t/\hbar}\sigma_x e^{-i\mathscr{H}_{\mathrm{A}}t/\hbar} \\ &= e^{i\omega_0 t\sigma_z/2}\sigma_x e^{-i\omega_0 t\sigma_z/2}.\end{aligned} \tag{5.19}$$

Expressions of this type already have been evaluated in Eq. (3.65). Therefore,

$$(\sigma_{\mathrm{H}})_x = \sigma_x \cos\omega_0 t - \sigma_y \sin\omega_0 t = \sigma^+ e^{i\omega_0 t} + \sigma^- e^{-i\omega_0 t}, \tag{5.20}$$

$$(\sigma_{\mathrm{H}})_y = \sigma_x \sin\omega_0 t + \sigma_y \cos\omega_0 t = \sigma^+ e^{-i\omega_0 t} + \sigma^- e^{i\omega_0 t},$$

$$\sigma_{\mathrm{H}}^+ = \sigma^+ e^{i\omega_0 t}, \qquad \sigma_{\mathrm{H}}^- = \sigma^- e^{-i\omega_0 t}, \qquad (\sigma_{\mathrm{H}})_z = \sigma_z, \tag{5.21}$$

$$\sigma_{\mathrm{H}}^+\sigma_{\mathrm{H}}^- = \sigma^+\sigma^-, \qquad \sigma_{\mathrm{H}}^-\sigma_{\mathrm{H}}^+ = \sigma^-\sigma^+. \tag{5.22}$$

The same relations apply to the interaction representation when $\mathscr{H} = \mathscr{H}_{\mathrm{A}} + V$. In particular,

$$\tilde{\sigma}^+(t) = \sigma^+ e^{i\omega_0 t}, \qquad \tilde{\sigma}^-(t) = \sigma^- e^{-i\omega_0 t}. \tag{5.23}$$

Based on the discussion in Section 3.2, it is seen that the transformations to the Heisenberg or interaction representations are associated closely with transformations to a rotating coordinate system.

The density matrix for the atomic two-level system must be of the same form as the density matrix for the spin-1/2 case. Referring to Eqs. (2.32) and (2.35), we write (bearing in mind the correspondence $|\alpha\rangle \leftrightarrow |\mathrm{e}\rangle$, $|\beta\rangle \leftrightarrow |\mathrm{g}\rangle$)

$$\rho = \begin{pmatrix} \langle \mathrm{e}|\rho|\mathrm{e}\rangle & \langle \mathrm{e}|\rho|\mathrm{g}\rangle \\ \langle \mathrm{g}|\rho|\mathrm{e}\rangle & \langle \mathrm{g}|\rho|\mathrm{g}\rangle \end{pmatrix} = \begin{pmatrix} \rho_{\mathrm{ee}} & \rho_{\mathrm{eg}} \\ \rho_{\mathrm{ge}} & \rho_{\mathrm{gg}} \end{pmatrix} \tag{5.24a}$$

$$= \frac{1}{2}\begin{pmatrix} 1 + \langle \sigma_z \rangle & \langle \sigma_x \rangle - i\langle \sigma_y \rangle \\ \langle \sigma_x \rangle + i\langle \sigma_y \rangle & 1 - \langle \sigma_z \rangle \end{pmatrix} = \frac{1}{2}(I + \langle \boldsymbol{\sigma} \rangle \cdot \boldsymbol{\sigma}), \tag{5.24b}$$

or

$$\rho = \begin{pmatrix} \langle \sigma^+\sigma^- \rangle & \langle \sigma^- \rangle \\ \langle \sigma^+ \rangle & \langle \sigma^-\sigma^+ \rangle \end{pmatrix} \tag{5.25a}$$

$$= \frac{1}{2}I + \left[\langle \sigma^+\sigma^- \rangle - \frac{1}{2}\right]\sigma_z + \langle \sigma^+ \rangle\sigma^- + \langle \sigma^- \rangle\sigma^+. \tag{5.25b}$$

As in Eq. (3.115), we introduce the quantities u_1, u_2, and u_3 defined by

$$\begin{aligned} \langle \sigma_x \rangle &= \rho_{\mathrm{eg}} + \rho_{\mathrm{ge}} \equiv u_1, \\ \langle \sigma_y \rangle &= i(\rho_{\mathrm{eg}} - \rho_{\mathrm{ge}}) \equiv u_2, \\ \langle \sigma_z \rangle &= \rho_{\mathrm{ee}} - \rho_{\mathrm{gg}} \equiv u_3, \end{aligned} \tag{5.26}$$

with

$$u_1^2 + u_2^2 + u_3^2 = 1, \tag{5.27}$$

in agreement with the general relation $\operatorname{Tr}\rho = 1$.

When the lower state $|\mathrm{g}\rangle$ is occupied and the upper state $|\mathrm{e}\rangle$ is vacant, $\langle \sigma_z \rangle = -1$, $\rho_{\mathrm{gg}} = 1$, and all other density matrix elements are zero. If the occupation of the states is reversed, $\langle \sigma_z \rangle = 1$ and the only nonvanishing matrix element is $\rho_{\mathrm{ee}} = 1$. Under thermal equilibrium both states are occupied, and the density matrix becomes

$$\rho_0 = \frac{1}{Z}\begin{pmatrix} e^{-\beta E_{\mathrm{e}}} & 0 \\ 0 & e^{-\beta E_{\mathrm{g}}} \end{pmatrix} = \frac{1}{Z}\begin{pmatrix} e^{-\beta\hbar\omega_0/2} & 0 \\ 0 & e^{\beta\hbar\omega_0/2} \end{pmatrix}, \tag{5.28}$$

where

$$\beta = \frac{1}{kT}, \qquad Z = e^{-\beta E_{\mathrm{e}}} + e^{-\beta E_{\mathrm{g}}}. \tag{5.29}$$

The analogy between the states $|\mathrm{e}\rangle$ and $|\mathrm{g}\rangle$ of a two-level atomic system on the one hand, and the state $|\alpha\rangle$ and $|\beta\rangle$ of a spin-1/2 system on the other hand (Section 2.2), enables us to extend the spin formalism to an assembly of N two-level atoms. Such an extension becomes important because in some circumstances N atoms can act cooperatively in their interaction with radiation, rather than as independent atoms. The characteristics of the emitted radiation for the two cases are significantly different (Section 5.6). In the present section, pursuing the spin analogy, we concentrate on the atomic states constructed on the basis of the rules governing the coupling of angular momenta. Starting with the case of two two-level atoms with the same resonant frequency ω_0, the atomic Hamiltonian is

$$\mathscr{H}_A = \frac{\hbar\omega_0}{2}(\sigma_{z1} + \sigma_{z2}), \tag{5.30}$$

and the four possible eigenstates and eigenvalues of $\mathscr{H}_A$ are

$|g_1g_2\rangle$: both atoms in their ground states; $E_A = -\hbar\omega$.

$|e_1e_2\rangle$: both atoms excited; $E_A = \hbar\omega$.

$|g_1e_2\rangle, |e_1g_2\rangle$: one atom in the ground state and one atom excited; $E_A = 0$.

Since $|g_1e_2\rangle$ and $|e_1g_2\rangle$ are degenerate, any linear combination of these states also will be an eigenstate of $\mathscr{H}_A$ with the same eigenvalue. Therefore, following the prescription shown in Eq. (2.56), we construct the *singlet Dicke state* [3, 4]

$$|SM\rangle = |00\rangle = \frac{1}{\sqrt{2}}[|g_1e_2\rangle - |e_1g_2\rangle], \tag{5.31}$$

and the *triplet Dicke* states

$$\begin{aligned} |SM\rangle &= |11\rangle = |e_1e_2\rangle, \\ |SM\rangle &= |10\rangle = \frac{1}{\sqrt{2}}[|g_1e_2 + |e_1g_2\rangle], \\ |SM\rangle &= |1-1\rangle = |g_1g_2\rangle. \end{aligned} \tag{5.32}$$

The states $|11\rangle$ and $|1-1\rangle$ correspond, respectively, to both atoms in their excited states and both atoms in their ground states. The singlet state $|00\rangle$ and the triplet state $|10\rangle$ both correspond to one atom in $|e\rangle$ and one in $|g\rangle$, but $|00\rangle$ is antisymmetric while $|10\rangle$ is symmetric under an interchange of atoms. Diagrams of the Dicke states are shown in Fig. 5.2.

For N two-level atoms, the Dicke states are designated by $|SM\rangle$ where S, also known as the *cooperation number*, and M are defined in Eq. (2.55). Note, also, that $M = (N_e - N_g)/2$ where $N_e + N_g = N$.

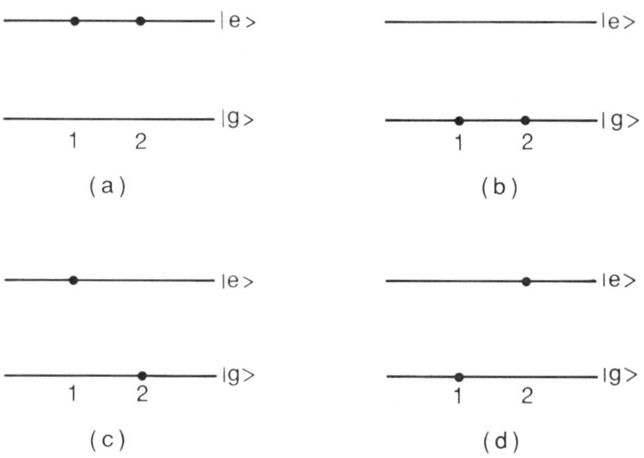

FIGURE 5.2 Two-atom Dicke states. (a) $|11\rangle = |e_1e_2\rangle$, both atoms in the upper state $|e\rangle$; (b) $|1-1\rangle = |g_1g_2\rangle$, both atoms in the lower state $|g\rangle$. The states $|11\rangle$ and $|1-1\rangle$ belong to the triplet manifold ($S = 1$). In (c) and (d) one atom is in $|g\rangle$ and one in $|e\rangle$. The symmetrical combination of (c) and (d) yields the third triplet state $|10\rangle = (1/2)^{1/2}[|g_1e_2\rangle + |e_1g_2\rangle]$ while the antisymmetrical combination is the singlet state $|00\rangle = (1/2)^{1/2}[|g_1e_2\rangle - |e_1g_2\rangle]$.

Several cases are of interest:

$|SM\rangle = |N/2\ N/2\rangle$. All atoms are in $|e\rangle$ states:

$|SM\rangle = |N/2\ 0\rangle$ (N even). Half of the atoms are in $|e\rangle$ and half in $|g\rangle$ states:

$|SM\rangle = |0\ 0\rangle$ (N even). Half of the atoms are in $|e\rangle$ and half in $|g\rangle$ states:

$|SM\rangle = |N/2\ -N/2\rangle$. All atoms are in $|g\rangle$ states:

$|SM\rangle = |N/2\ 1-(N/2)\rangle$. One atom in $|e\rangle$ and the rest in $|g\rangle$ states.

We shall resume the discussion of these states in connection with the phenomenon of superradiance (Section 5.6).

5.2 Semiclassical Equations of Motion

Owing to the smallness of atomic dimensions compared with optical wavelengths, most radiative interactions in atomic and molecular systems are described sufficiently well by the electric dipole approximation. We therefore take as our basic Hamiltonian

$$\mathscr{H} = \mathscr{H}_A + V(t), \tag{5.33}$$

where $\mathcal{H}_A$ for an atomic two-level system is given by Eq. (5.15) and $V(t)$ is the interaction term of Eq. (4.329):

$$V(t) = e\mathbf{r} \cdot \mathbf{E}(t) \equiv -\mathbf{d} \cdot \mathbf{E}(t). \tag{5.34}$$

In the semiclassical formulation, $\mathbf{E}(t)$ is the classical electric field associated with the radiation to which the atom is exposed. For a single electron, $\mathbf{r}$ is simply the position vector of the electron referred to the nucleus in the case of an atom, or to the center of positive charge when there is more than one nucleus; $\mathbf{d}$ is the dipole moment operator. For a multielectron system, a summation over all the electrons is required.

With the closure relation, an arbitrary operator A may be expressed in the form

$$A = \sum_{nm} |n\rangle\langle n|A|m\rangle\langle m|, \tag{5.36}$$

which reduces to

$$\begin{aligned} \mathbf{d} = {} & |\mathrm{g}\rangle\langle \mathrm{g}|\mathbf{d}|\mathrm{g}\rangle\langle \mathrm{g}| + |\mathrm{g}\rangle\langle \mathrm{g}|\mathbf{d}|\mathrm{e}\rangle\langle \mathrm{e}| \\ & + |\mathrm{e}\rangle\langle \mathrm{e}|\mathbf{d}|\mathrm{g}\rangle\langle \mathrm{g}| + |\mathrm{e}\rangle\langle \mathrm{e}|\mathbf{d}|\mathrm{e}\rangle\langle \mathrm{e}| \end{aligned} \tag{5.37}$$

for the dipole moment operator in the two-state basis consisting of the states $|\mathrm{g}\rangle$ and $|\mathrm{e}\rangle$. A further reduction results from the fact that $\mathbf{d}$ is an odd operator (i.e., it changes sign upon inversion in the origin), and the assumption that $|\mathrm{g}\rangle$ and $|\mathrm{e}\rangle$ have a definite parity (odd or even). The diagonal matrix elements then vanish. Further, if $|\mathrm{g}\rangle$ and $|\mathrm{e}\rangle$ have opposite parities

$$\mathbf{d} = |\mathrm{g}\rangle\langle \mathrm{g}|\mathbf{d}|\mathrm{e}\rangle\langle \mathrm{e}| + |\mathrm{e}\rangle\langle \mathrm{e}|\mathbf{d}|\mathrm{g}\rangle\langle \mathrm{g}| \tag{5.38}$$

or, with the definition (5.2),

$$\mathbf{d} = \langle \mathrm{g}|\mathbf{d}|\mathrm{e}\rangle\sigma^- + \langle \mathrm{e}|\mathbf{d}|\mathrm{g}\rangle\sigma^+ = \begin{pmatrix} 0 & \langle \mathrm{e}|\mathbf{d}|\mathrm{g}\rangle \\ \langle \mathrm{g}|\mathbf{d}|\mathrm{e}\rangle & 0 \end{pmatrix}. \tag{5.39}$$

Although the expectation values $\langle \mathrm{g}|\mathbf{d}|\mathrm{g}\rangle$ and $\langle \mathrm{e}|\mathbf{d}|\mathrm{e}\rangle$ vanish for reasons of symmetry, the argument does not apply to a superposition state $|\psi\rangle$ such as the one shown in Eq. (2.193). If $|\mathrm{g}\rangle$ and $|\mathrm{e}\rangle$ have opposite parities, there is no definite parity associated with $|\psi\rangle$, and the expectation value of the dipole moment operator in the state $|\psi\rangle$ does not vanish. In fact, one finds, with the help of Eq. (5.38), that

$$\langle \psi|\mathbf{d}|\psi\rangle = a_{\mathrm{g}}^* a_{\mathrm{e}}\langle \mathrm{g}|\mathbf{d}|\mathrm{e}\rangle + a_{\mathrm{e}}^* a_{\mathrm{g}}\langle \mathrm{e}|\mathbf{d}|\mathrm{g}\rangle. \tag{5.40a}$$

The same result may be achieved by means of the density matrix (Eq. (2.196)) and the matrix for $\mathbf{d}$ given by Eq. (5.39), since $\langle \psi|\mathbf{d}|\psi\rangle \equiv \langle \mathbf{d}\rangle = \mathrm{Tr}\{\rho\mathbf{d}\}$. Note that only off-diagonal elements of the density matrix appear in the expression for $\langle \mathbf{d}\rangle$. The macroscopic polarization $\mathbf{P}$ is defined by $\mathbf{P} = N\langle \mathbf{d}\rangle$ where N is the number of dipoles per unit volume. Consequently,

$$\mathbf{P} = N\langle\mathbf{d}\rangle = N\,\mathrm{Tr}\{\rho\mathbf{d}\} = N(\rho_{\mathrm{eg}}\mathbf{d}_{\mathrm{ge}} + \rho_{\mathrm{ge}}\mathbf{d}_{\mathrm{eg}}). \tag{5.40b}$$

Since a general electric field may be regarded as a superposition of plane waves, we confine our attention to the single wave

$$\begin{aligned}\mathbf{E}(t) = \hat{\boldsymbol{\varepsilon}}E(t) &= \frac{1}{2}\hat{\boldsymbol{\varepsilon}}[Ee^{-i\omega t} + E^*e^{i\omega t}]\\ &= \frac{1}{2}[\mathbf{E}e^{-i\omega t} + \mathbf{E}^*e^{i\omega t}],\end{aligned} \tag{5.41}$$

in which the spatial dependence has been suppressed. Combining Eqs. (5.39) and (5.41), the interaction term $V(t)$ assumes the form

$$\begin{aligned}V(t) = -\mathbf{d}\cdot\mathbf{E}(t) &= -\frac{1}{2}\mathbf{d}\cdot\hat{\boldsymbol{\varepsilon}}[Ee^{-\omega t} + E^*e^{i\omega t}]\\ &= -\frac{1}{2}[\sigma^-\langle \mathrm{g}|\mathbf{d}\cdot\hat{\boldsymbol{\varepsilon}}|\mathrm{e}\rangle Ee^{-i\omega t} + \sigma^+\langle \mathrm{e}|\mathbf{d}\cdot\hat{\boldsymbol{\varepsilon}}|\mathrm{g}\rangle Ee^{-i\omega t}\\ &\quad + \sigma^-\langle \mathrm{g}|\mathbf{d}\cdot\hat{\boldsymbol{\varepsilon}}|\mathrm{e}\rangle E^*e^{i\omega t} + \sigma^+\langle \mathrm{e}|\mathbf{d}\cdot\hat{\boldsymbol{\varepsilon}}|\mathrm{g}\rangle E^*e^{i\omega t}].\end{aligned} \tag{5.42}$$

It now will be shown that the rotating wave approximation (RWA), which appeared in the discussion of the magnetic two-level system (Section 3.1), also may be adopted in the atomic case. For this purpose, we transform the Hamiltonian of Eq. (5.33) via the transformation of Eq. (3.60) with the operator of Eq. (3.61). Referring to Eq. (3.65), it is readily verified that

$$\mathscr{H}' = \mathscr{H}'_{\mathrm{A}} + V', \tag{5.43a}$$

$$\begin{aligned}\mathscr{H}'_{\mathrm{A}} &= \frac{1}{2}\hbar\omega_0 e^{i\omega t\sigma_z/2}\sigma_z e^{-i\omega t\sigma_z/2} + i\hbar\frac{\partial O}{\partial t}O^\dagger\\ &= \frac{1}{2}\hbar(\omega_0 - \omega)\sigma_z = \frac{1}{2}\hbar\Delta\sigma_z,\end{aligned} \tag{5.43b}$$

$$\begin{aligned}V' = -\frac{1}{2}[&\sigma^-\langle \mathrm{g}|\mathbf{d}\cdot\hat{\boldsymbol{\varepsilon}}|\mathrm{e}\rangle Ee^{-2i\omega t} + \sigma^+\langle \mathrm{e}|d\cdot\hat{\boldsymbol{\varepsilon}}|\mathbf{g} > \mathbf{E}\\ &+\sigma^-\langle \mathrm{g}|\mathbf{d}\cdot\hat{\boldsymbol{\varepsilon}}|\mathrm{e}\rangle E^* + \sigma^+\langle \mathrm{e}|\mathbf{d}\cdot\hat{\boldsymbol{\varepsilon}}|\mathrm{g}\rangle E^*e^{2i\omega t}]\end{aligned} \tag{5.43c}$$

is the Hamiltonian for an atomic two-level system in a coordinate frame rotating at a frequency ω. The essence of the rotating wave approximation consists of the elimination of the high-frequency terms, as in the magnetic case. Omitting terms that depend on 2ω and reverting to the laboratory frame, the Hamiltonian reduces to

$$\mathscr{H} = \mathscr{H}_A + V(t), \tag{5.44a}$$

$$\mathscr{H}_A = \frac{\hbar}{2}\omega_0\sigma_z, \tag{5.44b}$$

$$\begin{aligned} V(t) &= -\frac{1}{2}[\sigma^+\langle e|\mathbf{d}\cdot\hat{\varepsilon}|g\rangle E e^{-i\omega t} + \sigma^-\langle g|\mathbf{d}\cdot\hat{\varepsilon}|e\rangle E^* e^{i\omega t}] \\ &= \hbar[q\sigma^+ e^{-i\omega t} + q^*\sigma^- e^{i\omega t}], \end{aligned}$$

where

$$\begin{aligned} \hbar q &= -\frac{E}{2}\langle e|\mathbf{d}\cdot\hat{\varepsilon}|g\rangle = -\frac{1}{2}\mathbf{d}_{eg}\cdot\mathbf{E}, \\ \hbar q^* &= -\frac{E^*}{2}\langle g|\mathbf{d}\cdot\hat{\varepsilon}|e\rangle = -\frac{1}{2}\mathbf{d}_{ge}\cdot\mathbf{E}^*. \end{aligned} \tag{5.45}$$

The quantities q and q^* act as coupling constants; the magnitude $|q|$ measures the strength of the coupling between the material system and the field.

An alternative expression for the Hamiltonian in the rotating wave approximation is obtained by converting σ^+ and σ^- to rectangular components:

$$\mathscr{H} = \frac{\hbar}{2}[\omega_0\sigma_z + (qe^{-i\omega t} + q^*e^{i\omega t})\sigma_x + i(qe^{-i\omega t} - q^*e^{i\omega t})\sigma_y]. \tag{5.46}$$

In most situations involving radiation interactions, it is possible to choose real eigenfunctions. The basic reason is associated with the fact that, in the absence of an external magnetic field, the Hamiltonian is invariant under time reversal. To probe this feature, we define the *time-reversal operator* T acting upon an arbitrary function $f(\mathbf{r}, t)$ by the relation

$$Tf(\mathbf{r}, t) = f^*(\mathbf{r}, -t). \tag{5.47}$$

If a time-independent Hamiltonian $\mathscr{H}(\mathbf{r})$ is a real function of $\mathbf{r}$,

$$T\mathscr{H}(\mathbf{r})\psi(\mathbf{r}, t) = \mathscr{H}^*(\mathbf{r})\psi^*(\mathbf{r}, -t) = \mathscr{H}(\mathbf{r})\psi(\mathbf{r}, -t) = \mathscr{H}(\mathbf{r})T\psi(\mathbf{r}, t). \tag{15.48}$$

Therefore,

$$T\mathscr{H}(\mathbf{r}) = \mathscr{H}(\mathbf{r})T \quad \text{or} \quad T\mathscr{H}(\mathbf{r})T^{-1} = \mathscr{H}(\mathbf{r}), \tag{15.49}$$

which indicates that a real Hamiltonian commutes with the time reversal operator or, alternatively, that such a Hamiltonian is invariant under a transformation by T. Let us now suppose that $\psi(\mathbf{r}, t)$ obeys the time-dependent

Schrödinger equation. Transforming the equation with the T-operator,

$$
\begin{aligned}
T\mathscr{H}(\mathbf{r})\psi(\mathbf{r},t) = \mathscr{H}(\mathbf{r})T\psi(\mathbf{r},t) = Ti\hbar\frac{\partial}{\partial t}\psi(\mathbf{r},t) &= -i\hbar T\frac{\partial}{\partial t}\psi(\mathbf{r},t) \\
&= -i\hbar\frac{\partial}{\partial(-t)}T\psi(\mathbf{r},t) \\
&= i\hbar\frac{\partial}{\partial t}T\psi(\mathbf{r},t).
\end{aligned}
\tag{5.50}
$$

Therefore, if $\psi(\mathbf{r},t)$ is a solution to the time-dependent Schrödinger equation, $T\psi(\mathbf{r},t)$ is also a solution. This statement is applicable to stationary states as well; namely, if $\psi(\mathbf{r})$ is an eigenfunction of a real Hamiltonian $\mathscr{H}(\mathbf{r})$ then $T\psi(\mathbf{r}) = \psi^*(\mathbf{r})$ is also an eigenfunction of $\mathscr{H}(\mathbf{r})$ with the same eigenvalue. Since we now have a two-fold degeneracy, any linear combination of $\psi(\mathbf{r})$ and $\psi^*(\mathbf{r})$, such as

$$
\Psi = \mathrm{a}\psi(\mathbf{r}) + \mathrm{b}\psi^*(\mathbf{r}),
\tag{5.51}
$$

where a and b are constants, will statisfy $\mathscr{H}(\mathbf{r})\Psi = E\Psi$. In particular, let $\psi(\mathbf{r}) = u + iv$. Then

$$
\begin{aligned}
\frac{\mathscr{H}(\mathbf{r})[\psi^*(\mathbf{r}) + \psi(\mathbf{r})]}{2} &= \mathscr{H}(\mathbf{r})u = Eu, \\
\frac{\mathscr{H}(\mathbf{r})i[\psi^*(\mathbf{r}) - \psi(\mathbf{r})]}{2} &= \mathscr{H}(\mathbf{r})v = Ev.
\end{aligned}
\tag{5.52}
$$

Thus, u and v, the real and imaginary parts of the eigenfunction $\psi(\mathbf{r})$, separately satisfy the Schrödinger equation, which means that when the Hamiltonian is real it is always possible to choose real eigenfunctions. An important violation of the reality condition on the Hamiltonian occurs in the presence of magnetic interactions because such interactions involve angular momentum operators that are imaginary (e.g., $L_z = -i\hbar\partial/\partial\phi$).

Since $\mathbf{d}\cdot\hat{\varepsilon}$ is a real operator, the matrix elements will be real in a basis in which $|g\rangle$ and $|e\rangle$ are real; also, if $E = E^*$, which merely implies an arbitrary choice of phase angle, then $q = q^*$ in Eq. (5.45). We often shall have occasion to invoke this property.

With the Hamiltonian (Eq. (5.43)), we have

$$
\begin{aligned}
&V_{\mathrm{gg}} = V_{\mathrm{ee}} = 0, \\
&V_{\mathrm{eg}} = \hbar \mathrm{q} e^{-i\omega t}, \qquad V_{\mathrm{ge}} = \hbar q^* e^{i\omega t}.
\end{aligned}
\tag{5.53}
$$

The equations of motion may be obtained directly from Eqs. (2.192), (5.25), and (5.53):

$$\dot{\rho}_{ee} = \frac{d}{dt}\langle\sigma^+\sigma^-\rangle = i(\mathrm{q}^* e^{i\omega t}\langle\sigma^-\rangle - \mathrm{q}e^{-i\omega t}\langle\sigma^+\rangle)$$

$$= i(\mathrm{q}^* e^{i\omega t}\rho_{eg} - \mathrm{q}e^{-i\omega t}\rho_{ge}) = -\dot{\rho}_{gg} = -\frac{d}{dt}\langle\sigma^-\sigma^+\rangle, \tag{5.54a}$$

$$\dot{\rho}_{eg} = \frac{d}{dt}\langle\sigma^-\rangle = -i\omega_0\langle\sigma^-\rangle + 2i\mathrm{q}e^{-i\omega t}\left(\langle\sigma^+\sigma^-\rangle - \frac{1}{2}\right)$$

$$= -i\omega_0\langle\sigma^-\rangle + i\mathrm{q}e^{-i\omega t}\langle\sigma_z\rangle$$

$$= -i\omega_0\rho_{eg} + i\mathrm{q}e^{-i\omega t}(\rho_{ee} - \rho_{gg}), \tag{5.54b}$$

$$\dot{\rho}_{ge} = \frac{d}{dt}\langle\sigma^+\rangle = i\omega_0\langle\sigma^+\rangle - 2i\mathrm{q}^* e^{i\omega t}\left(\langle\sigma^+\sigma^-\rangle - \frac{1}{2}\right)$$

$$= i\omega_0\langle\sigma^+\rangle - i\mathrm{q}^* e^{i\omega t}\langle\sigma_z\rangle$$

$$= i\omega_0\rho_{ge} - i\mathrm{q}^* e^{i\omega t}(\rho_{ee} - \rho_{gg}). \tag{5.54c}$$

The corresponding equations for the rectangular components are

$$\frac{d}{dt}\langle\sigma_x\rangle = \frac{d}{dt}(\langle\sigma^+\rangle + \langle\sigma^-\rangle) = -\omega_0\langle\sigma_y\rangle + i(\mathrm{q}e^{-i\omega t} - \mathrm{q}^* e^{i\omega t})\langle\sigma_z\rangle,$$

$$\frac{d}{dt}\langle\sigma_y\rangle = i\frac{d}{dt}(\langle\sigma^-\rangle - \langle\sigma^+\rangle) = \omega_0\langle\sigma_x\rangle - (\mathrm{q}e^{-i\omega t} + \mathrm{q}^* e^{i\omega t})\langle\sigma_z\rangle,$$

$$\frac{d}{dt}\langle\sigma_z\rangle = 2\frac{d}{dt}\langle\sigma^+\sigma^-\rangle \tag{5.55}$$

$$= (\mathrm{q}e^{-i\omega t} + \mathrm{q}^* e^{i\omega t})\langle\sigma_y\rangle - i(\mathrm{q}e^{-i\omega t} - \mathrm{q}^* e^{i\omega t})\langle\sigma_x\rangle.$$

In view of the connections between the Pauli and fermion operators expressed by Eqs. (5.13) and (5.14), Eqs. (5.54) may be written

$$\frac{d}{dt}c_e^\dagger c_e = i(\mathrm{q}^* e^{i\omega t}c_g^\dagger c_e - \mathrm{q}e^{-i\omega t}c_e^\dagger c_g) = -\frac{d}{dt}c_g^\dagger c_g,$$

$$\frac{d}{dt}c_g^\dagger c_e = -i\omega_0 c_g^\dagger c_e + i\mathrm{q}e^{-i\omega t}(c_e^\dagger c_e - c_g^\dagger c_g), \tag{5.56}$$

$$\frac{d}{dt}c_e^\dagger c_g = i\omega_0 c_e^\dagger c_g - i\mathrm{q}^* e^{i\omega t}(c_e^\dagger c_e - c_g^\dagger c_g).$$

Since these are operators relations, it is understood that they are in the Heisenberg representation.

Equations of motion also may be derived with respect to the rotating frame. Referring to Eq. (5.43) and omitting the high-frequency terms (rotating wave approximation), we have the time-independent Hamiltonian

$$(R)\qquad \begin{gathered}\mathscr{H}' = \mathscr{H}'_{\mathrm{A}} + V', \\ H'_{\mathrm{A}} = \frac{1}{2}\hbar\Delta\sigma_z, \qquad V' = \hbar(q\sigma^+ + q^*\sigma^-).\end{gathered} \tag{5.57}$$

The corresponding equations of motion are

$$\begin{aligned}\dot{\rho}_{\mathrm{ee}} &= \frac{d}{dt}\langle\sigma^+\sigma^-\rangle = i[q^*\langle\sigma^-\rangle - \mathrm{q}\sigma^+] \\ &= i[q^*\rho_{\mathrm{eg}} - \mathrm{q}\rho_{\mathrm{ge}}] = -\dot{\rho}_{\mathrm{gg}} = -\frac{d}{dt}\langle\sigma^-\sigma^+\rangle,\end{aligned}$$

$$(R)\qquad \begin{aligned}\dot{\rho}_{\mathrm{eg}} &= \frac{d}{dt}\langle\sigma^-\rangle = -i\Delta\langle\sigma^-\rangle + 2i\mathrm{q}\left(\langle\sigma^+\sigma^-\rangle - \frac{1}{2}\right) \\ &= -i\Delta\langle\sigma^-\rangle + i\mathrm{q}\langle\sigma_z\rangle \\ &= -i\Delta\rho_{\mathrm{eg}} + i\mathrm{q}(\rho_{\mathrm{ee}} - \rho_{\mathrm{gg}}), \\ \dot{\rho}_{\mathrm{ge}} &= \frac{d}{dt}\langle\sigma^+\rangle = i\Delta\langle\sigma^+\rangle - 2i\mathrm{q}^*\left(\langle\sigma^+\sigma^-\rangle - \frac{1}{2}\right) \\ &= i\Delta\langle\sigma^+\rangle - i\mathrm{q}^*\langle\sigma_z\rangle \\ &= i\Delta\rho_{\mathrm{ge}} - i\mathrm{q}^*(\rho_{\mathrm{ee}} - \rho_{\mathrm{gg}})\end{aligned} \tag{5.58}$$

or

$$(R)\qquad \begin{aligned}\frac{d}{dt}\langle\sigma_x\rangle &= \frac{d}{dt}(\langle\sigma^+\rangle + \langle\sigma^-\rangle) = -\Delta\langle\sigma_y\rangle + i(\mathrm{q} - \mathrm{q}^*)\langle\sigma_z\rangle, \\ \frac{d}{dt}\langle\sigma_y\rangle &= i\frac{d}{dt}(\langle\sigma^-\rangle - \langle\sigma^+\rangle) = \Delta\langle\sigma_x\rangle - (\mathrm{q} + \mathrm{q}^*)\langle\sigma_z\rangle, \\ \frac{d}{dt}\langle\sigma_z\rangle &= 2\frac{d}{dt}\langle\sigma^+\sigma^-\rangle = (\mathrm{q} + \mathrm{q}^*)\langle\sigma_y\rangle - i(\mathrm{q} - \mathrm{q}^*)\langle\sigma_x\rangle.\end{aligned} \tag{5.59}$$

Equations (5.58) may be written in matrix form similar to Eq. (2.198)

$$i\begin{pmatrix}\dot{\rho}_{\mathrm{gg}} \\ \dot{\rho}_{\mathrm{ge}} \\ \dot{\rho}_{\mathrm{eg}} \\ \dot{\rho}_{\mathrm{ee}}\end{pmatrix} = \begin{pmatrix}0 & -\mathrm{q} & \mathrm{q}^* & 0 \\ -q^* & -\Delta & 0 & \mathrm{q}^* \\ \mathrm{q} & 0 & \Delta & -\mathrm{q} \\ 0 & \mathrm{q} & -\mathrm{q}^* & 0\end{pmatrix}\begin{pmatrix}\rho_{\mathrm{gg}} \\ \rho_{\mathrm{ge}} \\ \rho_{\mathrm{eg}} \\ \rho_{\mathrm{ee}}\end{pmatrix}. \tag{5.60}$$

The eigenvalues, λ, of the matrix are evaluated easily; they are

$$\lambda = 0, 0, \pm\sqrt{\Delta^2 + 4|q|^2} \equiv \pm\Omega. \tag{5.61}$$

In the next section, we shall demonstrate that the frequency Ω is the Rabi frequency for the atomic two-level case.

For the quantities u_1, u_2, and u_3, defined in Eq. (5.26), the equations of motion based on Eq. (5.58), with $\mathrm{q} = \mathrm{q}^*$, are

(R)
$$\begin{aligned} \dot{u}_1 &= \dot{\rho}_{\mathrm{ge}} + \dot{\rho}_{\mathrm{eg}} = -\Delta u_2, \\ \dot{u}_2 &= i(\dot{\rho}_{\mathrm{eg}} - \dot{\rho}_{\mathrm{ge}}) = \Delta u_1 - 2\mathrm{q}u_3, \qquad (\Delta = \omega_0 - \omega) \\ \dot{u}_3 &= \dot{\rho}_{\mathrm{ee}} - \dot{\rho}_{\mathrm{gg}} = 2\mathrm{q}u_2, \end{aligned} \tag{5.62}$$

or

(R)
$$\dot{\boldsymbol{\beta}} = -\boldsymbol{\beta} \times \boldsymbol{\Omega} \tag{5.63}$$

where

$$\boldsymbol{\beta} = u_1\hat{\mathbf{i}} + u_2\hat{\mathbf{j}} + u_3\hat{\mathbf{k}}, \qquad \beta^2 = 1, \tag{5.64}$$

$$\boldsymbol{\Omega} = 2\mathrm{q}\hat{\mathbf{i}} + \Delta\hat{\mathbf{k}}, \qquad \Omega = \sqrt{\Delta^2 + 4|\mathrm{q}|^2}. \tag{5.65}$$

It is apparent that the atomic equations of motion (Eq. (5.62)) are of the same form as the corresponding Eqs. (3.27) in the magnetic case. The solutions for the magnetic case are given by Eqs. (3.32) and (3.33); hence, the solutions of Eq. (5.62), with initial conditions

$$\boldsymbol{\beta}(0) = u_3(0)\hat{\mathbf{k}}, \tag{5.66}$$

are

(R)
$$\begin{aligned} u_1(t) &= -u_3(0)\frac{2\mathrm{q}\,\Delta}{\Omega^2}(\cos\Omega t - 1), \qquad u_2(t) = -u_3(0)\frac{2\mathrm{q}}{\Omega}\sin\Omega t \\ u_3(t) &= u_3(0)\left[1 + \frac{4|\mathrm{q}|^2}{\Omega^2}(\cos\Omega t - 1)\right]. \end{aligned} \tag{5.67}$$

It follows, therefore, that the geometrical interpretation of the Bloch vector $\boldsymbol{\beta}$ must be the same in the two cases. In the magnetic case, the magnetic moment precesses in real space about an effective magnetic field (Fig. 3.3), the latter being the field observed in a rotating coordinate system. In the atomic case, on the other hand, there is no precession in real space; the geometrical interpretation of the Bloch vector $\boldsymbol{\beta}$ precessing about the torque vector $\boldsymbol{\Omega}$ (Fig. 5.3) is an abstraction that describes the time-development of the atomic operators. In the next section, we shall examine these features in greater detail. The analogy with the magnetic case also extends to the Bloch-Siegert shift associated with the high-frequency terms that were neglected in the rotating

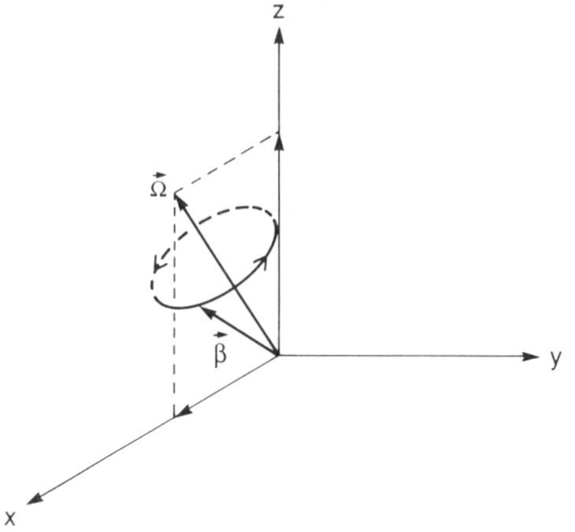

FIGURE 5.3 A two-level atom interacting with a radiation field can be interpreted geometrically as a precession of the Bloch vector $\boldsymbol{\beta}$ about the torque vector $\boldsymbol{\Omega}$ where the two vectors are defined by Eqs. (5.64) and (5.65), respectively. Note the close analogy with the magnetic case shown in Fig. 3.3.

wave approximation. Finally, we note that the remarks in Section 3.4 concerning the equations of motion in the Heisenberg representation apply in the atomic case as well.

5.3 Semiclassical Transition Probabilities

The fundamental quantity that determines the strength of the interaction between matter and electromagnetic radiation is the fine structure constant

$$\alpha = \frac{e^2}{4\pi\varepsilon_0 \hbar c} \simeq \frac{1}{137}. \tag{5.68}$$

Because $\alpha \ll 1$, perturbation expansions in the calculation of matter-radiation interactions are justifiable [5]. A weak field then may be defined as one in which the perturbation expansion need not be carried beyond the first (or, in some cases, possibly the second) order. Weak field conditions also may be described as those in which the initial state of the matter system (atom) and the population within that state are not changed significantly by the interaction with the field. Still another criterion for a weak interaction is the linearity of response to an applied field as, for example, when the induced polarization is proportional to the first power of the field. Nevertheless, one should bear in mind that at very high intensities, such as those available from pulsed lasers, the applied electric field can become comparable to the internal fields within the atom. Under such circumstances perturbation theory may not be applicable.

With most ordinary (thermal) sources the intensities are such that practically all atom-field interactions are weak. It is only with laser sources that sufficiently high intensities are available for nonlinear (strong) interactions to be observed easily.

We consider first an absorption process in a monochromatic radiation field by an atom with two infinitely sharp levels [6]. With the semiclassical Hamiltonian (Eq. (5.44)), the general Eq. (2.84) for the probability amplitudes (in RWA) reduces to

$$i\dot{c}_g(t) = \mathrm{q}^* c_e(t) e^{-i(\omega_0 - \omega)t}, \tag{5.69a}$$

$$i\dot{c}_e(t) = \mathrm{q} c_g(t) e^{i(\omega_0 - \omega)t}, \tag{5.69b}$$

where c_g and c_e (not to be confused with fermion operators) are the probability amplitudes of the states $|g\rangle$ and $|e\rangle$, respectively. Assuming the ground state population remains essentially constant for all time, $c_g(t) \simeq 1$. Hence, the solution to Eq. (5.695) is simply

$$c_e(t) = -i\mathrm{q} \int_0^t e^{i(\omega_0 - \omega)t'}\, dt' = \mathrm{q} \frac{1 - e^{i(\omega_0 - \omega)t}}{(\omega_0 - \omega),}, \tag{5.70}$$

as in Eq. (2.90), and the probability for absorption is

$$P(g \to e) \equiv P_{eg} = |c_e(t)|^2 = 4|\mathrm{q}|^2 F(\omega), \tag{5.71}$$

where

$$F(\omega) = \frac{\sin^2[\frac{1}{2}(\omega_0 - \omega)t]}{(\omega_0 - \omega)^2}. \tag{5.72a}$$

The function $F(\omega)$ is shown in Fig. 5.4. Since

$$\lim_{\omega \to \omega_0} F(\omega) = \frac{t^2}{4}, \tag{5.72b}$$

the transition probability at resonance is

$$P_{eg}(\omega = \omega_0) = |\mathrm{q}|^2 t^2. \tag{5.73}$$

In contrast to the classical case, there is a nonvanishing transition probability even when $\omega \neq \omega_0$, that is, even when the conservation of energy is not strictly observed. The transition probability falls off very rapidly as ω departs from ω_0, however.

The transition probability P_{eg} increases quadratically with time. One might suppose that we would arrive, ultimately, at the nonsensical result that $P_{eg} > 1$. But this would be inconsistent with the requirement that c_g remain

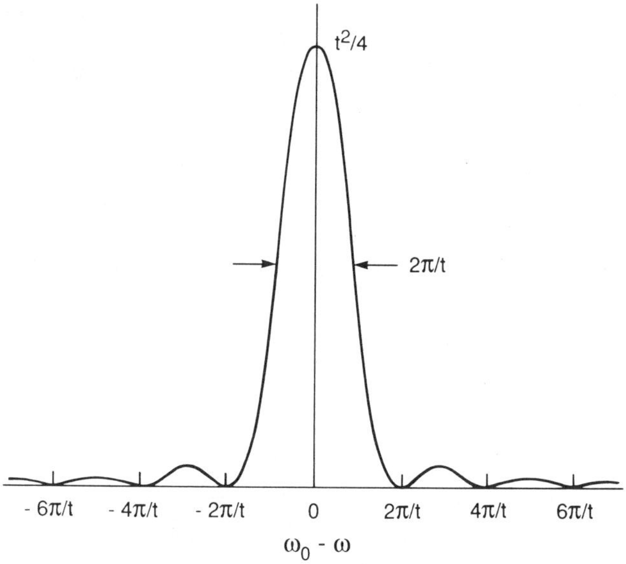

FIGURE 5.4 A plot of $F(\omega) = \sin^2[1/2(\omega_0 - \omega)t]/(\omega_0 - \omega)^2$ as a function of the detuning $\omega_0 - \omega$. The probability of absorption in a weak field is proportional to $F(\omega)$.

~ 1 for all time. It is necessary, therefore, to constrain the interaction time to satisfy the condition

$$|q|^2 t^2 \ll 1 \quad \text{or} \quad t \ll \frac{1}{|q|}. \tag{5.74}$$

But if the time is curtailed to some value T, a single frequency wave train automatically contains additional Fourier components spread out over a width $\Delta\omega \simeq 1/T$. What this means is that in a rigorous sense we cannot speak of excitation by a single frequency but rather by a distribution of frequencies. We shall return to these matters in Section 5.5 on the basis of a quantized field.

Another limitation on the time stems from the characteristics of $F(\omega)$. It may be verified that

$$\int_{-\infty}^{\infty} F(\omega)\, d\omega = \frac{\pi}{2} t, \tag{5.75}$$

with most of the area under the central peak. The total area $(\pi t/2)$ divided by the height of the central peak $(t^2/4)$ is equal approximately to the full width of the central peak at half maximum. Hence, the transition probability is effectively confined to a band width

$$|\omega_0 - \omega| = \frac{\pi t/2}{t^2/4} = \frac{2\pi}{t}, \tag{5.76}$$

consistent with the uncertainty principle. For typical spectral lines $|\omega_0 - \omega|/\omega_0 \ll 1$. Therefore,

$$t \gg \frac{2\pi}{\omega} \simeq \frac{2\pi}{\omega_0}, \tag{5.77}$$

that is, the interaction time t is much longer than the oscillation period of the radiation. In other words, a significant probability of atomic excitation occurs only after the elapse of sufficient time for the electromagnetic wave to have gone through many cycles of oscillation. Combining the two conditions of Eqs. (5.74) and (5.77), the condition for the validity of first-order perturbation theory for excitation by electromagnetic waves is

$$\frac{2\pi}{\omega_0} \ll t \ll \frac{1}{|\mathrm{q}|}. \tag{5.78}$$

For atomic transitions, this condition is satisfied easily since, typically, $2\pi/\omega_0$ is of the order of 10^{-16} s.

When $t \gg 2\pi/\omega_0$ and $\omega \simeq \omega_0$, the sharp central peak of $F(\omega)$ may be approximated by

$$\lim_{t\to\infty} F(\omega) = \frac{\pi t}{2}\delta(\omega_0 - \omega), \tag{5.79}$$

even though the upper bound on t is constrained by Eq. (5.74). Equation (5.71) for the transition probability becomes

$$P_{\mathrm{eg}} = |c_{\mathrm{e}}(t)|^2 = 2\pi|\mathrm{q}|^2 t\,\delta(\omega_0 - \omega), \tag{5.80}$$

and the transition probability per unit time is

$$W(\mathrm{g}\to\mathrm{e}) \equiv W_{\mathrm{eg}} = \frac{P_{\mathrm{eg}}}{t} = 2\pi|\mathrm{q}|^2\,\delta(\omega_0 - \omega) = \frac{\pi}{2\hbar^2}|\langle \mathrm{e}|\mathbf{d}\cdot\mathbf{E}|\mathrm{g}\rangle|^2\,\delta(\omega_0 - \omega)$$

$$\equiv \frac{2\pi}{\hbar^2}|\langle \mathrm{e}|V(t)|\mathrm{g}\rangle|^2\,\delta(\omega_0 - \omega). \tag{5.81}$$

The last expression is recognized from Eq. (2.91) as the Fermi Golden Rule.

We have shown that first-order perturbation theory is valid only when the transition probability $P_{\mathrm{eg}} \ll 1$, and as a consequence of this requirement, the time of interaction must be sufficiently short to satisfy Eq. (5.74). The question then arises as to how the system behaves when the interaction time is long, as under steady illumination, for example. To answer this question we shall calculate transition probabilities without recourse to perturbation theory.

Returning to Eqs. (5.69), we no longer assume $c_{\mathrm{g}}(t) \simeq 1$; instead, the ground state population will be permitted to vary in time. The initial conditions are then taken to be

$$c_{\mathrm{g}}(0) = 1, \qquad c_{\mathrm{e}}(0) = 0. \tag{5.82}$$

Now, let

$$c_g(t) = e^{i\mu t}. \tag{5.83}$$

After differentiating with respect to time and substituting in Eq. (5.69a), we get

$$c_e(t) = -\frac{\mu}{q^*} e^{i(\omega_0 - \omega + \mu)t}. \tag{5.84}$$

This expression is now differentiated with respect to time and compared with the time derivative of $c_e(t)$ from Eq. (5.69b). The result is

$$\mu^2 + (\omega_0 - \omega)\mu - |q|^2 = 0, \tag{5.85}$$

whose solutions are

$$\mu_\pm = -\frac{1}{2}(\omega_0 - \omega) \pm \frac{1}{2}\sqrt{(\omega_0 - \omega)^2 + 4|q|^2}, \tag{5.86}$$

with

$$\mu_+ - \mu_- = \sqrt{(\omega_0 - \omega)^2 + 4|q|^2} = \Omega, \qquad \mu_+\mu_- = |q|^2. \tag{5.87}$$

The probability amplitude $c_g(t)$ now becomes

$$c_g(t) = Ae^{i\mu_+ t} + Be^{i\mu_- t}, \tag{5.88}$$

and its time-derivative, when substituted in Eq. (5.69a), yields

$$c_e(t) = -\frac{1}{q^*}[A\mu_+ e^{i(\omega_0 - \omega + \mu_+)t} + B\mu_- e^{i(\omega_0 - \omega + \mu_-)t}]. \tag{5.89}$$

The initial conditions of Eq. (5.82) determine the constants A and B:

$$A = \frac{\mu_-}{\Omega}, \qquad B = \frac{\mu_+}{\Omega}, \tag{5.90}$$

and since

$$\omega_0 - \omega + \mu_+ = \frac{1}{2}(\omega_0 - \omega + \Omega), \qquad \omega_0 - \omega + \mu_- = \frac{1}{2}(\omega_0 - \omega - \Omega), \tag{5.91}$$

we have

$$c_e(t) = -i\frac{2q}{\Omega} e^{i(\omega_0 - \omega)t/2} \sin\frac{1}{2}\Omega t. \tag{5.92}$$

Hence, the occupation probability for the state $|e\rangle$ is

$$|c_e(t)|^2 = \frac{4|q|^2}{(\omega_0 - \omega)^2 + 4|q|^2} \sin^2\frac{1}{2}t\sqrt{(\omega_0 - \omega)^2 + 4|q|^2}$$

$$= \frac{\Omega_0^2}{\Delta^2 + \Omega_0^2} \sin^2\frac{1}{2}t\sqrt{\Delta^2 + \Omega_0^2} = \frac{\Omega_0^2}{\Omega^2} \sin^2\frac{1}{2}\Omega t = P_{eg}, \tag{5.93}$$

in which (with $q = q^*$)

$$\Omega_0 = 2\mathrm{q} = \frac{\langle \mathrm{e}|\mathbf{d}\cdot\mathbf{E}|\mathrm{g}\rangle}{\hbar}, \qquad \Omega^2 = \Omega_0^2 + \Delta^2, \qquad \Delta = \omega_0 - \omega. \tag{5.94}$$

Equation (5.93) is the Rabi formula for the transition probability in the atomic case; evidently, it is identical in form to Eq. (3.83) for the magnetic case. It is therefore legitimate to employ a common interpretation for both the atomic and magnetic cases, the latter illustrated in Figs. 3.5 and 3.6. P_{eg} oscillates (or nutates) between zero and Ω_0^2/Ω^2 with a period $T = 2\pi/\Omega$, while P_{ge} oscillates with the same period between one and $1 - \Omega_0^2/\Omega^2$. At resonance both probabilities oscillate between zero and one with a period of $2\pi/\Omega_0$ and a phase difference of $\pi/2$. What this means is that energy is exchanged periodically between the atom and the field; as the atom is excited to the higher level, the field loses energy, and when the atom reverts to the lower level, the field recoups its losses. Over a long time the average values of the occupation probabilities or the populations in the two levels equalize. The close connection between q, the coupling parameter, and the Rabi frequency Ω should be noted.

For the two-level system, the normalization condition on the probability amplitudes implies that the population in $|\mathrm{g}\rangle$ varies according to

$$|c_{\mathrm{g}}(t)|^2 = 1 - |c_{\mathrm{e}}(t)|^2 = 1 - \frac{\Omega_0}{\Omega^2}\sin^2\frac{1}{2}\Omega t. \tag{5.95}$$

At resonance, the probabilities are

$$|c_{\mathrm{e}}(t)|^2 = \sin^2\tfrac{1}{2}\Omega_0 t, \qquad |c_{\mathrm{g}}(t)|^2 = \cos^2\tfrac{1}{2}\Omega_0 t, \tag{5.96}$$

and the atomic wave function is

$$\psi(t) = |\mathrm{g}\rangle\cos\tfrac{1}{2}\Omega_0 t + |\mathrm{e}\rangle\sin\tfrac{1}{2}\Omega_0 t. \tag{5.97}$$

Similar conclusions can be drawn from the equations of motion (Eqs. (5.62). With $\Delta = 0$,

$$\dot{u}_1 = 0, \qquad \dot{u}_2 = -2\mathrm{q}u_3, \qquad \dot{u}_3 = 2\mathrm{q}u_2. \tag{5.98}$$

The solutions (also obtainable from Eq. (5.67)) under the initial condition of Eq. (5.66) are

$$u_1(t) = 0, \qquad u_2(t) = -u_3(0)\sin 2\mathrm{q}t, \qquad u_3(t) = u_3(0)\cos 2\mathrm{q}t, \tag{5.99}$$

which, of course, are identical in form to the magnetic Eqs. (3.30). When the atom is initially in $|\mathrm{g}\rangle$ (and $|\mathrm{e}\rangle$ is vacant), $u_3(0) = \rho_{\mathrm{ee}}(0) - \rho_{\mathrm{gg}}(0) = -1$. The orientations of the Bloch vector at $t = 0, \pi/4\mathrm{q} = \pi/2\Omega_0, \pi/2\mathrm{q} = \pi/\Omega_0$ are then

$$\boldsymbol{\beta}(0) = -\mathbf{k}, \qquad \boldsymbol{\beta}(\pi/2\Omega_0) = \mathbf{j}, \qquad \boldsymbol{\beta}(\pi/\Omega_0) = \mathbf{k}. \tag{5.100}$$

The Bloch vector $\boldsymbol{\beta}$ therefore describes a circle in the yz plane about the torque vector $\boldsymbol{\Omega}$, which lies along the x axis (see Fig. 3.2 for the analogous magnetic case). At $t = 0$, the $\boldsymbol{\beta}$ points in the negative z direction; after the elapse of a time $t = \pi/2\Omega_0$ (known as a $\pi/2$ pulse), $\boldsymbol{\beta}$ rotates into the positive y direction. Since $u_3 = \rho_{ee} - \rho_{gg} = 0$, the populations in the two states are equal or, in the case of a single atom, the probabilities of finding the atom in $|g\rangle$ or $|e\rangle$ are equal. The atomic wave function would then be written

$$\psi = \frac{1}{\sqrt{2}}(|g\rangle + |e\rangle), \tag{5.101}$$

which is recognized as a superposition state or, in another language, a *coherent* or *maximally mixed state*. The Bloch vector continues to rotate so that when $t = \pi/2q = \pi/\Omega_0$ (known as a π pulse), the Bloch vector has reversed itself and points in the positive z direction, which now corresponds to the state $|e\rangle$ being occupied and $|g\rangle$ vacant. Thus, the rotation of $\boldsymbol{\beta}$ through an angle of π (from $-\pi$ to 0) corresponds to the transition $|g\rangle \to |e\rangle$. The higher the frequency Ω_0, the stronger the atom-field interaction and the shorter the duration of the π pulse (as well as the $\pi/2$ pulse), and the shorter the time required for the transition $|g\rangle \to |e\rangle$.

Had we started with the atom initially in $|e\rangle$ (and $|g\rangle$ vacant), we would have $u_3(0) = \rho_{ee}(0) - \rho_{gg}(0) = 1$; $\boldsymbol{\beta}$ would then assume the orientations

$$\boldsymbol{\beta}(0) = \hat{\mathbf{k}}, \qquad \boldsymbol{\beta}\left(\frac{\pi}{2\Omega_0}\right) = -\hat{\mathbf{j}}, \qquad \boldsymbol{\beta}\left(\frac{\pi}{\Omega_0}\right) = -\hat{\mathbf{k}}. \tag{5.102}$$

Here, too, a π pulse reverses the Bloch vector with the interpretation that the atom has undergone a transition $|e\rangle \to |g\rangle$.

We also may examine the behavior of the density matrix elements. The initial conditions of Eq. (5.82) indicate that the state $|e\rangle$ is vacant at $t = 0$, i.e., $\rho_{ee}(0) = 0$. For $t > 0$, $\rho_{ee}(t)$ is determined by $|c_e(t)|^2$—the probability of a transition $|g\rangle \to |e\rangle$—which, at resonance, is given by Eq. (5.96). Therefore,

$$\begin{aligned} \rho_{ee} &= \sin^2 \tfrac{1}{2}\Omega_0 t, \\ \rho_{gg} &= 1 - \rho_{ee} = \cos^2 \tfrac{1}{2}\Omega_0 t. \end{aligned} \tag{5.103}$$

The off-diagonal density matrix elements (with $\Delta = 0$) now may be evaluated from the equations of motion (Eqs. (5.58)):

$$\begin{aligned} \dot{\rho}_{ge} &= -iq^*(\rho_{ee} - \rho_{gg}) = -iq^*(2\sin^2 \tfrac{1}{2}\Omega_0 t - 1), \\ \rho_{ge} &= -iq^* \int_0^t \left(2\sin^2 \frac{1}{2}\Omega_0 t - 1\right) dt = iq^* \frac{\sin \Omega_0 t}{\Omega_0} = \rho_{eg}^*. \end{aligned} \tag{5.104}$$

It is seen that, as the populations in the two states change under the influence of the radiation field, both diagonal and off-diagonal density matrix elements evolve in time.

With the initial condition of (Eq. (5.82), we saw that the effect of a π-pulse is to excite the atom to the state $|e\rangle$. Let this be regarded as the state of the atom at $t = 0$; i.e., we now replace Eq. (5.82) by $c_g(0) = 0$, $c_e(0) = 1$. It then is necessary merely to interchange g and e in the Rabi formula (Eq. (5.93)) and in the expression for the wave function (Eq. (5.97)). All subsequent formulas would similarly have the g and e indices interchanged. For weak radiative interactions, the Rabi formula merges with Eq. (5.71) derived on the basis of first-order perturbation theory. As the interaction increases in strength, however the Rabi formula maintains its validity whereas the perturbative treatment does not.

The off-resonance situation is more complicated; nevertheless, the atomic transition still can be interpreted as a rotation in an abstract space. Hence, in general, the temporal evolution of a two-level system can be described in terms of a Bloch vector $\boldsymbol{\beta}$ precessing about a torque vector $\boldsymbol{\Omega}$, in analogy with the classical picture of a magnetic moment precessing about a magnetic field.

We have concentrated on transitions induced by monochromatic plane waves, but it is a relatively simple matter to enlarge the discussion to include incident radiation spread over a narrow band in the vicinity of the resonance condition $\omega = \omega_0$. Referring to Eq. (5.81), the transition probability per unit time W_{eg} may be written

$$W_{eg} = \frac{\pi}{\varepsilon_0 c\hbar^2} |\langle e|\mathbf{d}\cdot\hat{\varepsilon}|g\rangle|^2 \mathrm{I}\, \delta(\omega_0 - \omega), \tag{5.105}$$

where

$$\mathbf{E} = \hat{\varepsilon} E, \qquad \mathrm{I} = \frac{1}{2}\varepsilon_0 c |E|^2. \tag{5.106}$$

To allow for a spectral width, we replace I, the incident intensity or time-averaged power flow density (energy crossing unit area in unit time), by $\mathrm{I}(\omega)\, d\omega$ and integrate over ω:

$$\begin{aligned} W_{eg} &= \frac{\pi}{\varepsilon_0 c\hbar^2} |\langle e|\mathbf{d}\cdot\hat{\varepsilon}|g\rangle|^2 \int_{-\infty}^{\infty} \mathrm{I}(\omega)\, \delta(\omega_0 - \omega)\, d\omega \\ &= \frac{\pi}{\varepsilon_0 c\hbar^2} |\langle e|\mathbf{d}\cdot\hat{\varepsilon}|g\rangle|^2 \mathrm{I}(\omega_0) \simeq \frac{\pi}{\varepsilon_0 c\hbar^2} |\langle e|\mathbf{d}\cdot\hat{\varepsilon}|g\rangle|^2 \mathrm{I}(\omega). \end{aligned} \tag{5.107}$$

It should be evident that the derivation of the transition probabilities given so far have been idealized in the sense that ever-present relaxation processes have not been taken into account. In particular, the semiclassical treatment employed in this section does not predict relaxation by spontaneous emission. The latter appears only when the classical radiation field is replaced by a quantized field (Section 5.5). Hence, the validity of the interpretation given above will depend on the relative strength of the radiative interaction, as measured by $\Omega_0 = 2q$, compared to the inverse relaxation times. As the strength of the radiative interaction is increased, relaxation effects play a diminishing role.

5.4 Quantized Hamiltonian, Equations of Motion

In the last two sections, we employed the semiclassical treatment of the interaction $V(t) = -\mathbf{d} \cdot \mathbf{E}(t)$ in which the dipole moment $\mathbf{d}$ was regarded as a quantum mechanical operator but $\mathbf{E}(t)$ retained its classical form. For many, but not all, cases, the semiclassical formulation is quite adequate. Nevertheless, we shall find that certain phenomena arising from the interaction between matter and radiation cannot be fully understood unless the electromagnetic field is quantized also.

Based on the discussion of the quantum mechanical gauge transformation, the quantized dipole Hamiltonian has the same general form as Eq. (5.34) with the important difference that the classical electric field is replaced by the quantized field. In the Schrödinger representation, the field has the form shown in Eq. (4.119b). Since $\mathbf{k} \cdot \mathbf{r} \ll 1$ in the dipole approximation, we write

$$\mathscr{H}_{\mathrm{AF}} = -\mathbf{d} \cdot \mathbf{E}(\mathbf{r}) = -i \sum_{\mathbf{k}\lambda} \sqrt{\frac{\hbar\omega_k}{2\varepsilon_0 V}}\, \mathbf{d} \cdot \hat{\boldsymbol{\varepsilon}}_{\mathbf{k}\lambda}(a_{\mathbf{k}\lambda} - a^{\dagger}_{\mathbf{k}\lambda}), \tag{5.107a}$$

$$= -d \cdot (\mathbf{E}^{(+)}(\mathbf{r}) + \mathbf{E}^{(-)}(\mathbf{r})), \tag{5.107b}$$

where $\mathbf{E}^{(+)}(\mathbf{r})$ and $\mathbf{E}^{(-)}(\mathbf{r})$ are defined in Eq. (4.121). Specializing to a two-level system interacting with a single mode of the field, we first replace $\mathbf{d}$ by the equivalent form of Eq. (5.39). Then

$$\mathscr{H}_{\mathrm{AF}} = \hbar(\mathrm{g}\sigma^{+} - \mathrm{g}^{*}\sigma^{-})(a - a^{\dagger}), \tag{5.108}$$

where

$$\hbar g = -i \sqrt{\frac{\hbar\omega}{2\varepsilon_0 V}} \langle \mathrm{e}|\mathbf{d} \cdot \hat{\varepsilon}|\mathrm{g}\rangle, \qquad \hbar \mathrm{g}^{*} = i \sqrt{\frac{\hbar\omega}{2\varepsilon_0 V}} \langle \mathrm{g}|\mathbf{d} \cdot \hat{\varepsilon}|\mathrm{e}\rangle. \tag{5.109}$$

The field operators a and $a^{\dagger}$ commute with the atomic operators σ^{+} and σ^{-} at all times, since the two kinds of operators belong to different spaces; hence, the order in which the operators appear in Eq. (5.108) is of no consequence.

The four terms in Eq. (5.108) may be displayed in the form of Feynman diagrams (Fig. 5.5). Examination of Eq. (5.108) or the diagrams reveals that

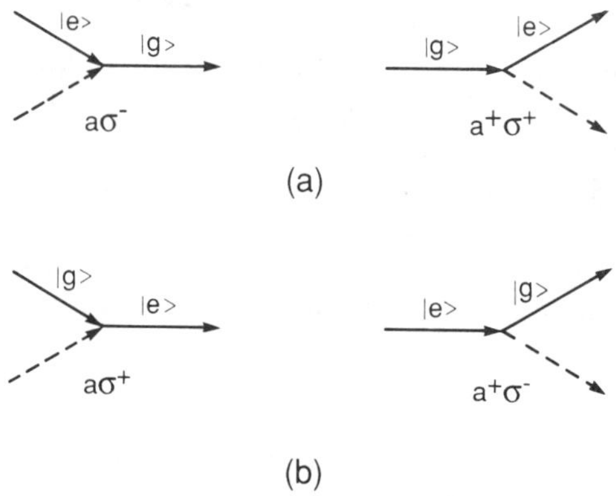

FIGURE 5.5 (a) Diagrams of the operator combinations $a\sigma^-$ and $a^\dagger\sigma^+$. With $E_e > E_g$, these combinations violate the conservation of energy; their elimination is equivalent to the rotating wave approximation. (b) Diagrams of the energy-conserving operator combinations $a\sigma^+$ and $a^\dagger\sigma^-$.

two of the four terms do not conserve energy. Thus, with $E_e > E_g$, the operator combination $a\sigma^-$ annihilates a photon and induces a transition from the higher energy state $|e\rangle$ to the lower energy state $|g\rangle$, in clear violation of energy conservation. A similar violation occurs for the combination $a^\dagger\sigma^+$, in which a photon is created during a transition $|g\rangle \to |e\rangle$. Retaining only those terms that conserve energy, the interaction Hamiltonian simplifies to

$$\mathscr{H}_{AF} = \hbar(ga\sigma^+ + g^*a^\dagger\sigma^-) = \frac{\hbar}{2}[(ga + g^*a^\dagger)\sigma_x + i(ga - g^*a^\dagger)\sigma_y]. \quad (5.110)$$

In the interaction representation, the atomic and field operators transform according to Eqs. (5.23) and (4.89), respectively. The Hamiltonian (Eq. (5.110)) is transformed then to

$$\begin{aligned}\tilde{\mathscr{H}}_{AF} &= \hbar(g\tilde{\sigma}^+ - g^*\tilde{\sigma}^-)(\tilde{a} - \tilde{a}^\dagger)\\ &= \hbar[g\sigma^+ a e^{i(\omega_0-\omega)t} - g\sigma^+ a^\dagger e^{i(\omega_0+\omega)t}\\ &\quad - g^*\sigma^- a e^{-i(\omega_0+\omega)t} + g^*\sigma^- a^\dagger e^{-i(\omega_0-\omega)t}]. \quad (5.111)\end{aligned}$$

It is seen that the energy-conserving terms contain exponentials with powers of $\pm i(\omega_0 - \omega)t$ while the powers in the exponentials associated with terms that do not conserve energy are $\pm i(\omega_0 - \omega)t$. Therefore, we may regard the elimination of terms in the quantum formulation that do not conserve energy as the analog of the elimination of the high-frequency terms in the semiclassical formulation. It then is appropriate to apply the terminology of the rotating wave approximation to the interaction Hamiltonian (Eq (5.110)). Indeed, in complete analogy with the magnetic case and the semiclassical formulation of the atomic case, the energy-nonconserving terms produce a Bloch-Siegert shift at sufficiently intense fields ($\sim 10^8$ W/cm^2).

We now may write the fully quantized Hamiltonian for a two-level atom interacting with a radiation field subject to the dipole and rotating wave approximations. The general form consists of three terms: an atomic term $\mathscr{H}_A$, a radiation term $\mathscr{H}_F$, and an interaction term $\mathscr{H}_{AF}$. The atomic Hamiltonian is given by Eq. (5.15). For the radiation field we take the general expression Eq. (4.97) without the zero-point energy; the interaction term is given by Eq. (5.110). Thus, in the Schrödinger representation, for a single mode we have the *Jaynes-Cummings* model [1]:

$$\mathscr{H} = \mathscr{H}_A + \mathscr{H}_F + \mathscr{H}_{AF'} \qquad \mathscr{H}_A = \frac{1}{2}\omega_0\sigma_z$$
$$\mathscr{H}_F = \hbar\omega a^\dagger a, \qquad \mathscr{H}_{AF} = \hbar(ga\sigma^+ + g^* a^\dagger \sigma^-). \tag{5.112}$$

For a multimode field,

$$\mathscr{H}_A = \frac{\hbar}{2}\omega_0\sigma_z, \qquad \mathscr{H}_F = \sum_j \hbar\omega_j a_j^\dagger a_j$$
$$\mathscr{H}_{AF} = \hbar\left[\sigma^+ \sum_j g_j a_j + \sigma^- \sum_j g_j^* a_j^\dagger\right]$$
$$= \hbar\left[c_e^\dagger c_g \sum_j g_j a_j + c_g^\dagger c_e \sum_j g_j^* a_j^\dagger\right], \tag{5.113}$$

in which the single index j signifies summation over $\mathbf{k}$ and λ.

The temporal change of the atomic and field operators in the Heisenberg representation is based on the general expression Eq. (2.107). Bearing in mind that atomic and field operators commute and referring to the commutator relations (Eq. (5.71)) (which are equally valid in the Heisenberg representation), one finds

$$\frac{d}{dt}(\sigma^+\sigma^-)_H = -i\left[\sigma_H^+ \sum_j g_j (a_H)_j - \sigma_H^- \sum_j g_j^\dagger (a_H^\dagger)_j\right],$$
$$= -\frac{d}{dt}(\sigma^-\sigma^+)_H,$$
$$\frac{d}{dt}\sigma_H^- = -i\omega_0\sigma_H^- + i\sigma_z \sum_j g_j (a_H)_j, \qquad \sigma_z = (\sigma_H)_z, \tag{5.114}$$
$$\frac{d}{dt}\sigma_H^+ = i\omega_0\sigma_H^+ - i\sigma_z \sum_j g_j^* (a_H^\dagger)_j,$$
$$\frac{d}{dt}(\sigma_H)_x = -\omega_0(\sigma_H)_y + i\sum_j [g_j (a_H)_j - g_j^* (a_H^\dagger)_j]\sigma_z,$$

$$\frac{d}{dt}(\sigma_{\rm H})_y = \omega_0(\sigma_{\rm H})_x - \sum_j [g_j(a_{\rm H})_j + g_j^*(a_{\rm H}^\dagger)_j]\sigma_z, \tag{5.115}$$

$$\begin{aligned}\frac{d}{dt}(\sigma_{\rm H})_z &= 2i\left[\sigma_{\rm H}^- \sum_j g_j^*(a_{\rm H}^\dagger)_j - \sigma_{\rm H}^+ \sum_j g_j(a_{\rm H})_j\right],\\ &= (\sigma_{\rm H})_y \sum_j [g_j(a_{\rm H})_j + g_j^*(a_{\rm H}^\dagger)_j]\\ &\quad - i(\sigma_{\rm H})_x \sum_j [g_j(a_{\rm H})_j - g_j^*(a_{\rm H}^\dagger)j].\end{aligned}$$

Equivalent relations may be written in terms of the fermion operators (in the Heisenberg representation):

$$\begin{aligned}\frac{d}{dt}c_{\rm e}^\dagger c_{\rm e} &= -ic_{\rm e}^\dagger c_{\rm g}\sum_j g_j(a_{\rm H})_j + ic_{\rm g}^\dagger c_{\rm e}\sum_j g_j^*(a_{\rm H}^\dagger)_j,\\ &= -\frac{d}{dt}c_{\rm g}^\dagger c_{\rm g},\end{aligned}$$

$$\frac{d}{dt}c_{\rm g}^\dagger c_{\rm e} = -i\omega_0 c_{\rm g}^\dagger c_{\rm e} + i(c_{\rm e}^\dagger c_{\rm e} - c_{\rm g}^\dagger c_{\rm g})\sum_j g_j(a_{\rm H})_j, \tag{5.116}$$

$$\frac{d}{dt}c_{\rm e}^\dagger c_{\rm g} = i\omega_0 c_{\rm e}^\dagger c_{\rm g} - i(c_{\rm e}^\dagger c_{\rm e} - c_{\rm g}^\dagger c_{\rm g})\sum_j g_j^*(a_{\rm H}^\dagger)_j.$$

By the same method, we also may obtain expressions for the time rate of change of the annihilation and creation operators. With the commutator relations of Eq. (4.96), the equations for a single mode interacting with a set of two-level atoms are

$$\frac{d}{dt}a_{\rm H} = -i\omega a_{\rm H} - i\sum_j g_j^*(\sigma_{\rm H}^-)_j,$$

$$\frac{d}{dt}a_{\rm H}^\dagger = i\omega a_{\rm H}^\dagger + i\sum_j g_j(\sigma_{\rm H}^+)_j. \tag{5.117}$$

One also may formulate equations of motion analogous to Eq. (5.62) for the case of an atom interacting with a quantized radiation field. Consider a basis set consisting of two states defined by

$$|{\rm I}\rangle = |{\rm g}\rangle|n\rangle \equiv |{\rm g}, n\rangle, \qquad |{\rm F}\rangle = |{\rm e}\rangle|n-1\rangle \equiv |{\rm e}, n-1\rangle. \tag{5.118}$$

In $|{\rm I}\rangle$ the atom is assumed to be in the state $|{\rm g}\rangle$ and the radiation field is in the state $|n\rangle$, that is, a single mode containing n photons. When the atom is elevated to the state $|{\rm e}\rangle$ and the number of photons reduced to $n-1$, the state of the system consisting of the atom and the field is $|F\rangle$. We now may compute matrix elements for the interaction. Dropping the terms in Eq. (5.111) that do not conserve energy,

$$\begin{aligned}\langle \mathrm{F}|\tilde{\mathscr{H}}_{\mathrm{AF}}|\mathrm{I}\rangle &= \hbar g e^{i\Delta t}\langle \mathrm{e}, n-1|\sigma^{+}a|\mathrm{g}, n\rangle \\ &\quad + \hbar g^{*} e^{-i\Delta t}\langle \mathrm{e}, n-1|\sigma^{-}a^{*}|\mathrm{g}, n\rangle \\ &= \hbar g e^{i\Delta t}\sqrt{n} = \langle \mathrm{I}|\tilde{\mathscr{H}}_{\mathrm{AF}}|\mathrm{F}\rangle^{*}, \qquad (\Delta = \omega_0 - \omega)\end{aligned}$$

and following the pattern established by Eq. (2.198),

$$\begin{aligned}i\hbar\frac{d}{dt}\tilde{\rho}_{\mathrm{II}} &= \langle \mathrm{I}|\tilde{\mathscr{H}}_{\mathrm{AF}}|\mathrm{F}\rangle\tilde{\rho}_{\mathrm{FI}} - \langle \mathrm{F}|\tilde{\mathscr{H}}_{\mathrm{AF}}|\mathrm{I}\rangle\tilde{\rho}_{\mathrm{IF}} \\ &= \hbar\sqrt{n}\,[g^{*}e^{-i\Delta t}\tilde{\rho}_{\mathrm{FI}} - g e^{i\Delta t}\tilde{\rho}_{\mathrm{IF}}], \\ i\hbar\frac{d}{dt}\tilde{\rho}_{\mathrm{FF}} &= \langle \mathrm{F}|\tilde{\mathscr{H}}_{\mathrm{AF}}|\mathrm{I}\rangle\tilde{\rho}_{\mathrm{IF}} - \langle \mathrm{I}|\tilde{\mathscr{H}}_{\mathrm{AF}}|\mathrm{F}\rangle\tilde{\rho}_{\mathrm{FI}} \\ &= \hbar\sqrt{n}\,[g e^{i\Delta t}\tilde{\rho}_{\mathrm{IF}} - g^{*}e^{-i\Delta t}\tilde{\rho}_{\mathrm{FI}}], \qquad (5.120) \\ i\hbar\frac{d}{dt}\tilde{\rho}_{\mathrm{IF}} &= \langle \mathrm{I}|\tilde{\mathscr{H}}_{\mathrm{AF}}|\mathrm{F}\rangle(\tilde{\rho}_{\mathrm{FF}} - \tilde{\rho}_{\mathrm{II}}) \\ &= \hbar\sqrt{n}\,g^{*}e^{-i\Delta t}(\tilde{\rho}_{\mathrm{FF}} - \tilde{\rho}_{\mathrm{II}}), \\ i\hbar\frac{d}{dt}\tilde{\rho}_{\mathrm{FI}} &= \langle \mathrm{F}|\tilde{\mathscr{H}}_{\mathrm{AF}}|\mathrm{I}\rangle(\tilde{\rho}_{\mathrm{II}} - \tilde{\rho}_{\mathrm{FF}}) \\ &= \hbar\sqrt{n}\,g e^{i\Delta t}(\tilde{\rho}_{\mathrm{II}} - \tilde{\rho}_{\mathrm{FF}}).\end{aligned}$$

We now define

$$g \equiv |g|e^{i\phi}, \qquad (5.121)$$

$$\begin{aligned}u_1 &= e^{i\phi}e^{i\Delta t}\tilde{\rho}_{\mathrm{IF}} + e^{-i\phi}e^{-i\Delta t}\tilde{\rho}_{\mathrm{FI}}, \\ u_2 &= i[e^{-i\phi}e^{-i\Delta t}\tilde{\rho}_{\mathrm{FI}} - e^{i\phi}e^{i\Delta t}\tilde{\rho}_{\mathrm{IF}}], \qquad (5.122) \\ u_3 &= \tilde{\rho}_{\mathrm{FF}} - \tilde{\rho}_{\mathrm{II}}.\end{aligned}$$

Note that u_1 and u_2 satisfy

$$\begin{aligned}u_1 + iu_2 &= 2e^{i\phi}e^{i\Delta t}\tilde{\rho}_{\mathrm{IF}}, \\ u_1 - iu_2 &= 2e^{-i\phi}e^{-i\Delta t}\tilde{\rho}_{\mathrm{FI}}. \qquad (5.123)\end{aligned}$$

These definitions, together with Eq. (5.120), yield

$$\dot{u}_1 = -\Delta u_2, \qquad \dot{u}_2 = \Delta u_1 - 2|g|\sqrt{n}\,u_3, \qquad \dot{u}_3 = 2|g|\sqrt{n}\,u_2, \quad (5.124)$$

which clearly are identical with Eqs. (5.62) when $|g|n^{1/2}$ is replaced by q. As in the semiclassical case, these equations lend themselves to a geometrical interpretation in which the tip of the Bloch vector, consisting of the components u_1, u_2, and u_3, moves on the surface of a sphere. Thus, when $u_3 = 1$ ($\boldsymbol{\beta} = \hat{\mathbf{k}}$), the atom-field system is in the state $|\mathrm{F}\rangle = |\mathrm{e}, n-1\rangle$ and a reversal of u_3 to -1 ($\boldsymbol{\beta} = -\hat{\mathbf{k}}$) corresponds to a transition to $|\mathrm{I}\rangle = |\mathrm{g}, n\rangle$.

5.5 One-Photon Transitions

Let us consider an atom initially in the state $|\mathrm{g}\rangle$ interacting with a single mode of a quantized radiation field containing n photons. If the atom is excited to the higher state $|\mathrm{e}\rangle$ while the field suffers a loss of one photon, the transition

$$|\mathrm{I}\rangle = |\mathrm{g}, n\rangle \to |\mathrm{F}\rangle = |\mathrm{e}, n-1\rangle \tag{5.125}$$

is regarded as an absorption process. On the other hand, if the initial state of the atom is $|\mathrm{e}\rangle$ and that of the field is $|n\rangle$, while the final state consists of an atom in $|\mathrm{g}\rangle$ and the field consists of an atom $|n+1\rangle$, the transition

$$|\mathrm{I}\rangle = |\mathrm{e}, n\rangle \to |\mathrm{F}\rangle = |\mathrm{g}, n+1\rangle \tag{5.126}$$

represents an emission process. Note that in this scheme the initial states for both absorption and emission contain n photons. The energies are

$$E_\mathrm{e} - E_\mathrm{g} = \hbar\omega_0. \tag{5.127}$$

$$\text{Absorption:}\quad E_\mathrm{I} = E_\mathrm{g} + n\hbar\omega, \qquad E_\mathrm{F} = E_\mathrm{e} + (n-1)\hbar\omega,$$
$$E_\mathrm{F} - E_\mathrm{I} = \hbar(\omega_0 - \omega). \tag{5.128}$$

$$\text{Emission:}\quad E_\mathrm{I} = E_\mathrm{e} + n\hbar\omega, \qquad E_\mathrm{F} = E_\mathrm{g} + (n+1)\hbar\omega,$$
$$E_\mathrm{F} - E_\mathrm{I} = -\hbar(\omega_0 - \omega). \tag{5.129}$$

At exact resonance ($\omega = \omega_0$), $E_\mathrm{F} = E_\mathrm{I}$ for absorption and emission.

For the two-level case the states $|\mathrm{I}\rangle$ and $|\mathrm{F}\rangle$ constitute a complete set and the time-dependent Schrödinger equation is expressed by

$$\begin{aligned} i\hbar\dot{c}_\mathrm{I}(t) &= \langle \mathrm{I}|\mathscr{H}_\mathrm{AF}|\mathrm{F}\rangle c_\mathrm{F}(t) e^{i(E_\mathrm{I} - E_\mathrm{F})t/\hbar}, \\ i\hbar\dot{c}_\mathrm{F}(t) &= \langle \mathrm{F}|\mathscr{H}_\mathrm{AF}|\mathrm{I}\rangle c_\mathrm{I}(t) e^{i(E_\mathrm{F} - E_\mathrm{I})t/\hbar}, \end{aligned} \tag{5.130}$$

in conformity with the general relations of Eq. (2.84). The normalization condition in this case is

$$|c_\mathrm{I}(t)|^2 + |c_\mathrm{F}(t)|^2 = 1, \tag{5.131}$$

and the matrix elements, with $\mathscr{H}_\mathrm{AF}$ given by Eq. (5.107), are

$$\begin{aligned} \text{Absorption:}\quad \langle \mathrm{F}|\mathscr{H}_\mathrm{AF}|\mathrm{I}\rangle_\mathrm{a} &= \hbar\langle \mathrm{e}, n-1|ga\sigma^+ + g^* a^\dagger \sigma^-|\mathrm{g}, n\rangle \\ &= \hbar\langle \mathrm{e}, n-1|ga\sigma^+|\mathrm{g}, n\rangle \\ &= \hbar g\sqrt{n} = \langle \mathrm{I}|\mathscr{H}_\mathrm{AF}|\mathrm{F}\rangle_\mathrm{a}^*. \end{aligned} \tag{5.132a}$$

$$\text{Emission:}\quad \langle \mathrm{F}|\mathscr{H}_{\mathrm{AF}}|\mathrm{I}\rangle_{\mathrm{e}} = \hbar\langle \mathrm{g}, n+1|\mathrm{g}a\sigma^{+} + \mathrm{g}^{*}a^{\dagger}\sigma^{-}|\mathrm{e}, n\rangle$$
$$= \hbar\langle \mathrm{g}, n+1|\mathrm{g}^{*}a^{\dagger}\sigma^{-}|\mathrm{e}, n\rangle$$
$$= \hbar\mathrm{g}^{*}\sqrt{n+1} = \langle \mathrm{I}|\mathscr{H}_{\mathrm{AF}}|\mathrm{F}\rangle_{\mathrm{e}}^{*}. \tag{5.132b}$$

In anticipation of results to be derived later in this section, the ratio of emission to absorption probabilities is given by the ratio of their respective absolute squared matrix elements. From Eq. (5.132) it is seen that

$$\frac{|\langle \mathrm{F}|\mathscr{H}_{\mathrm{AF}}|\mathrm{I}\rangle_{\mathrm{e}}|^{2}}{|\langle \mathrm{F}|\mathscr{H}_{\mathrm{AF}}|\mathrm{I}\rangle_{\mathrm{a}}|^{2}} = \frac{n+1}{n}. \tag{5.133}$$

For absorption and emission in weak fields, sufficiently accurate transition probabilities may be calculated on the basis of first-order perturbation theory. For absorption, the weak field assumption implies that the population in the initial atomic state remains essentially unchanged. We therefore assume

$$c_{\mathrm{I}}(0) = c_{\mathrm{I}}(t) = 1, \qquad c_{\mathrm{F}}(0) = 0. \tag{5.134}$$

The first-order calculation now proceeds as in Section 5.3. The probability amplitude in absorption is

$$c_{\mathrm{F}}(t) = -i\mathrm{g}\sqrt{n}\int_{0}^{t} e^{i(\omega_0-\omega)t'}\,dt' = \mathrm{g}\sqrt{n}\,\frac{1-e^{i(\omega_0-\omega)t}}{(\omega_0-\omega)}, \tag{5.135}$$

and the transition probability is

$$P(\mathrm{g}, n \to \mathrm{e}, n-1) \equiv P_a = |c_{\mathrm{F}}(t)|^{2} = 4|\mathrm{g}|^{2} n \frac{\sin^{2}[\frac{1}{2}(\omega_0-\omega)t]}{(\omega_0-\omega)^{2}}. \tag{5.136}$$

This is essentially the same expression as Eq. (5.71) for the semiclassical case; therefore, the same comments are applicable and we may write immediately the analog to the Golden Rule:

$$W(\mathrm{I}\to\mathrm{F}) \equiv W_{\mathrm{a}} = \frac{2\pi}{\hbar}|\langle \mathrm{F}|\mathscr{H}_{\mathrm{AF}}|\mathrm{I}\rangle|^{2}\,\delta(E_{\mathrm{F}}-E_{\mathrm{I}})$$
$$= \frac{2\pi}{\hbar^{2}}|\langle \mathrm{F}|\mathscr{H}_{\mathrm{AF}}|\mathrm{I}\rangle|^{2}\,\delta(\omega_0-\omega)$$
$$= 2\pi n|\mathrm{g}|^{2}\,\delta(\omega_0-\omega)$$
$$= \frac{\pi\omega n}{\varepsilon_0\hbar|}|\langle \mathrm{e}|\mathbf{d}\cdot\hat{\boldsymbol{\varepsilon}}|\mathrm{g}\rangle|^{2}\,\delta(\omega_0-\omega). \tag{5.137}$$

W_{a} is the absorption probability per unit time for the interaction of an atom with a single (plane polarized) mode of the radiation field. The field loses one photon and the atom is excited from the lower state $|\mathrm{g}\rangle$ to the higher state $|\mathrm{e}\rangle$.

To this point in the development, both atomic levels are assumed to be infinitely sharp. Here, too, it is seen that Eq. (5.137) becomes identical with Eq. (5.71) upon replacement of $|q|^2$ by $|g|^2 n$.

As has already been noted in Section 5.3, a more realistic approach must take into account the finite spectral width inherent in all light sources. We shall therefore replace the photon number n by a quasi-continuous number density $n(\omega)$ such that $n(\omega)\,d\omega$ is the number of photons in the interval $(\omega, \omega + d\omega)$. The energy density per unit frequency interval, $U(\omega)$, then will be

$$U(\omega) = \frac{n(\omega)\hbar\omega}{V} = \frac{\mathrm{I}(\omega)}{\mathrm{c}}, \tag{5.138}$$

in which V is the volume of the cavity and c is the velocity of light. $U(\omega)\,d\omega$ is the energy per unit volume of the cavity for photons lying in the interval $(\omega, \omega + d\omega)$ and $\mathrm{I}(\omega)\,d\omega$ is the energy crossing unit area in unit time (intensity) for photons in $(\omega, \omega + d\omega)$. Replacing n by $n(\omega)\,d\omega$ in Eq. (5.137) and integrating over frequency, we have

$$\begin{aligned} W_a &= \frac{\pi}{\varepsilon_0 \hbar V} |\langle \mathrm{e}|\mathbf{d}\cdot\hat{\varepsilon}|\mathrm{g}\rangle|^2 \int_{-\infty}^{\infty} \omega n(\omega)\,\delta(\omega_0 - \omega)\,d\omega \\ &= \frac{\pi}{\varepsilon_0 \hbar V} |\langle \mathrm{e}|\mathbf{d}\cdot\hat{\varepsilon}|\mathrm{g}\rangle|^2 \omega_0 n(\omega_0) = \frac{\pi}{\varepsilon_0 \hbar^2_{\mathrm{c}}} |<\mathrm{e}|\mathbf{d}\cdot\hat{\varepsilon}|\mathrm{g}>|^2 \mathrm{I}(\omega_0) \\ &= \frac{2\pi V}{\hbar \mathrm{c}\omega_0} |\mathrm{g}|^2 \mathrm{I}(\omega_0). \end{aligned} \tag{5.139}$$

The δ-function served the useful purpose of simplifying the calculation, but we must remember that the δ-function was introduced to approximate the narrow bandwidth in the vicinity of ω_0 where $P_a \neq 0$. Within this bandwidth, $\mathrm{I}(\omega)$ is a slowly varying function so that $\mathrm{I}(\omega)$ can be approximated by $\mathrm{I}(\omega_0)$; hence, the general expression for the absorption probability per unit time for radiation at the frequency ω is

$$W_{\mathrm{a}} = \frac{2\pi V}{\hbar \mathrm{c}\omega} |\mathrm{g}|^2 \mathrm{I}(\omega), \qquad \text{(finite spectral width).} \tag{5.140}$$

It is also instructive to note that W_{a} is proportional to the fine structure constant α defined by Eq. (5.68):

$$W_{\mathrm{a}} = \frac{4\pi^2\alpha}{\hbar} |\langle \mathrm{e}|\mathbf{r}\cdot\hat{\varepsilon}|\mathrm{g}\rangle|^2 \mathrm{I}(\omega), \qquad \text{(finite spectral width)} \tag{5.141}$$

in which the displacement vector $\mathbf{r}$ is related to the dipole moment operator $\mathbf{d}$ by $\mathbf{d} = -\mathrm{e}\mathbf{r}$. Were it not for the fact that $\alpha \ll 1$, the expressions derived by first-order perturbation theory would not be valid. A further point to note is that W_{a} is independent of time.

One may also derive an expression for the case where the radiation is distributed isotropically so that the absorbing atom sees radiation incident upon it from all directions. The number of cavity modes dN in a frequency interval $d\omega$ and propagating in an interval of solid angle $d\Omega$ is given by Eq. (4.42). If n is the average number of photons per mode within the interval dN,

$$n(\omega)\, d\omega = n dN, \tag{5.142}$$

and

$$\mathrm{I}(\omega) = \frac{\mathrm{c}n(\omega)\hbar\omega}{V} = \frac{\hbar n \omega^3\, d\Omega}{8\pi^3 \mathrm{c}^2}. \tag{5.143}$$

The absorption probability per unit time then takes the form

$$W_{\mathrm{a}} = \frac{\omega^2 n V}{4\pi^2 \mathrm{c}^3} |\mathrm{g}|^2\, d\Omega = \frac{\alpha \omega^3 n}{2\pi \mathrm{c}^2} |\langle \mathrm{e}|\mathbf{r}\cdot\hat{\varepsilon}|\mathrm{g}\rangle|^2\, d\Omega \qquad \text{(isotropic)}. \tag{5.144}$$

Comparing Eq. (5.144) with Eq. (5.137) reveals that the replacement of $\delta(\omega_0 - \omega)$ in Eq. (5.137) by $dN/d\omega$ from Eq. (4.42) yields Eq. (5.144). In other words, when discrete frequencies are replaced by a quasi-continuum, the δ-function in the Golden Rule is replaced by a mode density. This procedure is quite general and enables one to translate expressions pertaining to discrete modes into expressions applicable to cases where the bandwidth is continuous (though not too broad).

Equation (5.144) for the absorption probability per unit time refers to a specific polarization $\hat{\varepsilon}$ and a specific direction of propagation defined by the element of solid angle $d\Omega$. In many instances such details are of no interest, in

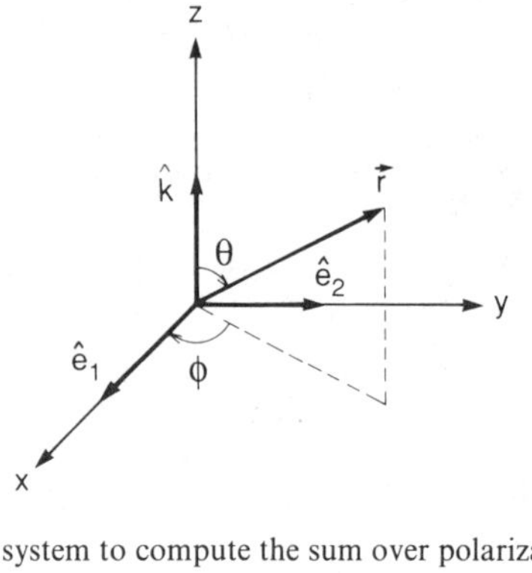

FIGURE 5.6 Coordinate system to compute the sum over polarizations and integration over solid angles.

which case it is necessary to sum over the polarizations and integrate over all solid angles. Referring to Fig. 5.6 it is seen that

$$\hat{\varepsilon}_1 \cdot \mathbf{r} = r \sin\theta \cos\phi, \qquad \hat{\varepsilon}_2 \cdot \mathbf{r} = r \sin\theta \sin\phi, \tag{5.145}$$

$$|\langle e|\hat{\varepsilon}_1 \cdot \mathbf{r}|g\rangle|^2 + |\langle e|\hat{\varepsilon}_2 \cdot \mathbf{r}|g\rangle|^2 = |\langle e|\mathbf{r}|g\rangle|^2 \sin^2\theta, \tag{5.146}$$

$$|\langle e|\mathbf{r}|g\rangle|^2 \int \sin^2\theta d\Omega = \frac{8\pi}{3}|\langle e|\mathbf{r}|g\rangle|^2. \tag{5.147}$$

Thus,

$$W_a = \frac{4\alpha\omega^3 n}{3c^2}|\langle e|\mathbf{r}|g\rangle|^2 \quad \text{(summed over polarizations and integrated over all directions).} \tag{5.148}$$

We now turn to the emission process and again invoke the Golden Rule. In view of Eq. (5.132b), the emission probability per unit time now reads

$$W(\mathrm{I} \to \mathrm{F}) \equiv W_e = 2\pi(n+1)|g|^2\,\delta(\omega_0 - \omega)$$
$$= \frac{\pi\omega(n+1)}{\varepsilon_0 \hbar V}|\langle g|\mathbf{d}\cdot\hat{\varepsilon}|e\rangle|^2\,\delta(\omega_0 - \omega). \tag{5.149}$$

Bearing in mind that the initial state $|\mathrm{I}\rangle = |e, n\rangle$ contains n photons, it is seen that there is a nonvanishing probability per unit time for emission, even when $n = 0$. The two terms in Eq. (5.149), one proportional to n and the other independent of n, are probability rates for *stimulated* (or *induced*) and *spontaneous* emission, respectively. That is,

$$\text{Stimulated:} \quad W_{st} = 2\pi n|g|^2\,\delta(\omega_0 - \omega)$$
$$= \frac{\pi\omega n}{\varepsilon_0 \hbar V}|\langle g|\mathbf{d}\cdot\hat{\varepsilon}|e\rangle|^2\,\delta(\omega_0 - \omega). \tag{5.150}$$

$$\text{Spontaneous:} \quad W_{sp} = 2\pi|g|^2\,\delta(\omega_0 - \omega)$$
$$= \frac{\pi\omega}{\varepsilon_0 \hbar V}|\langle g|\mathbf{d}\cdot\varepsilon|e\rangle|^2\,\delta(\omega_0 - \omega). \tag{5.151}$$

Spontaneous emission appears in the present calculations, which employ a quantized radiation field; it did not appear in the semiclassical calculations in which the field was treated classically. This is one of the basic distinguishing features between the two formalisms.

A comparison of Eq. (5.150) with Eq. (5.137) indicates that the probability rate for stimulated emission is identical with that for absorption when the initial states in the two cases contain the same number of photons; i.e.,

$$W_{st}(e, n \to g, n+1) = W_a(g, n \to e, n-1). \tag{5.152}$$

Hence, the probability rates for stimulated emission for the various cases treated previously are

$$W_{st} = \frac{4\pi^2\alpha}{\hbar}|\langle g|\mathbf{r}\cdot\hat{\varepsilon}|e\rangle|^2 I(\omega), \quad \text{(finite spectral width)} \tag{5.153a}$$

$$= \frac{\alpha\omega^3 n}{2\pi c^2}|\langle g|\mathbf{r}\cdot\hat{\varepsilon}|e\rangle|^2 d\Omega, \quad \text{(isotropic)} \tag{5.153b}$$

$$= \frac{4\alpha\omega^3 n}{3c^2}|\langle g|\mathbf{r}|e\rangle|^2 \quad \begin{array}{l}\text{(summed over polarizations and}\\ \text{integrated over all directions).}\end{array} \tag{5.153c}$$

In examining these equations, it is recalled that in the initial state $|I\rangle = |e, n\rangle$ (Eq. (5.126)), the radiation field is confined to a single mode with propagation vector **k** and polarization λ so that, more accurately, the photon-number state $|n\rangle$ should be written $|n_{\mathbf{k}\lambda}\rangle$. The final state $|F\rangle = |g, n+1\rangle$ is one in which a photon has been added to the *same mode*. This means that a photon emitted by a stimulated emission process has the same polarization, phase, and direction of propagation as the incident photons. Stimulated emission is said, therefore, to be a coherent process.

For spontaneous emission, the summation over polarizations and integration over all direction yields

$$W_{sp} = \frac{4\alpha\omega^3}{3c^2}|\langle g|\mathbf{r}|e\rangle|^2 \quad \begin{array}{l}\text{(summed over polarizations and}\\ \text{integrated over all directions).}\end{array} \tag{5.154}$$

Since incident photons are not required for spontaneous emission, the emitted photons are not constrained to any particular mode. Their directions of propagation, phases, and polarizations may range over all possible values. Spontaneous emission, therefore, is said to be an incoherent process. Further discussion of spontaneous emission will be resumed in the next section.

In the event that the states $|g\rangle$ and $|e\rangle$ have degeneracies d_g and d_e, respectively, the expressions given above need to be modified. The total transition probability per unit time for absorption is

$$W_a = \frac{4\alpha\omega^3 n}{3c^2 d_g}\sum_{ge}|\langle e|\mathbf{r}|g\rangle|^2, \tag{5.155}$$

in which all the transition probabilities have been added. But in view of the degeneracy d_g in the initial state, the probability of finding an atom in any one particular member of the degenerate manifold is $1/d_g$; hence, the appearance of d_g in the denominator of Eq. (5.155). The analogous expression for emission is

$$W_e = \frac{4\alpha(n+1)\omega^3}{3c^2 d_e}\sum_{ge}|\langle g|\mathbf{r}|e\rangle|^2. \tag{5.156}$$

It often is convenient to work with quantities that do not contain degeneracy factors and therefore are symmetric with respect to initial and final states. Such a quantity is the *line strength* $S(\mathrm{g}, \mathrm{e})$ defined by

$$S(\mathrm{g}, \mathrm{e}) = S(\mathrm{e}, \mathrm{g}) = \hbar c\alpha \sum_{\mathrm{ge}} |\langle \mathrm{g}|\mathbf{r}|\mathrm{e}\rangle|^2 = \frac{\mathrm{e}^2}{4\pi\varepsilon_0} \sum_{\mathrm{ge}} |\langle \mathrm{g}|\mathbf{r}|\mathrm{e}\rangle|^2. \tag{5.157}$$

The total transition probabilities per unit time then become

$$W_{\mathrm{a}} = \frac{4\omega^3 n}{3\hbar c^3 \mathrm{d_g}} S(\mathrm{g}, \mathrm{e}), \qquad W_{\mathrm{e}} = \frac{4\omega^3 (n+1)}{3\hbar c^3 \mathrm{d_e}} S(\mathrm{g}, \mathrm{e}). \tag{5.158}$$

The semiclassical Rabi formula for the transition probability in a two-level system was derived in Section 5.3. When the field is quantized, the Rabi formula acquires a slightly altered form, although the physical interpretation remains essentially the same. To derive the quantized version we revert to Eq. (5.130) with matrix elements for absorption as in Eq. (5.132a):

$$\dot{c}_{\mathrm{I}}(t) = -ig^* \sqrt{n}\, c_{\mathrm{F}}(t) e^{-i\Delta t}, \tag{5.159}$$

$$\dot{c}_{\mathrm{F}}(t) = ig\sqrt{n}\, c_{\mathrm{I}}(t) e^{i\Delta t}.$$

These equations have the same form as Eq. (5.69). Therefore, with initial conditions

$$c_{\mathrm{I}}(0) = 1, \qquad c_{\mathrm{F}}(0) = 0, \tag{5.160}$$

the derivation proceeds in precisely the same fashion as in Section starting with Eq. (5.83); the result is

$$P_{\mathrm{FI}} \equiv P(\mathrm{I} \to \mathrm{F}) = |c_{\mathrm{F}}(t)|^2 = \frac{\Omega_0^2}{\Omega^2} \sin^2 \frac{1}{2}\Omega t, \tag{5.161}$$

where

$$\Omega_2^0 = 4|g|^2 n, \qquad \Omega^2 = \Delta^2 + \Omega_0^2. \tag{5.162}$$

5.6 Spontaneous Emission, Rydberg States, Superradiance

In the previous section, it was shown that even in the absence of incident photons, a coupling between an atomic system in an excited state and the electromagnetic field still exists. The coupling gives rise to the emission of photons—a process called spontaneous emission to distinguish it from stimulated emission, which requires the presence of an incident beam.

Furthermore, a photon emitted by stimulated emission enters the same mode as that of the photon that stimulated the transition. But, since there are no incident photons in the case of spontaneous emission, the emitted photon can enter any available radiation mode allowed by the selection rules and conservation laws. It also was remarked in Section 5.5 that spontaneous emission does not appear in the standard version of semiclassical theory, but arises very naturally as soon as the electromagnetic field is quantized. Nevertheless, it should be mentioned that modified versions of semiclassical theory [7] can account for spontaneous emission.

Let us now investigate the spontaneous emission process in further detail [5, 8]. For the two-level system undergoing an emission process, the initial and final states of the photon-atom system, with the field in a single mode, were assumed to be $|\mathrm{I}\rangle = |\mathrm{e}, n\rangle$ and $|\mathrm{F}\rangle = |\mathrm{g}, n+1\rangle$, respectively, as shown in Eq. (5.126). These may be simplified by setting $n = 0$ since spontaneous emission is independent of the initial photon number; hence, the two states involved in spontaneous emission may be taken to be

$$|\mathrm{I}\rangle = |\mathrm{e}, 0\rangle, \qquad |\mathrm{F}\rangle = |\mathrm{g}, 1\rangle. \tag{5.163}$$

The field in $|\mathrm{I}\rangle$ is still regarded to be in a single mode, albeit there are no photons present, but the field in $|\mathrm{F}\rangle$ can no longer be confined to a single mode since the emitted photon can enter any one of an infinite number of possible modes with different directions of propagation, phases, and polarizations. It therefore will be necessary to take into account the fact that the state $|\mathrm{F}\rangle$ is not unique.

For absorption (and stimulated emission) it was possible to assume that $c_\mathrm{I}(0) = c_\mathrm{I}(t) = 1$ as in Eq. (5.134). But in the present case we must refrain from making such an assumption to allow for the possibility that the initial state is not stationary under spontaneous emission. Hence, we take

$$c_\mathrm{I}(0) = 1, \qquad c_\mathrm{F}(0) = 0, \qquad c_\mathrm{I}(t) \neq c_\mathrm{I}(0). \tag{5.164}$$

The physical picture thus assumes that at $t = 0$ the atom is in the upper (excited) state $|\mathrm{e}\rangle$ and the radiation field, assumed to consist of a single mode, is in the vacuum state $|0\rangle$. After the transition, the atom is found in the lower (ground) state $|\mathrm{g}\rangle$ and one photon has been created in an arbitrary mode. The description of spontaneous emission then must include all possible states $|\mathrm{F}\rangle$, which implies that Eqs. (5.130) must be modified to read

$$\begin{aligned} i\hbar\dot{c}_\mathrm{I}(t) &= \sum_{\mathrm{F}\neq\mathrm{I}} \langle\mathrm{I}|\mathscr{H}_{\mathrm{AF}}|\mathrm{F}\rangle c_\mathrm{F}(t) e^{i(\mathrm{E_I}-\mathrm{E_F})t/\hbar} \\ &= \sum_{\mathrm{F}\neq\mathrm{I}} \langle\mathrm{I}|\mathscr{H}_{\mathrm{AF}}|\mathrm{F}\rangle c_\mathrm{F}(t) e^{i(\omega_0-\omega)t}, \end{aligned} \tag{5.165a}$$

$$\begin{aligned} i\hbar\dot{c}_\mathrm{F}(t) &= \langle\mathrm{F}|\mathscr{H}_{\mathrm{AF}}|\mathrm{I}\rangle c_\mathrm{I}(t) e^{i(\mathrm{E_F}-\mathrm{E_I})t/\hbar} \\ &= \langle\mathrm{F}|\mathscr{H}_{\mathrm{AF}}|\mathrm{I}\rangle c_\mathrm{I}(t) e^{-i(\omega_0-\omega)t}. \end{aligned} \tag{5.165b}$$

Presently, we are concerned with the time-development of the state $|\mathrm{I}\rangle$; hence, we solve for $c_F(t)$ from Eq. (5.165b) and insert it into Eq. (5.165a):

$$\dot{c}_F(t) = -\frac{i}{\hbar}\langle \mathrm{F}|\mathscr{H}_{\mathrm{AF}}|\mathrm{I}\rangle \int_0^t c_I(t')e^{-i(\omega_0-\omega)t'}\,dt', \tag{5.166}$$

$$\dot{c}_I(t) = \frac{-1}{\hbar^2}\sum_{\mathrm{F}\neq \mathrm{I}} |\langle \mathrm{F}|\mathscr{H}_{\mathrm{AF}}|\mathrm{I}\rangle|^2 \int_0^t c_I(t')e^{i(\omega_0-\omega)(t-t')}\,dt'. \tag{5.167}$$

To solve this differential equation we note that the major contribution to the integral comes from the region where $t' = t$, which therefore permits the removal of $c_I(t') \simeq c_I(t)$ from the integral. Then

$$\dot{c}_I(t) = \frac{-1}{\hbar^2}\sum_{\mathrm{F}\neq \mathrm{I}} |\langle \mathrm{F}|\mathscr{H}_{\mathrm{AF}}|\mathrm{I}\rangle|^2 c_I(t) \int_0^t e^{i(\omega_0-\omega)\tau}\,d\tau, \tag{5.168}$$

in which

$$\tau = t - t'. \tag{5.169}$$

The integral in Eq. (5.168) appears to be identical with the integral in Eq. (5.70). In the latter, however, the upper limit is restricted as the subsequent discussion indicated, but in Eq. (5.168) there is no constraint to prevent the time from increasing indefinitely. Furthermore, the integrand is sharply peaked at $\omega \simeq \omega_0$; the upper limit then may be extended to infinity. Thus, since

$$\lim_{t\to\infty} \int_0^t e^{i(\omega_0-\omega)\tau}\,d\tau = iP\left(\frac{1}{\omega_0-\omega}\right) + \pi\,\delta(\omega_0-\omega), \tag{5.170}$$

according to Eq. (2.284), we have

$$\dot{c}_I(t) + \frac{\hbar\Gamma + i\Delta E}{2\hbar} c_I(t) = 0, \tag{5.171}$$

where

$$\frac{\hbar\Gamma}{2} = \frac{\pi}{\hbar}\sum_{\mathrm{F}\neq \mathrm{I}} |\langle \mathrm{F}|\mathscr{H}_{\mathrm{AF}}|\mathrm{I}\rangle|^2\,\delta(\omega_0-\omega), \tag{5.172}$$

$$\begin{aligned}\frac{\Delta E}{2} &= \frac{1}{\hbar}\sum_{\mathrm{F}\neq \mathrm{I}} |\langle \mathrm{F}|\mathscr{H}_{\mathrm{AF}}|\mathrm{I}\rangle|^2 P\left(\frac{1}{\omega_0-\omega}\right), \\ &= \frac{1}{\hbar}\sum_{\mathrm{F}\neq \mathrm{I}} |\langle \mathrm{F}|\mathscr{H}_{\mathrm{AF}}|\mathrm{I}\rangle|^2 \times \begin{cases} \dfrac{1}{\omega_0-\omega}, & \text{when } \omega_0-\omega \neq 0, \\ 0, & \text{when } \omega_0-\omega = 0. \end{cases}\end{aligned} \tag{5.173}$$

The solution to the differential equation (5.171) with the initial conditions of Eq. (5.164) gives

$$c_I(t) = e^{-\Gamma t/2}e^{-i\Delta E t/2\hbar}, \qquad |c_I(t)|^2 = e^{-\Gamma t}, \tag{5.174}$$

a result first obtained by Weisskopf and Wigner [9].

It may be noted that the same results are obtainable by means of the Green's function discussed in Section 2.10. Referring to Eq. (2.333), the replacements

$$V = \mathscr{H}_{AF}, \qquad x = E_I, \qquad E_n = E_F, \qquad |\psi_k\rangle = |I\rangle, \qquad |\psi_n\rangle = |F\rangle, \quad (5.175)$$

immediately yield the results in Eq. (5.174). The probability amplitude in Eq. (5.174) matches the probability amplitude in Eq. (2.33b) to within a (dynamical) phase factor, and the probabilities of finding the system in the initial state $|\psi_k\rangle = |I\rangle$ are identical. Another approach to spontaneous emission based on vacuum fluctuations is presented in Section 6.9.

We see, then, that the probability for the system to be found in the state $|I\rangle = |e, 0\rangle$, after a time t has elapsed, is a decreasing exponential with decay constant Γ, which means that an atom cannot reside in the excited state $|e\rangle$ for all time despite the absence of incident photons. The state has a finite lifetime that may be defined by

$$\tau = \frac{1}{\Gamma}, \quad (5.176)$$

and, in view of the Heisenberg uncertainty relation, $\hbar\Gamma$ is interpreted as a level width. In other words, as a consequence of spontaneous emission, the atomic state $|e\rangle$ cannot be an infinitely sharp state but must have a finite spread in energy equal to $\hbar\Gamma$. The level width arising from spontaneous emission, also known as the natural line width, is the minimum possible width, assuming all other broadening mechanisms have been eliminated. Examination of Eq. (5.172) indicates that Γ is just the expression, according to the Golden Rule, for the transition probability per unit time from a state $|I\rangle$ to all possible states $|F\rangle$. But we already have such an expression in Eq. (5.154) that gives the transition probability per unit time for spontaneous emission summed over polarizations and integrated over all angles. Therefore, at the transition frequency $\omega = \omega_0$,

$$\Gamma = \frac{1}{\tau} = W_{sp} = \frac{4\alpha\omega^3}{3c^2}|\langle g|\mathbf{r}|e\rangle|^2. \quad (5.177)$$

Indeed, Eq. (5.172) corroborates the fact that Γ is due mainly to resonant photons.

The interpretation of ΔE may be obtained by inserting $c_I(t)$ from Eq. (5.176) into Eq. (5.165b); then

$$i\hbar\dot{c}_F(t) = \langle F|\mathscr{H}_{AF}|I\rangle e^{-\Gamma t/2\hbar} e^{-i[(\omega_0 + \Delta E/2\hbar) - \omega]t}. \quad (5.178)$$

Evidently $\Delta E/2$ is a shift in the energy separation between the two atomic states, as we have already seen in connection with Eq. (2.336). Since, according to Eq. (5.173), $\Delta E = 0$ for $\omega = \omega_0$, only off-resonance photons contribute to ΔE. The computation of ΔE is far from trivial, however. Ordinary perturbation methods yield results much larger than experimental values. It is only when the computation proceeds on the basis of relativistic quantum electrodynamics with mass renormalization that correct results are achieved. Fortunately, in many cases of physical interest, ΔE is very small compared to $\hbar\omega_0$ so that it need not be taken into account.

Let us compare the transition probabilities derived from semiclassical theory with those from the fully quantized theory. In the former case, there is no spontaneous emission so that, in the absence of other interactions that may cause a decay of the excited state, one could assume that the two atomic states $|g\rangle$ and $|e\rangle$ were perfectly sharp. With this assumption, we deduced the Rabi formula (Eq. 5.93)) as well as Eq. (5.161), which indicated that the probability of finding the atom in one or the other state oscillated indefinitely. In the quantum mechanical version, however, we find that the probability of finding the atom in the state $|e\rangle$ decays irreversibly as a result of spontaneous emission; hence, the assumption that $|e\rangle$ is perfectly sharp is not tenable and Rabi oscillations cannot be sustained indefinitely. Only when the period of the Rabi oscillations is much shorter than the mean life of $|e\rangle$ is it possible for a number of (damped) oscillations to occur before the probability of finding the atom in $|e\rangle$ has effectively vanished. Further discussion of damping due to spontaneous emission and other mechanisms is considered in Chapter VI.

The calculation leading to Eq. (5.172) and (5.177) for the spontaneous emission rate assumes that all possible modes of the radiation field are available to the emitted photon. If the emitting atom is surrounded by a cavity with a limited number of modes at the transition frequency, the process of spontaneous emission will be inhibited [10–12], thereby increasing the lifetime of the excited state. Indeed, if the dimensions of the cavity are small compared to transition wavelength, there will be no spontaneous emission at all. On the other hand, a high Q (narrow bandwidth) cavity tuned to the transition frequency ω_0 will have a high mode density in the vicinity of ω_0. The spontaneous emission rate from an atom in such a cavity will be enhanced.

The experimental realization of these effects requires cavities whose dimensions are of the order of the wavelength of the radiation. For ordinary atomic transitions at optical frequencies, the cavity dimensions are far too small. One may take advantage, however, of the fact that atomic energy levels become more and more closely spaced as they approach the ionization limit. Consequently, the wavelength of the radiation associated with transitions between nearby levels near the ionization limit is much longer and can extend into the far infrared and even into the microwave region.

For large values of the principal quantum number n, the binding energy of a valence electron is approximately proportional to n^{-2}, and the energy separation between adjacent levels therefore scales as n^{-3}. These characteristics are precisely those of a hydrogen atom, hence, the basis for calling a highly excited atom a *Rydberg* atom. Indeed, the valence electron in a Rydberg atom with $n \sim 20$ or greater is so far from the nucleus and the core electrons that the entire system behaves like a giant hydrogen atom. With Rydberg atoms placed in high-Q cavities, experimenters [13–15] have been able to study field-atom interactions such as inhibited spontaneous emission, line shape, and effects arising from the coupling of an atom to a single mode of the field.

An inherent assumption in the discussion of spontaneous emission so far has been that in any assembly of atoms each atom acts independently with no effects arising from the presence of other atoms. The formalism then could concentrate on the behavior of a single atom and leads to the conclusion that the rate of spontaneous emission from an assembly of N atoms is simply N times the rate from a single atom. This conclusion and with it, the assumption of independence, was challenged by Dicke [3] who showed that atoms may become coupled by virtue of their interaction with a radiation field. Under those circumstances, the atoms behave in a collective, rather than an independent fashion, with the result that the spontaneous emission from N atoms is proportional to N^2 rather than N. This effect was named *superradiance*. Let us first illustrate how this comes about for $N = 2$.

We assume that the two atoms have the same resonant frequency ω_0, and that the system is confined to a region whose dimensions are small compared to a wavelength. Interactions other than photon-atom interactions (e.g., collisons) are ignored. In the rotating wave approximation, the Hamiltonian for the atomic system interacting with a single mode ($\omega = \omega_0$) of the radiation field is

$$\mathscr{H} = \mathscr{H}_A + \mathscr{H}_F + \mathscr{H}_{AF} \tag{5.179}$$

where

$$\mathscr{H}_A = \frac{\hbar\omega_0}{2}(\sigma_{z1} + \sigma_{z2}), \qquad \mathscr{H}_F = \hbar\omega_0 a^\dagger a,$$
$$\mathscr{H}_{AF} = \hbar|g|(\sigma_1^+ a + \sigma_2^+ a + \sigma_1^- a^\dagger + \sigma_2^- a^\dagger). \tag{5.180}$$

We now consider the interaction of the two atoms with the radiation field when the two atoms are (a) independent, (b) coupled into a singlet Dicke state, and (c) coupled into a triplet Dicke state. For the definitions of the Dicke states we refer to Section 5.1.

If the radiation field contains n photons, the zero-order states, that is, the eigenstates of $\mathscr{H}_A + \mathscr{H}_F$ are $|g_1g_2n\rangle$, $|e_1e_2n\rangle$, $|g_1e_2n\rangle$, and $|e_1g_2n\rangle$ with total energies (atom plus field) equal to $E_A + n\hbar\omega_0 = 0$; there are three states that satisfy this condition,

$$|g_1e_20\rangle, \qquad |g_2e_10\rangle, \qquad |g_1g_21\rangle. \tag{5.181}$$

The state $|g_1e_20\rangle$ corresponds to the case where the first atom is in the state $|g\rangle$ with energy $-\hbar\omega_0/2$, the second atom is in $|e\rangle$ with energy $\hbar\omega_0/2$, and the field mode is vacant. When the atoms are interchanged, the state of the total system becomes $|g_2e_10\rangle$. In $|g_1g_21\rangle$ the two atoms are in $|g\rangle$ states and the field mode contains one photon of energy $\hbar\omega_0$. The computation of matrix elements of $\mathscr{H}_{AF}$ is straightforward; the nonvanishing matrix elements are those that involve a change of one photon

$$\begin{aligned}\langle g_1e_20|\mathscr{H}_{AF}|g_1g_21\rangle &= \langle g_1g_21|\mathscr{H}_{AF}|g_1e_20\rangle \\ &= \langle g_2e_10|\mathscr{H}_{AF}|g_1g_21\rangle = \langle g_1g_21|\mathscr{H}_{AF}|g_2e_10\rangle = \hbar|g|. \end{aligned} \quad (5.182)$$

These matrix elements are in agreement with those developed in Section 5.5 for a single atom; hence, they correspond to the case in which the atoms are uncorrelated. Now consider the Dicke states

$$\begin{aligned}|\psi_a\rangle &= \frac{1}{\sqrt{2}}[|g_1e_20\rangle - |g_2e_10\rangle], \\ |\psi_s\rangle &= \frac{1}{\sqrt{2}}[|g_1e_20\rangle + |g_2e_10\rangle]. \end{aligned} \quad (5.183)$$

The state $|\psi_a\rangle$ is an antisymmetric (zero order) state of the total system; the two atoms reside in the singlet state $|00\rangle$ and the field mode is vacant. In $|\psi_s\rangle$ the atoms are in the triplet state $|10\rangle$. With the results from Eq. (5.182), one finds immediately that

$$M_a \equiv \langle g_1g_21|\mathscr{H}_{AF}|\psi_a\rangle = 0 \quad (5.184a)$$

$$M_s \equiv \langle g_1g_21|\mathscr{H}_{AF}|\psi_s\rangle = \sqrt{2}\,\hbar|g|. \quad (5.184b)$$

We now may compare the probability per unit time for spontaneous emission from a state in which one atom is in an excited state $|e\rangle$, the other atom is in the ground state $|g\rangle$, and the number of photons is $n = 0$, to a state in which both atoms are in their respective ground states $|g\rangle$ and the number of photons is $n = 1$. The following situations arise:

a. Independent case. The atom in $|e\rangle$ makes a transition to $|g\rangle$ and radiates a photon; the second atom in $|g\rangle$ remains in that state. The probability for the transition $|e_1g_2\rangle$ (or $|g_1e_2\rangle$) $\rightarrow |g_1g_2\rangle$ is proportional to $\hbar^2|g|^2$. The existence of the second atom in $|g\rangle$ does not contribute to the process.
b. Singlet Dicke case. The two atoms are initially coupled into the singlet state $|00\rangle$. To compute the rate of spontaneous emission for this case, we need the matrix element of $\mathscr{H}_{AF}$ for the transition $|\psi_a\rangle \rightarrow |g_1g_21\rangle$ which determines the probability for the emission of a photon. But according to Eq. (5.184a) the matrix element is zero which means that the excited atom has been prevented from radiating. This phenomenon is known as *subradiance*.

c. Triplet Dicke case. The two atoms are initially coupled into the triplet state $|10\rangle$. In this case, the spontaneous emission rate is proportional to the square of the matrix element of $\mathscr{H}_{\text{AF}}$ for the transition $|\psi_s\rangle \rightarrow |g_1 g_2 1\rangle$, i.e., proportional to $|M_s|^2$ which is equal to $2\hbar^2|g|^2$. Thus, an excited atom in the triplet state can radiate a photon with a probability that is twice that for the independent or single atom case; hence, the origin of the name superradiance.

For the more general case, it was shown by Dicke that the spontaneous radiation rate from an assembly of N atoms in a state $|SM\rangle$ is proportional to

$$W = (S + M)(S - M + 1). \tag{5.185}$$

Several cases are of interest:

$|SM\rangle = |N/2\; N/2\rangle$. All atoms are in excited states (complete inversion); $W = N$. This corresponds to N atoms radiating independently.

$|SM\rangle = |N/2\; 0\rangle$ (N even). Half of the atoms are in $|e\rangle$ and half in $|g\rangle$. For this case,

$$W = \frac{N}{2}\left(\frac{N}{2} + 1\right). \tag{5.186}$$

When $N \gg 1$, $W \simeq N^2/4$; that is, the radiation rate is proportional to the square of the number of atoms. The state $|N/2\; 0\rangle$, therefore, is superradiant, and the lifetime is shortened by a factor of N due to the enhanced emission rate.

$|SM\rangle = |00\rangle$ (N even). Here, too, half of the atoms are excited and half are in their ground states but $W = 0$. This state, therefore, is subradiant. The dramatic difference in radiation rates between $|00\rangle$ and $|N/2\; 0\rangle$, both of which have the same distribution of atoms in $|e\rangle$ and $|g\rangle$, is attributable to the difference in their symmetry properties.

$|SM\rangle = |N/2\; -N/2\rangle$. All atoms are in their ground states; $W = 0$. For obvious reasons this state does not radiate.

$|SM\rangle = |N/2\; 1 - N/2\rangle$. One atom is excited and all the rest are in their ground states; $W = N$. This is a remarkable result since it equates the radiation rate from a state in which only one atom is excited to the radiation rate from $|N/2\; N/2\rangle$ in which all atoms are excited.

When N atoms radiate at a rate proportional to N^2, it must mean that a coherence has been established among the atoms as if they were N classical oscillators all radiating with the same phase. One way the coherence can be brought about is by coherent irradiation; this is a straightforward process that will receive attention in connection with free induction decay and photon echoes (Section 6.5). The more subtle issue addressed in the modern literature [6, 16–21] is how an assembly of independent two-level atoms, all in the upper state initially, organize themselves through their common coupling to the radiation field, so as to radiate in a cooperative fashion. It is now customary to distinguish between *superradiance* and *superfluorescence*. Both are characterized by radiation proportional to N^2. In the former, a macroscopic polarization is established by external means, as by coherent irradiation. In the latter, a macroscopic polarization is created, where none existed initially, through the mutual interactions of the atoms and the field. Superfluorescence depends on the shape of the sample whose dimensions can be large compared to a wavelength. Experiments [22] have been conducted with pencil-shaped samples in order to confine the emission along a specific direction. Emission of a pulse with peak intensity proportional to N^2 was observed in a small solid angle around the pencil axis in both forward and backward directions.

5.7 Einstein Coefficients, Natural Line Shape

The probability per unit time for spontaneous emission is also known as the Einstein A coefficient. Referring to Eq. (5.177)

$$A = W_{sp} = \Gamma = \frac{1}{\tau} = \frac{4\alpha\omega^3}{3c^2}|\langle g|\mathbf{r}|e\rangle^2. \tag{5.187}$$

The Einstein B coefficient is related to the absorption (or stimulated emission) probability per unit time (Eq. (5.148) or (5.153c)) by the relation

$$BU(\omega) = W_a = W_{st} = \frac{4\alpha\omega^3 n}{3c^2}|\langle e|\mathbf{r}|g\rangle|^2, \tag{5.188}$$

in which $U(\omega)\,d\omega$ is the energy per unit volume for photons in the interval $(\omega, \omega + d\omega)$ summed over polarizations and integrated over all directions. To compute $U(\omega)\,d\omega$, we construct the product of (a) the two independent polarizations for each mode; (b) n, the average number of photons per mode within the interval $(\omega, \omega + d\omega)$; (c) $\hbar\omega$, the average photon energy in the interval and (d) dN, the number of cavity modes between ω and $\omega + d\omega$ integrated over all directions (Eq. (4.43)). The product divided by the cavity volume gives

$$U(\omega)\,d\omega = \frac{2n\hbar\omega}{V}\,dN(\omega) = \frac{n\hbar\omega^3}{\pi^2 c^3}\,d\omega. \tag{5.189}$$

With Eq. (5.189) inserted into Eq. (5.188), the Einstein B coefficient is found to be

$$B = \frac{4\pi^2 \alpha c}{3\hbar} |\langle e|\mathbf{r}|g\rangle|^2 = \frac{\pi e^2}{3\varepsilon_0 \hbar^2} |\langle e|\mathbf{r}|g\rangle|^2, \tag{5.190}$$

and the ratio of the two coefficients is

$$\frac{A}{B} = \frac{\hbar\omega^3}{\pi^2 c^3}. \tag{5.191}$$

Let us consider now the equilibrium situation in a two-level system exposed to a beam of radiation. Assume N_g and N_e are the number of atoms per unit volume in $|g\rangle$ and $|e\rangle$, respectively, with $N_g + N_e = N$ (Fig. 5.7). Under the influence of the radiation field, N_g will decrease by virtue of transitions $|g\rangle \rightarrow |e\rangle$ (absorption) and will increase as a result of transitions $|e\rangle \rightarrow |g\rangle$ (stimulated and spontaneous emission). Hence, the rate equations are

$$\begin{aligned} \frac{dN_g}{dt} = -\frac{dN_e}{dt} &= -N_g W_a + N_e(W_{st} + W_{sp}) \\ &= -N_g BU(\omega) + N_e[BU(\omega) + A]. \end{aligned} \tag{5.192}$$

Under the equilibrium conditions

$$\frac{dN_g}{dt} = -\frac{dN_e}{dt} = 0, \tag{5.193}$$

the difference in populations $N_g - N_e$ becomes

$$N_g - N_e = \frac{N}{1 + \dfrac{2BU(\omega)}{A}} = \frac{N}{1+S}, \tag{5.194}$$

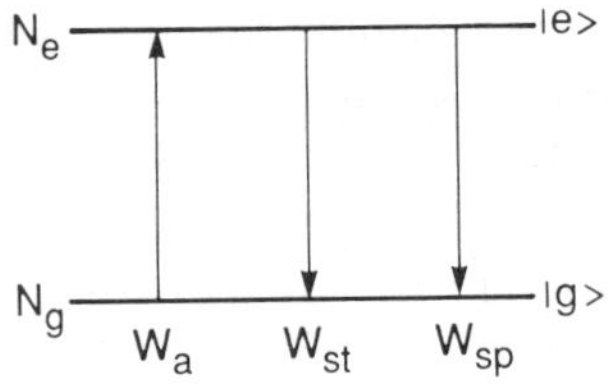

FIGURE 5.7 Absorption and emission processes in a two-level system. N_g and N_e are the populations in $|g\rangle$ and $|e\rangle$, respectively, W_a, W_{st} and W_{sp} are the probabilities per unit time for absorption, stimulated emission, and spontaneous emission.

where

$$S = \frac{2BU(\omega)}{A} = \frac{2\pi^2 c^3 U(\omega)}{\hbar\omega^3} = \frac{2\pi^2 c^2 I(\omega)}{\hbar\omega^3} \tag{5.195}$$

is known as the *saturation parameter*. (Caution: some authors define the saturation parameter to be $BU(\omega)/A$.) For a constant energy density or intensity, the saturation parameter is inversely proportional to ω^3. In the vicinity of resonance, $u \simeq \omega_0$; consequently, the ratio of stimulated to spontaneous transition rates fall rapidly with increasing separation between energy levels.

With the initial conditions $N_g(0) = N$, $N_e(0) = 0$, the solution to the differential Eq. (5.192) is

$$N_e(t) = \frac{NBU(\omega)}{A + 2BU(\omega)}(1 - e^{-[A+BU(\omega)]t}), \tag{5.196}$$

and as $t \to \infty$

$$N_e(\infty) = \frac{NBU(\omega)}{A + 2BU(\omega)} = \frac{NS}{2(1+S)}; \tag{5.197a}$$

also, since $\rho_{ee} = N_e/N$,

$$\rho_{ee} = \frac{S}{2(1+S)} \qquad (t \to \infty). \tag{5.197b}$$

We now may distinguish between two extreme cases when $t \to \infty$:
Unsaturated condition

$$S = 0, \quad N_e = 0, \quad N_g = N, \quad N_g - N_e = N, \quad \rho_{ee} = 0. \tag{5.198}$$

Totally saturated condition

$$S = \infty, \quad N_e = N/2, \quad N_g = N/2, \quad N_g - N_e = 0, \quad \rho_{ee} = \frac{1}{2}. \tag{5.199}$$

For ordinary (thermal) light sources, the intensities (in the optical region) are such that $A \gg BU(\omega)$; that is, spontaneous emission is far more probable than stimulated emission. In that case

$$N_e(\infty) \simeq \frac{NBU(\omega)}{A} = \frac{NS}{2}, \qquad (S \ll 1); \tag{5.200}$$

that is, the population in the excited state is a small fraction of the ground state population and is proportional to $U(\omega)$, the energy density of the light beam. With laser sources, however, stimulated emission can be comparable to, or greater than, spontaneous emission. If $A \ll BU(\omega)$,

$$N_e(\infty) \simeq \frac{N}{2} \qquad (S \gg 1). \tag{5.201}$$

We note that according to Eq. (5.197a) $N_e(\infty)$ cannot exceed $N/2$ (or N_g can never fall below $N/2$). But one should bear in mind that this restriction pertains to a two-level system. When more levels are present, the transition $|g\rangle \to |e\rangle$ may occur in a roundabout fashion; thus, if there is a level $|i\rangle$ lying above $|e\rangle$ (Fig. 5.8), we might arrive at $|e\rangle$ by the route $|g\rangle \to |i\rangle$ (absorption) followed by a radiative or nonradiative relaxation from $|i\rangle$ to $|e\rangle$. Under such circumstances N_e may exceed N_g. The condition $N_e > N_g$ is known as *population inversion.*

When degeneracies exist, the Einstein coefficients are modified in accordance with Eq. (5.150). The A coefficient becomes

$$A = \frac{4\omega^3}{3\hbar c^3 d_e} S(g, e). \tag{5.202}$$

For the B coefficient it is necessary to distinguish between absorption and emission. Thus,

$$B_{eg} = \frac{4\pi^2}{3\hbar^2 d_g} S(g, e), \qquad \text{(absorption)} \tag{5.203a}$$

$$B_{ge} = \frac{4\pi^2}{3\hbar^2 d_e} S(g, e) \qquad \text{(stimulated emission).} \tag{5.203b}$$

Note that

$$B_{ge} = \frac{\pi^2 c^3}{\hbar\omega^3} A, \qquad B_{eg} = \frac{d_e}{d_g} \frac{\pi^2 c^3}{\hbar\omega^3} A. \tag{5.204}$$

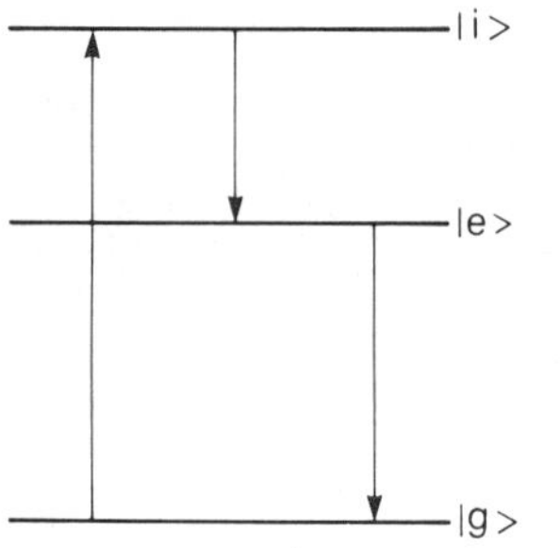

FIGURE 5.8 In a three-level system consisting of the states $|g\rangle$, $|e\rangle$, and $|i\rangle$, it is possible to create a population inversion in $|e\rangle$ relative to $|g\rangle$ by pumping the transition $|g\rangle \to |i\rangle$. A rapid relaxation (radiative or nonradiative) $|i\rangle \to |e\rangle$ allows the population in $|e\rangle$ to build up and eventually to exceed the population in $|g\rangle$.

Planck's radiation law was derived in Section 4.7. It is instructive to see how the same result may be obtained on the basis of the Einstein coefficients. When the system is in equilibrium, $dN_g/dt = dN_e/dt = 0$; hence,

$$N_g W_a = N_e(W_{st} + W_{sp}), \tag{5.205}$$

and in terms of the Einstein coefficients, with degeneracies included,

$$N_g B_{eg} U(\omega) = N_e[B_{ge} U(\omega) + A]. \tag{5.206}$$

But under thermal equilibrium the Boltzman distribution requires that

$$\frac{N_g}{N_e} = \frac{d_g}{d_e} e^{\beta\hbar\omega}, \tag{5.207}$$

in which $E_e - E_g = \hbar\omega_0 = \hbar\omega$. Combining Eqs. (5.206), (5.207) and (5.204), we solve for $U(\omega)$ at thermal equilibrium:

$$U(\omega) = \frac{\hbar\omega^3}{\pi^2 c^3} \frac{1}{e^{\beta\hbar\omega} - 1} \tag{5.208}$$

in agreement with Eq. (4.227) and independent of the degeneracies. The present derivation of Planck's law displays the crucial role of spontaneous emission. The derivation in Section 4.7, although it makes no explicit reference to spontaneous emission, nevertheless contains this feature implicitly because it is based on a quantized radiation field. Such a field automatically leads to the existence of spontaneous emission, as shown in Section 5.5.

When $U(\omega)$ from Eq. (5.208) is inserted into Eq. (5.191), we obtain the ratio of spontaneous to stimulated emission rates in a thermal source characterized by Eq. (5.208):

$$\frac{A}{BU(\omega)} = \frac{\hbar\omega^3}{\pi^2 c^3 U(\omega)} = e^{\beta\hbar\omega} - 1 = \frac{1}{\langle n \rangle}, \tag{5.209}$$

where $\langle n \rangle$ is the average number of photons at the equilibrium temperature, as in Eq. (4.217). As indicated in Section 4.7, $\langle n \rangle \ll 1$ in the optical region at room temperature; spontaneous emission is therefore dominant. But in the microwave region, with $\langle n \rangle \gg 1$, stimulated emission far exceeds spontaneous emission.

For two states $|g\rangle$ and $|e\rangle$, with $E_e - E_g = \hbar\omega_0$, the *oscillator strength* for absorption is defined by the dimensionless quantity

$$f_{eg} = \frac{2\varepsilon_0 m(E_e - E_g)}{\pi e^2} B_{eg} = \frac{8\pi\varepsilon_0 m\omega_0}{3\hbar e^2 d_g} S(g, e), \tag{5.210}$$

and for stimulated emission the oscillator strength is

$$f_{ge} = \frac{2\varepsilon_0 m(E_g - E_e)}{\pi e^2} B_{ge} = -\frac{8\pi\varepsilon_0 m\omega_0}{3\hbar e^2 d_e} S(g, e) = -\frac{d_g}{d_e} f_{eg}. \tag{5.211}$$

Oscillator strengths satisfy a number of sum rules, one of the most important being the Reiche-Thomas-Kuhn rule [23, 24]:

$$\sum_k f_{ki} = Z, \tag{5.212}$$

where f_{ki} is the oscillator strength for the transition $|i\rangle \rightarrow |k\rangle$ and Z is the total number of electrons. The summation extends over all levels $|k\rangle$ accessible from $|i\rangle$ by electric dipole transitions.

We have shown (Section 5.6) that spontaneous emission from an excited state is inextricably linked with a level width $\hbar\Gamma$. Transitions between such an excited state and the ground state, in either direction, must therefore exhibit a line shape having a finite width. Consider, first, the emission process of Eq. (5.126) in which, according to Eq. (5.174), the probability amplitude of the initial state, neglecting the energy shift, is

$$c_I(t) = e^{-\Gamma t/2}. \tag{2.213}$$

Following the general pattern of Eq. (2.84), we have, for the present case of a two-level system,

$$\begin{aligned} i\hbar\dot{c}_F(t) &= c_I(t)\langle F|\mathscr{H}_{AF}|I\rangle e^{i(E_F - E_I)t/\hbar} \\ &= \langle F|\mathscr{H}_{AF}|I\rangle e^{-[i(\omega_0 - \omega) + \Gamma/2]t}. \end{aligned} \tag{5.214}$$

The integration from 0 to t yields

$$c_F(t) = \frac{e^{-[i(\omega_0 - \omega) + \Gamma/2]t} - 1}{\hbar\left[(\omega_0 - \omega) - \dfrac{i\Gamma}{2}\right]}\langle F|\mathscr{H}_{AF}|I\rangle, \tag{5.215}$$

and, as $t \rightarrow \infty$,

$$|c_F(\infty)|^2 = \frac{|\langle F|\mathscr{H}_{AF}|I\rangle|^2}{\hbar^2\left[(\omega_0 - \omega)^2 + \dfrac{\Gamma^2}{4}\right]}. \tag{5.216}$$

Hence,

$$\begin{aligned} \sum_F |c_F(\infty)|^2 \delta(\omega_0 - \omega) &= \frac{\sum_F |\langle F|\mathscr{H}_{AF}|I\rangle|^2 \delta(\omega_0 - \omega)}{\hbar^2\left[(\omega_0 - \omega)^2 + \dfrac{\Gamma^2}{4}\right]} \\ &= \frac{\dfrac{\Gamma}{2\pi}}{(\omega_0 - \omega)^2 + \dfrac{\Gamma^2}{4}} \equiv L(\omega). \end{aligned} \tag{5.217}$$

The sum over the states $|F\rangle$ gives rise to photons with various polarizations and directions of propagation. The inclusion of the δ-function serves a two-fold purpose; it ensures that energy is conserved in the emission process (to the extent permitted by quantum mechanical uncertainties) and it enables us to make use of Eq. (5.172) in the derivation of the Lorentzian function $L(\omega)$. We are led, therefore, to conclude that the probability of finding a photon with frequency ω in the spectrum of spontaneously emitted photons in a two-level system with mean separation $\hbar\omega_0$, is given by $L(\omega)$ which is appreciable only when $\omega \simeq \omega_0$. In other words, the intensity distribution of spontaneously emitted radiation, usually known as the *natural* or *radiative line shape*, is the Lorentzian $L(\omega)$. As noted previously (Section 1.6), $L(\omega)$ is a normalized function and the full width at half maximum is Γ.

A similar situation must exist in absorption; that is, the absorption line is broadened due to the finite level width of the excited state. Although this conclusion is expected on general grounds, it may be shown explicitly by returning to the Schrödinger equation for the two-level system. The states in the absorption process are defined by Eq. (5.125). It is assumed, to first order, that for $|\mathrm{I}\rangle = |\mathrm{g}, n\rangle$

$$\mathrm{c_I}(0) = \mathrm{c_I}(t) = 1, \qquad \mathrm{c_F}(0) = 0. \tag{5.218}$$

The state $|\mathrm{F}\rangle = |\mathrm{e}, n-1\rangle$ is enhanced by transitions $|\mathrm{I}\rangle \to |\mathrm{F}\rangle$ and is depleted by transitions $|\mathrm{F}\rangle \to |\mathrm{I}\rangle$. Combining the two processes, the rate of change of the probability amplitude for the state $|F\rangle$ is

$$\begin{aligned} i\hbar\dot{\mathrm{c}}_\mathrm{F}(t) &= \mathrm{c_I}(t)\langle \mathrm{F}|\mathscr{H}_\mathrm{AF}|\mathrm{I}\rangle e^{i(\mathrm{E_F}-\mathrm{E_I})t/\hbar} - \frac{i\hbar\Gamma}{2}\mathrm{c_F}(t) \\ &= \langle \mathrm{F}|\mathscr{H}_\mathrm{AF}|\mathrm{I}\rangle e^{i(\omega_0-\omega)t} - \frac{i\hbar\Gamma}{2}\mathrm{c_F}(t), \end{aligned} \tag{5.219}$$

where the first term on the right represents the growth in the probability amplitude of $|\mathrm{F}\rangle$ due to absorption from $|\mathrm{I}\rangle$, and the second term on the right represents the loss due to spontaneous emission. The solution to the differential equation is

$$\mathrm{c_F}(t) = \mathrm{A}e^{-\Gamma t/2\hbar} + \frac{\langle \mathrm{F}|\mathscr{H}_\mathrm{AF}|\mathrm{I}\rangle e^{i(\omega_0-\omega)t}}{\dfrac{i\hbar\Gamma}{2} - \hbar(\omega_0-\omega)}, \tag{5.220}$$

in which the constant A is evaluated from the initial condition $\mathrm{c_F}(0) = 0$. For $t \to \infty$ it is readily verified that

$$|\mathrm{c_F}(\infty)|^2 = \frac{|\langle \mathrm{F}|\mathscr{H}_\mathrm{AF}|\mathrm{I}\rangle|^2}{\hbar^2\left[(\omega_0-\omega)^2 + \dfrac{\Gamma^2}{4}\right]}, \tag{5.221}$$

as in Eq. (5.216). Hence, the absorption line shape is also a Lorentzian with the same width as the corresponding emission line.

5.8 Cross Sections, Dipole Correlation Function

The absorption *cross section* is defined by

$$\sigma_a = \frac{P(\omega)}{I(\omega)}, \tag{5.222}$$

in which $P(\omega)\,d\omega$ is the absorbed power in the frequency interval $(\omega, \omega + d\omega)$ and $I(\omega)\,d\omega = cU(\omega)\,d\omega$ is the incident power crossing unit area for radiation in the same frequency interval. To obtain an explicit expression for σ_a we refer to the Einstein B coefficient (Eq. (5.203)) and write for the total absorbed power

$$P = \hbar\omega W_a = \hbar\omega B_{eg} U(\omega) = \frac{\hbar\omega B_{eg}}{c} \mathrm{I}(\omega) = \frac{d_e}{d_g}\frac{\pi^2 c^2}{\omega^2} \mathrm{AI}(\omega)$$
$$= \frac{d_e}{d_g}\frac{\pi^2 c^2}{\omega^2} \Gamma \mathrm{I}(\omega). \tag{5.223}$$

The absorbed power per unit frequency interval $P(\omega)$ is then obtained by multiplying the total absorbed power by the line shape $g(\omega)$,

$$P(\omega) = Pg(\omega) = \frac{d_e}{d_g}\frac{\pi^2 c^2}{\omega^2} \Gamma \mathrm{I}(\omega) g(\omega), \tag{5.224}$$

where $g(\omega)$ is a normalized line shape function defined by

$$\int_{-\infty}^{\infty} g(\omega)\,d\omega = 1, \tag{5.225}$$

as, for example, in the case of the Lorentzian function $L(\omega)$ (Eq. (5.217)). Hence, the absorption cross section is

$$\sigma_a(\omega) = \frac{d_e}{d_g}\frac{\pi^2 c^2}{\omega^2} \Gamma g(\omega). \tag{5.226}$$

For stimulated emission we must refer to the Einstein coefficient B_{ge} in place of B_{eg}; the cross section then becomes

$$\sigma_{st}(\omega) = \frac{\pi^2 c^2}{\omega^2} \Gamma g(\omega) = \frac{d_g}{d_e} \sigma_a(\omega). \tag{5.227}$$

If N_g and N_e are the number of atoms per unit volume in the states $|g\rangle$ and $|e\rangle$, respectively, and the atoms are contained in a thin slab of thickness dx, a narrow beam of radiation traversing the slab will suffer a change in intensity given by

$$-dI(\omega) = [N_g\sigma_a(\omega) - N_e\sigma_{st}(\omega)]I(\omega)\,dx. \tag{5.228}$$

The first term on the right represents the loss in intensity due to absorption and the second term on the right represents the gain in intensity due to stimulated emission. With relation (5.227),

$$-\frac{dI(\omega)}{I(\omega)} = \left(N_g - \frac{d_g}{d_e}N_e\right)\sigma_a(\omega)\,dx \equiv \mu(\omega)\,dx, \tag{5.229}$$

in which $\mu(\omega)$ is known as the *linear attenuation* or *absorption coefficient*. Note that the absorption coefficient vanishes when $N_g d_e = N_e d_g$. If $d_e = d_g$ we may refer to Eq. (5.194) to obtain the dependence of the absorption coefficient on the saturation parameter,

$$\mu(\omega) = \frac{\mu_0(\omega)}{1 + S}, \tag{5.230}$$

where $\mu_0(\omega)$ is the absorption coefficient when $S = 0$. In other words, absorption can occur only when there is a difference in the populations of the two states involved in the transition. At high intensities, the populations approach equality ($S = \infty$); hence, the medium becomes more transparent. This is the basis for so-called "hole burning" (Section 6.5). Spontaneous emission, on the other hand, depends only on the population of the excited state.

For low intensities of illumination, the effect of stimulated emission is negligible, and if, further, $N_e \ll N_g \simeq N$, Eq. (5.229) reduces to

$$-\frac{dI(\omega)}{I(\omega)} = N\sigma_a(\omega)\,dx = \mu(\omega)\,dx, \tag{5.231a}$$

or

$$I(\omega) = I_0(\omega)e^{-\sigma_a(\omega)Nx} = I_0(\omega)e^{-\mu(\omega)x}, \tag{5.231b}$$

where $I(\omega)$ is the incident intensity, $I(\omega)$ is the transmitted intensity, and x is the thickness of the slab in the direction of the beam. At higher intensities, N_g may decrease sufficiently to violate the condition $N_g \simeq N$. In that event, both N_g and N_e become functions of intensity and the simple exponential law (Eq. (5.231)) is no longer valid; the absorption then is said to be nonlinear.

Among the effects that arise as a result of nonlinear absorption is the phenomenon of optical bistability [25–28] observed in the transmission of light by an optical cavity filled with an absorptive medium. At low power levels, the transmitted intensity I_t is proportional to the incident intensity I_i, but at higher power levels—where saturation becomes important—I_t

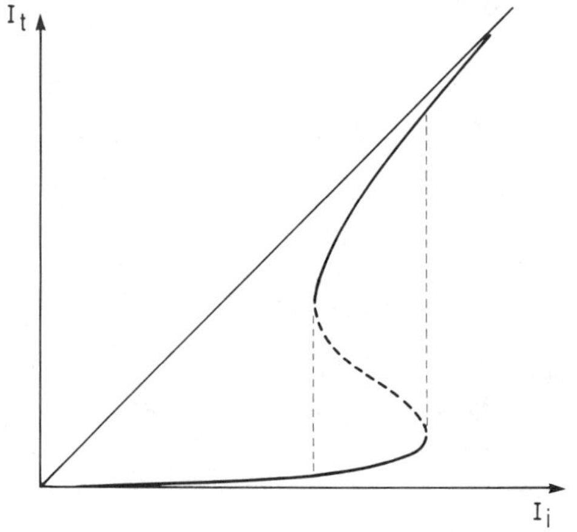

FIGURE 5.9 Optical bistability. In an optical cavity containing an absorbing medium, the transmitted light intensity I_t at low power levels is proportional to the incident intensity I_i. At high power levels I_t becomes a discontinuous function of I_i with two separate branches. The lower solid curve is applicable at low power levels but as I_i is increased beyond a threshold value, I_t jumps to the upper solid curve. An analogous jump occurs upon the return from the upper of the lower curve.

becomes a discontinuous function of I_i, as shown in Fig. 5.9. The curve of $I_t(I_i)$ has two branches, one for low and one for high transmission with a region between the two branches that is bistable. This behavior sets in above a threshold value of $\mu L/T$ where μ is the unsaturated absorption coefficient, L is the length of the cavity and T is the mirror transmissivity. The threshold value depends on several parameters, including the deviation of the incident frequency from both atomic and cavity resonance and the inhomogeneous linewidth of the absorbing medium. For the general case of bistability, there is also a contribution from the dispersion of the medium.

Under thermal equilibrium, the Boltzmann distribution prevails, in which case the condition $N_g > N_e d_g/d_e$ always holds. Nevertheless, as we have indicated previously, it is possible to create nonequilibrium conditions such that $N_g < N_e d_g/d_e$. This is the condition for population inversion when degeneracies are present. The right side of Eq. (5.229) then changes sign which means that the transmitted intensity is greater than the incident intensity. Alternatively, it is seen that $\mu(\omega)$ becomes negative: we then refer to $\mu(\omega)$ as a *gain constant*.

Next, we demonstrate that the absorption coefficient in the vicinity of resonance can be expressed in terms of a dipole correlation function. For this purpose, we refer to Eq. (5.137), the probability per unit time for the transition

$|g\rangle \to |e\rangle$. Multiplying by $\hbar\omega \simeq \hbar\omega_0$, we obtain the rate of energy loss

$$\hbar\omega W_a = \frac{\pi\omega^2 n}{\varepsilon_0 V} |\langle e|\mathbf{d}\cdot\hat{\varepsilon}|g\rangle|^2\,\delta(\omega_0 - \omega). \tag{5.232}$$

If there is more than one initial atomic state $|g\rangle$, that is, $|g\rangle$ is now regarded as a distribution of states, and the probability of finding the atom in a particular member of the distribution is ρ_g, the rate of energy loss must be

$$\sum_g \rho_g \hbar\omega W_a. \tag{5.233}$$

The final state $|e\rangle$ also may consist of a distribution of states, in which case the total power loss in all possible transitions $|g\rangle \to |e\rangle$ is

$$P_{eg} = \frac{\pi\omega^2 n}{\varepsilon_0 V} \sum_{ge} \rho_g |\langle g|\mathbf{d}\cdot\hat{\varepsilon}|e\rangle|^2\,\delta(\omega_0 - \omega). \tag{5.234}$$

In precisely the same fashion, the total rate of energy gain by the field as a result of stimulated emission is

$$P_{ge} = \frac{\pi\omega^2 n}{\varepsilon_0 V} \sum_{ge} \rho_e |\langle g|\mathbf{d}\cdot\hat{\varepsilon}|e\rangle|^2\,\delta(\omega_0 - \omega). \tag{5.235}$$

Therefore, the difference between the absorption and stimulated emission rates is the *net* absorption rate (power loss by the field)

$$P = P_{eg} - P_{ge} = \frac{\pi\omega^2 n}{\varepsilon_0 V} \sum_{ge} (\rho_g - \rho_e) |\langle g|\mathbf{d}\cdot\hat{\varepsilon}|e\rangle|^2\,\delta(\omega_0 - \omega). \tag{5.236}$$

If the atoms are in thermal equilibrium, ρ_g and ρ_e are related by the Boltzmann factor so that, with $\omega \simeq \omega_0$,

$$P = \frac{\pi\omega^2 n}{\varepsilon_0 V} (1 - e^{-\beta\hbar\omega}) \sum_{ge} \rho_g |\langle g|\mathbf{d}\cdot\hat{\varepsilon}|e\rangle|^2\,\delta(\omega_0 - \omega). \tag{5.237}$$

Let us now replace the δ-function by its integral representation

$$\rho_g |\langle g|\mathbf{d}\cdot\hat{\varepsilon}|e\rangle|^2\,\delta(\omega_0 - \omega) = \frac{1}{2\pi}\int_{-\infty}^{\infty} \rho_g |\langle g|d\cdot\varepsilon|e\rangle|^2 e^{i(\omega_0-\omega)t}\,dt \tag{5.238}$$

and the dipole operator by its interaction representation

$$\mathbf{d} = e^{-i\mathscr{H}_A t/\hbar}\,\mathbf{d}(t)\,e^{i\mathscr{H}_A t/\hbar}. \tag{5.239}$$

The operator $\mathscr{H}_A$ is the atomic Hamiltonian, that is, $\mathscr{H}_A|g\rangle = E_g|g\rangle$ and $\mathscr{H}_A|e\rangle = E_e|e\rangle$ with $E_e - E_g = \hbar\omega_0$. Hence, we now may write

$$\begin{aligned}|\langle g|\mathbf{d}\cdot\hat{\varepsilon}|e\rangle|^2 &= \langle g|\mathbf{d}\cdot\hat{\varepsilon}|e\rangle\langle e|\mathbf{d}\cdot\hat{\varepsilon}|g\rangle \\ &= \langle g|\mathbf{d}\cdot\hat{\varepsilon}|e\rangle\langle e|e^{-i\mathscr{H}_A t/\hbar}\tilde{\mathbf{d}}(t)\cdot\hat{\varepsilon}e^{i\mathscr{H}_A t/\hbar}|g\rangle \\ &= \langle g|\mathbf{d}\cdot\hat{\varepsilon}|e\rangle\langle e|\tilde{\mathbf{d}}(t)\cdot\hat{\varepsilon}|g\rangle e^{-i\omega_0 t}.\end{aligned} \tag{5.240}$$

Inserting Eqs. (5.238) and (5.240) into Eq. (5.237),

$$P = \frac{\omega^2 n}{2\varepsilon_0 V}(1 - e^{-\beta\hbar\omega})\sum_{ge}\int_{-\infty}^{\infty}\rho_g\langle g|\mathbf{d}\cdot\hat{\varepsilon}|e\rangle\langle e|\tilde{\mathbf{d}}(t)\cdot\hat{\varepsilon}|g\rangle e^{-i\omega t}\,dt \tag{5.241a}$$

$$= \frac{\omega^2 n}{2\varepsilon_0 V}(1 - e^{-\beta\hbar\omega})\sum_{g}\int_{-\infty}^{\infty}\rho_g\langle g|\mathbf{d}\cdot\hat{\varepsilon}\tilde{\mathbf{d}}(t)\cdot\hat{\varepsilon}|g\rangle e^{-i\omega t}\,dt \tag{5.241b}$$

$$= \frac{\omega^2 n}{2\varepsilon_0 V}(1 - e^{-\beta\hbar\omega})\int_{-\infty}^{\infty}\langle\mathbf{d}\cdot\hat{\varepsilon}\tilde{\mathbf{d}}(t)\cdot\hat{\varepsilon}\rangle e^{-i\omega t}\,dt. \tag{5.241c}$$

To obtain Eq. (5.241b) we used the closure relation on the states $|e\rangle$. The justification for this procedure stems from the fact that all the states $|e\rangle$ to which the transition probability is nonvanishing already are included in the sum of Eq. (5.241a). Therefore, even if the set $|e\rangle$ is not complete we may add other states to complete the set since the transition probability to the additional states will be zero. Finally, in Eq. (5.241c) we employed the definition of the ensemble average

$$\sum_g \rho_g\langle g|\mathbf{d}\cdot\hat{\varepsilon}\tilde{\mathbf{d}}(t)\cdot\hat{\varepsilon}|g\rangle \equiv \langle\mathbf{d}\cdot\hat{\varepsilon}\tilde{\mathbf{d}}(t)\cdot\hat{\varepsilon}\rangle. \tag{5.242}$$

If the incident photon intensity is

$$I = \frac{n\hbar\omega c}{V}, \tag{5.243}$$

the net absorption cross section is $\sigma = P/I$ and the absorption coefficient is

$$\mu = \sigma N = \frac{N\omega}{2\varepsilon_0 c\hbar}(1 - e^{-\beta\hbar\omega})\int_{-\infty}^{\infty}\langle\mathbf{d}\cdot\hat{\varepsilon}\tilde{\mathbf{d}}(t)\cdot\hat{\varepsilon}\rangle e^{-i\omega t}\,dt, \tag{5.244}$$

which indicates that the absorption coefficient is proportional to the Fourier transform of the dipole moment correlation function.

It often is sufficient to consider the special case of plane polarized light, say along the x direction. Then

$$\langle\mathbf{d}\cdot\hat{\varepsilon}\tilde{\mathbf{d}}(t)\cdot\varepsilon\rangle = e^2\langle x\tilde{x}(t)\rangle, \tag{5.245}$$

and with the help of the relations at the end of Section 2.8, the energy loss rate can be written

$$P = \frac{e^2\omega^2 n}{2\varepsilon_0 V}(1 - e^{-\beta\hbar\omega}) \int_{-\infty}^{\infty} \langle x\tilde{x}(t)\rangle e^{-i\omega t}\, dt \tag{5.246a}$$

$$= \frac{e^2\omega^2 n}{2\varepsilon_0 V}(1 - e^{-\beta\hbar\omega}) \int_{-\infty}^{\infty} \langle \tilde{x}(t)x\rangle e^{i\omega t}\, dt \tag{5.246b}$$

$$= \frac{e^2\omega^2 n}{2\varepsilon_0 V}(e^{\beta\hbar\omega} - 1) \int_{-\infty}^{\infty} \langle x\tilde{x}(t)\rangle e^{i\omega t}\, dt \tag{5.246c}$$

$$= \frac{\pi e^2\omega^2 n}{\varepsilon_0 V} \tan h\left(\frac{\beta\hbar\omega}{2}\right) R(\omega), \tag{5.246d}$$

where

$$R(\omega) = \frac{1}{2\pi} \int_{-\infty}^{\infty} \langle x\tilde{x}(t) + \tilde{x}(t)x\rangle e^{i\omega t}\, dt \tag{5.247}$$

is the Fourier transform of the symmetrized correlation function [29]. In view of Eq. (5.244), the absorption coefficient may be expressed similarly. It is evident that the correlation function of the dipole moment operator (or, equivalently, the displacement operator) is the crucial quantity in the determination of energy losses.

Also, for an isotropic medium such as a liquid or gas

$$\langle x\tilde{x}(t)\rangle = \langle y\tilde{y}(t)\rangle = \langle z\tilde{z}(t)\rangle = \frac{1}{3}\langle \mathbf{r} \cdot \tilde{\mathbf{r}}(t)\rangle. \tag{5.248}$$

5.9 Kramers-Heisenberg Cross Section

The general scattering process is one in which an incident photon of frequency ω_i, energy $\hbar\omega_i$, momentum $\hbar\mathbf{k}_i$ (or propagation vector $\mathbf{k}_i$), and polarization $\hat{\boldsymbol{\varepsilon}}_i$ is converted to a scattered photon of frequency ω_s, energy $\hbar\omega_s$, momentum $\hbar\mathbf{k}_s$, and polarization $\hat{\boldsymbol{\varepsilon}}_s$ while the scattering system undergoes a transition from one of its eigenstates to another. There are many kinds of scattering systems, e.g., bound systems of electrons (atoms, molecules, etc.), pressure waves, phonons, plasmons. Despite important differences in detail, the various processes have certain common features that are embedded in a general quantum mechanical expression [5].

Let the initial and final states $|\mathrm{I}\rangle$ and $|\mathrm{F}\rangle$ of the scatterer-photon system be defined by

$$|\mathrm{I}\rangle = |\mathrm{g}, n_i, n_s\rangle, \qquad |\mathrm{F}\rangle = |\mathrm{e}, n_i - 1, n_s + 1\rangle, \tag{5.249}$$

in which $|g\rangle$ and $|e\rangle$ are eigenstates of the scattering system in the absence of a radiation field; n_i and n_s are the photon numbers for single modes of the incident and scattered beams, respectively. Since the mode containing n_i photons is depleted by one photon while the mode with n_s photons is enhanced by one photon, the scattering is said to be a two-photon process. The energies of the two states (neglecting zero point energies) are

$$\begin{aligned} E_I &= E_g + n_i\hbar\omega_i + n_s\hbar\omega_s, \\ E_F &= E_e + (n_i - 1)\hbar\omega_i + (n_s + 1)\hbar\omega_s. \end{aligned} \tag{5.250}$$

For the general scattering process, ω_s may be smaller, equal to, or greater than ω_i, but to conserve energy we require that

$$E_I = E_T \quad \text{or} \quad E_e - E_g = \hbar\omega_i - \hbar\omega_s, \tag{5.251}$$

which means that E_e may be greater, equal to, or smaller than E_g.

In the dipole approximation, the interaction Hamiltonian

$$\mathscr{H}_{AF} = -\mathbf{d} \cdot \mathbf{E} \tag{5.252}$$

for a single mode, according to Eq. (5.107), is

$$\mathscr{H}_{AF} = \mathscr{H}_{AF}^{(+)} + \mathscr{H}_{AF}^{(-)}, \tag{5.253}$$

where

$$\mathscr{H}_{AF}^{(+)} = -i\sqrt{\frac{\hbar\omega}{2\varepsilon_0 V}}\,\mathbf{d}\cdot\hat{\varepsilon}a, \qquad \mathscr{H}_{AF}^{(-)} = i\sqrt{\frac{\hbar\omega}{2\varepsilon_0 V}}\,\mathbf{d}\cdot\hat{\varepsilon}a^\dagger. \tag{5.254}$$

It is readily verified that

$$\langle F|\mathscr{H}_{AF}^{(+)}|I\rangle = \langle F|\mathscr{H}_{AF}^{(-)}|I\rangle = 0, \tag{5.255}$$

a result to be expected since $\mathscr{H}_{AF}$ contains a and $a^\dagger$, which can annihilate or create only one photon in a particular mode but cannot simultaneously annihilate a photon in one mode and create a photon in another mode. We conclude, then, that $\mathscr{H}_{AF}$ cannot contribute to scattering in the first order of perturbation theory but may do so in higher order. In second order, it is necessary to evaluate matrix elements of the type

$$M = \sum_T \frac{\langle F|\mathscr{H}_{AF}|T\rangle\langle T|\mathscr{H}_{AF}|I\rangle}{E_I - E_T}, \tag{5.256}$$

where the intermediate states $|T\rangle$ are accessible by one-photon transitions from $|I\rangle$ and $|F\rangle$. This means that $|T\rangle$ has two possible forms:

$$|T\rangle = \begin{cases} |T_1\rangle = |t_1, n_i - 1, n_s\rangle, \\ |T_2\rangle = |t_2, n_i, n_s - 1\rangle, \end{cases} \tag{5.257}$$

in which $|t_1\rangle$ and $|t_2\rangle$ are eigenstates of the scattering system. Then

$$\begin{aligned} E_{\mathrm{T}_1} &= E_{t_1} + (n_\mathrm{i} - 1)\hbar\omega_\mathrm{i} + n_\mathrm{s}\hbar\omega_\mathrm{s}, \\ E_{\mathrm{T}_2} &= E_{t_2} + n_\mathrm{i}\hbar\omega_\mathrm{i} + (n_\mathrm{s} - 1)\hbar\omega_\mathrm{s}, \end{aligned} \tag{5.258}$$

$$\begin{aligned} E_\mathrm{I} - E_{\mathrm{T}_1} &= E_\mathrm{g} - E_{t_1} + \hbar\omega_\mathrm{i}, \\ E_\mathrm{I} - E_{\mathrm{T}_2} &= E_\mathrm{g} - E_{t_2} - \hbar\omega_\mathrm{s} = E_\mathrm{e} - E_{t_2} - \hbar\omega_\mathrm{i}, \end{aligned} \tag{5.259}$$

$$\begin{aligned} \langle \mathrm{F}|\mathscr{H}_\mathrm{AF}|\mathrm{T}_1\rangle &= \langle \mathrm{e}, n_\mathrm{i} - 1, n_\mathrm{s} + 1|\mathscr{H}_\mathrm{AF}|t_1, n_\mathrm{i} - 1, n_\mathrm{s}\rangle \\ &= \langle \mathrm{e}, n_\mathrm{s} + 1|\mathscr{H}^{(-)}_\mathrm{AF}|t^1, n_\mathrm{s}\rangle = \mathrm{i}\sqrt{\frac{(n_\mathrm{s}+1)\hbar\omega_\mathrm{s}}{2\varepsilon_0 V}}\langle \mathrm{e}|\mathbf{d}\cdot\hat{\boldsymbol{\varepsilon}}_\mathrm{s}|t^1\rangle, \end{aligned} \tag{5.260a}$$

$$\begin{aligned} \langle \mathrm{T}_1|\mathscr{H}_\mathrm{AF}|\mathrm{I}\rangle &= \langle t_1, n_\mathrm{i} - 1, n_\mathrm{s}|\mathscr{H}_\mathrm{AF}|\mathrm{g}, n_\mathrm{i}, n_\mathrm{s}\rangle \\ &= \langle t_1, n_\mathrm{i} - 1|\mathscr{H}^{(+)}_\mathrm{AF}|\mathrm{g}, n_\mathrm{i}\rangle = -\mathrm{i}\sqrt{\frac{n\hbar\omega_\mathrm{i}}{2\varepsilon_0 V}}\langle t_1|\mathbf{d}\cdot\hat{\boldsymbol{\varepsilon}}_\mathrm{i}|\mathrm{g}\rangle, \end{aligned} \tag{5.260b}$$

$$\begin{aligned} \langle \mathrm{F}|\mathscr{H}_\mathrm{AF}|\mathrm{T}_2\rangle &= \langle \mathrm{e}, n_\mathrm{i} - 1, n_\mathrm{s} + 1|\mathscr{H}_\mathrm{AF}|t_2, n_\mathrm{i}, n_\mathrm{s} + 1\rangle \\ &= \langle \mathrm{e}, n_\mathrm{i} - 1|\mathscr{H}^{(+)}_\mathrm{AF}|t_2, n_\mathrm{i}\rangle = -i\sqrt{\frac{n\hbar\omega_\mathrm{i}}{2\varepsilon_0 V}}\langle \mathrm{e}|\mathbf{d}\cdot\hat{\boldsymbol{\varepsilon}}_\mathrm{i}|t_2\rangle, \end{aligned} \tag{5.260c}$$

$$\begin{aligned} \langle \mathrm{T}_2|\mathscr{H}_\mathrm{AF}|\mathrm{I}\rangle &= \langle t_2, n_\mathrm{i}, n_\mathrm{s} + 1|\mathscr{H}_\mathrm{AF}|\mathrm{g}, n_\mathrm{i}, n_\mathrm{s}\rangle \\ &= \langle t_2, n_\mathrm{s} + 1|\mathscr{H}^{(-)}_\mathrm{AF}|\mathrm{g}, n_\mathrm{s}\rangle = \mathrm{i}\sqrt{\frac{(n_\mathrm{s}+1)\hbar\omega_\mathrm{s}}{2\varepsilon_0 V}}\langle t_2|\mathbf{d}\cdot\hat{\boldsymbol{\varepsilon}}_\mathrm{s}|\mathrm{g}\rangle. \end{aligned} \tag{5.260d}$$

With Eqs. (5.259) and (5.260) we obtain the two types of matrix elements

$$\begin{aligned} M_1 &= \sum_{\mathrm{T}_1} \frac{\langle \mathrm{F}|\mathscr{H}_\mathrm{AF}|\mathrm{T}_1\rangle\langle \mathrm{T}_1|\mathscr{H}_\mathrm{AF}|\mathrm{I}\rangle}{E_\mathrm{I} - E_{\mathrm{T}_1}} \\ &= \frac{\hbar}{2\varepsilon_0 V}(n_\mathrm{s}+1)n_\mathrm{i}\omega_\mathrm{s}\omega_\mathrm{i}\sum_{t_1}\frac{\langle \mathrm{e}|\mathbf{d}\cdot\hat{\boldsymbol{\varepsilon}}_\mathrm{s}|t_1\rangle\langle t_1\rangle\langle t_1|\mathbf{d}\cdot\hat{\boldsymbol{\varepsilon}}_\mathrm{i}|\mathrm{g}\rangle}{E_\mathrm{g} - E_{t_1} + \hbar\omega_\mathrm{i}}, \end{aligned} \tag{5.261}$$

$$\begin{aligned} M_2 &= \sum_{\mathrm{T}_2} \frac{\langle \mathrm{F}|\mathscr{H}_\mathrm{AF}|\mathrm{T}_2\rangle\langle \mathrm{T}_2|\mathscr{H}_\mathrm{AF}|\mathrm{I}\rangle}{E_\mathrm{I} - E_{\mathrm{T}_2}} \\ &= \frac{\hbar}{2\varepsilon_0 V}\sqrt{n_\mathrm{i}(n_\mathrm{s}+1)\omega_\mathrm{i}\omega_\mathrm{s}}\sum_{ts}\frac{\langle \mathrm{e}|\mathbf{d}\cdot\hat{\boldsymbol{\varepsilon}}_\mathrm{i}|t_2\rangle\langle t_2|\mathbf{d}\cdot\hat{\boldsymbol{\varepsilon}}_\mathrm{s}|\mathrm{g}\rangle}{E_\mathrm{g} - E_{t_2} - \hbar\omega_\mathrm{s}}, \end{aligned} \tag{5.262a}$$

$$= \frac{\hbar}{2\varepsilon_0 V}\sqrt{n_\mathrm{i}(n_\mathrm{s}+1)\omega_1\omega_\mathrm{s}}\sum_{t_2}\frac{\langle \mathrm{e}|\mathbf{d}\cdot\hat{\boldsymbol{\varepsilon}}_\mathrm{i}|t_2\rangle\langle t_2|\mathbf{d}\cdot\hat{\boldsymbol{\varepsilon}}_\mathrm{s}|\mathrm{g}\rangle}{E_\mathrm{e} - E_{t_2} - \hbar\omega_\mathrm{i}} \tag{5.262b}$$

The scattering process in second order may be pictured as a two-step process proceeding along two possible paths:

1. $|\mathrm{I}\rangle \rightarrow |\mathrm{T}_1\rangle \rightarrow |\mathrm{F}\rangle$,
2. $|\mathrm{I}\rangle \rightarrow |\mathrm{T}_2\rangle \rightarrow |\mathrm{F}\rangle$.

In the first path, the transition $|\mathrm{I}\rangle \rightarrow |\mathrm{T}_1\rangle$ involves the loss (absorption) of an incident photon accompanied by an atomic transition $|\mathrm{g}\rangle \rightarrow |t_1\rangle$; the transition $|\mathrm{T}_1\rangle \rightarrow |\mathrm{F}\rangle$ involves the gain (emission) of a scattered photon with an accompanying atomic transition $|t_1\rangle \rightarrow |\mathrm{e}\rangle$. In the second path, there is an emission of a scattered photon with an atomic transition $|\mathrm{g}\rangle \rightarrow |t_2\rangle$ in the step $|\mathrm{I}\rangle \rightarrow |\mathrm{T}_2\rangle$ and an absorption of an incident photon with an atomic transition $|t_2\rangle \rightarrow |\mathrm{e}\rangle$ in the step $|\mathrm{T}_2\rangle \rightarrow |\mathrm{F}\rangle$. In each path, the intermediate state differs from the initial states by just one photon. The two paths are symbolized by diagrams as in Fig. 5.10. The two paths are also said to be coherent, which implies that in calculating the transition probability $|\mathrm{I}\rangle \rightarrow |\mathrm{F}\rangle$, the matrix elements M_1 and M_2 are to summed before computing the absolute square rather than computing the sum of $|M_1|^2$ and $|M_2|^2$.

Let us now define α_{eg} by the relation

$$\alpha_{\mathrm{eg}} = \sum_t \left[\frac{\langle \mathrm{e}|\mathbf{d}\cdot\hat{\boldsymbol{\varepsilon}}_{\mathrm{s}}|t\rangle\langle t|\mathbf{d}\cdot\hat{\boldsymbol{\varepsilon}}_{\mathrm{i}}|\mathrm{g}\rangle}{E_{\mathrm{g}} - E_{\mathrm{t}} + \hbar\omega_{\mathrm{i}}} + \frac{\langle \mathrm{e}|\mathbf{d}\cdot\hat{\boldsymbol{\varepsilon}}_{\mathrm{i}}|t\rangle\langle t|\mathbf{d}\cdot\hat{\boldsymbol{\varepsilon}}_{\mathrm{s}}|\mathrm{g}\rangle}{E_{\mathrm{e}} - E_{\mathrm{t}} - \hbar\omega_{\mathrm{i}}} \right], \tag{5.263}$$

in which the general summation index t ranges over all possible states t_1 and t_2. In component form,

$$\mathbf{d}\cdot\hat{\boldsymbol{\varepsilon}}_{\mathrm{s}} = \sum_{\mu} d_{\mu}\varepsilon_{\mathrm{s}\mu'} \qquad \mathbf{d}\cdot\hat{\boldsymbol{\varepsilon}}_{\mathrm{i}} = \sum_{\nu} d_{\nu}\varepsilon_{i}^{\nu'}, \tag{5.264}$$

$$\mu, \nu = x, y, z.$$

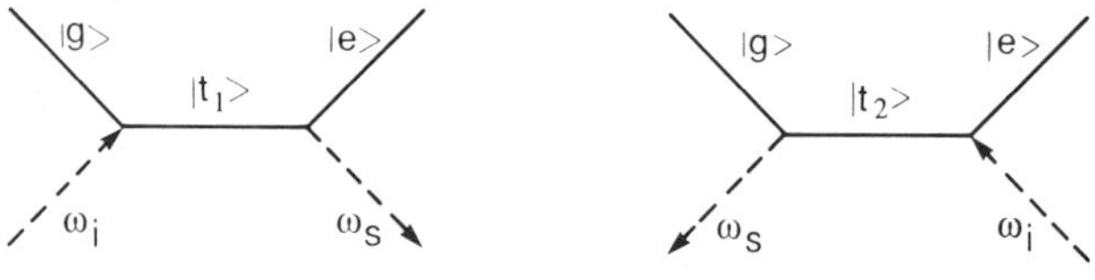

(a) (b)

FIGURE 5.10 Diagrammatic representation of scattering matrix elements. (a) $|\mathrm{I}\rangle \rightarrow |T_1\rangle \rightarrow |\mathrm{F}\rangle$ corresponds to $|\mathrm{g}, n_{\mathrm{i}}, n_{\mathrm{s}}\rangle \rightarrow |t_1, n_{\mathrm{i}} - 1, n_{\mathrm{s}}\rangle \rightarrow |\mathrm{e}, n_{\mathrm{i}} - 1, n_{\mathrm{s}} + 1\rangle$. In the step $|\mathrm{I}\rangle \rightarrow |T_1\rangle$ an incident photon (ω_{i}) is absorbed and in $|T_1\rangle \rightarrow |\mathrm{F}\rangle$ a scattered photon (ω_{s}) is emitted. (b) $|\mathrm{I}\rangle \rightarrow |T_2\rangle \rightarrow |\mathrm{F}\rangle$ corresponds to $|\mathrm{g}, n_{\mathrm{i}}, n_{\mathrm{s}}\rangle \rightarrow |t_2, n_{\mathrm{i}}, n_{\mathrm{s}} + 1\rangle \rightarrow |\mathrm{e}, n_{\mathrm{i}} - 1, n_{\mathrm{s}} + 1\rangle$. In $|\mathrm{I}\rangle \rightarrow |T_2\rangle$ a scattered photon (ω_{s}) is emitted and in $|T_2\rangle \rightarrow |\mathrm{F}\rangle$ an incident photon (ω_{i}) is absorbed.

The numerators in α_{eg} then can be replaced by

$$\langle e|\mathbf{d}\cdot\hat{\varepsilon}_s|t\rangle\langle t|\mathbf{d}\cdot\hat{\varepsilon}_i|g\rangle = \sum_{\mu\nu}\varepsilon_{s\mu}\langle e|d_\mu|t\rangle\langle t|d_\nu|g\rangle\varepsilon_{i\nu'}$$
$$\langle e|\mathbf{d}\cdot\hat{\varepsilon}_i|t\rangle\langle t|\mathbf{d}\cdot\hat{\varepsilon}_s|g\rangle = \sum_{\mu\nu}\varepsilon_{s\mu}\langle e|d_\nu|t\rangle\langle t|d_\mu|g\rangle\varepsilon_{i\nu'} \tag{5.265}$$

and α_{eg} can be written in the form

$$\alpha_{eg} = \sum_{\mu\nu}\varepsilon_{s\mu}\alpha_{\mu\nu}^{eg}\varepsilon_{i\nu'} \tag{5.266}$$

where

$$\alpha_{\mu\nu}^{eg} = \sum_t\left[\frac{\langle e|d_\mu|t\rangle\langle t|d_\nu|g\rangle}{E_g - E_t + \hbar\omega_i} + \frac{\langle e|d_\nu|t\rangle\langle t|d_\mu|g\rangle}{E_e - E_t - \hbar\omega_i}\right] \tag{5.267}$$

is defined as the *polarizability* (or *scattering*) tensor. Note that the polarizability has dimensions of an electric dipole moment induced by a unit field which is simply a volume.

In terms of α_{eg},

$$M = M_1 + M_2$$
$$= \frac{\hbar}{2\varepsilon_0 V}\sqrt{n_i(n_s+1)\omega_i\omega_s}\,\alpha_{eg}. \tag{5.268}$$

We now may use the Golden Rule to compute the transition probability per unit time:

$$W_{FI} \equiv W(I \to F)$$
$$= \frac{2\pi}{\hbar}|M|^2\rho_F(E), \tag{5.269}$$

in which, according to Eq. (4.42),

$$\rho_F(E) = \frac{1}{\hbar}\frac{dN}{d\omega_s}$$
$$= \frac{V\omega_s^2\,d\Omega_s}{\hbar(2\pi c)^3} \tag{5.270}$$

is the density of final states for scattered photons; $d\Omega_s$ is an element of solid angle around $\mathbf{k}_s$, the propagation vector of the scattered radiation. Combining Eqs. (5.268) to (5.270),

$$W_{FI} = \frac{n_i(n_s+1)\omega_i\omega_s^3\,d\Omega_s}{16\varepsilon_0^2\pi^2c^3V}|\alpha_{eg}|^2. \tag{5.271}$$

The scattering process also may be expressed in terms of a differential scattering cross section defined by

$$\frac{d\sigma}{d\Omega} = \frac{\text{number of scattered photons/(second)(steradian)}}{\text{number of incident photons/(second)(cm}^2\text{)}} \tag{5.272}$$

$$\equiv \frac{\mathrm{I}}{\mathrm{I}_0} = \frac{(W_{\mathrm{FI}})/d\Omega_{\mathrm{s}}}{n_{\mathrm{i}}c/V} = \frac{(n_{\mathrm{s}}+1)\omega_{\mathrm{i}}\omega_{\mathrm{s}}^2}{16\varepsilon_0^2\pi^4 c^4}|\alpha_{\mathrm{eg}}|^2 \tag{5.273a}$$

$$= \frac{r_0^2(n_s+1)\omega_{\mathrm{i}}\omega_{\mathrm{s}}^3 m^2}{e^4}|\alpha_{\mathrm{eg}}|^2 \tag{5.273b}$$

$$= \frac{r_0^2(n_{\mathrm{s}}+1)\omega_{\mathrm{i}}\omega_{\mathrm{s}}^3 m^2}{e^4}\left|\sum_{\mu\nu}\varepsilon_{\mathrm{s}\mu}\alpha_{\mu\nu}^{\mathrm{eg}}\varepsilon_{\nu}\right|^2, \tag{5.273c}$$

in which m is the mass of the electron and

$$r_0 = \frac{e^2}{4\pi\varepsilon_0 mc^2} = 2.8179 \times 10^{-15}\ \mathrm{m} \tag{5.274}$$

is the classical radius of the electron. Alternatively, we may write

$$\frac{d\sigma}{d\Omega} = r_0^2(n_{\mathrm{s}}+1)\omega_{\mathrm{i}}\omega_{\mathrm{s}}^3 m^2\left|\sum_t\left[\frac{\langle \mathrm{e}|\mathbf{r}\cdot\hat{\varepsilon}_{\mathrm{s}}|t\rangle\langle t|\mathbf{r}\cdot\hat{\varepsilon}_{\mathrm{i}}|\mathrm{g}\rangle}{E_{\mathrm{g}}-E_{\mathrm{t}}+\hbar\omega_{\mathrm{i}}} + \frac{\langle \mathrm{e}|\mathbf{r}\cdot\hat{\varepsilon}_{\mathrm{i}}|t\rangle\langle t|\mathbf{r}\cdot\hat{\varepsilon}_{\mathrm{s}}|\mathrm{g}\rangle}{E_{\mathrm{e}}-E_{\mathrm{t}}-\hbar\omega_{\mathrm{i}}}\right]\right|^2. \tag{5.275}$$

With the *Kramers-Heisenberg* formula (Eq. (5.273) or (5.275)), it is possible, in principle, to compute the cross section for the conversion of an incident photon to a scattered photon propagating in the direction $d\Omega$ while the scattering system has undergone a transition from $|\mathrm{g}\rangle$ to $|\mathrm{e}\rangle$. To make literal use of the formula, however, all the intermediate states, i.e., all the eigenstates of the scattering system that participate in one-photon transitions must be known. Since such information is rarely available, it is necessary to resort to one approximation or another. In some cases, the sum in Eq. (5.275) may be evaluated by averaging procedures; in other cases, significant contributions to the sum arise from only one or a small number of terms whose denominators are very small, i.e., near a resonance.

The state with energy $E_{\mathrm{g}} + \hbar\omega_{\mathrm{i}}$ is called a *nonresonant state* if it does not coincide with an eigenstate of the scattering system, similarly for states with energy $E_{\mathrm{e}} - \hbar\omega_{\mathrm{i}}$. Such states also are said to be *virtual* [30] since they are not observable. When the frequency of the incident radiation is far removed from any resonance, effects of damping are small and generally may be ignored since none of the denominators in the scattering formula are close to zero. When a material is irradiated with light in the vicinity of an absorption band, however, the energy denominators for this band become so small that their effect dominates the sum in Eq. (5.275). It is then no longer possible to ignore damping; hence, the polarizability tensor (Eq. (5.267) is written

$$\alpha_{\mu\nu}^{\text{eg}} = \sum_t \left[\frac{\langle e|d_\mu|t\rangle\langle t|d_\nu|g\rangle}{\dfrac{E_g - E_t + \hbar\omega + i\hbar\Gamma_t}{2}} + \frac{\langle e|d_\nu|t\rangle\langle t|d_\mu|g\rangle}{\dfrac{E_e - E_t - \hbar\omega + i\hbar\Gamma_t}{2}} \right], \tag{5.276}$$

where Γ_t is the damping constant for the state $|t\rangle$.

The factor $(n_s + 1)$ in Eq. (5.275) has an interpretation similar to that in one-photon emission. If $n_s = 0$ in the initial state $|I\rangle$, we have *spontaneous scattering* whose probability per unit time, according to Eq. (5.271), is proportional to n_i, the number of incident photons. When $n_s \neq 0$, we have, in addition to spontaneous scattering, a term proportional to the product $n_i n_s$. The scattering cross-section associated with the latter requires the presence of scattered photons; that is, the scattered photons stimulate the production of other scattered photons. For this reason, the process is known as *stimulated scattering*. Clearly, stimulated scattering can occur only after the production of scattered photons by spontaneous scattering (assuming no scattered photons have been injected into the system). At ordinary intensities, stimulated scattering is usually negligible because n_s is small, but leads to important nonlinear effects at higher intensities (see Section 7.5).

The matrix elements in the differential cross section (Eq. (5.275)) contain the dipole moment (or displacement) operator which is an odd parity operator. Then, if the states $|g\rangle$, $|t\rangle$, and $|e\rangle$ have a definite parity (odd or even), $|g\rangle$ and $|t\rangle$ must have opposite parity, similarly for $|t\rangle$ and $|e\rangle$. Consequently, the cross section is subject to the selection rule that $|g\rangle$ and $|e\rangle$ must both have odd parity or both have even parity (odd-odd or even-even transition). In view of the dipole selection rule for single photon absorption, which requires that $|g\rangle$ and $|e\rangle$ have opposite parity (odd-even or even-odd transition), states which absorb strongly tend to scatter weakly and vice versa.

An alternative derivation of the Kramers-Heisenberg formula may be obtained on the basis of the interaction Hamiltonian, Eq. (4.328). We write

$$\mathscr{H}_{\text{int}} = \mathscr{H}_1 + \mathscr{H}_2 \tag{5.277a}$$

where

$$\mathscr{H}_1 = \frac{e}{m}\mathbf{p}\cdot\mathbf{A}, \qquad \mathscr{H}_2 = \frac{e^2}{2m}A^2. \tag{5.277b}$$

For a single mode, the vector potential (Eq. (4.117)) in the dipole approximation reduces to

$$\mathbf{A}(\mathbf{r}) = \sqrt{\frac{\hbar}{2\varepsilon_0\omega V}}\,\hat{\varepsilon}(a + a^\dagger). \tag{5.278}$$

We consider, first, the contribution of $\mathscr{H}_1$. Let

$$\mathscr{H}_1 = \mathscr{H}_1^{(+)} + \mathscr{H}_1^{(-)}, \tag{5.279a}$$

where

$$\mathscr{H}_1^{(+)} = \frac{e}{m}\sqrt{\frac{\hbar}{2\varepsilon_0\omega V}}(\mathbf{p}\cdot\hat{\varepsilon})a, \qquad \mathscr{H}_1^{(-)} = \frac{e}{m}\sqrt{\frac{\hbar}{2\varepsilon_0\omega V}}(\mathbf{p}\cdot\hat{\varepsilon})a^\dagger. \tag{5.279b}$$

With $|\mathrm{I}\rangle$ and $|\mathrm{F}\rangle$ defined by Eq. (5.249),

$$\langle \mathrm{F}|\mathscr{H}_1^{(+)}|\mathrm{I}\rangle = \langle \mathrm{F}|\mathscr{H}_1^{(-)}|\mathrm{I}\rangle = 0. \tag{5.280}$$

Hence, there is no contribution from $\mathscr{H}_1$ in first order as we have found previously with the Hamiltonian (Eq. (252)). It is necessary, therefore, to proceed to second order.

In second order, the matrix element is of the same form as Eq. (5.256) but with $\mathscr{H}_{\mathrm{AF}}$ replaced by $\mathscr{H}_1$; the intermediate states T_1 and T_2 are the same as those in Eq. (5.257) and their energies are given by Eq. (5.258); the energies E_{I} and E_{F} are the same as in Eq. (5.250). Thus,

$$\begin{aligned}\langle \mathrm{F}|\mathscr{H}_1|\mathrm{T}_1\rangle &= \langle \mathrm{e}, n_{\mathrm{i}} - 1, n_{\mathrm{s}} + 1|\mathscr{H}_1|t_i, n_{\mathrm{i}} - 1, n_{\mathrm{s}}\rangle \\ &= \langle \mathrm{e}, n_{\mathrm{s}} + 1|\mathscr{H}_1^{(-)}|\mathrm{g}, n_{\mathrm{s}}\rangle \\ &= \frac{e}{m}\sqrt{\frac{\hbar(n_{\mathrm{s}}+1)}{2\varepsilon_0\omega_{\mathrm{s}}V}}\langle \mathrm{e}|\mathbf{p}\cdot\hat{\varepsilon}_{\mathrm{s}}|t_1\rangle,\end{aligned} \tag{5.281a}$$

$$\begin{aligned}\langle \mathrm{T}_1|\mathscr{H}_1|\mathrm{I}\rangle &= \langle t_1, n_{\mathrm{i}} - 1, n_{\mathrm{s}}|\mathscr{H}_1|\mathrm{g}, n, n_{\mathrm{s}}\rangle \\ &= \langle t_1, n_{\mathrm{i}} - 1|\mathscr{H}_1^{(+)}|\mathrm{g}, n_{\mathrm{i}}\rangle \\ &= \frac{e}{m}\sqrt{\frac{\hbar n_{\mathrm{i}}}{2\varepsilon_0\omega_{\mathrm{i}}V}}\langle t_1|\mathbf{p}\cdot\hat{\varepsilon}_{\mathrm{i}}|\mathrm{g}\rangle,\end{aligned} \tag{5.281b}$$

$$\begin{aligned}\langle \mathrm{F}|\mathscr{H}_1|\mathrm{T}_2\rangle &= \langle \mathrm{e}, n_{\mathrm{i}} - 1, n_{\mathrm{s}} + 1|\mathscr{H}_1|t_2, n_{\mathrm{i}}, n_{\mathrm{s}} + 1\rangle \\ &= \langle \mathrm{e}, n_{\mathrm{i}} - 1|\mathscr{H}_1^{(+)}|t_2, n_{\mathrm{i}}\rangle \\ &= \frac{e}{m}\sqrt{\frac{\hbar n_{\mathrm{i}}}{2\varepsilon_0\omega_{\mathrm{i}}V}}\langle \mathrm{e}|\mathbf{p}\cdot\hat{\varepsilon}_{\mathrm{i}}|t_2\rangle,\end{aligned} \tag{5.281c}$$

$$\begin{aligned}\langle \mathrm{T}_2|\mathscr{H}_1|\mathrm{I}\rangle &= \langle t_2, n_{\mathrm{i}}, n_{\mathrm{s}} + 1|\mathscr{H}_1|\mathrm{g}, n_1, n_{\mathrm{s}}\rangle \\ &= \langle t_2, n_{\mathrm{s}} + 1|\mathscr{H}_1^{(-)}|\mathrm{g}, n_{\mathrm{s}}\rangle \\ &= \frac{e}{m}\sqrt{\frac{\hbar(n_{\mathrm{s}}+1)}{2\varepsilon_0\omega_{\mathrm{s}}V}}\langle t_2|\mathbf{p}\cdot\hat{\varepsilon}_{\mathrm{s}}|\mathrm{g}\rangle.\end{aligned} \tag{5.281d}$$

We now may write the second order matrix elements N_1 and N_2 corresponding to the paths via T_1 and T_2:

$$N_1 = \sum_{T_1} \frac{\langle F|\mathscr{H}_1|T_1\rangle\langle T_1|\mathscr{H}_1|I\rangle}{E_I - E_{T_1}}$$

$$= \frac{e^2\hbar}{2\varepsilon_0 m^2 V} \sum_{t_1} \sqrt{\frac{(n_s + 1)n_i}{\omega_s\omega_i}} \langle e|\mathbf{p}\cdot\hat{\varepsilon}_s|t_1\rangle\langle t_1|\mathbf{p}\cdot\hat{\varepsilon}_i|g\rangle, \qquad (5.282a)$$

$$N_2 = \sum_{T_2} \frac{\langle F|\mathscr{H}_1|T_2\rangle\langle T_2|\mathscr{H}_1|I\rangle}{E_I - E_{T_2}}$$

$$= \frac{e^2\hbar}{2\varepsilon_0 m^2 V} \sum_{t_2} \sqrt{\frac{n_i(n_s + 1)}{\omega_i\omega_s}} \langle e|\mathbf{p}\cdot\hat{\varepsilon}_i|t_2\rangle\langle t_2|\mathbf{p}\cdot\hat{\varepsilon}_s|g\rangle. \qquad (5.282b)$$

Thus, the complete second order matrix element is

$$N = N_1 + N_2$$

$$= \frac{e^2\hbar}{2\varepsilon_0 m^2 V} \sqrt{\frac{n_i(n_s + 1)}{\omega_i\omega_s}} \sum_t \left[\frac{\langle e|\mathbf{p}\cdot\hat{\varepsilon}_s|t\rangle\langle t|\mathbf{p}\cdot\hat{\varepsilon}_i|g\rangle}{E_g - E_t + \hbar\omega_i}\right.$$

$$\left. + \frac{\langle e|\mathbf{p}\cdot\hat{\varepsilon}_i|t\rangle\langle t|\mathbf{p}\cdot\hat{\varepsilon}_s|g\rangle}{E_e - E_t - \hbar\omega_i}\right], \qquad (5.283)$$

in which, again the general summation index t replaces t_1 and t_2.

We now consider the contribution of $\mathscr{H}_2$ in Eq. (5.277b) to the scattering process. Since $\mathscr{H}_2$ contains bilinear combinations of creation and annihilation operators, there is a nonvanishing contribution in first order. The matrix elements of the four bilinear combinations are

$$\begin{aligned}
\langle n_i - 1, n_s + 1|a_i a_s|n_i, n_s\rangle &= 0, \\
\langle n_i - 1, n_s + 1|a_i^\dagger a_s^\dagger|n_i, n_s\rangle &= 0, \\
\langle n_i - 1, n_s + 1|a_i^\dagger a_s|n_i, n_s\rangle &= 0, \\
\langle n_i - 1, n_s + 1|a_i a_s^\dagger|n_i, n_s\rangle &= \sqrt{n_i(n_s + 1)}.
\end{aligned} \qquad (5.284)$$

The evaluation of $F|\mathscr{H}_2|I\rangle$ follows immediately but it is necessary to take cognizance of the fact that the vector potential, as written in Eq. (4.117) is a sum over all modes. For $\mathscr{H}_1$, matrix elements involving specific modes, such as those corresponding to incident and scattered photons, appear just once in the sum. But $\mathscr{H}_2$, which depends quadratically on the vector potential, is a double sum over the photon modes; nonvanishing matrix elements therefore appear twice. With this in mind, we have

$$\langle \mathrm{F}|\mathscr{H}_2|\mathrm{I}\rangle = 2\langle \mathrm{F}|\frac{e^2}{2m}A^2|\mathrm{I}\rangle = 2\langle \mathrm{e}, n_\mathrm{i} - 1, n_\mathrm{s} + 1|\frac{e^2}{2m}A^2|\mathrm{g}, n_\mathrm{i}, n_\mathrm{s}\rangle$$
$$= \frac{e^2\hbar}{2\varepsilon_0 mV}\sqrt{\frac{n_\mathrm{i}(n_\mathrm{s}+1)}{\omega_\mathrm{i}\omega_\mathrm{s}}}(\hat{\boldsymbol{\varepsilon}}_\mathrm{i}\cdot\hat{\boldsymbol{\varepsilon}}_\mathrm{s})\,\delta_{\mathrm{eg}}. \tag{5.285}$$

Here, too, the first and second order terms are coherent, which means that they are to be added before squaring. Hence, the total scattering matrix element is

$$\langle \mathrm{F}|\mathscr{H}_2|\mathrm{I}\rangle + \sum_\mathrm{T}\frac{\langle \mathrm{F}|\mathscr{H}_1|\mathrm{T}\rangle\langle \mathrm{T}|\mathscr{H}_1|\mathrm{I}\rangle}{E_\mathrm{I} - E_\mathrm{T}} = \frac{e^2\hbar}{2\varepsilon_0 mV}\sqrt{\frac{n_\mathrm{i}(n_\mathrm{s}+1)}{\omega_\mathrm{i}\omega_\mathrm{s}}}\left\{(\hat{\varepsilon}_\mathrm{i}\cdot\hat{\varepsilon}_\mathrm{s})\,\delta_{\mathrm{eg}} + \frac{1}{m}\sum_t\left[\frac{\langle \mathrm{e}|\mathbf{p}\cdot\hat{\boldsymbol{\varepsilon}}_\mathrm{s}|t\rangle\langle t|\mathbf{p}\cdot\hat{\boldsymbol{\varepsilon}}_\mathrm{i}\rangle}{E_\mathrm{g} - E_t + \hbar\omega_\mathrm{i}} + \frac{\langle \mathrm{e}|\mathbf{p}\cdot\hat{\boldsymbol{\varepsilon}}_\mathrm{i}|t\rangle\langle t|\mathbf{p}\cdot\hat{\boldsymbol{\varepsilon}}_\mathrm{s}|\mathrm{g}\rangle}{E_\mathrm{e} - E_t - \hbar\omega_\mathrm{i}}\right]\right\}. \tag{5.286}$$

Following the procedure leading to Eq. (5.275) the differential cross section is

$$\frac{d\sigma}{d\Omega} = r_0^2\frac{\omega_\mathrm{s}}{\omega_\mathrm{i}}(n_\mathrm{s}+1)\left|(\hat{\boldsymbol{\varepsilon}}_\mathrm{i}\cdot\hat{\boldsymbol{\varepsilon}}_\mathrm{s})\,\delta_{\mathrm{eg}} + \frac{1}{m}\sum_t\left[\frac{\langle \mathrm{e}|\mathbf{p}\cdot\hat{\boldsymbol{\varepsilon}}_\mathrm{s}|t\rangle\langle t|\mathbf{p}\cdot\hat{\boldsymbol{\varepsilon}}_\mathrm{i}|\mathrm{g}\rangle}{E_\mathrm{g} - E_t + \hbar\omega_\mathrm{i}} + \frac{\langle \mathrm{e}|\mathbf{p}\cdot\hat{\boldsymbol{\varepsilon}}_\mathrm{i}|t\rangle\langle t|\mathbf{p}\cdot\hat{\boldsymbol{\varepsilon}}_\mathrm{s}|\mathrm{g}\rangle}{E_\mathrm{e} - E_t - \hbar\omega_\mathrm{i}}\right]\right|^2. \tag{5.287}$$

This form of the differential scattering cross section is based on matrix elements of the momentum operator whereas Eq. (5.275) contains matrix elements of the dipole moment operator. The two forms are equivalent and one can be transformed into the other. Also, as in Eq. (5.276), damping terms are required in the vicinity of a resonance.

5.10 Rayleigh and Thomson Scattering

The energy of the scattered radiation may or may not be the same as the energy of the incident radiation. If the two are the same, the state of the scattering system must remain unchanged and the scattering process is said to be elastic. Setting $|\mathrm{g}\rangle = |\mathrm{e}\rangle$ and $\omega_\mathrm{i} = \omega_\mathrm{s}$, the spontaneous differential cross section, according to Eq. (5.275), is

$$\frac{d\sigma}{d\Omega} = r_0^2\omega^4 m^2\left|\sum_t\left[\frac{\langle \mathrm{g}|\mathbf{r}\cdot\hat{\boldsymbol{\varepsilon}}_\mathrm{s}|t\rangle\langle t|\mathbf{r}\cdot\hat{\boldsymbol{\varepsilon}}|\mathrm{g}\rangle}{E_\mathrm{g} - E_t + \hbar\omega} + \frac{\langle \mathrm{g}|\mathbf{r}\cdot\hat{\boldsymbol{\varepsilon}}|t\rangle\langle t|\mathbf{r}\cdot\hat{\boldsymbol{\varepsilon}}_\mathrm{s}|\mathrm{g}\rangle}{E_\mathrm{g} - E_t - \hbar\omega}\right]\right|^2. \tag{5.288}$$

When $\hbar\omega \ll |E_g - E_t|$, the scattering is known as *Rayleigh scattering* with a cross section

$$\frac{d\sigma}{d\Omega} = r_0^2\omega^4 m^2 \left| \sum_t \frac{\langle g|\mathbf{r}\cdot\hat{\varepsilon}_s|t\rangle\langle t|\mathbf{r}\cdot\hat{\varepsilon}|g\rangle + \langle g|\mathbf{r}\cdot\hat{\varepsilon}|t\rangle\langle t|\mathbf{r}\cdot\hat{\varepsilon}_s|g\rangle}{E_g - E_t} \right|^2, \quad (5.289)$$

which yields the familiar result that at long wavelengths the Rayleigh scattering cross section is inversely proportional to the fourth power of the wavelength.

In the vicinity of a resonance, by which we mean that there exists a state $|t\rangle$ such that $E_g - E_t = -\hbar\omega_t \simeq -\hbar\omega$, the second term in the square brackets of Eq. (5.288) may be neglected. Of the remaining terms in the sum one will be dominant, and the denominator for that term will be close to zero. In such cases, it is necessary to include a damping constant as was done in Eq. (5.276). Thus, we now have

$$\frac{d\sigma}{d\Omega} = \frac{r_0^2\omega^4 m^2}{\hbar^2}\left[\frac{|\langle g|\mathbf{r}\cdot\hat{\varepsilon}_s|t\rangle|^2|\langle t|\mathbf{r}\cdot\hat{\varepsilon}|g\rangle|}{(\omega-\omega_t)^2 + \dfrac{\Gamma^2}{4}}\right], \quad (5.290)$$

in which it is assumed that Γ is a radiative damping constant associated with the spontaneous emission from the excited state as in Eq. (5.187). After integrating over all solid angles and summing over the two polarizations, a factor of $8\pi/3$ is obtained as has been shown previously (Section 5.5). Another factor of 1/3 appears as a result of averaging over the random orientations in an assembly of atoms. We then have for the total cross section

$$\sigma = \frac{8\pi}{9} r_0^2\omega^4 m^2 \frac{|\langle t|\mathbf{r}|g\rangle|^4}{\hbar^2\left[(\omega-\omega_t)^2 + \dfrac{\Gamma^2}{4}\right]} \quad (5.291)$$

The matrix element and the radiative decay constant are related by Eq. (5.187); hence, for the present case

$$|\langle t|r|g\rangle|^2 = \frac{3c^2}{4\alpha\omega^3}\Gamma. \quad (5.292)$$

With this relation and the expressions for r_0 and α from Eqs. (5.274) and (5.68), respectively,

$$\sigma = \frac{e^2\omega}{3\varepsilon_0\hbar c}|\langle t|\mathbf{r}|g\rangle|^2 \frac{\dfrac{\Gamma}{2}}{(\omega-\omega_t)^2 + \dfrac{\Gamma^2}{4}}. \quad (5.293)$$

The matrix element may be replaced by Eq. (5.292) to obtain the resonant cross section

$$\sigma = 2\pi \frac{c^2}{\omega^2} = \frac{\lambda^2}{2\pi}, \tag{5.294}$$

where λ is the wavelength of the incident radiation. In the visible region, $\sigma \simeq 5 \times 10^{-10}\ \mathrm{cm}^2$.

When $\omega = \omega_s$ and $\hbar\omega \gg |E_g - E_t|$, we have another example of elastic scattering known as *Thomson* scattering. For this case, it is more convenient to refer to Eq. (5.287) for the differential cross section since now the terms in the sum over intermediate states are small and may be neglected. The remaining term, which is entirely due to the interaction Hamiltonian $\mathscr{H}_2$ in Eq. (5.277b), yields

$$\frac{d\sigma}{d\Omega} = r_0^2(\hat{\boldsymbol{\varepsilon}}_i \cdot \boldsymbol{\varepsilon}_s^2)\,\delta_{ge}. \tag{5.295}$$

For a many-electron system, $d\sigma/d\Omega$ is multiplied by Z^2. Since the atomic states do not appear in the cross section, Thomson scattering may be regarded as the scattering of photons from free electrons.

If the incident beam is polarized, so is the scattered beam, and the differential cross section depends on the polarization of the latter. Referring to Fig. 5.11, $\mathbf{k}_i$ and $\mathbf{k}_s$ are unit vectors in the direction of propagation of the incident and scattered beams, respectively. The incident beam is assumed to propagate in the z direction and to be polarized along x; the scattered beam propagates in the (θ, ϕ) direction. Then

$$\frac{d\sigma}{d\Omega} = \begin{cases} 0, & \hat{\boldsymbol{\varepsilon}}_s \text{ perpendicular to } \hat{\boldsymbol{\varepsilon}}_i, \quad (5.296a) \\ r_0^2(1 - \sin^2\theta\cos^2\phi), & \hat{\varepsilon}_s \text{ lies in the } (\hat{\varepsilon}_i, \hat{k}_s) \text{ plane.} \quad (5.296b) \end{cases}$$

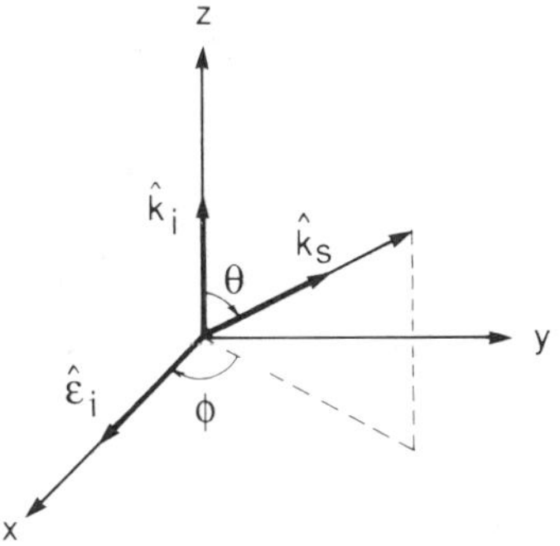

FIGURE 5.11 Thomson scattering geometry. $\mathbf{k}_i$ and $\mathbf{k}_s$ are unit propagation vectors for the incident and scattered beams. $\boldsymbol{\varepsilon}_i$ is the polarization vector of the incident beam.

Alternatively, $\boldsymbol{\varepsilon}_s$ may be resolved into two components $\boldsymbol{\varepsilon}_{s1}$ and $\boldsymbol{\varepsilon}_{s2}$ with $\boldsymbol{\varepsilon}_{s1}$ perpendicular to the $(\mathbf{k}_i, \mathbf{k}_s)$ plane and $\boldsymbol{\varepsilon}_{s2}$ perpendicular to the $(\boldsymbol{\varepsilon}_{s1}, \mathbf{k}_s)$ plane. Then

$$\frac{d\sigma}{d\Omega} = \begin{cases} r_0^2 \sin^2\phi, & \hat{\boldsymbol{\varepsilon}}_{s1} \text{ polarization}, \quad (5.297a) \\ r_0^2 \cos^2\theta \cos^2\phi, & \hat{\boldsymbol{\varepsilon}}_{s2} \text{ polarization}. \quad (5.297b) \end{cases}$$

A detector that is insensitive to polarization will record the sum of Eqs. (5.297a) and (5.297b) which is identical with Eq. (5.296b).

When the incident beam is unpolarized, the differential cross section is obtained by averaging Eq. (5.296b) over ϕ,

$$\frac{d\sigma}{d\Omega} = \frac{1}{2} r_0^2 (1 + \cos^2\theta). \quad (5.298a)$$

Finally, when Eq. (5.298a) is integrated over solid angle, one obtains the total cross section under conditions where the incident light is unpolarized and the detector sensitivity is independent of polarization:

$$\sigma = \frac{8\pi}{3} r_0^2. \quad (5.298b)$$

5.11 Spontaneous Raman Scattering

A beam of light propagating in a transparent medium (gas, liquid, or solid) generally will be partially scattered. Most of the scattered light has the same frequency as the incident light (elastic or Rayleigh scattering), although the direction of propagation may differ from that of the incident light. A smaller fraction of the scattered light differs both in frequency and direction of propagation from the incident light. Such scattering has its origin in inelastic interactions of the incident light with internal degrees of freedom of the medium. A common example, but by no means the only one, is the vibrational Raman effect in which a molecule of the medium gains or losses a vibrational quantum (typically with the energy of 100–4000 cm^{-1}) through interaction with the incident light. If the molecule gains a vibrational quantum (by a transition to the next higher vibrational state), the scattered photon energy must be reduced to conserve energy, in which case $\hbar\omega_s < \hbar\omega_i$ where ω_i and ω_s are the frequencies of the incident and scattered light, respectively. This is known as a *Stokes shift* (Fig. 5.12a). On the other hand, should the molecule be in an excited state, it may lose a vibrational quantum (by a transition to the next lower vibrational state); the scattered photon energy will then be increased, i.e. $\hbar\omega_s > \hbar\omega_i$. This is known as an *anti-Stokes shift* (Fig. 5.12b) and is generally much weaker than the Stokes shift simply because of the lower

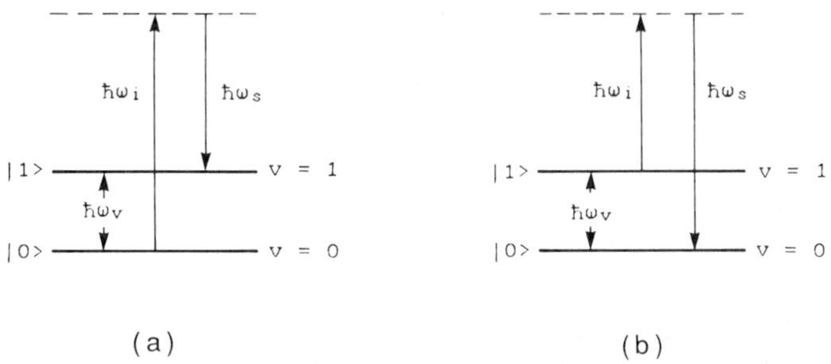

FIGURE 5.12 Vibrational Raman scattering involving the interchange of a single vibrational quantum of energy $\hbar\omega_v$. ω_i and ω_s are the frequencies of the incident and scattered radiation. The broken lines represent virtual (unobserved) levels that may or may not coincide with an eigenstate of the molecule; in the former case, one refers to the process as resonance Raman scattering. (a) Stokes shift in which $\omega_i = \omega_s + \omega_v$ and the molecular transition is $|0\rangle \rightarrow |1\rangle$ where $|0\rangle$ and $|1\rangle$ refer to the lowest and first-excited vibrational levels; (b) anti-Stokes shift in which $\omega_i = \omega_s - \omega_v$ and the molecular transition is $|1\rangle \rightarrow |0\rangle$.

population of molecules residing in excited states. Transitions involving more than one vibrational quantum are also possible. The difference $|\omega_i - \omega_s|$ for either the Stokes or anti-Stokes shift is independent of ω, and is solely a property of the medium.

In addition to the vibrational Raman effect, there are other types of Raman effects, e.g., (a) rotational Raman in which a rotational quantum (5–400 cm^{-1}) is absorbed or emitted; (b) electronic Raman involving an electronic transition (1–2 eV); (c) Raman scattering by phonons, plasmons, excitions, and others.

Classical considerations cannot provide a complete theory of scattering; ultimately, one must resort to quantum mechanics. Nevertheless, the simple physical picture that emerges from classical theory is correct in its gross aspects and serves as a useful introduction to the subject.

When a molecule is placed in a weak electric field, the electrons and nuclei undergo a displacement resulting in the induction of a net dipole moment proportional to the molecule's polarizability and to the applied electric field. If the incident field oscillates, the induced dipole moment also is forced into oscillation and acts as a source of scattered radiation. Both nuclei and electrons contribute to the scattered radiation but the nuclei, because of their large mass, scatter very little at wavelengths corresponding to the visible region of the spectrum. The direct nuclear contribution then may be neglected; however, there is an indirect effect of the nuclei associated with the dependence of the electronic polarizability on the instantaneous position of the nuclei. Therefore, as the nuclei vibrate, the polarizability does not remain constant but also becomes (weakly) time-dependent. The oscillation frequencies of the induced dipole moment then depend not only on the applied field but also, to a small extent, on the nuclear motion. It is because of the latter effect that the scattered radiation contains frequency components not found in the incident field.

Let us now consider a light wave with a real electric field

$$\mathbf{E}(t) = \mathbf{E}\cos\omega t, \tag{5.299}$$

incident on a molecule with polarizability α but with no permanent dipole moment. The *induced* dipole moment $\mathbf{D}(t)$ will then be

$$\mathbf{D}(t) = \varepsilon_0 \alpha \mathbf{E}(t). \tag{5.300}$$

We may picture the polarizability as a measure of the deformation of the electron cloud of the molecule in response to the external field. Since $\mathbf{D}(t)$ and $\mathbf{E}(t)$ are not necessarily colinear, α is a second rank tensor in the general case; that is, the response of the molecule for different directions of the applied field relative to the molecule's internal axes may not be the same.

If Q_k is a vibrational coordinate associated with a normal mode such that $Q_k = 0$ at the equilibrium position, we may expand α in a Taylor series about $Q_k = 0$. For most cases, an expansion to first order is sufficient; thus,

$$\alpha(Q_i) = \alpha_0 + \left(\frac{\partial\alpha}{\partial Q_k}\right)_0 Q_k. \tag{5.301}$$

For small amplitudes of vibration, the harmonic approximation gives

$$Q_k = Q_{k0}\cos(\omega_k t + \delta_k). \tag{5.302}$$

Combining Eq. (5.299) with Eq. (5.301),

$$\mathbf{D}(t) = \varepsilon_0\left[\alpha_0 + \left(\frac{\partial\alpha}{\partial Q_k}\right)_0 Q_{k0}\cos(\omega_k t + \delta_k)\right]\mathbf{E}\cos\omega t, \tag{5.303}$$

where the second term represents the modulation of the induced dipole moment by the nuclear vibrations. Further details may be displayed by writing Eq. (5.303) in the form

$$\mathbf{D}(t) = \varepsilon_0\alpha_0\mathbf{E}\cos\omega t + \varepsilon_0\frac{1}{2}\left(\frac{\partial\alpha}{\partial Q_k}\right)_0 Q_{k0}\mathbf{E}(\cos[(\omega - \omega_k)t - \delta_k] + \cos[(\omega + \omega_k)t + \delta_k]\}. \tag{5.304}$$

It now is evident that the oscillation of the induced dipole moment, which is the source of the scattered radiation, contains three frequencies: the frequency of the incident radiation at ω; the Stokes frequency at $\omega - \omega_k$; and the anti-Stokes frequency at $\omega + \omega_k$. The scattering at ω accounts for Rayleigh scattering and the scattering at $\omega \pm \omega_k$ represents Raman scattering. It also is observed that the phase of the Rayleigh scattered radiation is the same as that of the incident radiation. Rayleigh scattering therefore, is, said to be coherent. On the other hand, there is a phase shift of $\pm\delta_k$ in Raman scattering due to the phase of the normal vibration Q_k relative to the incident field. But the phase shift varies from one scattering molecule to another; spontaneous Raman scattering therefore, is, incoherent. This means that Raman scattering from two independent sources cannot produce interference; hence, the total scattering rate is found by adding the rates from each source.

In the vibrational Raman effect, the interaction between an incident field and a vibrating molecule leads to a transfer of energy between a molecule and a radiation field. The eigenstates of the scattering system are therefore molecular eigenstates whose description is most often facilitated by the *adiabatic* (Born-Oppenheimer) approximation. Because of the large ratio of nuclear to electronic mass, the coupling between electronic and nuclear motions is weak. For a fixed nuclear configuration, the electronic wave function depends only on the electronic coordinates, but with changes in nuclear configuration, which occur at relatively slow rates, the electronic wave function adjusts itself almost immediately (adiabatically). The total molecular wave function in the adiabatic approximation, therefore, is factored into a product of an electronic and a nuclear wave function

$$\Psi(Q, q) = \psi(Q, q)v(Q), \tag{5.305}$$

where Q and q denote the collection of nuclear and electronic coordinates, respectively; $\psi(Q, q)$ is an electronic wave function that depends primarily on electronic coordinates and to a much lesser extent (parametrically) on nuclear coordinates; $v(Q)$ is a vibrational wave function of the nuclei.

We now shall consider the following molecular states and their wave functions (Fig. 5.13) for an ensemble of identical noninteracting molecules:

$$\begin{aligned}
\psi_{\mathrm{g}}(Q, q)v_{\mathrm{g}\gamma}(Q) &\equiv |\mathrm{g}\gamma\rangle = \text{initial state, energy } E_{\mathrm{g}\gamma},\\
\psi_{\mathrm{t}}(Q, q)v_{\mathrm{t}\tau}(Q) &\equiv |\mathrm{t}\tau\rangle = \text{intermediate state, energy } E_{\mathrm{t}\tau},\\
\psi_{\mathrm{g}}(Q, q)v_{\mathrm{g}\varepsilon}(Q) &\equiv |\mathrm{g}\varepsilon\rangle = \text{final state, energy } E_{\mathrm{g}\varepsilon}.
\end{aligned} \tag{5.306}$$

The assumption implicit in these expressions is that the initial and final states have the same electronic wave function $\psi_{\mathrm{g}}(Q, q)$; that is, the electrons are initially in the ground state and remain in the ground state after the scattering event. This is a reasonable assumption because at ordinary temperatures a molecular system is practically always in the lowest electronic state. But the vibrational states before and after the scattering are not the same; thus, $|\mathrm{g}\gamma\rangle$ contains the vibrational wave function $v_{\mathrm{g}\gamma}(Q)$ and $|g\varepsilon\rangle$ contains $v_{\mathrm{g}\varepsilon}(Q)$. The intermediate state $|t\tau\rangle$ is a vibronic state in which both electronic and vibrational wave functions differ from those in $|\mathrm{g}\gamma\rangle$ and $|\mathrm{g}\varepsilon\rangle$. However, $|t\tau\rangle$ is constrained to states such that single photon, electric dipole transitions

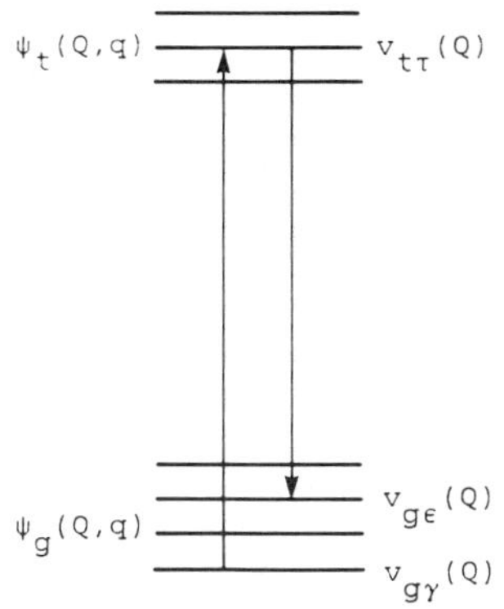

FIGURE 5.13 Vibrational Raman effect in the Born-Oppenheimer approximation. $v_{g\gamma}(Q)$ and $v_{g\varepsilon}(Q)$ are vibrational states in the ground electronic state $\psi_g(Q,q)$; $v_{t\tau}(Q)$ is a vibrational state in an excited electronic state $\psi_t(Q,q)$. It is assumed that the molecule stays in the ground electronic state before and after scattering. The only changes occur in the vibrational states: the initial vibronic wave function is $\psi_g(Q,q)v_{g\gamma}(Q)$, and the final vibronic wave function, after scattering, is $\psi_g(Q,q)v_{g\varepsilon}(Q)$.

between $|g\gamma\rangle$ and $|t\tau\rangle$ and between $|g\varepsilon\rangle$ and $|t\tau\rangle$ are allowed. We note further that $v_{g\gamma}$ and $v_{g\varepsilon}$ are orthogonal since they belong to the same electronic state, but they are not orthogonal to $v_{t\tau}$, which belongs to a different electronic state.

The wave functions (Eq. (5.306)) now may be inserted into the expression for the polarizability (Eq. (5.267)) with the replacement $E_g - E_t + \hbar\omega_i = E_{g\gamma} - E_{t\tau} - \hbar\omega_i$, $E_g - E_t - \hbar\omega_s = E_{g\varepsilon} - E_{t\tau} - \hbar\omega_i$ and the inclusion of damping factors as in Eq. (5.276):

$$\alpha_{\mu\nu}^{g\varepsilon,g\gamma} = \sum_{t\tau}\left[\frac{\langle g\varepsilon|d_\mu|t\tau\rangle\langle t\tau|d_\nu|g\gamma\rangle}{E_{g\gamma} - E_{t\tau} + \dfrac{\hbar\omega_i + i\hbar\Gamma_{t\tau}}{2}} + \frac{\langle g\varepsilon|d_\nu|t\tau\rangle\langle t\tau|d_\mu|g\gamma\rangle}{E_{g\varepsilon} - E_{t\tau} - \dfrac{\hbar\omega_i + i\hbar\Gamma_{t\tau}}{2}}\right]. \quad (5.307)$$

Since $E_{g\gamma}$ and $E_{g\varepsilon}$ correspond to the same electronic energy, their difference is equal to the energy of one or at most, a few vibrational quanta. (In the case of Brillouin scattering, acoustical phonons are absorbed or emitted.) But differences in energy between electronic states are far greater than the differences between vibrational states. Then, barring the possibility of an exact resonance, which would cause one of the denominators to vanish, we may set

$$E_{g\gamma} - E_{t\tau} \simeq E_{g\varepsilon} - E_{t\tau} \simeq E_g - E_t, \quad (5.308)$$

where E_g and E_t are electronic energies. By virtue of this approximation, vibrational energies have been eliminated from the denominators. One now may invoke the closure relation

$$\sum_{\tau}|v_{t\tau}\rangle\langle v_{t\tau}| = 1 \quad (5.309)$$

applied to the vibrational states belonging to the intermediate electronic state. With Eqs. (5.308) and (5.309) the polarizability tensor becomes

$$\alpha_{\mu\nu}^{g\varepsilon,g\gamma} = \sum_{t}\left[\frac{\langle g\varepsilon|d_\mu|t\rangle\langle t|d_\nu|g\gamma\rangle}{E_g - E_t + \hbar\omega_i + \dfrac{i\hbar\Gamma_t}{2}} + \frac{\langle g\varepsilon|d_\nu|t\rangle\langle t|d_\mu|g\gamma\rangle}{E_g - E_t - \hbar\omega_i + \dfrac{i\hbar\Gamma_t}{2}}\right]$$

$$= \left\langle \varepsilon \left| \sum_{r}\left[\frac{\langle g|d_\mu|t\rangle\langle t|d_\nu|g\rangle}{E_g - E_t + \hbar\omega_i + \dfrac{i\hbar\Gamma_t}{2}} + \frac{\langle g|d_\nu|t\rangle\langle t|d_\mu|g\rangle}{E_g - E_t - \hbar\omega_i + \dfrac{i\hbar\Gamma_t}{2}}\right] \right| \gamma \right\rangle$$

$$= \langle \varepsilon|\alpha_{\mu\nu}^{gg}|\gamma\rangle \equiv \langle v_{g\varepsilon}|\alpha_{\mu\nu}^{gg}|v_{g\gamma}\rangle. \tag{5.310}$$

In addition to the parity selection rule (Section 5.9), the polarizability tensor is constrained by another selection rule based on a theorem from group theory. For the group to which the molecule belongs, the theorem states that if

$$M = \langle \psi_\alpha^{(i)}|Q_\beta^{(j)}|\phi_\gamma^{(k)}\rangle \tag{5.311}$$

is a matrix element where $\psi_\alpha^{(i)}$, $Q_\beta^{(j)}$, and $\phi_\gamma^{(k)}$ transform according to the irreducible representations $\Gamma^{(i)}$, $\Gamma^{(j)}$, $\Gamma^{(k)}$, respectively, then $M = 0$ unless the direct product $\Gamma^{(i)*} \times \Gamma^{(j)} \times \Gamma^{(k)}$ contains the unit (totally symmetric) representation. (It is, of course, possible for M to vanish for other reasons.) The application of this theorem to Eq. (5.310) requires that $\Gamma(\alpha_{\mu\nu}^{gg})$ be contained in the reduction of $\Gamma(\varepsilon) \times \Gamma(\gamma)$.

Because of the weak coupling between electrons and nuclei, higher order approximations may be obtained by expanding electronic wave functions around the ground state equilibrium configuration of the nuclei. This is known as a *Herzberg-Teller expansion.* To see the essential features of this method, consider a system of electrons and nuclei interacting entirely through Coulombic interactions. The Hamiltonian for such a system is

$$\mathscr{H} = \sum_\alpha \frac{P_\alpha^2}{2M_\alpha} + \sum_i \frac{p_i^2}{2m} + V(Q_a, q_i), \tag{5.312}$$

in which $P_{\alpha_2}^2/2M_\alpha$ is the kinetic energy operator for a nucleus of mass M_α, $p_i/2m$ is kinetic energy operator for the ith electron of mass m, Q_α is a set of nuclear coordinates, q_i is a set of electronic coordinates, and

$$V(Q_\alpha, q_i) = \sum_{i<j} \frac{e^2}{r_i^j} + \sum_{\alpha<\beta} \frac{Z_\alpha Z_\beta e^2}{r_{\alpha\beta}} - \sum_{i\alpha} \frac{Z_\alpha e^2}{r_{i\alpha}}. \tag{5.313}$$

$V(Q_\alpha, q_i)$ is the potential energy of the entire system of electrons and nuclei; it consists of all the Coulomb interactions between all pairs of charges. The first term in Eq. (5.313) is the sum of the Coulomb interactions between all pairs of electrons, the second is the same for all pairs of nuclei with atomic numbers Z_α, Z_β, and the third term is the sum of the Coulomb interactions between nuclei and electrons.

The electronic Hamiltonian $\mathscr{H}_e$ is defined as the Hamiltonian (Eq. (5.312)) without the nuclear kinetic energy term, i.e., with the nuclei clamped in fixed positions:

$$\mathscr{H}_e = \sum_i \frac{p_i^2}{2m} + V(Q_\alpha, q_i). \tag{5.314}$$

Since $\mathscr{H}_e$ depends on the nuclear coordinates through $V(Q_\alpha, q_i)$, we may expand $\mathscr{H}_e$ about the equilibrium nuclear configuration labelled $Q_\alpha = 0$. Then, to first order,

$$\mathscr{H}_e = \mathscr{H}_0 + \mathscr{H}', \tag{5.315}$$

where $\mathscr{H}_0$ refers to $Q_\alpha = 0$, and

$$\mathscr{H}' = \left(\frac{\partial \mathscr{H}_e}{\partial Q_\alpha}\right)_0 Q_\alpha. \tag{5.316}$$

If $\psi_g^{(0)}(Q, q) \equiv \psi_g^{(0)}$ is an eigenfunction of $\mathscr{H}_0$ with eigenvalue $E_g^{(0)}$, the Herzberg-Teller expansion to first order is

$$\psi_g^{(1)} = \psi_g^{(0)} + \sum_\alpha \sum_{n \neq g} Q_\alpha V_{gn} \psi_n^{(0)}, \tag{5.317}$$

where

$$V_{gn} = \frac{\left\langle \psi_g^{(0)} \middle| \left(\frac{\partial \mathscr{H}_e}{\partial Q_\alpha}\right)_0 \middle| \psi_n^{(0)} \right\rangle}{E_n^{(0)} - E_g^{(0)}}. \tag{5.318}$$

The same treatment is applied to the intermediate states; the corrected wave functions are then inserted into the expression for the polarizability.

It is important to recognize the intrinsic difference between *fluorescence* and Raman scattering. The former is a two-step process in which a photon is absorbed and after a measurable time delay, typically on the order of 10^{-8} seconds, a second photon is emitted with an energy that is never larger (and usually smaller) than that of the absorbed photon. The two steps are statistically independent and individually observable. Therefore, fluorescence can occur only within an absorption band of the molecule. Each of the two transitions in fluorescence is a first order, one-photon radiative process with a probability proportional to the absolute square of the matrix element pertaining to each transition. The probability for the occurrence of the total process is the product of the probabilities for absorption and emission.

Raman scattering, on the other hand, is a single (two-photon) radiative process that cannot be decomposed into individual observable steps. Even though the formalism of second-order perturbation theory employs a stepwise procedure, the intermediate states are not separable in time and are not detectable. The energies of the incident and scattered photons bear no relation to energy differences in the molecule so that Raman scattering can occur at any energy of incident light and is not confined to any particular band.

When the frequency of the incident light corresponds to a molecular transition, one or more energy denominators in Eq. (5.110) approaches zero, and the intensity of the Raman spectrum increases strongly. We then have the case of *resonance Raman* scattering, which shares the common feature with fluorescence that, in both cases, the irradiation occurs within an absorption band of the scatterer. In such cases, the distinction between fluorescence and resonance Raman scattering becomes less sharp [31–34]. When the incident light is pulsed, it often is possible to distinguish between the two types of phenomena by examining the line shape of the emitted radiation. In fluorescence, the excited state created by the absorption process exhibits a decay associated with its lifetime. Therefore, after the pulse is turned off the emitted light will contain an exponentially decaying tail, characteristic of the excited state. In resonance Raman scattering, the relaxation of the excited state is not observed and the scattered light mainly follows the time-dependence of the incident light; i.e., emission stops when the pulse of incident radiation stops.

5.12 First-Order Coherence Functions

In the classical description of light, the visibility (or contrast) of interference fringes, in an arrangement such as Young's double slit experiment (Fig. 5.14) or a Michelson interferometer, is regarded as a measure of the coherence of the light beams [35]. In these experiments, typically, a light beam from a single source is divided into two equal parts; each part propagates along a separate path until they are recombined to produce the interference pattern. Since light from ordinary (thermal) sources exhibits rapid phase (as well as intensity) fluctuations, two beams from independent sources cannot be made to inter-

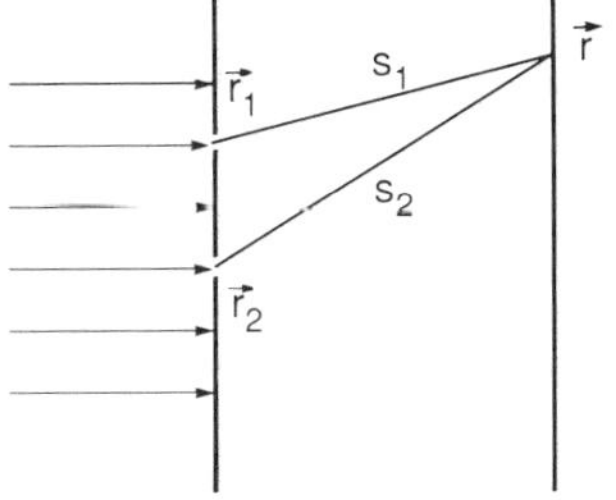

FIGURE 5.14 Schematic drawing of Young's double slit experiment.

fere. (We remark, parenthetically, that interference has been observed in light beams from two independent *laser* sources).

If the spectral bandwidth of the source is $\Delta\omega$ and the path difference between the two light beams is Δs, interference is observed only when Δs is less than or approximately equal to $c/\Delta\omega$ where c is the velocity of light. The maximum path difference

$$\Delta s_c \simeq \frac{c}{\Delta\omega} \tag{5.319}$$

is known as the *coherence length* and

$$\Delta t_c = \frac{\Delta s_c}{c} \simeq \frac{1}{\Delta\omega} \tag{5.320}$$

is the *coherence time*. Hence, the criterion for fringes to be visible is

$$\Delta t_c \Delta\omega \simeq 1. \tag{5.321}$$

Deeper analysis reveals that there are various orders of coherence and that fringe visibility is a measure only of the lowest order. More generally, it is found that two beams with identical intensities, spectral properties and polarizations nevertheless may differ in their statistical aspects. Moreover, there are also nonclassical light beams whose statistics can be described only in a quantum mechanical context. Such considerations acquire special importance in nonlinear optics.

We begin the discussion of first-order coherence functions with a classical description of interference. Referring to Fig. 5.14, let $\mathbf{r}_1$, $\mathbf{r}_2$, and $\mathbf{r}$ be the position vectors (coordinates) of slit 1, slit 2, and the detector, respectively. The complex scalar (or single component) electric field $E(\mathbf{r}t)$ at the position of the detector at time t is a linear superposition of the fields at the two slits at the earlier times $t_1 = t - s_1/c$ and $t_2 = t - s_2/c$, where s_1 and s_2 are the distances from the slits to the detector and c is the velocity of light. Thus,

$$E(\mathbf{r}t) = k_1 E(\mathbf{r}_1 t_1) + k_2 E(\mathbf{r}_2 t_2) \tag{5.322}$$

where the proportionality constants k_1 and k_2 depend on the path length. We shall ignore all complications arising from such effects as diffraction, finite size of the source, and slits.

It now will be assumed that $E(\mathbf{r}_1 t_1)$ and $E(\mathbf{r}_2 t_2)$ are subject to stationary, ergodic fluctuations of amplitude and phase arising from the large number of radiating atoms subject to various random processes such as spontaneous emission and thermal motion. We also shall assume that the response time of the detector is long compared to the correlation time of the fluctuations; the detector then will measure an average intensity proportional to the time average of $|E(\mathbf{r}t)|^2$:

$$I(\mathbf{r}) = K\langle|E(\mathbf{r}t)|^2\rangle. \tag{5.323}$$

Substituting Eq. (5.322) into Eq. (5.323)

$$I(\mathbf{r}) = K(|\mathrm{k}_1|^2\langle|E(\mathbf{r}_1t_1)|^2\rangle + |\mathrm{k}_2|^2\langle|E(\mathbf{r}_2t_2)|^2\rangle + \mathrm{k}_1\mathrm{k}_2^*\langle E(\mathbf{r}_1t_1)E^*(\mathbf{r}_2t_2)\rangle + \mathrm{k}_1^*\mathrm{k}_2\langle E^*(\mathbf{r}_1t_1)E(\mathbf{r}_2t_2)\rangle). \tag{5.324}$$

In view of the ergodic assumption, the angular brackets may be interpreted either as averages over time or as ensemble averages; in classical applications, the time average is generally preferred. Letting

$$I_1 = K|\mathrm{k}_1|^2\langle|E(\mathbf{r}_1t_1)|^2\rangle, \qquad I_2 = K|\mathrm{k}_2|^2\langle|E(\mathbf{r}_2t_2)|^2\rangle, \tag{5.325}$$

we have

$$\begin{aligned} I(\mathbf{r}) &= I_1 + I_2 + 2K\,\mathrm{Re}[\mathrm{k}_1^*\mathrm{k}_2\langle E^*(\mathbf{r}_1t_1)E(\mathbf{r}_2t_2)\rangle] \\ &= I_1 + I_2 + 2K\,\mathrm{Re}[\mathrm{k}_1^*\mathrm{k}_2\langle E^*(x_1)E(x_2)\rangle]. \end{aligned} \tag{5.326}$$

In the second expression, the space-time point (rt) has been replaced by (x).

We now introduce the *first-order coherence function*, also known as the *mutual coherence function*:

$$g^{(1)}(x_1, x_2) = \frac{\langle E^*(x_1)E(x_2)\rangle}{\sqrt{\langle|E(x_1)|^2\rangle\langle|E(x_2)|^2\rangle}}. \tag{5.327}$$

With

$$\mathrm{k}_1 = |\mathrm{k}_1|e^{i\phi_1}, \qquad \mathrm{k}_2 = |\mathrm{k}_2|e^{i\phi_2}, \tag{5.328}$$

$$g^{(1)}(x_1, x_2) = |g^{(1)}(x_1, x_2)|e^{i\Theta_{12}}, \tag{5.329}$$

the average intensity at the detector acquires the form

$$I(\mathbf{r}) = I_1 + I_2 + 2\sqrt{I_1I_2}\,|g^{(1)}(x_1, x_2)|\cos(\Theta_{12} - \phi). \tag{5.340}$$

In this expression, $\phi = \phi_1 - \phi_2$ is the phase difference arising from the path difference from the two slits to the detector. Note, however, as we have indicated previously, that the path difference must not exceed the coherence length. Θ_{12} is the phase difference between the fields $E(\mathbf{r}_1t_1)$ and $E(\mathbf{r}_2t_2)$ at the slits.

Equation (5.340) describes the spatial interference pattern. To see the effect of the first-order coherence function on the fringe pattern, we define the fringe contrast (or visibility)

$$\eta = \frac{I(\mathbf{r})_{\max} - I(\mathbf{r})_{\min}}{I(\mathbf{r})_{\max} + I(\mathbf{r})_{\min}} \tag{5.341}$$

When $|g^{(1)}(x_1,x_2)| = 0$, the intensity at the detector is simply the sum of the intensities; $I(\mathbf{r})_{max} = I(\mathbf{r})_{min}$ and $\eta = 0$. There is then no interference and the two beams are said to be incoherent. Since $g^{(1)}(x_1,x_2)$ is a normalized correlation function, the maximum value of $|g^{(1)}(x_1,x_2)|$ is 1, so that the maximum fringe contrast is

$$\eta = \frac{2\sqrt{I_1 I_2}}{I_1 + I_2}. \tag{5.342}$$

Taking $I_1 = I_2 = I$, we get $I(\mathbf{r})_{max} = 4I$, $I(\mathbf{r})_{min} = 0$, $\eta = 1$. The fringe visibility for several values of $|g^{(1)}(x_1,x_2)|$ is show in Fig. 5.15.

In summary, the coherence properties of a light beam are characterized by the electric field correlation function $g^{(1)}(x_1,x_2)$. Its absolute value may be used as a measure of coherence:

$$\begin{aligned} |g^{(1)}(x_1,x_2)| &= 0 && \text{incoherent} \\ 0 < |g^{(1)}(x_1,x_2)| &< 1 && \text{partially coherent} \\ |g^{(1)}(x_1,x_2)| &= 1 && \text{first-order coherent.} \end{aligned} \tag{5.343}$$

Further, it is useful to observe that when

$$\langle E^*(x_1)E(x_2)\rangle = \langle E^*(x_1)\rangle\langle E(x_2)\rangle, \tag{5.344}$$

Eq. (5.327) gives $|g^{(1)}(x_1,x_2)| = 1$. Consequently, the factorization property (Eq. (5.344)) may be used as the condition for optical coherence. It may be shown that condition (5.344) is both necessary and sufficient.

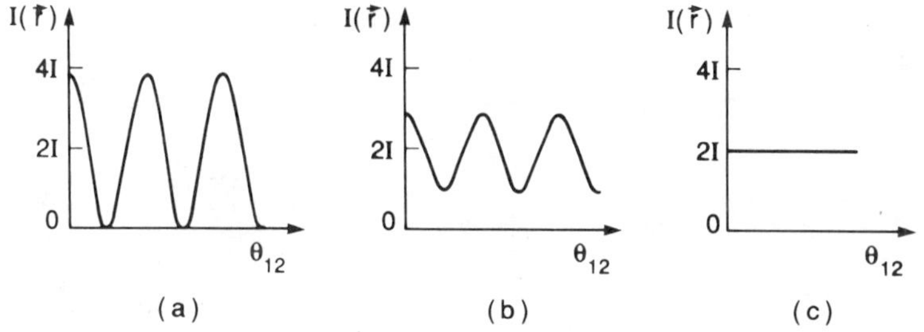

FIGURE 5.15 Effect of the mutual coherence function on the fringe contrast η (defined by Eq. (5.341)) for two interfering beams of equal intensity I. The intensity at the detector is $I(\mathbf{r})$.

a. $|g^{(1)}(x_1,x_2)| = 1$, $I(\mathbf{r})\text{max} = 4I$, $I(\mathbf{r})\text{min} = 0$, $\eta = 1$.

b. $|g^{(1)}(x_1,x_2)| = 0.5$, $I(\mathbf{r})\text{max} = 3I$, $I(\mathbf{r})\text{min} = I$, $\eta = 0.5$.

c. $|g^{(1)}(x_1,x_2)| = 0$, $I(\mathbf{r})\text{max} = 2I = I(\mathbf{r})\text{min}$, $\eta = 0$.

As a first example, consider a monochromatic plane wave propagating in the z direction. The fields $E(\mathbf{x}_1)$ and $E(\mathbf{x}_2)$ are

$$E(zt_1) = E_0 e^{i(kz - \omega t_1)}, \qquad E(zt_2) = E_0 e^{i(kz - \omega t_2)}, \tag{5.345}$$

which then yield

$$\langle E^*(zt_1)E(zt_2)\rangle = E_0^2 e^{-i\omega(t_2 - t_1)} = E_0^2 e^{-i\omega\tau}, \tag{5.346}$$

$$g^{(1)}(x_1, x_2) = g^{(1)}(\tau) = e^{-i\omega\tau} = g^{(1)}(-\tau)^*, \tag{5.347}$$

$$|g^{(1)}(\tau)| = 1, \tag{5.348}$$

where $\tau = t_2 - t_1 = (s_2 - s_1)/c$. A more realistic example is one in which the radiation has a Lorentzian power spectrum centered at a frequency ω_0 as, for example, in a collision-broadened light source. Under the assumption that the fluctuations in the light source are stationary and ergodic, we may use the Wiener-Khinchine theorem to derive the correlation function. Referring to Eqs. (1.106) and (1.109) and again assuming propagation in the z direction,

$$\langle E^*(zt_1)E(zt_2)\rangle = E_0^2 e^{-i\omega_0\tau - |\tau|/\tau_c}, \tag{5.349}$$

$$g^{(1)}(\tau) = e^{-i\omega_0\tau - |\tau|/\tau_c}, \qquad 0 \le |g^{(1)}(\tau)| \le 1. \tag{5.350}$$

Here the correlation time τ_c may be interpreted as a mean time between collisions. Since $|g^{(1)}(\tau)| \to 0$ as $\tau \to \infty$, the beam becomes more and more incoherent as the path difference increases, resulting in an exponentially diminishing fringe visibility. Only when $|\tau| \ll \tau_c$ does the light approach first-order coherence. If the spectrum is broadened, the result is a smaller value of τ_c and a more rapid diminution of fringe visibility with increasing path difference. A light beam with a correlation function of the form of Eq. (5.349) is said to be *chaotic*, although the general chaotic light beam may have a power spectrum other than Lorentzian as, for example, Gaussian, which could arise as a result of Doppler broadening (Section 6.5). Since the Fourier transform of a Gaussian is also a Gaussian, the relations analogous to Eqs. (5.349) and (5.350) are

$$\langle E^*(zt_1)E(zt_2)\rangle = E_0^2 e^{-i\omega_0 t} \exp\left[-\frac{1}{2}\frac{\tau^2}{\tau_c^2}\right] \tag{5.351}$$

where $1/\tau_c$ is the mean square spread in the Gaussian line shape and

$$g^{(1)}(\tau) = e^{-i\omega_0 t} \exp\left[-\frac{1}{2}\frac{\tau^2}{\tau_c^2}\right], \qquad 0 \le |g^{(1)}(\tau)| \le 1. \tag{5.352}$$

Other cases are treated by Loudon [36].

We now turn to the quantum formulation [37–40] beginning with first-order coherence. All devices that measure the intensity of a light beam depend on the absorption of photons that induce an observable response in the absorbing system. The quantum mechanical description of the absorption process therefore serves as a convenient starting point for the discussion of quantum coherence functions. Although there are numerous examples of photon detectors such as photographic media, photodiodes, phototubes, and photomultipliers, it is sufficient for our purpose to consider an ideal broad-band detector consisting of a single atom that becomes ionized in the act of absorbing a photon. The electron detached from the atom makes its presence known as an electrical pulse. The pulses may be counted and subjected to statistical analysis from which one derives the statistical properties of the light beams.

The interaction Hamiltonian in the dipole approximation is given by Eq. (4.329), and the matrix element for absorption between an initial state $|\mathrm{I}\rangle$ and a final state $|\mathrm{F}\rangle$ is

$$\langle \mathrm{F}|\mathscr{H}_{\mathrm{AF}}|\mathrm{I}\rangle = \langle \mathrm{F}|-\mathbf{d}\cdot\mathbf{E}(\mathbf{r})|\mathrm{I}\rangle = \langle \mathrm{F}|-\mathbf{d}\cdot\mathbf{E}^{(+)}(\mathbf{r})|\mathrm{I}\rangle. \tag{5.353}$$

Here $\mathbf{E}(\mathbf{r})$ is the general quantized electric field (Eq. (4.119b)) in the Schrödinger representation with $\mathbf{k}\cdot\mathbf{r} \ll 1$; $\mathbf{E}^{(+)}(\mathbf{r})$ is the component containing the annihilation operator (Eq. (4.121a)) and is therefore the operator responsible for absorption; $\mathbf{d}$ is the dipole moment operator. The initial state $|\mathrm{I}\rangle$ consists of an atom in the ground state $|\mathrm{g}\rangle$ and a radiation field in a general state $|\mathrm{i}\rangle$ that is not necessarily a photon number state or a product of such states. The final state $|\mathrm{F}\rangle$ consists of an atomic state $|\mathrm{e}\rangle$ in the continuum and a radiation field in a general state $|\mathrm{f}\rangle$. Since it is assumed that the atom is ionized, a good approximation to $|\mathrm{e}\rangle$ far from the ionization limit is the free electron state

$$|\mathrm{e}\rangle = \frac{1}{V^{1/2}} e^{i\mathbf{q}\cdot\mathbf{r}}, \tag{5.354}$$

where $\hbar\mathbf{q}$ is the momentum of the detached electron and the wave function is normalized in the volume V. Thus, in zero order, $|\mathrm{I}\rangle = |\mathrm{g}\rangle|\mathrm{i}\rangle \equiv |\mathrm{g},\mathrm{i}\rangle$ and $|\mathrm{F}\rangle = |\mathrm{e}\rangle|\mathrm{f}\rangle \equiv |\mathrm{e},\mathrm{f}\rangle$.

The dipole moment operator $\mathbf{d}$ acts only on the atomic states while $\mathbf{E}^{(+)}$ acts only on the states of the field; hence, the matrix element (Eq. (5.353)) may be factored:

$$\langle \mathrm{F}|\mathscr{H}_{\mathrm{AF}}|\mathrm{I}\rangle = -\langle \mathrm{e}|\mathbf{d}|\mathrm{g}\rangle \cdot \langle \mathrm{f}|\mathbf{E}^{(+)}(\mathbf{r})|\mathrm{i}\rangle. \tag{5.355}$$

Since the transition probability is proportional to $|\langle \mathrm{F}|\mathscr{H}_{\mathrm{AF}}|\mathrm{I}\rangle|^2$, the probability that the field will undergo a transition from the state $|\mathrm{i}\rangle$ to the state $|\mathrm{f}\rangle$ is proportional to

$$|\langle \mathrm{f}|\mathbf{E}^{(+)}(\mathbf{r})|\mathrm{i}\rangle|^2. \tag{5.356}$$

In practice, the final state $|f\rangle$ of the radiation field is rarely, if ever, observed; it then is necessary to sum the transition probability over all possible final states to which transitions from $|i\rangle$ are allowed. Using an argument we invoked previously (Section 5.8), the set of states $|f\rangle$ may be regarded as complete since one may always add states to which transitions are not allowed in order to satisfy the completeness requirement. Then

$$\sum_{f}|\langle f|\mathbf{E}^{(+)}(\mathbf{r})|i\rangle|^2 = \sum_{f}\langle i|\mathbf{E}^{(-)}(\mathbf{r})|f\rangle\cdot\langle f|\mathbf{E}^{(+)}(\mathbf{r})|i\rangle$$
$$= \langle i|\mathbf{E}^{(-)}(\mathbf{r})\cdot\mathbf{E}^{(+)}(\mathbf{r})|i\rangle. \tag{5.357}$$

Most often the initial state is a mixed state and cannot be prepared with infinite precision; it is more realistic to resort to average values. We therefore regard the transition probability or the output of the one-atom detector to be proportional to

$$\{\langle i|\mathbf{E}^{(-)}(\mathbf{r})\cdot\mathbf{E}^{(+)}(\mathbf{r})|i\rangle\}_{av} = \sum_{i} P_i\langle i|\mathbf{E}^{(-)}(\mathbf{r})\cdot\mathbf{E}^{(+)}(\mathbf{r})|i\rangle$$
$$= \mathrm{Tr}\{\rho\mathbf{E}^{(-)}(\mathbf{r})\cdot\mathbf{E}^{(+)}(\mathbf{r})\}, \tag{5.358}$$

consistent with the definition of the ensemble average (Eq. (2.22)). Note, incidentally, that the operators in Eq. (5.358) appear in normal order and the reason they do so is a direct consequence of having chosen a photon detector that annihilates photons. Transforming to the Heisenberg representation allows this expression to be generalized to include dependence on time; also, to conform to previous usage, we will consider the electric field operators as scalar quantities (or single components of a vector field). We then define

$$G^{(1)}(x,x) = \mathrm{Tr}\{\rho E_{\mathrm{H}}^{(-)}(x)E_{\mathrm{H}}^{(+)}(x)\}, \qquad (x) = (\mathbf{r}t), \tag{5.359}$$

which is a special case of the more general first-order correlation function

$$G^{(1)}(x_1,x_2) = \mathrm{Tr}\{\rho E_{\mathrm{H}}^{(-)}(x_1)E_{\mathrm{H}}^{(+)}(x_2)\}. \tag{5.360}$$

Since $E_{\mathrm{H}}^{(-)}$ and $E_{\mathrm{H}}^{(+)}$ do not commute, the order in which they appear in the correlation function is significant—in fact, the operators must be written in normal order (Section 4.3)—whereas in the classical case, the order in which the fields are written in a product such as $\langle E^*(x_1)E(x_2)\rangle$ is immaterial.

We now define the first-order coherence function by analogy with the classical expression of Eq. (5.327)

$$g^{(1)}(x_1,x_2) = \frac{G^{(1)}(x_1,x_2)}{[G^{(1)}(x_1,x_1)G^{(1)}(x_2,x_2)]^{1/2}}, \tag{5.361}$$

where

$$G^{(1)}(x_1, x_1) = \langle E_{\mathrm{H}}^{(-)}(x_1) E_{\mathrm{H}}^{(+)}(x_1)\rangle$$
$$= \mathrm{Tr}\{\rho E_{\mathrm{H}}^{(-)}(x_1) E_{\mathrm{H}}^{(+)}(x_1)\}, \tag{5.362a}$$
$$G^{(1)}(x_2, x_2) = \langle E_{\mathrm{H}}^{(-)}(x_2) E_{\mathrm{H}}^{(+)}(x_2)\rangle$$
$$= \mathrm{Tr}\{\rho E_{\mathrm{H}}^{(-)}(x_2) E_{\mathrm{H}}^{(+)}(x_2)\}, \tag{5.362b}$$
$$G^{(1)}(x_1, x_2) = \langle E_{\mathrm{H}}^{(-)}(x_1) E_{\mathrm{H}}^{(+)}(x_2)\rangle$$
$$= \mathrm{Tr}\{\rho E_{\mathrm{H}}^{(-)}(x_1) E_{\mathrm{H}}^{(+)}(x_2)\}. \tag{5.362c}$$

$G^{(1)}(x_i, x_i)$ is proportional to the counting rate of an ideal photodetector or the average intensity at the space-time point x_i. The intensity due to the superposition of photons at a point in space is governed by the coherence function $g^{(1)}(x_1, x_2)$ which satisfies the condition

$$0 \leq |g^{(1)}(x_1, x_2)| \leq 1. \tag{5.363}$$

First-order coherence corresponds to $|g^{(1)}(x_1, x_2)| = 1$; partial coherence to $0 < |g^{(1)}(x_1, x_2)| < 1$; and complete incoherence to $|g^{(1)}(x_1, x_2)| = 0$. These relations are identical with Eq. (5.342) for the classical case. Thus, the counting rate of an ideal detector in a Young's interference experiment is proportional to

$$G^{(1)}(x_1, x_1) + G^{(1)}(x_2, x_2) + 2\,\mathrm{Re}\, G^{(1)}(x_1, x_2). \tag{5.364}$$

Several properties of the correlation functions follow directly from the definitions in Eq. (5.362):

$$G^{(1)}(x_2, x_1) = [G^{(1)}(x_1, x_2)]^*. \tag{5.365}$$
$$G^{(1)}(x_1, x_2) \geq 0, \qquad G^{(1)}(x_2, x_2) \geq 0, \tag{5.366}$$
$$G^{(1)}(x_1, x_1) G^{(1)}(x_2, x_2) \geq |G^{(1)}(x_1, x_2)|^2. \tag{5.367}$$

In the last expression, the equality sign corresponds to maximum fringe visibility. To evaluate the correlation functions, we note that

$$E_{\mathrm{H}}^{(+)}(x) = E_{\mathrm{H}}^{(+)}(\mathbf{r}t) = \mathrm{i}Ka\mathrm{e}^{\mathrm{i}(\mathbf{k}\cdot\mathbf{r} - \omega t)},$$
$$E_{\mathrm{H}}^{(-)}(x) = E_{\mathrm{H}}^{(-)}(\mathbf{r}t) = -\mathrm{i}Ka^{\dagger}\mathrm{e}^{-\mathrm{i}(\mathbf{k}\cdot\mathbf{r} - \omega t)}, \tag{5.368}$$
$$K = \sqrt{\frac{\hbar\omega}{2\varepsilon_0 V}}.$$

For a photon number state $|n\rangle$, we have $\langle n|a^\dagger a|n\rangle = n$, $\langle n|a|n\rangle = \langle n|a^\dagger|n\rangle = 0$, or

$$\langle n|E_{\rm H}^{(+)}(x)|n\rangle = \langle n|E_{\rm H}^{(-)}(x)|n\rangle = 0, \tag{5.369a}$$

$$\langle n|E_{\rm H}^{(-)}(x)E_{\rm H}^{(+)}(x)|n\rangle = K^2 n, \tag{5.369b}$$

$$\langle n|E_{\rm H}^{(-)}(x_1)E_{\rm H}^{(+)}(x_2)|n\rangle = K^2 n \exp[\mathrm{i}\{\mathbf{k}\cdot(\mathbf{r}_2-\mathbf{r}_1)-\omega(t_2-t_1)\}]. \tag{5.369c}$$

Therefore,

$$|g^{(1)}(x_1, x_2)| = 1. \tag{5.370}$$

For a coherent state $|\alpha\rangle$,

$$\langle\alpha|a^\dagger a|\alpha\rangle = |\alpha|^2, \qquad \langle\alpha|a|\alpha\rangle = \alpha, \qquad \langle\alpha|a^\dagger|\alpha\rangle = \alpha^*, \tag{5.371}$$

and

$$\begin{aligned}
\langle\alpha|E_{\rm H}^{(-)}(x)|\alpha\rangle &= K\alpha^* \mathrm{e}^{-\mathrm{i}(\mathbf{k}\cdot\mathbf{r}-\omega t)},\\
\langle\alpha|E_{\rm H}^{(+)}(x)|\alpha\rangle &= K\alpha \mathrm{e}^{\mathrm{i}(\mathbf{k}\cdot\mathbf{r}-\omega t)},\\
\langle\alpha|E_{\rm H}^{(-)}(x)E_{\rm H}^{(+)}(x)|\alpha\rangle &= K^2|\alpha|^2,\\
\langle\alpha|E_{\rm H}^{(-)}(x_1)E_{\rm H}^{(+)}(x_2)|\alpha\rangle &= K^2|\alpha|^2 \exp[\mathrm{i}\{\mathbf{k}\cdot(r_2-r_1)-\omega(t_2-t_1)\}],
\end{aligned} \tag{5.372}$$

which yields the same result as Eq. (5.370).

5.13 Higher-Order Coherence Functions, Photon Statistics

The definition (Eq. (5.327)) of the first-order coherence function may be extended formally to second and higher order. Just as the first-order coherence function acquired a fruitful interpretation on the basis of Young's interference experiment, however, it will be shown that the second-order coherence function can be analogously interpreted in terms of the Hanbury Brown-Twiss experiment [41, 42]. There are several versions of the experiment—the arrangement shown in Fig. 5.16 is sufficient for the presentation of the basic ideas. A half-transparent mirror divides a light beam into two equal parts that necessarily have the same intensity fluctuations. The two separate beams are directed toward two photodetectors P_1 and P_2 acting as photon counters. The two counters and the two arms of the apparatus are identical except for the presence of a time-delay device in front of one counter. Thus, when viewed from the source, P_1 and P_2 are superimposed. The outputs from the counters are combined by a coincidence counter, and since the beam in one arm can

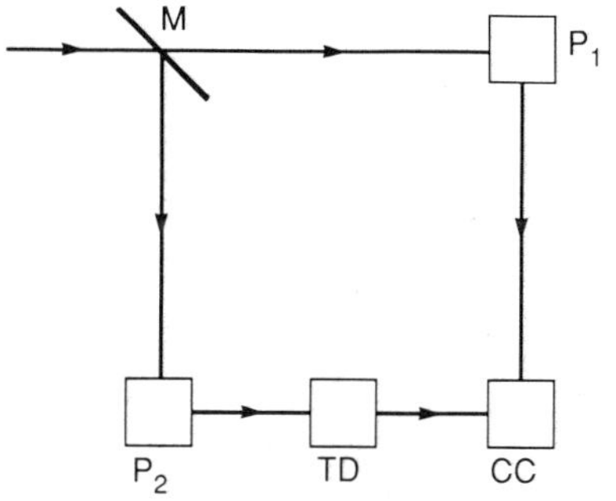

FIGURE 5.16 Schematic diagram of the Hanbury Brown-Twiss experiment. A beam of light entering from the left is split into two equal parts by a semitransparent mirror M. The output from photodetector P_1 proceeds to a coincidence counter CC; the output from photodetector P_2 is delayed by the variable time delay TD and then proceeds to the coincidence counter.

be delayed, one obtains a measurement of the delayed coincidence rate. Specifically, a delayed coincidence is said to occur when one detector registers a count (corresponding to the arrival of a photon) at the time t and the other detector registers a count at the time $t + \tau$. If the delay time τ is smaller than the coherence time τ_c of the intensity fluctuations, such an experiment is capable of providing information on the statistics of the light beam.

Although the apparatus bears some resemblance to the Michelson interferometer, it should be noted that, in the latter, the two separate beams are combined before detection so that their relative phase is preserved. In the Hanbury Brown-Twiss arrangement, on the other hand, the detectors measure intensities of separate beams with the consequent loss of all phase information. We note further that, whereas in Young's experiment, as well as in the Michelson interferometer, the crucial quantity is the correlation in the electric field fluctuations, the crucial quantity in the Hanbury Brown-Twiss experiment is the correlation in the *intensity* fluctuations.

The coincidence counting rate is proportional to

$$R = \lim_{T \to \infty} \frac{1}{2T} \int_{-T}^{T} I(t)I(t + \tau)\, dt, \tag{5.373}$$

where $I(t)$ and $I(t + \tau)$ are the intensities at the two detectors. By our assumption of stationary ergodic fluctuations, the average over time may be equated to the correlation function; thus,

$$R = \langle I(t)I(t + \tau) \rangle. \tag{5.374}$$

Assuming the fields are stationary, R is a function of τ, independent of t. If $\langle I \rangle$ is the same average intensity at each detector, the probability of a delayed coincidence is

$$g^{(2)}(\tau) = \frac{\langle I(t)I(t+\tau)\rangle}{\langle I\rangle^2} = \frac{\langle E^*(t)E^*(t+\tau)E(t+\tau)E(t)\rangle}{\langle E^*(t)E(t)\rangle^2} \qquad (5.375)$$

We take this expression to be the definition of the second-order coherence function.

A more elaborate form of the Hanbury Brown-Twiss experiment permits the counters to be displaced from one another. It then becomes possible to measure correlations in both time and space and the definition of the second-order coherence function may be enlarged to

$$\begin{aligned} g^{(2)}(x_1, x_2; x_2, x_1) &= \frac{\langle I(x_1)I(x_2)\rangle}{\langle I(x_1)\rangle\langle I(x_2)\rangle} \\ &= \frac{\langle E^*(x_1)E^*(x_2)E(x_2)E(x_1)\rangle}{\langle |E(x_1)|^2\rangle\langle |E(x_2)|^2\rangle}. \end{aligned} \qquad (5.376)$$

The order in which the fields are written is immaterial, but nevertheless follows a generally accepted convention. If the two conditions

$$|g^{(1)}(x_1, x_2)| = 1, \qquad g^{(2)}(x_1, x_2; x_2, x_1) = 1, \qquad (5.377)$$

are satisfied simultaneously, the light beam is said to have second-order coherence at x_1 and x_2.

For plane-polarized beams propagating in the z-direction, definition (5.376) reduces to

$$g^{(2)}(x_1, x_2; x_2, x_1) = g^{(2)}(z_1t_1, z_2t_2; z_2t_2, z_1t_1) = g^{(2)}(\tau), \qquad (5.378)$$

where

$$\tau = (t_2 - t_1) - \frac{(z_2 - z_1)}{c}. \qquad (5.379)$$

For a beam whose intensity $I(t) = I(t + \tau) = I_0$ where I_0 is a constant, it is evident that $g^{(2)}(t) = 1$.

To obtain the quantum mechanical second-order coherence function, it is necessary merely to repeat the arguments of the previous section but extend those arguments to the detection (by absorption) of two photons. Thus, in place of Eq. (5.356) we write (for scalar fields)

$$|\langle \mathrm{f}|E^{(+)}(\mathbf{r}_2)E^{(+)}(\mathbf{r}_1)|\mathrm{i}\rangle|^2, \qquad (5.380)$$

which, after summation over the complete set of final states, leads to the extension of Eq. (5.358):

$$\begin{aligned} &\{\langle \mathrm{i}|E^{(-)}(\mathbf{r}_1)E^{(-)}(\mathbf{r}_2)E^{(+)}(\mathbf{r}_2)E^{(+)}(\mathbf{r}_1)|\mathrm{i}\rangle\}_{\mathrm{av}} \\ &\quad = \mathrm{Tr}\{\rho E^{(-)}(\mathbf{r}_1)E^{(-)}(\mathbf{r}_2)E^{(+)}(\mathbf{r}_2)E^{(+)}(\mathbf{r}_1)\}. \end{aligned} \qquad (5.381)$$

The normalized quantum mechanical second-order coherence function then may be defined by

$$g^{(2)}(x_1, x_2; x_2, x_1) = \frac{G^{(2)}(x_1, x_2; x_2, x_1)}{G^{(1)}(x_1, x_1)G^{(1)}(x_2, x_2)}, \tag{5.382}$$

where

$$\begin{aligned} G^{(2)}(x_1, x_2; x_2, x_1) &= \langle E_{\mathrm{H}}^{(-)}(x_1)E_{\mathrm{H}}^{(-)}(x_2)E_{\mathrm{H}}^{(+)}(x_2)E_{\mathrm{H}}^{(+)}(x_1)\rangle \\ &= \mathrm{Tr}\{\rho E_{\mathrm{H}}^{(-)}(x_1)E_{\mathrm{H}}^{(-)}(x_2)E_{\mathrm{H}}^{(+)}(x_2)E_{\mathrm{H}}^{(+)}(x_1)\} \end{aligned} \tag{5.383}$$

is interpreted as the ensemble average of $I(x_1)I(x_2)$ as in the classical case. Note that the ordering of two operators with the same sign in the superscript is unimportant, but that normal order must be preserved as in the first-order function; that is, operators that create photons stand to the left of operators that annihilate photons. Here, too, the basic reason for the appearance of normal ordering is associated with the fact that the experiments on which the correlation functions are based involve detectors that annihilate photons.

The second-order coherence function (Eq. (5.382)), as in the corresponding classical case, may be related to the Hanbury-Brown and Twiss experiment in that $g^{(2)}(x_1, x_2; x_2, x_1)$ is the joint probability that a photon is detected at position $\mathbf{r}_1$ and time t_1 and that another photon is detected at position $\mathbf{r}_2$ and time t_2. At a fixed position in space, $g^{(2)}(x_1, x_2; x_2, x_1) = g^{(2)}(t_1, t_2; t_2, t_1)$, and under stationary conditions $g^{(2)}$ will depend only on the difference $\tau = t_2 - t_1$. We then interpret

$$\begin{aligned} g^{(2)}(\tau) &= \frac{G^{(2)}(t, t+\tau; t+\tau, t)}{G^{(1)}(t, t)G^{(1)}(t+\tau, t+\tau)} \\ &= \frac{\langle E_{\mathrm{H}}^{(-)}(t)E_{\mathrm{H}}^{(-)}(t+\tau)E_{\mathrm{H}}^{(+)}(t+\tau)E^{(+)}(t)\rangle}{\langle E_{\mathrm{H}}^{(-)}(t)E_{\mathrm{H}}^{(+)}(t)\rangle\langle E_{\mathrm{H}}^{(-)}(t+\tau)E_{\mathrm{H}}^{(+)}(t+\tau)\rangle} \end{aligned} \tag{5.384}$$

as the conditional probability that a photon is detected at $t + \tau$ given that a photon had been detected at the time t, i.e., the probability of a delayed coincidence.

On the basis of Eq. (4.121), the quantized field for a single mode may be written as the sum of the two components

$$E_{\mathrm{H}}^{(+)}(x) = \mathrm{f}(x)a, \qquad E_{\mathrm{H}}^{(-)}(x) = \mathrm{f}^*(x)a^\dagger \tag{5.385a}$$

$$\mathrm{f}(x) = \mathrm{i}\sqrt{\frac{\hbar\omega}{2\varepsilon_0 V}}\,\mathrm{e}^{\mathrm{i}(\mathbf{k}\cdot\mathbf{r}-\omega t)}. \tag{5.385b}$$

Then

$$\begin{aligned} G^{(1)}(x_1, x_1) = |\mathrm{f}(x_1|^2\langle a^\dagger a\rangle, \qquad G^{(1)}(x_2, x_2) = |\mathrm{f}(x_2)|^2\langle a^\dagger a\rangle, \\ G^{(2)}(x_1, x_2; x_2, x_1) = |\mathrm{f}(x_1)|^2|\mathrm{f}(x_2)|^2\langle a^\dagger a^\dagger aa\rangle. \end{aligned} \tag{5.386}$$

For a photon number state $|n\rangle$,

$$\langle n|a^\dagger a|n\rangle = n \qquad \langle n|a^\dagger a^\dagger aa|n\rangle = n(n-1); \tag{5.387}$$

thus,

$$g^{(2)}(\tau) = \frac{n-1}{n}\begin{cases} <1 & n \text{ finite} \\ =1 & n \to \infty \end{cases} \tag{5.388}$$

independent of τ.

For a coherent state $|\alpha\rangle$,

$$\langle \alpha|a^\dagger a|\alpha\rangle = |\alpha|^2, \qquad \langle \alpha|a^\dagger a^\dagger aa|\alpha\rangle = |\alpha|^4, \tag{5.389}$$

so that

$$g^{(2)}(\tau) = 1. \tag{5.390}$$

This means that the probability of a delayed coincidence is independent of the delay time. The second photon is equally likely to arrive at the counter at any time, zero to infinity, after the first. Also, it is evident from Eq. (5.389) that, in a coherent state, the second-order correlation function is factorable

$$G^{(2)}(x_1, x_2; x_2, x_1) = G^{(1)}(x_1, x_1)G^{(1)}(x_2, x_2). \tag{5.391}$$

Let us now calculate $G^{(2)}$ for a chaotic field. The method is based on a generating function [38, 43] defined by

$$\mathrm{F}[\xi(x)] = \mathrm{Tr}\left\{\rho \exp\left[\int \xi(x)E_{\mathrm{H}}^{(-)}(x)\, d^4x\right]\exp\left[-\int \xi^*(x)E_{\mathrm{H}}^{(+)}(x)\, d^4x\right]\right\} \tag{5.392}$$

where $d^4x = d\mathbf{r}\, dt$, that is, the four-dimensional differential element in space-time. The functional derivatives of $\mathrm{F}[\xi(x)]$ with respect to $\xi(x)$ and $-\xi^*(x)$ at $\xi(x) = 0$ yield correlation functions to various orders. For example,

$$\left.\frac{\delta^2 \mathrm{F}}{\delta\xi(x_1)\,\delta[-\xi^*(x_2)]}\right|_{\xi(x)=0} = \mathrm{Tr}\{\rho E_{\mathrm{H}}^{(-)}(x_1)E_{\mathrm{H}}^{(+)}(x_2)\}$$
$$= G^{(1)}(x_1, x_2) \tag{5.393}$$

This result is obtained by first differentiating with respect to $\xi(x_1)$; this brings down $E_{\mathrm{H}}^{(-)}(x_1)$. A subsequent differentiation with respect to $-\xi^*(x_2)$ brings down $E_{\mathrm{H}}^{(+)}(x_2)$. Finally, setting $\xi(x) = 0$ yields Eq. (5.593). Following the same procedure we find

$$\left.\frac{\delta^4 \mathrm{F}}{\delta\xi(x_1)\,\delta\xi(x_2)\,\delta[-\xi^*(x_3)]\,\delta[-\xi^*(x_4)]}\right|_{\xi(x)=0}$$
$$= \mathrm{Tr}\{\rho E_{\mathrm{H}}^{(-)}(x_1)E_{\mathrm{H}}^{(-)}(x_2)E_{\mathrm{H}}^{(+)}(x_3)E_{\mathrm{H}}^{(+)}(x_4)\}$$
$$= G^{(2)}(x_1, x_2, x_3, x_4). \tag{5.394}$$

Referring to Eq. (5.385),

$$\int \xi(x)E_{\mathrm{H}}^{(-)}(x)\,d^4x = \int \xi(x)f^*(x)a^\dagger\,d^4x \equiv \lambda a^\dagger,$$

$$\int -\xi^*(x)E_{\mathrm{H}}^{(+)}(x)\,d^4x = \int -\xi^*(x)f(x)a\,d^4x \equiv -\lambda^* a, \tag{5.395}$$

and

$$\mathrm{F}[\xi(x)] = \mathrm{Tr}\{\rho e^{\lambda a^\dagger} e^{-\lambda^* a}\}.$$

It is evident that $\mathrm{F}[\xi(x)]$ is a normally ordered operator similar to $\chi_n(\lambda)$, defined by Eq. (4.186b). We therefore may proceed, as in Eq. (4.252), to write

$$\mathrm{F}[\xi(x)] = \int P(\alpha)e^{\lambda\alpha^* - \lambda^*\alpha}\,d^2\alpha. \tag{5.397}$$

At this stage we invoke the chaotic property of the field by inserting for $P(\alpha)$ the expression obtained in Eq. (4.258). The generating function for a chaotic field then becomes

$$\mathrm{F}[\xi(x)]_{\mathrm{ch}} = \frac{1}{\pi\langle \mathrm{n}\rangle}\int \exp\left[-\frac{|\alpha|^2}{\langle \mathrm{n}\rangle}\right]e^{\lambda\alpha^* - \lambda^*\alpha}\,d^2\alpha. \tag{5.398}$$

From the discussion in Section 4.6, we recognize this expression to be the Fourier transform of a Gaussian. The evaluation is accomplished with the help of Eq. (4.256):

$$\mathrm{F}[\xi(x) = \exp[-\langle \mathrm{n}\rangle|\lambda|^2]. \tag{5.399}$$

But

$$\begin{aligned}
\langle \mathrm{n}\rangle|\lambda|^2 &= \mathrm{Tr}\{\rho a^\dagger a\}\int d^4x\xi(x)f^*(x)\int d^4x'\xi^*(x')f(x')\\
&= \mathrm{Tr}\{\rho\int d^4x\xi(x)f^*(x)a^\dagger\int d^4x'\xi^*(x')f(x')a\}\\
&= \mathrm{Tr}\{\rho\int \xi(x)E_{\mathrm{H}}^{(-)}(x)E_{\mathrm{H}}^{(+)}(x')\xi^*(x')\,d^4x\,d^4x'\}\\
&= \int \xi(x)\,\mathrm{Tr}\{\rho E_{\mathrm{H}}^{(-)}(x)E_{\mathrm{H}}^{(+)}(x')\}\xi^*(x')\,d^4x\,d^4x'\\
&= \int \xi(x)\mathrm{G}^{(1)}(x,x')\xi^*(x')\,d^4x\,d^4x'
\end{aligned} \tag{5.400}$$

and

$$\mathrm{F}[\xi(x)]_{\mathrm{ch}} = \exp\left[-\int \xi(x)G^{(1)}(x,x')\xi^*(x')\,d^4x\,d^4x'\right]. \tag{5.401}$$

This expression may be inserted into Eq. (5.394) to compute the second-order correlation function. Nonvanishing contributions to the derivative are obtained only when $x = x_1$ or x_2 and $x' = x_3$ or x_4. Therefore,

$$\begin{aligned} G^{(2)}&(x_1, x_2; x_2, x_1) \\ &= G^{(1)}(x_1, x_3)G^{(1)}(x_2, x_4) + G^{(1)}(x_1, x_4)G^{(1)}(x_2, x_3) \end{aligned} \tag{5.402}$$

or, upon setting $x_1 = x_3$ and $x_2 = x_4$,

$$\begin{aligned} G^{(2)}&(x_1, x_2; x_2, x_1) \\ &= G^{(1)}(x_1, x_1)G^{(1)}(x_2, x_2) + G^{(1)}(x_1, x_2)G^{(1)}(x_2, x_1) \\ &= G^{(1)}(x_1, x_1)G^{(1)}(x_2, x_2) + |G^{(1)}(x_1, x_2)|^2 \end{aligned} \tag{5.403}$$

where Eq. (5.365) has been invoked. Employing the definition of the first-order coherence function stated in Eq. (5.351),

$$G^{(2)}(x_1, x_2; x_2, x_1) = G^{(1)}(x_1, x_1)G^{(1)}(x_2, x_2)[1 + |g^{(1)}(x_1, x_2)|^2] \tag{5.404}$$

and, from Eq. (5.382), the second-order coherence function for a chaotic beam is

$$g^{(2)}(x_1, x_2; x_2, x_1) = 1 + |g^{(1)}(x_1, x_2)|^2 \geq 1. \tag{5.405}$$

Owing to the fact that $|g^{(1)}(x_1, x_2)|$ is confined to the interval (0, 1) as indicated by Eq. (5.343), the value of $g^{(2)}(x_1, x_2; x_2, x_1)$ lies between 1 and 2. Note the important fact that, for a chaotic beam, the second-order coherence function is determined by the coherence function in first order. It may be shown [36] that the quantum and classical results for a chaotic light beam are the same.

For a beam from a collision-broadened source we have an expression for $g^{(1)}(\tau)$ from Eq. (5.350); thus,

$$g^{(2)}(\tau) = e^{-2|\tau|/\tau_c} + 1, \tag{5.406a}$$

$$g^{(2)}(0) = 2, \qquad g^{(2)}(|\tau| \gg \tau_c) = 1. \tag{5.406b}$$

Similarly, for the Gaussian line shape of Eq. (5.352),

$$g^{(2)}(\tau) = 1 + \exp\left[-\frac{\tau^2}{\tau_c^2}\right] \tag{5.407a}$$

$$g^{(2)}(0) = 2, \qquad g^{(2)}(\tau \gg \tau_c) = 1. \tag{5.407b}$$

The extension of Eq. (5.382) to nth order [37–40, 44] is given by

$$g^{(n)}(x_1,\ldots,x_n;\,x_n,\ldots,x_1)=\frac{G^{(n)}(x_1,\ldots,x_n,\ldots,x_1)}{G^{(1)}(x_1,x_1)\cdots G^{(1)}(x_n,x_n)} \tag{5.408}$$

where the definition of $G^{(n)}(x_1,\ldots,x_n;\,x_n,\ldots x_1)$ is simply an extension of Eq. (5.383). Since intensities and coincidence counting rates are real, positive quantities,

$$G^{(n)}(x_1,\ldots,x_n;\,x_n,\ldots,x_1)\geq 0. \tag{5.409}$$

A radiation field is said to be nth order coherent if

$$|g^{(n)}(x_1,\ldots,x_n;\,x_n,\ldots,x_1)|=1 \tag{5.410}$$

for all $n \geq 1$ and fully coherent if $n \to \infty$. For this to be the case, a necessary and sufficient condition is for $G^{(n)}$ to be factorable, i.e.,

$$G^{(n)}(x_1,\ldots,x_n;\,x_n,\ldots,x_1)=G^{(1)}(x_1,x_1)\cdots G^{(1)}(x_n,x_n), \tag{5.411}$$

which is automatically satisfied by coherent states.

Let us return to the second-order coherence functions (as a function of τ) for different states of light. Our previous discussion indicated that $g^{(2)}(\tau) < 1$ for a light beam in a photon-number state (n finite); $g^{(2)}(\tau) = 1$ in a coherent state; and $g^{(2)}(\tau) = 2(\tau = 0)$, $g^{(2)}(\tau) = 1(\tau \gg \text{coherence time } (\tau_c))$, for the chaotic beam. These cases exemplify three classes of light beams (Fig. 5.17) characterized by

$$\begin{aligned} &g^{(2)}(\tau)=1, && \text{independent of } \tau \\ &g^{(2)}(\tau)=\begin{cases}2, & \tau\to 0\\ 1, & \tau\gg\tau_c\end{cases} \\ &g^{(2)}(\tau)<1, && \text{all } \tau \end{aligned} \tag{5.412}$$

As indicated previously, when $g^{(2)} = 1$, the photons arrive at random, i.e., in an apparatus such as that of Hanbury Brown-Twiss, the coincidence rate is independent of the delay time τ. The counts registered by the detectors are statistically independent and the probability distribution is therefore Poissonian (Section 1.3). A coherent state, such as the state of the radiation from a single-mode laser, fits this description as can be seen from the property of Eq. (4.145). The classical counterpart of the coherent state, we saw, is a classical radiation field without fluctuations (noise-free).

When $g^{(2)}(\tau) > 1$, as in the case of a chaotic (thermal) beam when τ is close to zero ($\tau \ll \tau_c$), the probability of registering two counts within a time interval τ is increased relative to the random case. The photon distribution is then said to be *bunched*. This means that the delayed coincidence rate exhibits a peak

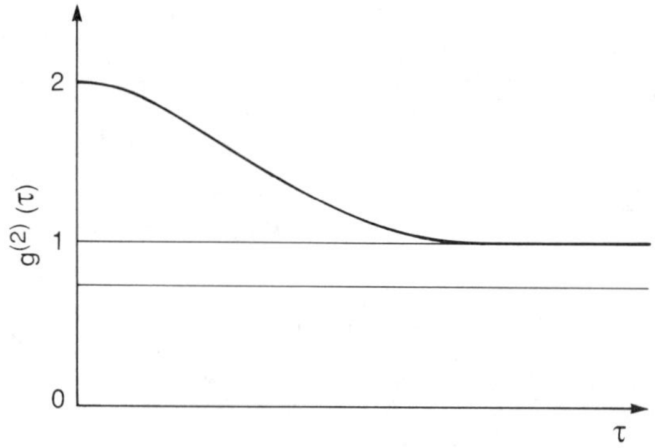

FIGURE 5.17 Classes of light beams distinguished by their second-order coherence function $g^{(2)}(\tau)$ where τ is the delay time in the arrival of successive photons. When $g^{(2)}(\tau) = 1$, independent of τ, photons arrive at random and the coincidence rate in a Hanbury Brown-Twiss experiment is independent of delay time. The counting statistics are Poissonian and correspond to light in a coherent state. When $g^{(2)}(\tau) < 1$ for all τ, the coincidence rate at any value of τ is less than for the random case. The statistics are subpoissonian and the light is said to be antibunched. For a chaotic or thermal beam, $g^{(2)}(\tau) = 2$ when $\tau = 0$ and decreases to 1 when $\tau \gg \tau_c$ is the correlation time. Since the coincidence rate in the vicinity of $\tau \simeq 0$ is greater than for the random case, a chaotic beam is said to be bunched. The beam reverts to the random case with the disappearance of bunching when $\tau \gg \tau_c$.

at $\tau = 0$ and decreases monotonically for increasing τ. After τ appreciably exceeds the coherence time, the fluctuations average to zero and the counting rate approaches a constant value as in the random case.

Finally, when $g^{(2)}(\tau) < 1$, as in the case of a photon-number state with a finite value of n, we have the opposite situation, in which the probability of registering two counts in a time τ is less than for the random case. This is then an *antibunched* distribution [45–50]. A special case of antibunching is one in which $g^{(2)}(\tau) \to 0$ as $\tau \to 0$, implying that the coincidence rate at $\tau = 0$ may be smaller than the rate at $\tau > 0$. An extreme case of antibunching occurs in the emission from a single atom [49]. After the first photon has been emitted (and absorbed by the detector) the atom reverts to the ground state and cannot emit a second photon until it is re-excited. Since the excitation process is not instantaneous but consumes a time of the order of the inverse Rabi frequency, there can be no coincidences for $\tau = 0$. Squeezed states (Section 4.9) also may exhibit antibunching; whether they do or not depends on the specific choice of parameters.

Another way of characterizing these distributions is in terms of an experiment in which a light beam impinges on a single detector. The number of counts in a time interval T is recorded, and the experiment is repeated may times (for the same value of T). Such data determine the probability distribution from which one obtains the difference between the variance $(\Delta n)^2$ of the photon number (or the counts registered by the detector) and the mean $\langle n \rangle$. If Δ is defined as the difference between the variance and the mean, i.e.,

$$\Delta \equiv (\Delta n)^4 - \langle n \rangle = \langle n^2 \rangle - \langle n \rangle^2 - \langle n \rangle, \tag{5.413}$$

the three possible types of distributions are

$$\begin{aligned} &\Delta = 0, \text{ random (or Poissonian) distribution,} \\ &\Delta > 0, \text{ bunched distribution,} \\ &\Delta < 0, \text{ antibunched distribution.} \end{aligned} \tag{5.414}$$

The first two cases—random and bunched—with $g^{(2)}(\tau) \geq 1$, $\Delta \geq 0$, are regarded as classical states, while the antibunched case is nonclassical. The nomenclature stems from the close connection between coherent states and classical fields. The antibunched case, however, with so-called sub-Poissonian statistics, is a strictly quantum mechanical manifestation and has no counterpart in classical physics.

References

[1] E. T. Jaynes and F. W. Cummings, *Proc. IEEE.* **51**, 89(1963).
[2] R. P. Feynman, F. L. Vernon, Jr., and R. W. Hellwarth, *App. Phys.* **28**, (1957).
[3] R. H. Dicke, *Phys. Rev.* **93**, 99(1954)
[4] F. T. Arecchi, E. Courtens, R. G. Gilmore, and H. Thomas, *Phys. Rev. A***6**, 2211(1972).
[5] W. Heitler, *The Quantum Theory of Light.* Oxford Press, Oxford, 1954.
[6] L. Allen and J. H. Eberly, *Optical Resonance and Two-Level Atoms.* J. Wiley, New York, 1975.
[7] E. T. Jaynes, *In Proc. Third Rochester Conf. on Coherence and Quantum Optics.* (L. Mandel and E. Wolf, eds.), Plenum, 1973.
[8] J. J. Sakurai, *Advanced Quantum Mechanics.* Addison-Wesley, Reading, Mass., 1967.
[9] V. Weisskopf and E. Wigner, Z. Physik. **63**, 54(1930).
[10] E. M. Purcell, *Phys. Rev.* **69**, 681(1946).
[11] D. Kleppner, *Phys. Rev. Lett.* **47**, 233(1981).
[12] J. J. Sanchez-Mondragon, N. B. Narozhny, and J. H. Eberly, *Phys. Rev. Lett.* **51,** 550(1983).
[13] R. G. Hulet, E. S. Hilfer and D. Kleppner, Phys. Rev. Lett. **55**, 2137(1985)
[14] S. Haroche, In *Methods of Laser Spectroscopy.* (Y. Prior, A. Ben-Reuven, and M. Rosenbluth eds.), Plenum, 1986.
[15] C. Rempe and H. Walther, *In Methods of Laser Spectroscopy.* (Y. Prior, A Ben-Reuven, and M. Rosenbluth, eds.), Plenum, New York, 1986.
[16] I. R. Senitzky, *Phys. Rev.* **111**, 3(1958).
[17] P. W. Milonni and P. L. Knight, *Phys. Rev.* A**10**, 1096(1974).
[18] G. S. Agarwal, *In Springer Tracts in Modern Physics.* (G. Hohler, ed.), Springer-Verlag, Berlin, 1974.
[19] M. F. H. Schuurmans and D. Polder, *In Laser Spectroscopy IV.* (H. Walther and K. W. Rothe, eds.), Springer-Verlag, Berlin, 1979.
[20] M. F. H. Schuurmans, Q. F. H. Vrehen, D. Polder, and H. M. Gibbs, *Adv. Atom. Mol. Phys.* **17**, 167(1981).

[21] F. W. Cummings and A. Dorri, *Phys. Rev.* **A28**, 2282(1983).
[22] Q. H. F. Vrehen and H. M. Gibbs, *In Dissipative Systems in Quantum Optics.* (R. Bonifacio, ed.), Springer-Verlag, Berlin, 1982.
[23] F. Reiche and W. Thomas, *Naturwiss.* **13**, 627(1925).
[24] W. Kuhn, *Z. Phys.* **33**, 408(1925).
[25] F. Casagrande and L. A. Lugiato *In Quantum Optics.* (C. A. Engelbrecht, ed.), Springer-Verlag, Berlin, 1982.
[26] R. Bonifacio and L. A. Lugiato, *In Dissipative Systems in Quantum Optics.* (R. Bonifacio, ed.), Springer-Verlag, Berlin, 1982.
[27] S. L. McCall and H. M. Gibbs, *In Dissipative Systems in Quantum Optics.* (R. Bonifacio, ed.), Springer-Verlag, Berlin, 1982.
[28] P. D. Drummond and D. F. Walls, *J. Phys. A*: *Math. Gen* **13**, 725(1980).
[29] J. H. Van Vleck and D. L. Huber, *Rev. Mod. Phys.* **49**, 939(1977).
[30] O. S. Mortensen and E. N. Svendsen, *J. Chem. Phys.* **74**, 3185(1981).
[31] J. Behringer, *Mol. Spec.* **2**, 100(1974).
[32] J. Behringer, *J. Raman. Spec.* **2**, 275(1974).
[33] Y. R. Shen, *Phys. Rev.* B**9**, 622(1974).
[34] H. J. Kimble and L. Mandel, *Phys. Rev.* **A13**, 2123(1976).
[35] L. Mandel and E. Wolf, *Rev. Mod. Phys.* **37**, 231(1965).
[36] R. Loudon, *The Quantum Theory of Light.* Second Edition, Clarendon, Oxford, 1983.
[37] R. J. Glauber, *Phys. Rev.* **130**, 2529(1963).
[38] R. J. Glauber, *In Quantum Optics and Electronics.* (C. DeWitt, A. Blandin, and C. Cohen-Tannoudji, eds.), Gordon and Breach, New York, 1965.
[39] R. J. Glauber, *In Quantum Optics.* (R. J. Glauber, ed.), Academic Press, New York, 1969.
[40] R. J. Glauber, *In Laser Handbook.* (F. T. Arecchi and E. O. Schulz-Dubois, eds.), North-Holland, Amsterdam, 1972.
[41] R. Hanbury Brown and R. Q. Twiss, *Proc. Roy. Soc.* A**242**, 300(1957).
[42] R. Hanbury Brown and R. Q. Twiss, *Proc. Roy. Soc.* A**243**, 291(1958).
[43] H. M. Nussenzveig, *Introduction to Quantum Optics.* Gordon and Breach, New York, 1973.
[44] M. Schubert and B. Wilhelmi, *In Progress in Optics XVII.* (E. Wolf, ed.), North-Holland, Amsterdam, 1980.
[45] D. Stoler, *Phys. Rev. Lett.* **33**, 1397(1974).
[46] H. J. Kimble, M. Dagenais, and L. Mandel, *Phys. Rev. Lett.* **39**, 691(1977).
[47] R. Loudon, *Rep. Prog. Phys.* **43**, 913(1980).
[48] H. Paul, *Rev. Mod Phys.* **54**, 1061(1982).
[49] H. J. Carmichael, *Phys. Rev. Lett.* **55**, 2790(1985).
[50] F. T. Arecchi and V. Degiorgio, *In Laser Handbook.* (F. T. Arecchi and E. O. Schulz-Dubois, eds.), North-Holland, Amsterdam, 1972.

VI Reservoir Theory and Damping

In Section 3.5 the equations describing the motion of magnetic moments (or spins) in an external, time-varying magnetic field were modified by the addition of terms containing the relaxation constants T_1 and T_2. This is an example in which a *dynamical system* is influenced by a *heat bath* or *reservoir*. The dynamical system consists of the magnetic moments in an external magnetic field and the reservoir is the physical system with which the dynamical system ultimately comes into equilibrium, for example, the molecules in a gas or liquid or a crystal lattice to which the spins may be coupled. Similarly, in the dynamical system consisting of a set of atoms interacting with a radiation field, the gas molecules with which the atoms collide or a crystal lattice in which the atoms are embedded act as reservoirs. In still another example, the dynamical system consists of excited atoms and the reservoir consists of empty modes of the radiation field (vacuum). The system-reservoir interaction gives rise to spontaneous emission. These examples and many others are manifestations of the general statement that real physical systems do not exist in complete isolation; they are said to be *open* because they are embedded in an environment with which the system interacts to a greater or lesser extent. The process whereby a dynamical system progresses toward equilibrium (steady state) with its reservoir is known as *relaxation* or *damping*. A sudden change in some parameter of the system will disturb the equilibrium; in that event the system relaxes until a new state of equilibrium is achieved.

The dynamical system is characterized by a finite (usually, small) number of degrees of freedom and discrete energy states while the reservoir is a large complex system with many (possibly infinite) degrees of freedom and a broadband quasicontinuous spectrum. To solve the equations of motion for the total system (dynamical system plus reservoir) not only would be difficult, if not impossible, but also would be wasteful in effort since our primary interest is in the dynamical system. The specifics of the reservoir, apart from their influence on the dynamical system, are totally irrelevant. There are two basic methods that attempt to avoid this problem. One is the *phenomenological* approach in which reservoir effects are taken into account by means of one or more empirical constants. We resorted to this method in Section 3.5 to obtain the Bloch equations. The second approach attempts to represent the reservoir by a model capable of describing the effect of the reservoir on the temporal development of the dynamical system as it approaches equilibrium. An essential requirement is that a solution for the detailed behavior of the total system shall not be necessary.

6.1 Reservoir Interaction and the Master Equation

Models for different experimental situations have been constructed; in these models, a common (but not necessarily unique) set of assumptions are the following:

1. The fluctuations in the reservoir are stationary and Markovian.

2. The correlation time τ_c of the reservoir fluctuations in of the order of the mean fluctuation period.

3. Ensemble averages of reservoir operators are zero.

4. The coupling between the dynamical system and the reservoir is sufficiently weak and the reservoir is a sufficiently large system that the effect of the coupling on the reservoir is negligible; it is assumed therefore that the reservoir remains in thermal equilibrium at a constant temperature at all times. The coupling has a stronger effect on the dynamical system, however, and ultimately brings it into equilibrium with the reservoir.

5. The reservoir induces Markovian fluctuations in the dynamical system.

6. The relaxation time constant $1/\gamma$, which governs the rate at which the dynamical system approaches equilibrium with the reservoir, is much longer than the correlation time τ_c.

The results that emerge from this model are embodied in a set of equations called the *master equations.* They describe the time development of the density operator associated with the dynamical system under the influence of the reservoir [1–9]. Specifically, for our purpose, one of the important applications of the master equations arises in the investigation of damping when light interacts with matter. We shall find the Langevin and Fokker-Planck equations appearing in these discussions (Sections 6.7 and 6.8) in close analogy with their role in Brownian motion.

We now shall derive the relevant equations by means of the reduced density matrix method described in Section 2.11. Consider a dynamical system (S) with Hamiltonian $\mathscr{H}_s$ and a reservoir (R) with Hamiltonian $\mathscr{H}_r$. Assuming that S is initially prepared in a stationary, nonequilibrium state and is uncoupled from R, the total Hamiltonian is $\mathscr{H} = \mathscr{H}_s + \mathscr{H}_r$. If the coupling between S and R is "turned on" at $t = 0$, the total Hamiltonian for $t > 0$ becomes

$$\mathscr{H} = \mathscr{H}_s + \mathscr{H}_r + V = \mathscr{H}_0 + V, \tag{6.1}$$

where V represents the coupling between the dynamical system and the reservoir. For the total system (R coupled to S), the temporal development of the density operator $\tilde{\rho}$ in the interaction representation is determined by Eq. (2.168)

$$\dot{\tilde{\rho}}(t) = -\frac{i}{\hbar}[\tilde{V}(t), \rho(0)] - \frac{1}{\hbar^2}\int_0^t dt'[\tilde{V}(t), [\tilde{V}(t'), \tilde{\rho}(t')]]. \tag{6.2}$$

Since S is a subsystem of the total system RS, the density operator for the dynamical system is obtained by a trace operation over the reservoir variables as in Eq. (2.364), i.e.,

$$\tilde{\rho}_s(t) = \mathrm{Tr}_r\, \tilde{\rho}(t). \tag{6.3}$$

To proceed with the computation it is necessary to provide a specific form for the coupling term V. Previously, we encountered interactions of the form $-\mathbf{m}\cdot\mathbf{B}$ (Eq. (3.36)) in the magnetic case and $-\mathbf{d}\cdot\mathbf{E}$ (Eq. (5.34)) in the atomic case. These forms suggest that a sufficiently general form for V may be written

$$V = \sum_i S_i R_i, \tag{6.4}$$

where S_i are operators belonging to the dynamical system and R_i are operators belonging to the reservoir. It is further assumed that all S_i commute with all R_i but not necessarily among themselves. Combining Eq. (6.2) with Eq. (6.4) we have

$$\dot{\tilde{\rho}}_s(t) = \mathrm{Tr_r}\,\dot{\tilde{\rho}}(t) = -\frac{i}{\hbar}\mathrm{Tr_r}\left[\sum_i \tilde{S}_i(t)\tilde{R}_i(t), \rho(0)\right]$$
$$-\frac{1}{\hbar^2}\mathrm{Tr_r}\int_0^t dt'\left[\sum_i \tilde{S}_i(t)\tilde{R}_i(t), \left[\sum_j \tilde{S}_j(t')\tilde{R}_j(t'), \tilde{\rho}(t')\right]\right], \tag{6.5}$$

where

$$\tilde{S}_i(t) = e^{i\mathscr{H}_0 t/\hbar} S_i e^{-i\mathscr{H}_0 t/\hbar} = e^{i\mathscr{H}_s t/\hbar} S_i e^{-i\mathscr{H}_s t/\hbar},$$
$$\tilde{R}_i(t) = e^{i\mathscr{H}_0 t/\hbar} R_i e^{-i\mathscr{H}_0 t/\hbar} = e^{i\mathscr{H}_r t/\hbar} R_i e^{-i\mathscr{H}_r t/\hbar}. \tag{6.6}$$

Assuming that the interaction between R and S is turned on at $t = 0$, then according to our previous discussion (Section 2.11) and Eq. (2.349), the density operator $\rho(0)$ may be factored into

$$\rho(0) = \rho_s(0)\rho_r(0). \tag{6.7}$$

But the reservoir has been assumed to remain in thermal equilibrium at all times; hence,

$$\rho_r(0) \equiv \rho_r = \frac{e^{-\beta\mathscr{H}_r}}{Z}, \tag{6.8}$$

as in Eq. (2.226). The first term in Eq. (6.5) now may be evaluated

$$\mathrm{Tr_r}\left[\sum_i \tilde{S}_i(t)\tilde{R}_i(t), \rho(0)\right]$$
$$= \mathrm{Tr_r}\sum_i [\tilde{S}_i(t)\tilde{R}_i(t)\rho_s(0)\rho_r - \rho_s(0)\rho_r\tilde{S}_i(t)\tilde{R}_i(t)]$$
$$= \sum_i [\tilde{S}_i(t)\rho_s(0)\mathrm{Tr_r}\{\tilde{R}_i(t)\rho_r\} - \rho_s(0)\tilde{S}_i(t)\mathrm{Tr_r}\{\rho_r\tilde{R}_i(t)\}]. \tag{6.9}$$

But

$$\mathrm{Tr_r}\{\tilde{R}_i(t)\rho_r\} = \mathrm{Tr_r}\{\rho_r\tilde{R}_i(t)\} = \langle\tilde{R}_i(t)\rangle = 0 \tag{6.10}$$

as a result of the rapid fluctuations in the reservoir (Assumption 3). Therefore, the first term in Eq. (6.5) vanishes.

Turning now to the second term of Eq. (6.5), a first approximation may be obtained by replacing $\tilde{\rho}(t')$ by $\rho_s(0)\rho_r$; the Markovian character of the fluctuations (Assumption 1), however, suggests that a better approximation is achieved by writing

$$\tilde{\rho}(t') = \tilde{\rho}_s(t)\rho_r, \tag{6.11}$$

since then $\dot{\tilde{\rho}}_s(t)$ will depend on the present value $\tilde{\rho}_s(t)$ rather than on $\tilde{\rho}_s(t')$ where t' varies from 0 to ∞. After expanding the double commutator in

(Eq. (6.5)) we now have

$$\dot{\tilde{\rho}}_s(t) = -\frac{1}{\hbar^2}\int_0^t dt' \sum_{ij}[\tilde{S}_i(t)\tilde{S}_j(t')\tilde{\rho}_s(t)\mathrm{Tr_r}\{\tilde{R}_i(t)\tilde{R}_j(t')\rho_r\}$$
$$- \tilde{S}_i(t)\tilde{\rho}_s(t)\tilde{S}_j(t')\mathrm{Tr_r}\{\tilde{R}_i(t)\rho_r\tilde{R}_j(t')\}$$
$$- \tilde{S}_j(t')\tilde{\rho}_s(t)\tilde{S}_i(t)\mathrm{Tr_r}\{\tilde{R}_j(t')\rho_r\tilde{R}_i(t)\}$$
$$+ \tilde{\rho}_s(t)\tilde{S}_j(t')\tilde{S}_i(t)\mathrm{Tr_r}\{\rho_r\tilde{R}_j(t')\tilde{R}_i(t)\}], \tag{6.12}$$

which may be simplified by (a) recalling the invariance property of the trace to cyclic permutations of the operators, (b) replacing each trace by an ensemble average, and (c) invoking the assumption that the fluctuations are stationary. Then, with

$$\tau = t - t', \qquad \tilde{R}_i(0) = R_i, \qquad \tilde{R}_j(0) = R_j,$$
$$\mathrm{Tr_r}\{\tilde{R}_i(t)\tilde{R}_j(t')\rho_r = \langle\tilde{R}_i(t)\tilde{R}_j(t')\rangle = \langle\tilde{R}_i(\tau)R_j\rangle, \tag{6.13a}$$
$$\mathrm{Tr_r}\{\tilde{R}_i(t)\rho_r\tilde{R}_j(t')\} = \mathrm{Tr_r}\{\tilde{R}_j(t')\tilde{R}_i(t)\rho_r\}$$
$$= \langle\tilde{R}_j(t')\tilde{R}_i(t)\rangle = \langle R_j\tilde{R}_i(\tau)\rangle, \tag{6.13b}$$
$$\mathrm{Tr_r}\{\tilde{R}_j(t')\rho_r\tilde{R}_i(t)\} = \mathrm{Tr_r}\{\tilde{R}_i(t)\tilde{R}_j(t')\rho_r\}$$
$$= \langle\tilde{R}_i(t)\tilde{R}_j(t')\rangle = \langle\tilde{R}_i(\tau)R_j\rangle, \tag{6.13c}$$
$$\mathrm{Tr_r}\{\rho_r\tilde{R}_j(t')\tilde{R}_i(t)\} = \langle\tilde{R}_j(t')\tilde{R}_i(t)\rangle = \langle R_j\tilde{R}_i(\tau)\rangle. \tag{6.13d}$$

With these expressions, Eq. (6.12) now reads

$$\dot{\tilde{\rho}}_s(t) = -\frac{1}{\hbar^2}\sum_{ij}\int_0^t d\tau([\tilde{S}_i(t), \tilde{S}_j(t-\tau)\tilde{\rho}_s(t)]\langle\tilde{R}_i(\tau)R_j\rangle$$
$$- [\tilde{S}_i(t), \tilde{\rho}_s(t)\tilde{S}_j(t-\tau)]\langle R_j\tilde{R}_i(\tau)\rangle). \tag{6.14}$$

The assumption that the dynamical system and the reservoir are characterized by two widely different time constants, $1/\gamma$ and τ_c, respectively, with $1/\gamma \gg \tau_c$, means that during a time interval on the order of τ_c, the dynamical system remains essentially unchanged. But the correlation functions are nonvanishing only during a time interval on the order of τ_c. Therefore, to a good approximation, the upper limit may be extended to infinity, resulting in the definition of the *coarse-grained derivative*

$$\dot{\tilde{\rho}}_s(t) = -\frac{1}{\hbar^2}\sum_{ij}\int_0^\infty d\tau([\tilde{S}_i(t), \tilde{S}_j(t-\tau)\tilde{\rho}_s(t)]\langle\tilde{R}_i(\tau)R_j\rangle$$
$$- [\tilde{S}_i(t), \tilde{\rho}_s(t)\tilde{S}_j(t-\tau)]\langle R_j\tilde{R}_i(\tau)\rangle). \tag{6.15}$$

This is a fundamental relation; it shows explicitly that the effect of the reservoir on the dynamical system is contained entirely in the correlation functions of the reservoir operators. All other operators that appear in the expression belong to the dynamical system.

The next step in the development of the master equations requires the evaluation of matrix elements of the terms in Eq. (6.15). In a basis set consisting of the eigenstates of $\mathscr{H}_s$, we may write

$$\begin{aligned}\langle k|\tilde{S}_i(t)\tilde{S}_j(t-\tau)\tilde{\rho}_s(t)|l\rangle &= \sum_{nm}\langle k|\tilde{S}_i(t)|n\rangle\langle n|\tilde{S}_j(t-\tau)|m\rangle\langle m|\tilde{\rho}_s(\mathrm{t}|l\rangle \\ &= \sum_{nm}\langle k|S_i|n\rangle\langle n|S_j|m\rangle\langle m|\tilde{\rho}_s(t)|l\rangle e^{i\omega_{kn}t}e^{i\omega_{nm}(t-\tau)}. \quad (6.16)\end{aligned}$$

In the second equation, $\tilde{S}_i(t)$ and $\tilde{S}_j(t-\tau)$ have been converted to the Schrödinger representation where, for example,

$$\begin{aligned}\langle k|\tilde{S}_i(t)|n\rangle &= \langle k|e^{i\mathscr{H}_s t/\hbar}S_i e^{-i\mathscr{H}_s t/\hbar}|n\rangle \\ &= \langle k|S_i|n\rangle e^{i(E_k-E_n)t/\hbar} \\ &= \langle k|S_i|n\rangle e^{i\omega_{kn}t}. \quad (6.17)\end{aligned}$$

Equivalent forms of Eq. (6.16) are given by

$$\begin{aligned}&\langle k|\tilde{S}_i(t)\tilde{S}_j(t-\tau)\tilde{\rho}_s(t)|l\rangle \\ &= \sum_{nm}\langle k|S_i|n\rangle\langle n|S_j|m\rangle\langle m|\tilde{\rho}_s(t)|l\rangle e^{i\omega_{km}t}e^{-i\omega_{nm}\tau} \quad (6.18a) \\ &= \sum_{nm}\langle m|\tilde{\rho}_s(t)|n\rangle\,\delta_{ln}\sum_r\langle k|S_i|r\rangle\langle r|S_j|m\rangle e^{-i\omega_{rm}\tau}e^{i(\omega_{km}+\omega_{nl})t}. \quad (6.18b)\end{aligned}$$

Following the same procedure,

$$\begin{aligned}&\langle k|\tilde{S}_j(t-\tau)\tilde{\rho}_s(t)\tilde{S}_i(t)|l\rangle \quad (6.19a) \\ &= \sum_{nm}\langle k|S_j|n\rangle\langle n|\tilde{\rho}_s(t)|m\rangle\langle m|S_i|l\rangle e^{i(\omega_{kn}+\omega_{ml})t}e^{-i\omega_{kn}\tau} \quad (6.19b) \\ &= \sum_{nm}\langle k|S_j|m\rangle\langle m|\rho_s(t)|n\rangle\langle n|S_i|l\rangle e^{i(\omega_{km}+\omega_{nl})t}e^{-i\omega_{km}\tau},\end{aligned}$$

in which the last equation was obtained by an interchange of the indices m and n. Combining Eq. (6.18b) and (6.19b) the matrix element of the first commutator in Eq. (6.15) is

$$\begin{aligned}\langle k|[\tilde{S}_i(t),\tilde{S}_j(t-\tau)\tilde{\rho}_s(t)]|l\rangle = \sum_{nm}\langle m|\tilde{\rho}_s(t)|n\rangle\Bigg(\delta_{ln}\sum_r\langle k|S_i|r\rangle\langle r|S_j|m\rangle e^{-i\omega_{rm}\tau} \\ - \langle n|S_i|l\rangle\langle k|S_j|m\rangle e^{-i\omega_{km}\tau}\Bigg)e^{i(\omega_{km}+\omega_{nl})t}. \quad (6.20)\end{aligned}$$

The second commutator in Eq. (6.15) is evaluated in similar fashion

$$\langle k|[\tilde{S}_i(t), \tilde{\rho}_s(t)\tilde{S}_j(t-\tau)]|l\rangle = \sum_{nm} \langle m|\tilde{\rho}_s(t)|n\rangle \Big(\langle n|\, S_j|l\rangle\langle k|S_i|m\rangle e^{-i\omega_{nl}\tau}$$

$$- \delta_{km} \sum_r \langle n|S_j|r\rangle\langle r|S_i|l\rangle e^{-i\omega_{nr}\tau} \Big) e^{i(\omega_{km}+\omega_{nl})\tau}. \tag{6.21}$$

The two expressions (Eqs. (6.20) and (6.21)) may be inserted into Eq. (6.15) to obtain the matrix element of the density operator, but before doing so it is convenient to introduce the notation

$$\Gamma^+_{nlkm} = \frac{1}{\hbar^2} \sum_{ij} \langle n|S_i|l\rangle\langle k|S_j|m\rangle \int_0^\infty d\tau e^{-i\omega_{km}\tau} \langle \tilde{R}_i(\tau) R_j\rangle, \tag{6.22a}$$

$$\Gamma^-_{nlkm} = \frac{1}{\hbar^2} \sum_{ij} \langle n|S_j|l\rangle\langle k|S_i|m\rangle \int_0^\infty d\tau e^{-i\omega_{nl}\tau} \langle R_j \tilde{R}_i(\tau)\rangle. \tag{6.22b}$$

Putting it all together,

$$\langle k|\dot{\tilde{\rho}}_s(t)|l\rangle = \sum_{nm} \langle m|\tilde{\rho}_s(t)|n\rangle \Big(-\delta_{ln} \sum_r \Gamma^+_{krrm} + \Gamma^+_{nlkm}$$

$$+ \Gamma^-_{nlkm} - \delta_{km} \sum_r \Gamma^-_{nrrl} \Big) e^{i(\omega_{km}+\omega_{nl})t} \tag{6.23a}$$

$$= \sum_{nm} \langle m|\tilde{\rho}_s(t)|n\rangle R_{klmn} e^{i(\omega_{km}+\omega_{nl})t}, \tag{6.23b}$$

where the *relaxation matrix* R_{klmn} is defined by

$$R_{klmn} = -\delta_{ln} \sum_r \Gamma^+_{krrm} + \Gamma^+_{nlkm} + \Gamma^-_{nlkm} - \delta_{km} \sum_r \Gamma^-_{nrrl}. \tag{6.24}$$

In the secular approximation, the high frequency terms are eliminated, keeping only those for which

$$\omega_{km} + \omega_{nl} = 0, \tag{6.25}$$

namely, when

$$\begin{aligned} &\text{a.} \quad k = m, \qquad l = n, \qquad k \neq l, \\ &\text{b.} \quad k = l, \qquad m = n, \qquad k \neq n, \\ &\text{c.} \quad k = l = n = m. \end{aligned} \tag{6.26}$$

The equations that arise as a result of the three conditions are

$$\text{a.} \quad \langle k|\dot{\tilde{\rho}}_s(t)|l\rangle = \langle k|\tilde{\rho}_s(t)|l\rangle \Big(-\sum_r \Gamma^+_{krrk} + \Gamma^+_{llkk} + \Gamma^-_{llkk} - \sum_r \Gamma^-_{lrrl} \Big), \tag{6.27}$$

b. $$\langle k|\dot{\tilde{\rho}}_s(t)|k\rangle = \sum_{n\neq k} \langle n|\tilde{\rho}_s(t)|n\rangle(\Gamma^+_{nkkn} + \Gamma^-_{nkkn}), \tag{6.28}$$

c. $$\langle k|\dot{\tilde{\rho}}_s(t)|k\rangle = \langle k|\rho_s(t)|k\rangle\left(-\sum_r \Gamma^+_{krrk} + \Gamma_{kkkk} + \Gamma^-_{kkkk} - \sum_r \Gamma^-_{krrk}\right). \tag{6.29}$$

Equation (6.29) is a special case of Eq. (6.27) when $k = l$, so it need not be considered separately. Equation (6.28) can be written

$$\langle k|\dot{\tilde{\rho}}_s(t)|l\rangle = \delta_{kl} \sum_{n\neq k} \langle n|\tilde{\rho}_s(t)|n\rangle(\Gamma^+_{nkkn} + \Gamma^-_{nkkn}), \tag{6.30}$$

which then may be combined with Eq. (6.27) to give

$$\langle k|\dot{\tilde{\rho}}_s(t)|l\rangle = \delta_{kl} \sum_{n\neq k} \langle n|\tilde{\rho}_s(t)|n\rangle(\Gamma^+_{nkkn} + \Gamma^-_{nkkn})$$
$$+ \langle k|\tilde{\rho}_s(t)|l\rangle\left(-\sum_r \Gamma^+_{krrk} + \Gamma^+_{llkk} + \Gamma^-_{llkk} - \sum_r \Gamma^-_{lrrl}\right). \tag{6.31}$$

We now define

$$W_{kn} = \Gamma^+_{nkkn} + \Gamma^-_{nkkn}, \tag{6.32}$$

$$\Gamma_{kl} = \sum_r (\Gamma^+_{krrk} + \Gamma^-_{lrrl}) - \Gamma^+_{llkk} - \Gamma^-_{llkk}. \tag{6.33}$$

Then

$$\langle k|\dot{\tilde{\rho}}_s(t)|l\rangle = \delta_{kl} \sum_{n\neq k} \langle n|\tilde{\rho}_s(t)|n\rangle W_{kn} - \Gamma_{kl}\langle k|\tilde{\rho}_s(t)|l\rangle. \tag{6.34}$$

Since the first term is multiplied by δ_{kl}, we may insert freely the factor $\exp(i\omega_{kl}t)$. Upon conversion of the right side to the Schrödinger representation,

$$\langle k|\dot{\tilde{\rho}}_s(t)|l\rangle = \delta_{kl} \sum_{n\neq k} \langle n|\rho_s(t)|n\rangle W_{kn} e^{i\omega_{kl}t} - \Gamma_{kl}\langle k|\rho_s(t)|l\rangle e^{i\omega_{kl}t}. \tag{6.35}$$

Finally, the conversion of the left side to the Schrödinger representation, accomplished with the help of Eq. (2.165), yields

$$\langle k|\dot{\tilde{\rho}}_s(t)|l\rangle = -\frac{i}{\hbar}\langle k|[\mathscr{H}_0, \rho_s(t)]|l\rangle + \langle k|e^{-i\mathscr{H}_0 t/\hbar}\dot{\tilde{\rho}}_s(t)e^{i\mathscr{H}_0 t/\hbar}|l\rangle \tag{6.36a}$$

$$= -\frac{i}{\hbar}\langle k|[\mathscr{H}_0, \rho_s(t)]|l\rangle + \langle k|\dot{\tilde{\rho}}_s(t)|l\rangle e^{-i\omega_{kl}t}. \tag{6.36b}$$

We now have the *master equation*

$$\langle k|\dot{\rho}_s(t)|l\rangle = -\frac{i}{\hbar}\langle k|[\mathscr{H}_s, \rho_s(t)]|l\rangle + \delta_{kl} \sum_{n\neq k} \langle n|\rho_s(t)|n\rangle W_{kn} - \Gamma_{kl}\langle k|\rho_s(t)|l\rangle. \tag{6.37}$$

The first term pertains to the motion of the dynamical system without coupling to the reservoir (free motion) while the second and third terms are consequences of the reservoir interactions. For diagonal elements, the first term vanishes and Γ_{kk}, according to the definitions (6.32) and (6.33), reduces

$$\begin{aligned}\Gamma_{kk} &= \sum_r (\Gamma^+_{krrk} + \Gamma^-_{krrk}) - \Gamma^+_{kkkk} - \Gamma^-_{kkkk} \\ &= \sum_{r|k} (\Gamma^+_{krrk} + \Gamma^-_{krrk}) = \sum_{n\neq k} (\Gamma^+_{knnk} + \Gamma^-_{knnk}) \\ &= \sum_{n\neq k} W_{nk}. \end{aligned} \tag{6.38}$$

When the terms corresponding to the free motion are omitted,

$$\begin{aligned}\langle k|\dot{\rho}_s(t)|k\rangle &= \sum_{n\neq k} (\langle n|\rho_s(t)|n\rangle W_{kn} - \langle k|\rho_s(t)|k\rangle W_{nk}) \\ &= -\langle k|\rho_s(t)|k\rangle \Gamma_{kk} + \sum_{n\neq k} \langle n|\rho_s(t)|n\rangle W_{kn}. \end{aligned} \tag{6.39}$$

This is the master equation for diagonal matrix elements of the density operator (also known as the Pauli master equation) for a dynamical system coupled to a reservoir. The interpretation of Eq. (6.39) follows immediately from the recognition that a matrix element such as $\langle k|\rho_s(t)|k\rangle$ represents the probability of finding the dynamical system in the eigenstate $|k\rangle$ (or the population in the state $|k\rangle$). The terms $\langle n|\rho_s(t)|n\rangle W_{kn}$ contribute to an increase in $\langle k|\rho_s(t)|k\rangle$; therefore, W_{kn} is interpretable as the probability per unit time for a transition from $|n\rangle$ to $|k\rangle$. Similarly, the terms $\langle k|\rho_s(t)|k\rangle W_{nk}$, which appear with a minus sign, act in the reverse direction to produce a decrease in the population of $|k\rangle$; hence, W_{nk} is the probability per unit time for a transition $|k\rangle$ to $|n\rangle$. Thus,

$$W_{kn} = W(n \to k), \qquad W_{nk} = W(k \to n), \tag{6.40}$$

and Γ_{kk}, defined by Eq. (6.38), is the population decay constant associated with the lifetime of the state $|k\rangle$. For the off-diagonal matrix elements in Eq. (6.37), without the free motion,

$$\langle k|\dot{\rho}_s(t)|l\rangle = -\Gamma_{kl}\langle k|\rho_s(t)|l\rangle. \tag{6.41}$$

Thus, Γ_{kl} is simply a damping or relaxation constant, which will later be identified with phase relaxation.

In summary, the model we have adopted for a dynamical system coupled to a reservoir leads to master equations which describe the time-rate of change of the density matrix elements associated with the dynamical system in terms of easily interpretable constants whose dependence on the reservoir is entirely contained in the correlation functions of the reservoir fluctuations. The master equations are used widely in discussions of relaxation and damping effects, population fluctuations, atomic correlations, photon statistics, and other phenomena involving statistical considerations [11].

For subsequent calculations, we adopt the notation

$$\rho(t) = \rho_s(t), \tag{6.42}$$

$$\left(\frac{\partial \rho_{kk}}{\partial t}\right)_{\text{damp}} = -\Gamma_{kk}\rho_{kk} + \sum_{n\neq k} \rho_{nn} W_{kn}, \tag{6.43a}$$

$$\left(\frac{\partial \rho_{kl}}{\partial t}\right)_{\text{damp}} = -\Gamma_{kl}\rho_{kl}. \tag{6.43b}$$

As in Eqs. (6.39) and (6.41), the equation of motion for the density operator is then written

$$\dot{\rho}(t) = -\frac{i}{\hbar}[\mathscr{H}_0 + V(t), \rho(t)] + \left(\frac{\partial \rho}{\partial t}\right)_{\text{damp}}, \tag{6.44}$$

in which $\mathscr{H}_0 + V(t)$ is the Hamiltonian of the system without reservoir interactions and $(\partial\rho/\partial t)_{\text{damp}}$ contains the reservoir effects. The general solution to Eq. (6.44) can be expanded into a series

$$\rho(t) = \rho^{(0)}(t) + \rho^{(1)}(t) + \rho^{(2)}(t) + \cdots \tag{6.45}$$

where

$$\begin{aligned}
i\hbar\dot{\rho}^{(0)} &= [\mathscr{H}_0, \rho^{(0)}] + i\hbar\left(\frac{\partial \rho^{(0)}}{\partial t}\right)_{\text{damp}},\\
i\hbar\dot{\rho}^{(1)} &= [\mathscr{H}_0, \rho^{(1)}] + [V(t), \rho^{(0)}] + i\hbar\left(\frac{\partial \rho^{(1)}}{\partial t}\right)_{\text{damp}},\\
&\;\vdots\\
i\hbar\dot{\rho}^{(n)} &= [\mathscr{H}_0, \rho^{(n)}] + [V(t), p^{(n-1)}] + i\hbar\left(\frac{\partial \rho^{(n)}}{\partial t}\right)_{\text{damp}}.
\end{aligned} \tag{6.46}$$

The equations for the matrix elements in the notation of (6.43) are

$$\begin{aligned}
i\hbar\dot{\rho}_{kl}^{(0)} &= [\mathscr{H}_0, \rho^{(0)}]_{kl} - i\hbar\Gamma_{kl}\rho_{kl}^{(0)},\\
i\hbar\dot{\rho}_{kl}^{(1)} &= [\mathscr{H}_0, \rho^{(1)}]_{kl} + [V(t), \rho^{(0)}]_{kl} - i\hbar\Gamma_{kl}\rho_{kl}^{(1)},\\
&\;\vdots\\
i\hbar\dot{\rho}_{kl}^{(n)} &= [\mathscr{H}_0, \rho^{(n)}]_{kl} + [V(t), \rho^{(n-1)}]_{kl} - i\hbar\Gamma_{kl}\rho_{kl}^{(n)}.
\end{aligned} \tag{6.47}$$

When spontaneous emission (Section 5.6) is the only mechanism by which the system tends to return to equilibrium, the damping constants are related by

$$\Gamma_{kl} = \frac{(\Gamma_{kk} + \Gamma_{ll})}{2}, \tag{6.48}$$

where Γ_{kk} and Γ_{ll} are known as *population decay constants* and are related to the lifetimes of the states $|k\rangle$ and $|l\rangle$ by

$$\tau_k = \frac{\pi}{\Gamma_{kk}}, \qquad \tau_1 = \frac{\pi}{\Gamma_{ll}}. \tag{6.49}$$

(Another approach to damping due to spontaneous emission is discussed in Section 6.9.) In the presence of additional damping mechanisms such as collision processes, phonon interactions, intramolecular processes, lattice interactions, and various other processes,

$$\Gamma_{kl} = \frac{(\Gamma_{kk} + \Gamma_{ll})}{2} + \Gamma'_{kl}, \tag{6.50}$$

in which Γ'_{kl} is known as a *pure dephasing constant*, i.e., a constant associated with processes that disturb the phase of the wave functions but do not alter the populations of the states.

6.2 Density Matrix with Damping

In Section 4.11 we developed expressions for the matrix elements of the density operator for interactions with monochromatic fields when the atomic system was initially in thermal equilibrium. No account was taken of damping effects, however; this now may be accomplished on the basis of the equations in (6.47). The procedure will be illustrated by the computation of the first- and second-order terms in Eq. (6.47). Adopting a basis set consisting of the eigenstates of $\mathscr{H}_0$,

$$i\hbar\dot{\rho}^{(1)}_{kl}(t) = [\mathscr{H}_0, \rho^{(1)}(t)]_{kl} + [V(t), \rho^{(0)}(t)]_{kl} - i\hbar\Gamma_{kl}\rho^{(1)}(t), \tag{6.51}$$

in which $\rho^{(0)}$ is the thermal density operator ρ_0 and

$$\begin{aligned}[\mathscr{H}_0, \rho^{(1)}(t)]_{kl} &= \langle k|\mathscr{H}_0\rho^{(1)}(t)|l\rangle - \langle k|\rho^{(1)}(t)\mathscr{H}_0|l\rangle \\ &= (E_k - E_l)\langle k|\rho^{(1)}(t)|l\rangle \equiv \hbar\omega_{kl}\rho^{(1)}_{kl}(t). \end{aligned} \tag{6.52}$$

In the present basis, the off-diagonal elements of ρ_0 are all zero; hence,

$$\begin{aligned}[V(t), \rho^{(0)}]_{kl} &= [V(t), \rho_0]_{kl} \\ &= \sum_j [V_{kj}(t)\langle j|\rho_0|l\rangle - \langle k|\rho_0|j\rangle V_{jl}(t)] \\ &= V_{kl}(t)\rho^0_{ll} - \rho^0_{kk}V_{kl}(t), \end{aligned} \tag{6.53}$$

with

$$\rho_{ll}^0 \equiv \langle l|\rho_0|l\rangle, \qquad \rho_{kk}^0 \equiv \langle k|\rho_0|k\rangle. \tag{6.54}$$

We now assume $V_{kl}(t)$ to have the form of Eq. (4.333). Equation (6.51) then may be written

$$\dot{\rho}_{kl}^{(1)}(t) = -i\omega_{kl}\rho_{kl}^{(1)}(t) - \frac{i}{\hbar}(\rho_{ll}^0 - \rho_{kk}^0)v_{kl}(\omega)e^{-i\omega t} - \Gamma_{kl}\rho_{kl}^{(1)}(t). \tag{6.55}$$

To solve this equation let

$$\rho_{kl}^{(1)}(t) = Ae^{-i\omega t}. \tag{6.56}$$

After substitution in Eq. (6.55), it is found that

$$A = \frac{v_{kl}(\omega)}{\hbar(\omega_{kl} - \omega - i\Gamma_{kl})}(\rho_{kk}^0 - \rho_{ll}^0). \tag{6.57}$$

Then, in place of Eq. (4.337) we now have

$$\rho_{kl}^{(1)}(t) = \frac{1}{\hbar}(\rho_{kk}^0 - \rho_{ll}^0)\frac{v_{kl}(\omega)e^{-i\omega t}}{\omega_{kl} - \omega - i\Gamma_{kl}}, \tag{6.58}$$

and upon inserting the definitions of $v_{kl}(\omega)$ from Eq. (4.332),

$$\begin{aligned}\rho_{kl}^{(1)}(t) &= \frac{1}{\hbar}(\rho_{ll}^0 - \rho_{kk}^0)\frac{\mathbf{d}_{kl}\cdot\mathbf{E}(\omega)e^{-i\omega t}}{\omega_{kl} - \omega - i\Gamma_{kl}},\\ &= \frac{1}{\hbar}(\rho_{ll}^0 - \rho_{kk}^0)\sum_\alpha \frac{d_{kl}^\alpha E_\alpha(\omega)e^{-i\omega t}}{\omega_{kl} - \omega - i\Gamma_{kl}},\end{aligned} \tag{6.59}$$

where $\alpha = x, y, z$.

To extend the procedure the second order, we refer once more to Eq. (6.47) which enables us to write

$$\begin{aligned}\dot{\rho}_{kl}^{(2)}(t) &= -\frac{i}{\hbar}[\mathscr{H}_0, \rho^{(2)}(t)]_{kl} - \frac{i}{\hbar}[V(t), \rho^{(1)}(t)]_{kl} - \Gamma_{l}\rho_{kl}^{(2)}(t),\\ &= -i\omega_{kl}\rho_{kl}^{(2)}(t) - \frac{i}{\hbar}\sum_r [V_{kr}(t)\rho_{rl}^{(1)}(t) - \rho_{kr}^{(1)}(t)V_{rl}(t)] - \Gamma_{kl}\rho_{kl}^{(2)}(t).\end{aligned} \tag{6.60}$$

For a second order process it is necessary to consider two monochromatic fields and their corresponding interactions as in Eq. (4.339) or Eq. (4.341). As in Section 4.11 we concentrate on those terms that contain the time factor $\exp[-i(\omega_1 - \omega_2)t]$. Then, by a simple extension of Eq. (6.58) or (4.338),

$$\rho_{rl}^{(1)}(t) = \frac{1}{\hbar}(\rho_{rr}^0 - \rho_{ll}^0)\left[\frac{v_{rl}(\omega_1)e^{-i\omega_1 t}}{\omega_{rl} - \omega_1 - i\Gamma_{rl}} + \frac{v_{rl}^*(\omega_2)e^{i\omega_2 t}}{\omega_{rl} + \omega_2 - i\Gamma_{rl}}\right]. \tag{6.61}$$

We then have

$$\sum_r V_{kr}(t)\rho_{rl}^{(1)}(t) = \frac{1}{\hbar}\sum_r(\rho_{rr}^0 - \rho_{ll}^0)\left[\frac{v_{kr}(\omega_1)v_{rl}^*(\omega_2)}{\omega_{rl}+\omega_2 - i\Gamma_{rl}} + \frac{v_{kr}^*(\omega_2)v_{rl}(\omega_1)}{\omega_{rl}-\omega_1 - i\Gamma_{rl}}\right]e^{-i(\omega_1-\omega_2)t}. \tag{6.62}$$

Similarly,

$$\sum_r \rho_{kr}^{(1)}(t)V_{rl}(t) = \frac{1}{\hbar}\sum_r(\rho_{kk}^0 - \rho_{rr}^0)\left[\frac{v_{kr}(\omega_1)v_{rl}^*(\omega_2)}{\omega_{kr}-\omega_1 - i\Gamma_{kr}} + \frac{v_{kr}^*(\omega_2)v_{rl}(\omega_1)}{\omega_{kr}+\omega_2 - i\Gamma_{kr}}\right]e^{-i(\omega_1-\omega_2)t}. \tag{6.63}$$

When these expressions are inserted into Eq. (6.60) it is apparent that the solution must be of the form

$$\rho_{kl}^{(2)}(t) = Ae^{-i(\omega_1-\omega_2)t}, \tag{6.64}$$

that is,

$$\begin{aligned}\rho_{kl}^{(2)}(t) = \left(\frac{1}{\hbar}\right)^2 \frac{e^{-i(\omega_1-\omega_2)t}}{\omega_{kl}-\omega_1+\omega_2 - i\Gamma_{kl}}\sum\Bigg[&\frac{v_{kr}(\omega_1)v_{rl}^*(\omega_2)}{\omega_{rl}+\omega_2 - i\Gamma_{rl}}\rho_{ll}^0 \\ &- \frac{v_{kr}(\omega_1)v_{rl}^*(\omega_2)}{\omega_{rl}+\omega_2 - i\Gamma_{rl}}\rho_{rr}^0 - \frac{v_{kr}(\omega_1)v_{rl}^*(\omega_2)}{\omega_{kr}-\omega_1 - i\Gamma_{kr}}\rho_{rr}^0 \\ &+ \frac{v_{kr}(\omega_1)v_{rl}^*(\omega_2)}{\omega_{kr}-\omega_1 - i\Gamma_{kr}}\rho_{kk}^0 + \frac{v_{kr}^*(\omega_2)v_{rl}(\omega_1)}{\omega_{rl}-\omega_1 - i\Gamma_{rl}}\rho_{ll}^0 \\ &- \frac{v_{kr}^*(\omega_2)v_{rl}(\omega_1)}{\omega_{rl}-\omega_1 - i\Gamma_{rl}}\rho_{rr}^0 - \frac{v_{kr}^*(\omega_2)vrl(\omega_1)}{\omega_{kr}+\omega_2 - i\Gamma_{kr}}\rho_{rr}^0 \\ &+ \frac{v_{kr}^*(\omega_2)v_{rl}(\omega_1)}{\omega_{kr}+\omega_2 - i\Gamma_{kr}}\rho_{kk}^0\Bigg].\end{aligned} \tag{6.65}$$

This expression replaces the sum of the eight terms contained in Eqs. (4.347) and (4.349). The computation of first- and second-order matrix elements indicates that damping effects may be taken into account by simply replacing a frequency ω_{ab} by $\omega_{ab} - i\Gamma_{ab}$. This property extends to higher order.

Since

$$\begin{aligned} v_{kr}(\omega_1) &= -\sum_\alpha d_{kr}^\alpha E_\alpha, \\ v_{kr}^*(\omega_2) &= -\sum_\beta d_{kr}^\beta E_\beta^*, \end{aligned} \tag{6.66}$$

with analogous relations for $v_{rl}(\omega_1)$ and $v^*_{rl}(\omega_2)$, we may rewrite Eq. (6.65) in the form

$$\rho_{kl}^{(2)}(t) = \left(\frac{1}{\hbar}\right)^2 \frac{e^{-i}(\omega_1 - \omega_2)t}{\omega_{kl} - \omega_1 + \omega_2 - i\Gamma_{kl}} \sum_r \sum_{\alpha\beta} E_\alpha(\omega_1)E^*_\beta(\omega_2)$$
$$\times \left[\frac{d^\alpha_{kr} d^\beta_{rl}}{\omega_{rl} + \omega_2 - i\Gamma_{rl}}(\rho^0_{ll} - \rho^0_{rr}) - \frac{d^\alpha_{kr} d^\beta_{rl}}{\omega_{kr} - \omega_1 - i\Gamma_{kr}}(\rho^0_{rr} - \rho^0_{kk})\right.$$
$$\left. + \frac{d^\beta_{kr} d^\alpha_{rl}}{\omega_{rl} - \omega_1 - i\Gamma_{rl}}(\rho^0_{ll} - \rho^0_{rr}) - \frac{d^\beta_{kr} d^\alpha_{rl}}{\omega_{kr} + \omega_2 - i\Gamma_{kr}}(\rho^0_{rr} - \rho^0_{kk})\right]. \quad (6.67)$$

The Feynman diagrams shown in Fig. 4.6 are still valid with the replacement of ω_{ab} by $\omega_{ab} - i\Gamma_{ab}$.

To obtain matrix elements of the density operator with the oscillatory factor $\exp[-i(\omega_1 + \omega_2)t]$, it is necessary merely to change the sign of ω_2 in all the terms

$$\rho_{kl}^{(2)}(t) = \left(\frac{1}{\hbar}\right)^2 \frac{e^{-i(\omega_1 + \omega_2)t}}{\omega_{kl} - \omega_1 - \omega_2 - i\Gamma_{kl}} \sum_r \sum_{\alpha\beta} E_\alpha(\omega_1)E_\beta(\omega_2)$$
$$\times \left[\frac{d^\alpha_{kr} d^\beta_{rl}}{\omega_{rl} - \omega_2 - i\Gamma_{rl}}(\rho^0_{ll} - \rho^0_{rr}) - \frac{d^\alpha_{kr} d^\beta_{rl}}{\omega_{kr} - \omega_1 - i\Gamma_{kr}}(\rho^0_{rr} - \rho^0_{kk})\right.$$
$$\left. + \frac{d^\beta_{kr} d^\alpha_{rl}}{\omega_{rl} - \omega_1 - i\Gamma_{rl}}(\rho^0_{ll} - \rho^0_{rr}) - \frac{d^\beta_{kr} d^\alpha_{rl}}{\omega_{kr} - \omega_2 - i\Gamma_{kr}}(\rho^0_{rr} - \rho^0_{kk})\right]. \quad (6.68)$$

6.3 Two-Level System with Damping

Referring to Eq. (6.43), the terms arising from the master equation for an atomic two-level system are

$$(\dot{\rho}_{ee})_{damp} = \rho_{gg} W_{eg} - \rho_{ee} W_{ge} = -(\dot{\rho}_{gg})_{damp},$$
$$(\dot{\rho}_{eg})_{damp} = -\Gamma_{eg}\rho_{eg}, \qquad (\dot{\rho}_{ge})_{damp} = -\Gamma_{ge}\rho_{ge}, \quad (6.69)$$

with the transition probabilities per unit time defined by

$$W_{eg} \equiv W(g \to e), \qquad W_{ge} \equiv W(e \to g). \quad (6.70)$$

We now set

$$W_{eg} + W_{ge} \equiv \frac{1}{T_1}, \qquad \Gamma_{eg} = \Gamma_{ge} \equiv \frac{1}{T_2}, \quad (6.71)$$

and since

$$\mathrm{Tr}\,\rho = \rho_{gg} + \rho_{ee} = 1, \quad (6.72)$$

we may write

$$(\dot{\rho}_{ee})_{damp} = \rho_{gg} W_{eg} - \rho_{ee} W_{ge} = W_{eg} - (W_{eg} + W_{ge})\rho_{ee}$$

$$= W_{eg} - \frac{1}{T_1}\rho_{ee}, \tag{6.73a}$$

$$(\dot{\rho}_{gg})_{damp} = \rho_{ee} W_{ge} - \rho_{gg} W_{eg} = W_{ge} - (W_{eg} + W_{ge})\rho_{gg} \tag{6.73b}$$

$$= W_{ge} - \frac{1}{T_1}\rho_{gg},$$

$$(\dot{\rho}_{eg})_{damp} = -\frac{1}{T_2}\rho_{eg}, \qquad (\dot{\rho}_{ge})_{damp} = -\frac{1}{T_2}\rho_{ge}, \tag{6.73c}$$

whose solutions are

$$[\rho_{ee}(t)]_{damp} = \rho_{ee}(0)e^{-t/T_1} + T_1 W_{eg}(1 - e^{-t/T_1})$$

$$= 1 - [\rho_{gg}(t)]_{damp}. \tag{6.74}$$

In the limit $t \to \infty$,

$$[\rho_{ee}(\infty)]_{damp} = T_1 W_{eg}, \qquad [\rho_{gg}(\infty)]_{damp} = T_1 W_{ge}. \tag{6.75}$$

Under thermal equilibrium

$$(\dot{\rho}_{ee})_{damp} = \rho_{gg} W_{eg} - \rho_{ee} W_{ge} = 0, \tag{6.76}$$

or

$$\frac{\rho_{ee}}{\rho_{gg}} = \frac{W_{eg}}{W_{ge}}. \tag{6.77}$$

Also, the ratio of the populations in the two states is equal to the ratio of the corresponding Boltzmann factors; thus,

$$\frac{\rho_{ee}}{\rho_{gg}} = \frac{e^{-\beta E_e}}{e^{-\beta E_g}} = e^{-\beta\hbar\omega_0} = \frac{W_{eg}}{W_{ge}}. \tag{6.78}$$

But since $\rho_{ee} + \rho_{gg} = 1$, we may solve for ρ_{ee} and ρ_{gg} in terms of W_{eg} and W_{ge}

$$\rho_{ee} = T_1 W_{eg}, \qquad \rho_{gg} = T_1 W_{ge}. \tag{6.79}$$

Comparing these relation with Eq. (6.75), it may be concluded that the reservoir interactions ultimately bring the atomic system into thermal equilibrium with the reservoir at a rate governed by the relaxation time T_1. Furthermore, from Eq. (6.78) it is seen that for $\beta > 0$, that is, for any finite temperature, the probability per unit time, W_{ge}, for a transition from the upper state $|e\rangle$ to the lower state $|g\rangle$ is greater than the probability per unit time for the reverse transition. Therefore, if the system initially had an excess of population in the upper state (by virtue of some prior excitation process) then, given sufficient time, the system relaxes to thermal equilibrium at the temperature of the reservoir.

It is clear from Eq. (6.69) that the off-diagonal elements decay to zero with a time constant equal to T_2.

It may be remarked that the notation T_1 and T_2 for the diagonal and off-diagonal relaxation times is not arbitrary. In Section 3.5 we encountered the longitudinal and transverse relaxation times T_1 and T_2 in connection with the Bloch equations for the magnetization. The similarity in notation is intended to suggest a similarity in interpretation. This will become clearer in the next section where the Bloch equations for the optical case are introduced.

Let us now examine the combined effects of an external radiation field and a reservoir acting on a two-level atom. For this purpose, we combine Eqs. (5.54) for the free motion (atom plus field) with the damping equations (Eqs. (6.69)):

$$\dot{\rho}_{ee} = i[q^* e^{i\omega t}\rho_{eg} - q e^{-i\omega t}\rho_{ge}] + \rho_{gg} W_{eg} - \rho_{ee} W_{ge}, \tag{6.80a}$$

$$= -\dot{\rho}_{gg},$$

$$\dot{\rho}_{eg} = -i\omega_0\rho_{eg} + iq e^{-i\omega t}(\rho_{ee} - \rho_{gg}) - \Gamma_{eg}\rho_{eg}, \tag{6.80b}$$

$$\dot{\rho}_{ge} = i\omega_0\rho_{ge} - iq^* e^{i\omega t}(\rho_{ee} - \rho_{gg}) - \Gamma_{ge}\rho_{ge}. \tag{6.80c}$$

Assuming a solution of the form

$$\rho_{eg}(t) = \rho^{eg}(0)e^{-i\omega t}, \tag{6.81}$$

and substituting into Eq. (6.80b) one obtains, with $\Gamma_{ge} = \Gamma_{eg}$,

$$\rho_{eg}(t) = \frac{q(\rho_{ee} - \rho_{gg})}{\omega_0 - \omega - i\Gamma_{eg}} e^{-i\omega t} = \rho^*_{ge}(t). \tag{6.82}$$

These expressions, when inserted into Eq. (6.80a), lead to

$$\dot{\rho}_{gg}(t) = (W + W_{ge})\rho_{ee} - (W + W_{eg})\rho_{gg},$$

$$\dot{\rho}_{ee}(t) = (W + W_{eg})\rho_{gg} - (W + W_{ge})\rho_{ee}, \tag{6.83}$$

where

$$W = \frac{2|q|^2\Gamma_{eg}}{(\omega_0 - \omega)^2 + \Gamma^2_{eg}} = \frac{2|q|^2 T_2}{1 + (\omega_0 - \omega)^2 T_2^2}. \tag{6.84}$$

The quantity W is clearly a transition probability per unit time associated with the external radiation field. By virtue of the equality $W_a = W_{st}$ for absorption and stimulated emission, as shown in Eq. (5.152), the field induces transitions in either direction, that is, $|g\rangle \to |e\rangle$ and $|e\rangle \to |g\rangle$, and the probability per unit time is the same for the two directions. On the other hand, as has been indicated in connection with Eq. (6.78), the transition probabilities per unit time, W_{eg} and W_{ge}, associated with the reservoir interactions, depend on the direction of the transition. Thus, the external field tries to equalize the populations in the two levels while the relaxation mechanisms associated with the reservoir try to establish a thermal (Boltzmann) distribution. The presence of both a constant external field and a reservoir ultimately leads to a new equilibrium in which the population in the upper state $|e\rangle$ is higher than it would be if it were in thermal equilibrium (in the absence of a radiation field) but nevertheless smaller than the population in $|g\rangle$. The net effect in the new equilibrium is to transfer energy from the radiation field to the atomic population from which it is drained to the reservoir. In weak fields, the new equilibrium is not far from thermal equilibrium; but with increasing field amplitude, the populations in the two states approach equality—a condition we have previously (Section 5.7) associated with saturation and a diminished net absorption.

As mentioned previously (Section 5.7), the impossibility of population inversion (population in an excited state greater than the population in a level at lower energy) is specific to the system under discussion, namely, a two-level atom interacting with a radiation field and a reservoir at thermal equilibrium. Population inversion can be produced and sustained in many other systems (e.g., a three-level system).

Equation (6.82) was derived explicitly for the two-level system. It is, of course, a special case of the first-order term of Eq. (6.58). If we set

$$|e\rangle = |k\rangle, \qquad |g\rangle = |l\rangle, \qquad \omega_0 = \omega_{eg} = \omega_{kl}, \qquad \Gamma = \Gamma_{eg} = \Gamma_{kl}, \tag{6.85}$$

$$v_{kl}(\omega) = v_{eg}(\omega) = -\frac{1}{2}\mathbf{d}_{eg} \cdot \mathbf{E} = \hbar q, \tag{6.86}$$

$$\rho^0_{kk} = \rho_{ee}, \qquad \rho^0_{ll} = \rho_{gg}, \qquad \rho^{(1)}_{kl}(t) = \rho_{eg}(t), \tag{6.87}$$

and consider only the term that varies as $\exp(-i\omega t)$, Eq. 16.58) gives

$$\rho^{(1)}_{kl}(t) = (\rho^0_{kk} - \rho^0_{ll})\frac{v_{kl}(\omega)e^{-i\omega t}}{\hbar(\omega_{kl} - \omega - i\Gamma_{kl})}. \tag{6.88}$$

For the two-level system, this becomes identical with Eq. (6.82).

We have seen that an atomic system, under the combined influence of a thermal reservoir and a radiation field, approaches equilibrium wherein the populations in $|e\rangle$ and $|g\rangle$ are no longer functions of time. Therefore, in a weak field, the Rabi transition probability (Eq. (5.93)) is meaningful only over a time $t \ll T_1$ where T_1, according to Eq. (6.74), is the characteristic time for the populations to approach thermal equilibrium.

Since the general effect of the reservoir interactions is to introduce decay constants, Eq. (2.82) for the probability amplitudes, with damping constants included, is written

$$i\hbar\dot{a}_l(t) = E_l a_l(t) + \sum_k a_k(t)\phi_l(\mathbf{r})|V(t)|\phi_k(\mathbf{r})\rangle - i\frac{\Gamma_l}{2}a_l(t). \tag{6.89}$$

For a two-level system, with $V_{gg} = V_{ee} = 0$,

$$\begin{aligned} i\hbar\dot{a}_g &= E_g a_g + a_1 V_{ge} - i\hbar\frac{\Gamma_g}{2}a_g, \\ i\hbar\dot{a}_e &= E_e a_e + a_g V_{eg} - i\hbar\frac{\Gamma_e}{2}a_e. \end{aligned} \tag{6.90}$$

Similarly, Eq. (2.84) with decay constants becomes

$$i\hbar\dot{c}_l(t) = \sum_k c_k(t)\langle\phi_l|V(t)|\phi_k\rangle e^{i\omega_{lk}t} - i\hbar\frac{\Gamma_l}{2}c_l(t), \tag{6.91}$$

and the two-level equations are

$$\begin{aligned} i\hbar\dot{c}_g &= c_e V_{ge} e^{-i\omega_0 t} - i\hbar\frac{\Gamma_g}{2}c_g, \\ i\hbar\dot{c}_e &= c_g V_{eg} e^{i\omega_0 t} - i\hbar\frac{\Gamma_e}{2}c_e. \end{aligned} \tag{6.92}$$

Specifically, in place of Eq. (5.69), we now have

$$\begin{aligned} i\dot{c}_g(t) &= q^* c_e(t)e^{-i(\omega_0-\omega)t} - \frac{i\Gamma_g}{2}c_g(t), \\ i\dot{c}_e(t) &= q c_g(t)e^{i(\omega_0-\omega)t} - \frac{i\Gamma_e}{2}c_e(t), \end{aligned} \tag{6.93}$$

In the event that $\Gamma_g = \Gamma_e \equiv \Gamma$ let

$$c'_g = c_g e^{-\Gamma t/2}, \qquad c'_e = c_e e^{-\Gamma t/2}. \tag{6.94}$$

Equations (6.93) then transform to

$$\begin{aligned} i\dot{c}'_g(t) &= q^* c'_e(t)e^{-i(\omega_0-\omega)t}, \\ i\dot{c}'_e(t) &= q c'_g(t)e^{i(\omega_0-\omega)t}, \end{aligned} \tag{6.95}$$

which are identical in form to Eq. (5.69). Solving these as was done previously,

$$|c'_e(t)|^2 = 4|q|^2 F(\omega), \tag{6.96a}$$

or

$$|c_e(t)|^2 = 4|q|^2 F(\omega)e^{-\Gamma t} = \frac{\sin^2[\frac{1}{2}(\omega_0 - \omega)t]}{(\omega_0 - \omega)^2}e^{-\Gamma t}. \tag{6.96b}$$

Hence, the effect of the damping constants is to superimpose an exponential decay on the transition probability.

The computation in Section 5.3 for the Rabi frequency and the transition probability now may be repeated with the inclusion of the damping terms. Starting again with

$$c_g(t) = e^{i\mu t}, \tag{6.97}$$

as in Eq. (5.83), it is found that μ satisfies the quadratic equation

$$\mu^2 + \left(\omega_0 - \omega - i\frac{\Gamma_g + \Gamma_e}{2}\right)\mu - \frac{i\Gamma_g}{2}(\omega_0 - \omega) - |q|^2 - \frac{\Gamma_e\Gamma_g}{4} = 0, \tag{6.98}$$

whose solutions are

$$\mu_\pm = -\frac{1}{2}(\omega_0 - \omega - i\Gamma') \pm \frac{1}{2}\sqrt{(\omega_0 - \omega - i\Gamma'')^2 + 4|q|^2} \tag{6.99}$$

where

$$\Gamma' = \frac{(\Gamma_e + \Gamma_g)}{2},$$

$$\Gamma'' = \frac{(\Gamma_e - \Gamma_g)}{2}. \tag{6.100}$$

The Rabi frequency is seen to be altered to

$$\Omega = \mu_+ - \mu_- = \sqrt{(\omega_0 - \omega - i\Gamma'')^2 + 4|q|^2}, \tag{6.101}$$

and the transition probability to

$$|c_e(t)|^2 = \frac{4|q|^2}{\Omega^2} e^{-\Gamma' t} \sin^2 \frac{1}{2}\Omega t. \tag{6.102}$$

In comparison with Eq. (5.93), the transition probability in the presence of damping no longer oscillates indefinitely but ultimately vanishes (Fig. 6.1). The

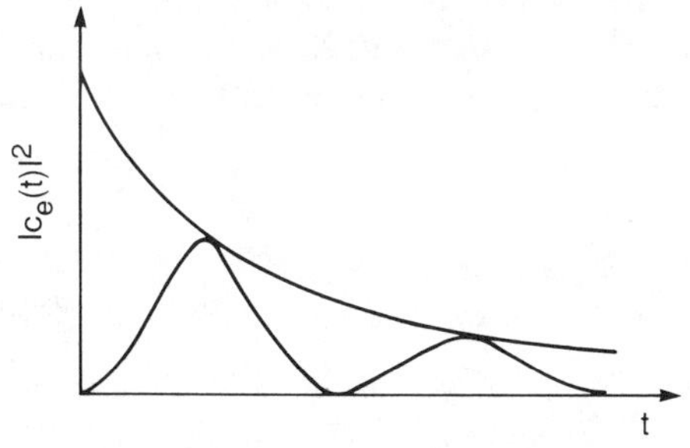

FIGURE 6.1 Rabi oscillations in the presence of damping.

system reaches a steady state determined by the magnitudes of the Rabi frequency and damping constants.

We also may modify Eqs. (2.192) by the addition of damping constants. To accomplish this in a consistent manner, we refer to Eq. (2.196), which relates the density matrix to the matrix of the coefficients a_g and a_e whose evolution in time is governed by Eq. (6.90). As an illustration of the computation, we have $\rho_{gg} = |a_g|^2$ and

$$\begin{aligned}\dot{\rho}_{gg} &= a_g\dot{a}_g^* + \dot{a}_g a_g^* \\ &= a_g\left[\frac{i}{\hbar}E_g a_g^* + \frac{i}{\hbar}a_e^* V_{eg} - \frac{\Gamma_g}{2}a_g^*\right] + \left[-\frac{i}{\hbar}E_g a_g - \frac{i}{\hbar}a_e V_{ge} - \frac{\Gamma_g}{2}a_g\right]a_g^* \\ &= -\frac{i}{\hbar}V_{ge}\rho_{eg} + \frac{i}{\hbar}V_{eg}\rho_{ge} - \Gamma_g\rho_{gg}. \end{aligned} \tag{6.103a}$$

For the other components of the density matrix one gets

$$\dot{\rho}_{ee} = -\frac{i}{\hbar}V_{eg}\rho_{ge} + \frac{i}{\hbar}V_{ge}\rho_{eg} - \Gamma_e\rho_{ee}, \tag{6.103b}$$

$$\dot{\rho}_{ge} = i\omega_0\rho_{ge} - \frac{i}{\hbar}V_{ge}(\rho_{ee} - \rho_{gg}) - \Gamma_{ge}\rho_{ge}, \tag{6.103c}$$

$$\dot{\rho}_{eg} = -i\omega_0\rho_{eg} + \frac{i}{\hbar}V_{eg}(\rho_{ee} - \rho_{gg}) - \Gamma_{ge}\rho_{eg}, \tag{6.103d}$$

where

$$\Gamma_{ge} = \frac{1}{2}(\Gamma_g + \Gamma_e). \tag{6.104}$$

These equations hold for a two-level system influenced by an arbitrary interaction V under the assumption that $|g\rangle$ and $|e\rangle$ can decay but are not pumped. Equations (6.80) are special cases for an interaction represented by the coupling constants q and q*. If the populations in $|g\rangle$ and $|e\rangle$ are pumped at rates λ_g and λ_e, such terms can be added on the right side of Eq. (6.103a) and (6.103b), respectively. Finally, we note that if Γ is written in the form

$$\Gamma = \begin{pmatrix}\Gamma_e & 0 \\ 0 & \Gamma_g\end{pmatrix}, \tag{6.105}$$

the matrix corresponding to the damping terms in Eq. (6.103) is

$$\frac{1}{2}(\Gamma_\rho + \rho\Gamma) = \begin{pmatrix}\Gamma_e\rho_{ee} & \Gamma_{eg}\rho_{eg} \\ \Gamma_{ge}\rho_{ge} & \Gamma_g\rho_{gg}\end{pmatrix}. \tag{6.106}$$

6.4 Optical Bloch Equations

We have noted previously (Section 5.1) the close formal resemblance between a two-level atomic system and a spin-1/2 system. It then would be reasonable to expect that the Bloch equations of magnetic resonance Eqs. (3.121) also would be applicable to the atomic case. In fact, the atomic Bloch vector has already been introduced in Eq. (5.64) so that it remains only to incorporate the effects of damping. For this purpose we return to Eq. (6.73c) which yields

$$
\begin{aligned}
(\dot{u}_1)_{\text{damp}} &= (\dot{\rho}_{\text{ge}})_{\text{damp}} + (\dot{\rho}_{\text{eg}})_{\text{damp}} = -\frac{1}{\text{T}_2}u_1, \\
(\dot{u}_2)_{\text{damp}} &= i(\dot{\rho}_{\text{eg}})_{\text{damp}} - i(\dot{\rho}_{\text{ge}})_{\text{damp}} = -\frac{1}{\text{T}_2}u_2.
\end{aligned} \tag{6.107}
$$

Following the same procedure leading to Eqs. (6.73a) and (6.73b), we have

$$
(\dot{u}_3)_{\text{damp}} = (\dot{\rho}_{\text{ee}})_{\text{damp}} - (\dot{\rho}_{\text{gg}})_{\text{damp}} = W_{\text{eg}} - W_{\text{ge}} - \frac{1}{\text{T}_1}u_3. \tag{6.108}
$$

But under thermal equilibrium the populations of the two states $|\text{g}\rangle$ and $|\text{e}\rangle$ remain unchanged; therefore, if u_3^0 is the value of the inversion u_3 at thermal equilibrium,

$$
(\dot{u}_3)_{\text{damp}} = W_{\text{eg}} - W_{\text{ge}} - \frac{1}{\text{T}_1}u_3^0 = 0. \tag{6.109}
$$

Thus,

$$
(\dot{u}_3)_{\text{damp}} = -\frac{u_3 - u_3^0}{\text{T}_1}. \tag{6.110}
$$

The reservoir Eqs. (6.107) and (6.110) now may be combined with Eq. (5.62) for the free motion of the system to obtain the *optical Bloch equations*:

$$
\dot{u}_1 + \Delta u_2 + \frac{u_1}{\text{T}_2} = 0, \tag{6.111a}
$$

(*R*)
$$
\dot{u}_2 - \Delta u_1 + 2\text{q}u_3 + \frac{u_2}{\text{T}_2} = 0, \qquad (\Delta = \omega_0 - \omega) \tag{6.111b}
$$

$$
\dot{u}_3 - 2\text{q}u_2 + \frac{u_3 - u_3^0}{\text{T}_1} = 0. \tag{6.111c}
$$

As in the magnetic case, T_1 and T_2 are the longitudinal and transverse relaxation times, respectively.

There are no general solutions to the Bloch equations; a few special cases of physical interest are the following [12–15]:

1. Steady state

$$\dot{u}_1 = \dot{u}_2 = \dot{u}_3 = 0. \tag{6.112}$$

The solutions are analogous to Eq. (3.126):

$$(R) \qquad \begin{aligned} u_1 &= u_3^0 \frac{2\mathrm{q}\mathrm{T}_2^2\Delta}{D}, \\ u_2 &= -u_3^0 \frac{2\mathrm{q}\mathrm{T}_2}{D}, \\ u_3 &= u_3^0 \frac{1 + \mathrm{T}_2^2\Delta^2}{D}, \end{aligned} \tag{6.113}$$

where

$$D = 1 + \mathrm{T}_2^2\Delta^2 + 4|\mathrm{q}|^2\mathrm{T}_1\mathrm{T}_2. \tag{6.114}$$

2. Quasi-steady state

$$\dot{u}_1 = \dot{u}_2 = 0, \qquad \dot{u}_3 \neq 0. \tag{6.115}$$

From Eqs. (6.111a) and (6.111b),

$$u_1 = \frac{2\mathrm{q}\mathrm{T}_2^2\Delta}{1 + \mathrm{T}_2^2\Delta^2} u_3, \qquad u_2 = -\frac{2\mathrm{q}\mathrm{T}_2}{1 + \mathrm{T}_2^2\Delta^2} u_3, \tag{6.116}$$

and from Eq. (6.111c),

$$\dot{u}_3 + \frac{4|\mathrm{q}|^2\mathrm{T}_2}{1 + \mathrm{T}_2^2\Delta^2} u_3 + \frac{u_3 - u_3^0}{\mathrm{T}_1} = 0. \tag{6.117}$$

Solving for u_3, one obtains

$$u_3(t) = \frac{1 + \mathrm{T}_2^2\Delta^2}{D} u_3^0 + \left(u_3(0) - \frac{1 + \mathrm{T}_2^2\Delta^2}{D} u_3^0\right) \exp\left[\frac{-\dfrac{D}{\mathrm{T}_1}}{1 + \mathrm{T}_2^2\Delta^2} t\right], \tag{6.118a}$$

$$= \frac{1 + \mathrm{T}_2^2\Delta^2}{D} u_3^0, \qquad (t \to \infty) \tag{6.118b}$$

consistent with the steady-state expression for u_3 in Eq. (6.113). Under extreme saturation ($D \gg 1 + \mathrm{T}_2^2\Delta^2$), $u_3 \to 0$.

3. $2\mathrm{q} \gg 1/\mathrm{T}_1$, $2\mathrm{q} \gg 1/\mathrm{T}_2$. The Bloch equations reduce to

$$\dot{u}_1 + \Delta u_2 = 0, \qquad \dot{u}_2 - \Delta u_1 + 2\mathrm{q}u_3 = 0, \qquad \dot{u}_3 - 2\mathrm{q}u_2 = 0. \tag{6.119}$$

With initial components $u_1(0)$, $u_2(0)$, and $u_3(0)$, the solutions are analogous to Eq. (3.22):

(R)

$$u_1(t) = u_1(0)\frac{4|\mathrm{q}|^2 + \Delta^2 \cos \Omega t}{\Omega^2} - u_2(0)\frac{\Delta}{\Omega}\sin \Omega t - u_3(0)\frac{2\mathrm{q}\,\Delta}{\Omega^2}(\cos \Omega t - 1),$$

$$u_2(t) = u_1(0)\frac{\Delta}{\Omega}\sin \Omega t + u_2(0)\cos \Omega t - u_3(0)\frac{2\mathrm{q}}{\Omega}\sin \Omega t, \tag{6.120}$$

$$u_3(t) = -u_1(0)\frac{2\mathrm{q}\,\Delta}{\Omega^2}(\cos \Omega t - 1) + u_2(0)\frac{2\mathrm{q}}{\Omega}\sin \Omega t + u_3(0)\left[1 + \frac{4|\mathrm{q}|^2}{\Omega^2}(\cos \Omega t - 1)\right],$$

where

$$\Omega^2 = \Delta^2 + 4|\mathrm{q}|^2 = \Delta^2 + \Omega_0^2. \tag{6.121}$$

For an initial state with $\rho_{\mathrm{gg}}(0) = 1$, $\rho_{\mathrm{ee}}(0) = 0$,

$$u_1(0) = u_2(0) = 0, \qquad u_3(0) = -1. \tag{6.122}$$

Further, if $\Delta = 0$, the equation for $u_3(t)$ gives $u_3(t) = -\cos \Omega_0 t$ or $u_3(\pi/\Omega_0) = 1$, i.e., $\rho_{\mathrm{gg}}(\pi/\Omega_0) = 0$, $\rho_{\mathrm{ee}}(\pi/\Omega_0) = 1$. In other words, a π-pulse reverses the Bloch vector, consistent with the Rabi transition probability (Section 5.3).

4. $\Delta \gg 2\mathrm{q}$. This situation arises when the coupling between the radiation field and the atom is very weak as would be the case, for example, when the field and the atom are far off resonance. The Bloch equations then become

$$\dot{u}_1 + \Delta u_2 + \frac{u_1}{\mathrm{T}_2} = 0, \qquad \dot{u}_2 - \Delta u_1 + \frac{u_2}{\mathrm{T}_2} = 0,$$

$$\dot{u}_3 + \frac{(u_3 - u_3^0)}{\mathrm{T}_1} = 0, \tag{6.123}$$

and the solutions, with initial components $u_1(0)$, $u_2(0)$, and $u_3(0)$, are

(R)

$$u_1(t) = [u_1(0)\cos \Delta t - u_2(0)\sin \Delta t]e^{-t/T_2},$$

$$u_2(t) = [u_1(0)\sin \Delta t + u_2(0)\cos \Delta t]e^{-t/T_2}, \tag{6.124}$$

$$u_3(t) = [u_3(0) - u_3^0]e^{-t/T_1} + u_3^0.$$

Under these conditions, u_1 and u_2 are decoupled from u_3 and

$$u_1(\infty) = u_2(\infty) = 0, \qquad u_3(\infty) = u_3^0. \tag{6.125}$$

Equations (6.124) indicate the significance of T_1 and T_2. The longitudinal component u_3 approaches the equilibrium value u_3^0 at a rate determined by T_1. The component u_3, defined in Eq. (5.26), is the difference between the population in $|e\rangle$ and the population in $|g\rangle$ and is, therefore, a measure of the energy of the atomic system. The relaxation time T_1 then must reflect the rate of energy exchange between the atoms and the reservoir leading to thermal equilibrium. The transverse components u_1 and u_2, which depend entirely on off-diagonal density matrix elements, ultimately decay to zero at a rate governed by T_2. The relaxation of u_1 and u_2 is associated with a loss of coherence among the atoms due to dephasing processes, which in the atomic case refer to the phases of the wave functions. Thus, in a hypothetical example, suppose that initially a group of atoms are all in the state $|e\rangle$ and that the phases of all the wave functions are alike. In the course of time, at a rate determined by T_2, the phases will become randomized. We see, then, that the physical interpretation of T_1 and T_2 is essentially the same in both the magnetic and atomic cases. Indeed, this analogy has led to fruitful experiments (Section 6.5) yielding direct measurements of T_1 and T_2.

5. $T_1 = T_2 = T$. With the definition

$$D' = 1 + T^2\Delta^2 + 4|q|^2T^2 \tag{6.126}$$

and with initial components $u_1(0)$, $u_2(0)$, and $u_3(0)$, the solutions are

$$\begin{aligned}
u_1(t) &= e^{-t/T}\left\{u_1(0) - \Delta\left[u_2(0) + \frac{2qTu_3^0}{D'}\right]\frac{\sin\Omega t}{\Omega}\right.\\
&\quad + \Delta\left[\Delta u_1(0) - 2qu_3(0) + \frac{2qu_3^0}{D'}\right]\frac{\cos\Omega t - 1}{\Omega^2}\\
&\quad \left. - \frac{2q\,\Delta T^2 u_3^0}{D'}\right\} + \frac{2q\,\Delta T^2 u_3^0}{D'},\\
u_2(t) &= e^{-t/T}\left\{\left[u_2(0) + \frac{2qTu_3^0}{D'}\right]\cos\Omega t\right.\\
&\quad \left. + \left[\Delta u_1(0) - 2qu_3(0) + \frac{2qu_3^0}{D'}\right]\frac{\sin\Omega t}{\Omega}\right\} - \frac{2qTu_3^0}{D'},\\
u_3(t) &= e^{-t/T}\left\{u_3(0) - u_3^0 + 2q\left[u_2(0) + \frac{2qTu_3^0}{D'}\right]\frac{\sin\Omega t}{\Omega}\right.\\
&\quad - 2q\left[\Delta u_1(0) - 2qu_3(0) + \frac{2qu_3^0}{D'}\right]\frac{\cos\Omega t - 1}{\Omega^2}\\
&\quad \left. + \frac{4|q|^2u_3^0}{D'}\right\} + u_3^0\left(1 - \frac{4|q|^2}{D'}\right).
\end{aligned} \qquad (R) \tag{6.127}$$

6.5 Line Shapes, Lamb Dip, Photon Echoes

Our attention been concentrated so far on line broadening due to spontaneous emission (radiative broadening). There are many other effects, however, that lead to the broadening of spectral lines so that, under experimental conditions, the observed line width may be many times larger than the radiative line width. The latter, since it is an inherent property of the radiation interaction, is the smallest possible line width. Much effort has been expended in recent years to reduce nonradiative broadening and a great deal of progress has been achieved.

Broadening effects may be grouped into two general categories—*homogeneous* and *inhomogeneous* effects. The former apply to the case of identical atoms with identical transition frequencies and line widths. The latter apply to the case in which there are differences among the atoms in regard to transition frequencies or probabilities of emission and absorption or other features. The line shape in homogeneous broadening is typically Lorentzian. Broadening due to spontaneous emission with linewidth Γ already has been shown to belong to this category. Another important contribution to homogeneous broadening arises from elastic and inelastic collisions between atoms. Classically, this is understood as arising from an interruption of the electromagnetic wave train emitted by an atom as a result of a collision with other atoms. The effect is to alter the phase of the wave in a random manner. When the time between collisions is shorter than the lifetime of the excited atoms, as is typically the case at atmospheric pressure and room temperature, the wave train is broken into many short sections whose Fourier transforms may be many times broader than the radiative linewidth. The quantum mechanical picture is not much different. The excited atom can relax to the ground state not only by spontaneous emission but also by transfering its excitation energy to another atom in the course of a collision. The existence of an alternative relaxation channel shortens the atom's lifetime and therefore results in line broadening.

We have seen (Section 5.7) that the saturation parameter is proportional to the intensity of the radiation. With an increase in intensity, the resulting increase in saturation also contributes to line broadening—a phenomenon known as *power broadening*. For a homogeneous (Lorentzian) profile, the major effect of saturation is an increase in the linewidth.

An inhomogeneously broadened spectrum consists of a superposition of many individual, homogeneously-broadened (typically Lorentzian) lines which merge into a single broadened line (Fig. 6.2). The line shape is often Gaussian with a width many times greater than the homogenous width. One of the most important mechanisms for inhomogeneous broadening is the

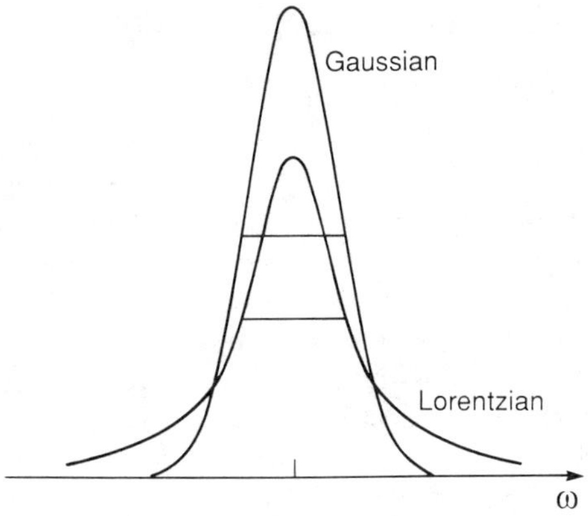

FIGURE 6.2 Comparison of a Gaussian and a Lorentzian line shape both having the same width at half-maximum.

Doppler effect. In a gas, for example, the Maxwellian distribution of velocities results in a spread of frequencies due to the various Doppler shifts, even though the atoms, in their own rest frames, have the same transition frequency. The Doppler width at a temperature T in a gas consisting of molecules of mass M is

$$\Delta\omega_{\mathrm{D}} = \frac{\omega_0}{\mathrm{c}}\left(\frac{8\mathrm{k}T\ln 2}{M}\right)^{1/2} \equiv (8\ln 2)^{1/2}\,\delta \tag{6.128}$$

where $\Delta\omega_{\mathrm{D}}$ is the full width at half maximum, c is the velocity of light, k is the Boltzmann constant, and ω_0 is the center frequency. The normalized Gaussian line shape (Section 1.6) is

$$G(\omega) = \frac{1}{\delta(2\pi)^{1/2}}\exp\left[-\frac{(\omega_0-\omega)^2}{2\,\delta^2}\right], \tag{6.129}$$

where

$$\delta^2 = \frac{\omega_0^2\mathrm{k}T}{M_{\mathrm{c}}^2} \tag{6.130}$$

is the variance or δ is the root-mean-square spread. A comparison between a Lorentzian and a Gaussian line shape, both normalized and both having the same width at half-maximum, is shown in Fig. 6.3. The Lorentzian has longer tails and falls off more slowly from the central peak.

In gases at room temperature, the Doppler width in the optical region exceeds the radiative linewidth by about two orders of magnitude and is generally much larger than the collision linewidth. The total line shape

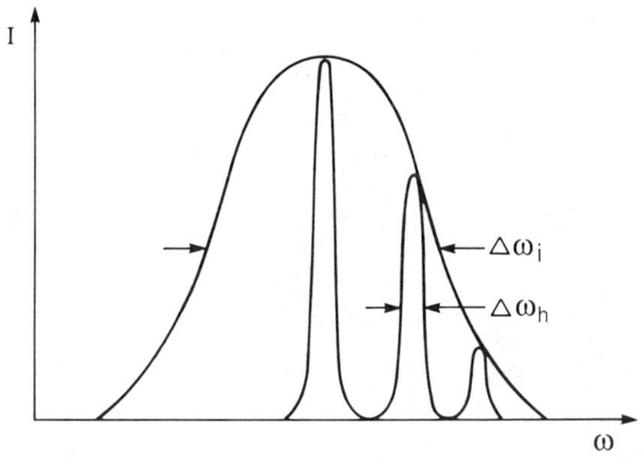

FIGURE 6.3 When many homogeneously broadened lines of width $\Delta\omega_h$ are sufficiently close to one another, they merge into a single inhomogeneous line of width $\Delta\omega_i$.

resulting from the convolution of Lorentzian line shapes (e.g., spontaneous emission) situated at different frequencies and a Gaussian line shape (e.g., Doppler broadening) is called a *Voigt profile.* There are many other causes for inhomogeneous broadening, for example, impurities and strains in solids that may give rise to inhomogeneous fields. One also may mention *Stark* and *resonance broadening.* The former occurs in a collision between an atom and a charged particle. In the course of the collision, the atom is subjected to a fluctuating electric field that perturbs its energy levels. Resonance broadening occurs when two identical atoms collide. When the two atoms are in close proximity they form a quasi-molecule with totally different energy levels.

Saturation of an inhomogeneously broadened lineshape is the basis for an important high-resolution spectroscopic application known as *saturation spectoscopy* [16, 17], which offers the possibility of extracting the homogeneous lineshape under circumstances where it is well hidden within the much broader inhomogeneous lineshape. Closely spaced lines buried in the inhomogeneous profile are thereby revealed. We shall describe the main ideas with reference to inhomogeneous broadening arising from the Doppler effect.

Consider a sample containing molecules with a Maxwellian velocity distribution. If the sample is irradiated by a beam of frequency ω propagating in the positive z direction, a molecule moving away from the beam with velocity v_z will encounter a (first-order) Doppler-shifted frequency $\omega(1 - v_z/c)$. If the molecular resonance frequency in the rest frame is ω_0, excitation will occur only within that group of molecules that satisfy $\omega_0 \simeq \omega(1 - v_z/c)$. Assuming the incoming beam is of sufficient intensity, the transition can be saturated; since the absorption coefficient of saturated molecules is reduced,

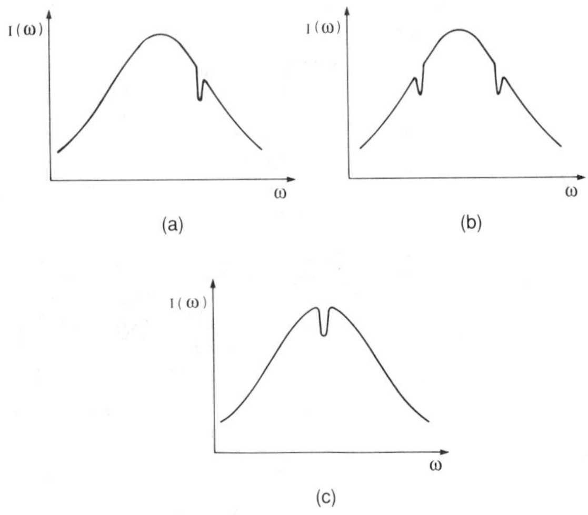

FIGURE 6.4 Hole burning by saturation in an inhomogeneous line shape with center frequency at ω_0. (a) Hole at a frequency $\omega = \omega_0/(1 - v_z/c)$. (b) Symmetrical holes produced by two counterpropagating beams at the frequency ω. (c) Single hole at $\omega = \omega_0$.

the inhomogeneous lineshape will reveal a dip or "hole" (Fig. 6.4a) when examined by a suitable probe beam. With counterpropagating beams of the same frequency ω, two groups of molecules—one with velocity v_z and the other with $-v_z$—will be resonant. There then will be two holes located in symmetrical positions (Fig. 6.4b). Clearly, when $\omega = \omega_0$ only those molecules with $v_z = 0$ will be excited, and a single hole will then be located exactly at the center of the inhomogeneous distribution (Fig. 6.4c). To a first approximation, the shape of the center hole corresponds to the homogeneous absorption lineshape.

It is also possible for an atom or molecule to interact with a radiation field by two-photon absorption (Section 7.4). In a process of this kind, absorption can occur (under certain restrictions) when the transition frequency ω_0 is equal to $\omega_1 + \omega_2$ where ω_1 and ω_2 are two frequencies contained in the incident beam. A special case is one in which $\omega_1 = \omega_2 = \omega$ so that $\omega_0 = 2\omega$. This process, too, may be exploited to advantage in saturation spectroscopy. Consider a molecule with a two-photon absorption line at $2\omega_0$. Since $\omega(1 + v_z/c) + \omega(1 - v_z/c) = 2\omega$, this means that for two counterpropagating beams each at the frequency ω, every molecule, regardless of the magnitude or direction of v_z, sees one photon shifted to a higher frequency and one photon shifted to a lower frequency by the same amount so that the sum of the two frequencies remains at 2ω. Therefore, in contrast to the one-photon absorption case where only certain groups of molecules can be excited, two-photon excitation applies to all molecules. By this method and its extensions, the effects of Doppler broadening have been largely eliminated in many cases.

With the development of laser source capable of generating coherent light beams, phenomena analogous to the magnetic case have been observed in the optical region. These include optical nutation, optical free induction decay, photon echoes, quantum beats, and others [13, 18–20]. Coherent optical spectroscopy, as the field is generally known, differs in important respects from conventional (incoherent) spectroscopy such as absorption, fluorescence, or Raman. In the latter cases, the exciting radiation disturbs the equilibrium populations in the ground and excited states and, when the excitation is removed, the return to equilibrium is accompanied by incoherent radiation. As we have seen previously, the most general description of an incoherent process is in terms of an appropriate correlation function. Coherent optical spectroscopy, on the other hand, finds its most natural form of expression in the Bloch equations.

We now shall describe the production of photon echoes [13, 21, 22]—a process closely analogous to spin echoes described qualitatively in Section 3.5. There are numerous applications employing photon echoes; the method is particularly powerful in the study of homogeneous optical line shapes. Our presentation will be confined to the simplest version that exhibits the basic physical principles. In a typical atomic photon-echo experiment, a sample of atoms is irradiated by two successive laser pulses, one at $t = 0$ and the other at $t = \tau$. At a time $t = 2\tau$, a pulse of radiation (the echo) is spontaneously emitted by the sample (see Fig. 3.9 for the pulse sequence).

It is assumed that a gas of atoms (or molecules) is irradiated under steady state conditions by a highly monochromatic laser beam and that within the velocity distribution of the atoms there is a subgroup in near resonance with the laser beam. If the radiation is sufficiently intense, each resonant atom will oscillate between the ground and excited states at the Rabi flopping frequency and will alternately absorb and emit radiation—a process that has been called *nutation*. In consequence of such excitation, the atomic wave functions become linear superpositions of the ground and excited states. This feature already has been noted in Section 5.3, where it was shown that after a $\pi/2$ pulse a two-level atom, initially in the ground state $|g\rangle$ evolves into the superposition state (Eq. (5.101)). This implies that the two-level atoms distribute themselves equally between the states $|g\rangle$ and $|e\rangle$. As noted previouly (Section 5.2), the expectation value of the electric dipole moment operator in the superposition state is nonzero. Each atom therefore acquires a net (electric) dipole moment.

To proceed quantitatively, we refer to the solutions of the Bloch equations in Section 6.4. Specifically, it is assumed that 2q is much larger than the reciprocal of any relaxation time; hence, the relevant solutions are those given by Eq. (6.120). If the atomic system is prepared so that $u_1(0) = u_2(0) = 0$, the equations at $t = t_1$ reduce to

$$u_1(t_1) = -u_3(0)\frac{2\mathrm{q}\,\Delta}{\Omega^2}(\cos\Omega t_1 - 1),$$

$$u_2(t_1) = -u_3(0)\frac{2\mathrm{q}}{\Omega}\sin\Omega t_1, \tag{6.131}$$

$$u_3(t_1) = u_3(0)\left[1 + \frac{4|\mathrm{q}|^2}{\Omega^2}(\cos\Omega t_1 - 1)\right].$$

The essential physics is retained under the simplifying assumption that resonance conditions prevail, $\Delta = 0$; then

$$\begin{aligned} u_1(t_1) &= 0, \\ u_2(t_1) &= -u_3(0)\sin 2\mathrm{q}t_1, \qquad (\Omega = 2\mathrm{q}) \\ u_3(t_1) &= u_3(0)\cos 2\mathrm{q}t_1, \end{aligned} \tag{6.132}$$

which indicate that the Bloch vector rotates about the x-axis, that is, about the direction of the torque vector. For atoms initially in the ground state $|\mathrm{g}\rangle$, the initial condition is $u_3(0) = -1$ (Fig. 6.5a). If $2\mathrm{q}t_1 = \pi/2$,

$$u_1(t_1) = 0, \qquad u_2(t_1) = -u_3(0) = 1, \qquad u_3(t_1) = 0. \tag{6.133}$$

At the end of the $\pi/2$-pulse, the Bloch vector lies along the positive y axis.

At $t = t_1$, the atomic system is moved rapidly (in a time less than T_2) far out of resonance with the light beam. This is achieved commonly by means of the Stark effect [13] in which a dc electric field is employed to shift the atomic levels. As far as the resonant atoms are concerned, the effect is the same as if the beam had been turned off altogether. In the Stark-shifted system, however, another group of atoms within the velocity distribution now become resonant as a consequence of the Doppler effect and, as far as they are concerned, the beam has just been turned on. The second group therefore responds in the same manner as the first group did before the Stark field was applied, that is, they begin to nutate. Let us now follow the events associated with the first group of atoms that have been shifted out of resonance. For $t > t_1$, the Bloch vector or the net dipole moment remains in the xy plane and precesses about the z axis in the counterclockwise sense. This motion results in the emission of radiation known as the *free induction decay signal* (FID). It is a coherent emission and propagates in the same direction as the laser beam.

The relevant form of the Bloch equations after the application of the Stark field is given then by Eq. (6.123) with the initial conditions of Eq. (6.133).

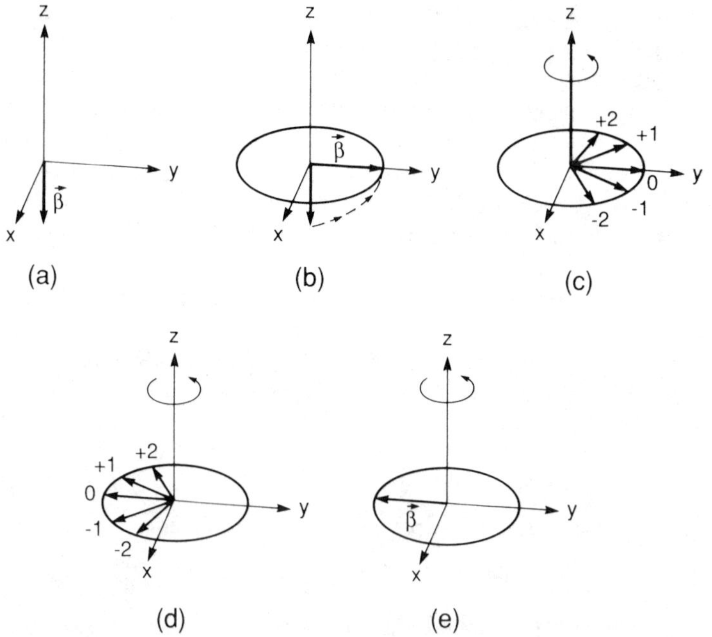

FIGURE 6.5 Development of a photon echo. (a) For atoms initially in the ground state, the Bloch vector $\boldsymbol{\beta}$ consists of one component $u_3 = -1$. (b) After a $\pi/2$ pulse, $\boldsymbol{\beta}$ becomes $u_2 = 1$. (c) Inhomogeneities cause individual atomic dipoles to fan out. (d) A π pulse rotates the atomic dipoles about the x axis resulting in the reversal of relative phases among the dipoles. (e) Precessional motion continues in the same sense allowing the dipoles to rephase.

Referring to Eq. (6.124), the solutions at $t = t_2 > t_1$ with $u_3^0 = 0$ are

$$\begin{aligned} u_1(t_2) &= u_3(0)\sin\Delta_s(t_2 - t_1)e^{-(t_2-t_1)/T_2}, \\ u_2(t_2) &= -u_3(0)\cos\Delta_s(t_2 - t_1)e^{-(t_2-t_1)/T_2}, \\ u_3(t_2) &= 0, \end{aligned} \tag{6.134}$$

where Δ_s is the detuning resulting from the Stark shift. Owing to the presence of various inhomogeneities such as the Doppler shifts for atoms traveling with different velocities, the individual atomic dipole moments precess at slightly different frequencies (Fig. 6.5c). They soon get out of phase with one another—in a time roughly equal to the reciprocal of $\Delta\omega_{\text{inh}}$, the inhomogeneous linewidth—resulting in the ultimate vanishing of the net dipole moment and the associated free induction signal.

While the first velocity group is undergoing free induction decay, the second velocity group is nutating. Hence, with a periodic pulsed dc field, the two velocity groups alternately exchange roles, and the net effect is a modulation of the transmitted laser beam.

It should be remarked that the radiation associated with free induction decay is a result of a collective atomic motion somewhat reminiscent of the radiation associated with Dicke states. There is a fundamental difference, however. The superposition state responsible for free induction decay has a net dipole moment induced by the external optical field whereas there is no net dipole moment in Dicke states. In the latter, atomic correlations arise from the fact that each atom is influenced by the fields produced by all the other atoms. The radiation occurs spontaneously as a consequence of these correlations.

After the free induction signal has effectively ceases at $t = t_2$, the light beam is brought back into resonance ($\Delta = 0$) with the atomic system. We then may invoke Eq. (6.120) once more, but now the initial conditions are given by Eq. (6.134) and the duration of the pulse is such that $2q(t_3 - t_2) = \pi$. The resulting solutions are

$$\begin{aligned} u_1(t_3) &= u_3(0)\sin\Delta_s(t_2 - t_1)e^{-(t_2-t_1)/T_2}, \\ u_2(t_3) &= u_2(t_2)\cos 2q(t_3 - t_2) - u_3(t_2)\sin 2q(t_3 - t_2) \\ &= -u_2(t_2) = u_3(0)\cos\Delta_s(t_2 - t_1)e^{-(t_2-t_1)/T_2}, \\ u_3(t_3) &= u_2(t_2)\sin 2q(t_3 - t_2) + u_3(t_2)\cos 2q(t_3 - t_2) \\ &= -u_3(t_2) = 0. \end{aligned} \tag{6.135}$$

The application of a π-pulse after the decay of the free induction signal has the same effect as in the magnetic case. It reverses the orientation of the atomic dipole moments (Fig. 6.5d) but, most importantly, the precessional motion continues in the *same* sense. After the π-pulse, the light beam is detuned again thereby removing the excitation. Equation (6.123) is again applicable; hence, for $t > t_3$,

$$\begin{aligned} u_1 &= [u_1(t_3)\cos\Delta_s(t - t_3) - u_2(t_3)\sin\Delta_s(t - t_3)]e^{-(t-t_3)/T_2}, \\ &= u_3(0)[\sin\Delta_s(t_2 - t_1)\cos\Delta_s(t - t_3) \\ &\qquad - \cos\Delta_s(t_2 - t_1)\sin\Delta_s(t - t_3)]e^{-[(t_2-t_1)+(t-t_3)]/T_2}, \\ &= u_3(0)\sin\Delta_s[(t_2 - t_1) - (t - t_3)]e^{-[(t_2-t_1)+(t-t_3)/T_2}, \\ u_2(t) &= [u_1(t_3)\sin\Delta_s(t - t_3) + u_2(t_3)\cos\Delta_s(t - t_3)]e^{-(t-t_3)/T_2}, \\ &= u_3(0)[\sin\Delta_s(t_2 - t_1)\sin\Delta_s(t - t_3) \\ &\qquad + \cos\Delta_s(t_2 - t_1)\cos\Delta_s(t - t_3)]e^{-[(t_2-t_1)+(t-t_3)]/T_2}, \\ &= u_3(0)\cos\Delta_s[(t_2 - t_1) - (t - t_3)]e^{-[(t_2-t_1)+(t-t_3)]/T_2}, \\ u_3(t) &= 0. \end{aligned} \tag{6.136}$$

When the transverse relaxation time T_2 is long compared to $(t_2 - t_1) + (t - t_3)$

$$u_1(t) = -u_3(0)\sin\Delta_s(t - 2\tau), \qquad u_2(t) = u_3(0)\cos\Delta_s(t - 2\tau),$$

where

$$2\tau = t_3 + t_2 - t_1.$$

Thus, when $t = 2\tau$, i.e., when $t - t_3 = t_2 - t_1$, all atoms are back in phase, and the Bloch vector again consists of the single component $u_2(t) = u_3(0)$ (Fig. 6.5e) as in Eq. (6.133) after the $\pi/2$ pulse. We note, however, that the direction of $u_2(t)$ is opposite to the direction of $u_2(t_1)$. In other words, the dephasing of the elementary dipoles has been reversed and the net dipole moment has been reconstituted. This occurs at a time such that the interval of free precession after the π-pulse is equal to the interval between the end of the $\pi/2$-pulse and the onset of the π-pulse. A burst of radiation is emitted again (the echo) which decays exponentially with the time constant T_2. For an assembly of N atoms, the radiated intensity of the echo, as well as the free induction signal, is proportional to N^2. This is a consequence of the initial preparation of the system whereby each atom evolves into a superposition state, as indicated previously.

We have noted already that T_2 must be long compared to τ for photon echoes to be observed. The same constraint must be applied to T_1, which governs the decay of the population in the upper state (e.g., by spontaneous emission or collisions), since the echo amplitude decreases with the same time constant. This implies that the light pulse must have sufficient intensity to populate the upper state before there is any appreciable decay. Thus, as a general statement, τ must be small compared to T_1 and T_2.

It is important to bear in mind that the remarkable phenomenon of phase reversal after a system has been dephased by inhomogeneous mechanisms does not apply to homogeneous broadening. Dephasing due to the latter is not reversible.

The Bloch equations, in conjunction with the Maxwell equations for propagation in a medium, lead to the remarkable conclusion that circumstances exist such that a light pulse of a special shape can propagate through an absorbing medium without suffering attenuation or change in shape [23–25]. The special pulse shape, known as the hyperbolic secant pulse, is described by a field $E(z, t)$ proportional to

$$\operatorname{sech}\left[\frac{1}{\tau}\left(t - \frac{z}{v}\right)\right] \tag{6.137}$$

where τ is the pulse width and v is the velocity in the medium. For this self-induced transparency to occur, the center frequency of the pulse must be in resonance with the atoms of the medium and τ must be short compared to T_1 and T_2. The process then may be pictured as occurring in two steps: (1) energy is absorbed from the leading edge of the pulse, thereby exciting the medium, and (2) subsequent emission by the medium returns the energy to the trailing edge in such a manner as to keep the pulse shape intact. It is as if each atom has been exposed to a 2π pulse in the rotating frame with the result that there is no net interchange of energy between the light pulse and the medium in which it propagates. The propagation velocity of the pulse, however, is considerably reduced.

6.6 Damped Oscillator—Reservoir Formulation

The intimate connection between the modes of a radiation field and an assembly of simple harmonic oscillators suggests that the damped oscillator is a useful model for the damping of a single mode of the field [26–28]. In the present section, we shall develop the formalism for a single oscillator (the dynamical system) with Hamiltonian

$$\mathscr{H}_s = \hbar\omega a^\dagger a \tag{6.138}$$

damped by an assembly of other oscillators (the reservoir) with Hamiltonian

$$\mathscr{H}_r = \sum_k \hbar\omega_k b_k^\dagger b_k \tag{6.139}$$

through the interaction

$$V = \hbar\left(a\sum_k \lambda_k b_k^\dagger + a^\dagger \sum_k \lambda_k^* b_k \right), \tag{6.140}$$

in which λ_k and λ_k^* are coupling constants whose detailed form need not be specified. The total Hamiltonian $\mathscr{H}$ is the sum of $\mathscr{H}_s$, $\mathscr{H}_r$, and V. The results of this computation then are directly applicable to a radiation mode in a cavity damped by the multitude of other cavity modes.

The method of calculation employed in this section is based on the general formalism of Section 6.1 where the main objective was to derive the equation of motion for the density operator associated with a dynamical system when the latter was in contact with a reservoir. We return to Eq. (6.15). After interchanging the indices i and j in the second term,

$$\dot{\tilde{\rho}}_s(t) = \frac{1}{\hbar^2}\sum_{ij}\int_0^\infty d\tau([\tilde{S}_j(t-\tau)\rho_s(t), \tilde{S}_i(t)]\langle \tilde{R}_i(\tau)\tilde{R}_j\rangle$$
$$+ [\tilde{S}_j(\tau), \tilde{\rho}_s(t)\tilde{S}_i(t-\tau)]\langle R_i\tilde{R}_j(\tau)\rangle) \tag{6.141}$$

in which the variables S and R are related by the interaction (6.4). For the present case,

$$V = S_1 R_1 + S_2 R_2 \tag{6.142}$$

where, according to Eq. (6.140) and $\mathscr{H}_0 = \mathscr{H}_s + \mathscr{H}_r$,

$$\begin{aligned} S_1 &= \hbar a, \qquad S_2 = \hbar a^\dagger, \\ \tilde{S}_1(t) &= S_1 e^{-i\omega t}, \qquad \tilde{S}_2(t) = S_2^{i\omega t}, \end{aligned} \tag{6.143}$$

$$\begin{aligned} R_1 &= \sum_k \lambda_k b_k^\dagger, \qquad R_2 = \sum_k \lambda_k^* b_k, \\ \tilde{R}_1(t) &= \sum_k \lambda_k b_k^\dagger e^{i\omega_k t}, \qquad \tilde{R}_2(t) = \sum_k \lambda_k^* b_k e^{-i\omega_k t}. \end{aligned} \tag{6.144}$$

In view of these relations we assume, quite generally, that

$$\tilde{S}_k(t) = S_k e^{i\omega_k t}. \tag{6.145}$$

Then

$$\begin{aligned} \dot{\tilde{\rho}}_s(t) = \frac{1}{\hbar^2} \sum_{ij} \Big(&[S_j \tilde{\rho}_s(t), S_i] e^{i(\omega_i + \omega_j)t} \int_0^\infty e^{-i\omega_j \tau} \langle \tilde{R}_i(\tau) R_j \rangle \, d\tau \\ &+ [S_j, \tilde{\rho}_s(t) S_i] e^{i(\omega_i + \omega_j)t} \int_0^\infty e^{-i\omega_i \tau} \langle R_i \tilde{R}_j(\tau) \rangle \, d\tau. \end{aligned} \tag{6.146}$$

In the secular approximation, only the slowly varying features of the density operator are of interest; in that case, the sum over i and j is confined to those terms for which

$$\omega_i + \omega_j = 0. \tag{6.147}$$

Thus,

$$\dot{\tilde{\rho}}_s(t) = \sum_{ij} ([S_j \tilde{\rho}_s(t), S_i] A_{ij} + [S_j, \tilde{\rho}_s(t) S_i] B_{ij}), \tag{6.148}$$

where

$$A_{ij} = \frac{1}{\hbar^2} \int_0^\infty e^{-i\omega_j \tau} \langle \tilde{R}_i(\tau) R_j \rangle \, d\tau, \tag{6.149a}$$

$$B_{ij} = \frac{1}{\hbar^2} \int_0^\infty e^{-i\omega_i \tau} \langle R_i \tilde{R}_j(\tau) \rangle \, d\tau. \tag{6.149b}$$

Under the special conditions that

$$R_j^\dagger(\tau) = R_i(\tau), \tag{6.150}$$

as exemplified by R_1 and R_2 in Eq. (6.144), and noting that for any operator A

$$\mathrm{Tr}\{A^\dagger\} = [\mathrm{Tr}\{A\}]^*, \tag{6.151}$$

we have

$$\langle \tilde{R}_i(\tau)R_j\rangle^* = [\mathrm{Tr}_r\{\rho_r\tilde{R}_j^\dagger(\tau)R_j\}]^* = \mathrm{Tr}_r\{(\rho_r\tilde{R}_j^\dagger(\tau)R_j)^\dagger\}$$
$$= \mathrm{Tr}_r\{R_j^\dagger\tilde{R}_j(\tau)\rho_r\} = \mathrm{Tr}_r\{R_i\tilde{R}_j(\tau)\rho_r\} = \langle R_i\tilde{R}_j(\tau)\rangle. \quad (6.152)$$

As a consequence of Eqs. (6.147) and (6.152), it is seen that

$$A_{ij}^* = \frac{1}{\hbar^2}\int_0^\infty e^{-i\omega_i\tau}\langle\tilde{R}_i(\tau)R_j\rangle^*\,d\tau = B_{ij}. \quad (6.153)$$

The assumption that the reservoir is in thermal equilibrium at all times permits us to evaluate the correlation functions in a basis in which ρ_r is diagonal:

$$\langle\tilde{R}_i(\tau)R_j\rangle = \mathrm{Tr}_r\{\tilde{R}_i(\tau)R_j\rho_r\} = \sum_n\langle n|\tilde{R}_i(\tau)R_j\rho_r|n\rangle$$
$$= \sum_{nm}\langle n|\tilde{R}_i(\tau)|m\rangle\langle m|R_j|n\rangle\langle n|\rho_r|n\rangle$$
$$= \sum_{nm}\langle n|R_i|m\rangle\langle m|R_j|n\rangle\langle n|\rho_r|n\rangle e^{i\omega_{nm}\tau}. \quad (6.154)$$

Thus,

$$A_{ij} = \frac{1}{\hbar^2}\sum_{nm}\langle n|R_i|m\rangle\langle m|R_j|n\rangle\langle n|\rho_r|n\rangle\int_0^\infty e^{i(\omega_{nm}-\omega_j)\tau}\,d\tau, \quad (6.155)$$

and by the same procedure,

$$A_{ji} = \frac{1}{\hbar^2}\sum_{nm}\langle n|R_j|m\rangle\langle m|R_i|n\rangle\langle n|\rho_r|n\rangle\int_0^\infty e^{i(\omega_{nm}-\omega_i)\tau}\,d\tau, \quad (6.156a)$$

$$= \frac{1}{\hbar^2}\sum_{nm}\langle m|R_j|n\rangle\langle n|R_i|m\rangle\langle m|\rho_r|m\rangle\int_0^\infty e^{i(\omega_{mn}-\omega_i)\tau}\,d\tau, \quad (6.156b)$$

in which the last expression results from an interchange of n and m in Eq. (6.156a). The integrals are evaluated according to Eq. (2.584); however, we shall neglect the principal value that corresponds to a small energy shift. In any case, the energy shift can be incorporated into the energies of the states $|n\rangle$ and $|m\rangle$. Then, on account of Eq. (6.147), we have

$$\int_0^\infty e^{i(\omega_{mn}-\omega_i)\tau}\,d\tau = \pi\delta(\omega_{mn}-\omega_i) = \pi\delta(-\omega_{nm}+\omega_j) = \pi\delta(\omega_{nm}-\omega_j)$$
$$= \int_0^\infty e^{i(\omega_{nm}-\omega_j)\tau}\,d\tau, \quad (6.157)$$

and

$$\frac{A_{ij}}{A_{ji}} = \frac{\langle n|\rho_r|n\rangle}{\langle m|\rho_r|m\rangle} = e^{\beta\hbar\omega_{mn}} = e^{\beta\hbar\omega_i} = e^{-\beta\hbar\omega_j}. \quad (6.158)$$

It is further noted that the approximation of the integral by the δ-function alone implies that A_{ij} is real which, according to Eq. (6.153), means that $B_{ij} = A_{ij}$.

For the system described by Eq. (6.143), Eq. (6.148) becomes [8]

$$\begin{aligned}\dot{\tilde{\rho}}_s(t) &= \hbar[a^\dagger\tilde{\rho}_s(t), a]A_{12} + \hbar[a\tilde{\rho}_s(t), a^\dagger]A_{21} \\ &\quad + \hbar[a^\dagger, \tilde{\rho}_s(t)a]B_{12} + \hbar[a, \tilde{\rho}_s(t)a^\dagger]B_{21} \\ &= \hbar([a^\dagger\tilde{\rho}_s(t), a] + [a^\dagger, \rho_s(t)a])A_{12} \\ &\quad + \hbar([a\tilde{\rho}_s(t), a^\dagger] + [a, \tilde{\rho}_s(t)a^\dagger])A_{21}. \end{aligned} \tag{6.159}$$

For most applications, it is desirable to convert this relation into the Schrödinger representation. Referring to Eq. (2.165) and noting that

$$e^{-i\mathscr{H}_{st}/\hbar}ae^{i\mathscr{H}_{st}/\hbar} = ae^{i\omega t}, \qquad e^{-i\mathscr{H}_{st}/\hbar}a^\dagger e^{i\mathscr{H}_{st}/\hbar} = a^\dagger e^{-i\omega t}, \tag{6.160}$$

we have

$$\begin{aligned}\dot{\rho}_s(t) &= -\frac{i}{\hbar}[\mathscr{H}_s, \rho_s] + \hbar([a^\dagger\rho_s, a] + [a^\dagger, \rho_s a])A_{12} \\ &\quad + \hbar([a\rho_s, a^\dagger] + [a, \rho_s a^\dagger])A_{21} \\ &= -i\omega(a^\dagger a\rho_s - \rho_s a^\dagger a) - \hbar A_{12}(aa^\dagger\rho_s - 2a^\dagger\rho_s a + \rho_s aa^\dagger) \\ &\quad - \hbar A_{21}(a^\dagger a\rho_s - 2a\rho_s a^\dagger + \rho_s a^\dagger a). \end{aligned} \tag{6.161}$$

This is the equation of motion for the density operator ρ_s belonging to the dynamical system (oscillator or radiation mode) in contact with a reservoir (other oscillators or cavity modes). All the operators belong to the dynamical system; the effect of the reservoir is contained in the constants A_{12} and A_{21}. We are now in position to calculate various ensemble averages, for example,

$$\frac{d}{dt}\langle a^\dagger\rangle = \frac{d}{dt}\,\mathrm{Tr}\{a^\dagger\rho_s\} = \mathrm{Tr}\{a^\dagger\dot{\rho}_s(t)\}. \tag{6.162}$$

Inserting Eq. (6.161) into Eq. (6.162), the resulting terms are simplified by means of the commutation relation $[a, a^\dagger] = 1$. We then have

$$\begin{aligned}\mathrm{Tr}\{a^\dagger a^\dagger a\rho_s - a^\dagger\rho_s a^\dagger a\} &= \mathrm{Tr}\{-a^\dagger\rho_s\} = -\langle a^\dagger\rangle \\ \mathrm{Tr}\{2a^\dagger a^\dagger\rho_s a - a^\dagger aa^\dagger\rho_s - a^\dagger\rho_s aa^\dagger\} &= \mathrm{Tr}\{a^\dagger\rho_s\} = \langle a^\dagger\rangle \\ \mathrm{Tr}\{2a^\dagger a\rho_s a^\dagger - a^\dagger a^\dagger a\rho_s - a^\dagger\rho_s a^\dagger a\} &= \mathrm{Tr}\{-a^\dagger\rho_s\} = -\langle a^\dagger\rangle. \end{aligned} \tag{6.163}$$

Consequently,

$$\frac{d}{dt}\langle a^\dagger\rangle = [i\omega + \hbar(\mathrm{A}_{12} - \mathrm{A}_{21})]\langle a^\dagger\rangle. \tag{6.164}$$

In similar fashion, one finds

$$\frac{d}{dt}\langle a\rangle = [-i\omega + \hbar(\mathrm{A}_{12} - \mathrm{A}_{21})]\langle a\rangle, \tag{6.165}$$

$$\frac{d}{dt}\langle a^\dagger a\rangle = 2\hbar[\mathrm{A}_{12} + (\mathrm{A}_{12} - \mathrm{A}_{21})\langle a^\dagger a\rangle], \tag{6.166}$$

$$\frac{d}{dt}\langle aa^\dagger\rangle = 2\hbar[\mathrm{A}_{21} + (\mathrm{A}_{12} - \mathrm{A}_{21})\langle aa^\dagger\rangle], \tag{6.167}$$

$$\frac{d}{dt}\langle [a, a^\dagger]\rangle = 2\hbar(\mathrm{A}_{21} - \mathrm{A}_{12}) + 2\hbar(\mathrm{A}_{12} - \mathrm{A}_{21})\langle [a, a^\dagger]\rangle. \tag{6.168}$$

Solving these equations, one gets

$$\langle a^\dagger\rangle = \langle a^\dagger\rangle_0 e^{[i\omega + \hbar(\mathrm{A}_{12} - \mathrm{A}_{21})]t}, \tag{6.169}$$

$$\langle a\rangle = \langle a\rangle_0 e^{[-i\omega + \hbar(\mathrm{A}_{12} - \mathrm{A}_{21})]t}, \tag{6.170}$$

$$\langle a^\dagger a\rangle = \langle a^\dagger a\rangle_0 e^{2\hbar(\mathrm{A}_{12} - \mathrm{A}_{21})t} + \frac{\mathrm{A}_{12}}{\mathrm{A}_{21} - \mathrm{A}_{12}}[1 - e^{2\hbar(\mathrm{A}_{12} - \mathrm{A}_{21})t}], \tag{6.171}$$

$$\langle aa^\dagger\rangle = \langle aa^\dagger\rangle_0 e^{2\hbar(\mathrm{A}_{12} - \mathrm{A}_{21})t} + \frac{\mathrm{A}_{21}}{\mathrm{A}_{21} - \mathrm{A}_{12}}[1 - e^{2\hbar(\mathrm{A}_{12} - \mathrm{A}_{21})t}], \tag{6.172}$$

$$\langle [a, a^\dagger]\rangle = \langle [a, a^\dagger]\rangle_0 e^{2\hbar(\mathrm{A}_{12} - \mathrm{A}_{21})t} + [1 - e^{2\hbar(\mathrm{A}_{12} - \mathrm{A}_{21})t}], \tag{6.173}$$

where $\langle\cdots\rangle_0 \equiv \langle\cdots\rangle_{t=0}$. With the initial condition

$$\langle [a, a^\dagger]\rangle_0 = 1, \tag{6.174}$$

we have

$$\langle [a, a^\dagger]\rangle = 1. \tag{6.175}$$

This is a significant result that will be rederived and interpreted in the next section.

6.7 Damped Oscillator—Langevin Formulation

We now turn to another formulation for the damping of a single radiation mode in a cavity (the dynamical system) by the aggregate of other modes in the cavity (the reservoir). Instead of focusing on the density operator and its equation of motion as was done in the previous section, we now concentrate on the time development of the operators (in the Heisenberg representation) that characterize the dynamical system [1, 11, 28]. As usual, the modes will be represented by oscillators. The total Hamiltonian is

$$\mathscr{H} = \mathscr{H}_s + \mathscr{H}_r + V \tag{6.176}$$

where $\mathscr{H}_s$, $\mathscr{H}_r$, and V are given by (Eq. (6.138), (6.139), and (6.140), (6.176), respectively. Consistent with the general formalism, it is assumed that the reservoir oscillators are in thermal equilibrium and that the system operators a and $a^\dagger$ commute with all reservoir operators b_k, $b_k^\dagger$. However, a and $a^\dagger$ as well as b_k and b_k^* satisfy the harmonic oscillator commutation rules.

The equations of motion of the operators must satisfy the Heisenberg Eq. (2.102):

$$\begin{aligned} i\hbar\dot{a}_{\mathrm{H}}(t) &= [a_{\mathrm{H}}, \mathscr{H}], \\ &= \Bigg[a_{\mathrm{H}}, \hbar\omega a_{\mathrm{H}}^\dagger a_{\mathrm{H}} + \sum_k \hbar\omega_k (b_k^\dagger)_{\mathrm{H}}(b_k)_{\mathrm{H}} \\ &\quad + \hbar\Bigg(a_{\mathrm{H}}\sum_k \lambda_k (b_k^\dagger)_{\mathrm{H}} + a_{\mathrm{H}}^\dagger \sum_k \lambda_k^*(b_k)_{\mathrm{H}}\Bigg)\Bigg], \\ &= \hbar\omega a_{\mathrm{H}} + \hbar\sum_k \lambda_k^*(b_k)_{\mathrm{H}}. \end{aligned} \tag{6.177a}$$

Similarly,

$$i\hbar\dot{a}_{\mathrm{H}}^\dagger(t) = -\hbar\omega a_{\mathrm{H}}^\dagger - \hbar\sum_k \lambda_k (b_k^\dagger)_{\mathrm{H}}, \tag{6.177b}$$

$$i\hbar(\dot{b}_k)_{\mathrm{H}}(t) = \hbar\omega_k (b_k)_{\mathrm{H}} + \hbar\lambda_k a_{\mathrm{H}}, \tag{6.177c}$$

$$i\hbar(\dot{b}_k^\dagger) = -\hbar\omega_k (b_k^\dagger)_{\mathrm{H}} - \hbar\lambda_k^* a_{\mathrm{H}}^\dagger. \tag{6.177d}$$

To solve these equations for the time-development of the system operators, we begin by integrating Eq. (6.177c):

$$(b_k)_{\mathrm{H}}(t) = -i\lambda_k \int_{t_0}^{t} dt' e^{-i\omega_k(t-t')} a_{\mathrm{H}}(t') + e^{-i\omega_k(t-t_0)}(b_k)_{\mathrm{H}}(t_0). \tag{6.178}$$

One then may insert Eq. (6.178) into Eq. (6.177a), with $t_0 = 0$, to obtain the integro-differential equation

$$\dot{a}_{\mathrm{H}}(t) + i\omega a_{\mathrm{H}}(t) + \sum_k |\lambda_k|^2 \int_0^t dt' e^{-i\omega_k(t-t')} a_{\mathrm{H}}(t') = -i\sum_k \lambda_k^* e^{-i\omega_k t} bk(0). \tag{6.179}$$

At optical frequencies, the modes in a macroscopic cavity are densely packed, i.e., the number of modes per unit energy interval is very large. It is possible then to treat the term containing the integral by converting the sum over k into an integral in which the discrete frequencies ω_k are replaced by a continuous frequency variable Ω:

$$\sum_k |\lambda_k|^2 e^{-i\omega_k(t-t')} \to \int_{-\infty}^{\infty} d\Omega \lambda^2(\Omega) e^{-i\Omega(t-t')}. \tag{6.180}$$

The major contribution to the interaction between the radiation mode and reservoir modes will arise from those modes whose frequencies are close to that of the radiation mode. In this narrow frequency range, $\lambda^2(\Omega)$ remains essentially constant. Then

$$\int_{-\infty}^{\infty} d\Omega \lambda^2(\Omega) e^{-i\Omega(t-t')} \simeq \lambda^2(\Omega) \int_{-\infty}^{\infty} d\Omega e^{-i\Omega(t-t')},$$
$$= 2\pi\lambda^2(\Omega)\,\delta(t-t'), \tag{6.181}$$

and

$$\sum_k |\lambda_k|^2 \int_0^t dt' e^{-\omega_k(t-t')} a_{\mathrm{H}}(t') \simeq 2\pi\lambda^2(\Omega) \int_0^t dt' a_{\mathrm{H}}(t')\,\delta(t-t'). \tag{6.182}$$

The sharp peak at $t \simeq t'$ allows one to extend the upper limit to infinity; also, we may set

$$\int_0^{\infty} f(x')\,\delta(x-x')\,dx' = \frac{1}{2} f(x). \tag{6.183}$$

This expression is consistent with Eq. (2.287) without the principle value term [1] whose neglect corresponds to the neglect of a small frequency shift, which, as noted previously, could be taken into account by an adjustment of the radiation mode frequency. Thus,

$$\sum_k |\lambda_k|^2 \int_0^t dt' e^{-i\omega_k(t-t')} a_{\mathrm{H}}(t') = \pi\lambda^2(\Omega) a_{\mathrm{H}}(t) \equiv \gamma a_{\mathrm{H}}(t). \tag{6.184}$$

Equation (6.179) now has been simplified to

$$\dot{a}_{\mathrm{H}}(t) + i\omega a_{\mathrm{H}}(t) + \gamma a_{\mathrm{H}}(t) = -i\sum_k \lambda_k^* e^{-i\omega_k t} b_k \equiv F(t), \tag{6.185a}$$

and the Hermitian conjugate yields the corresponding equation for the creation operator

$$\dot{a}_{\mathrm{H}}^\dagger(t) - i\omega a_{\mathrm{H}}^\dagger(t) + \gamma a_{\mathrm{H}}^\dagger(t) - i\sum_k \lambda_k e^{i\omega_k t} b_k^\dagger \equiv F^\dagger(t). \tag{6.185b}$$

Since reservoir operators are assumed to fluctuate, $F(t)$ and $F(t')$—which are solely dependent on the reservoir operators b_k and $b_k^\dagger$—must also fluctuate. The structure of these equations therefore suggests that they are of the Langevin type, analogous to the equation for Brownian motion (with a driving term), and that γ is to be interpreted as a damping constant. $F(t)$ and $F^\dagger(t)$ are also called *Langevin noise sources.*

The oscillator modes have been assumed to be in thermal equilibrium and since

$$\langle n_k|b_k|n_k\rangle = 0, \qquad \langle n_k|b_k^\dagger|n_k\rangle = 0, \tag{6.186}$$

where $|n_k\rangle$ is an eigenstate of the oscillator number operator $N_k = k_k^\dagger b_k$. It follows that

$$\langle b_k\rangle = \mathrm{Tr}_r\{\rho_0 b_k\} = 0, \qquad \langle b_k^\dagger\rangle = \mathrm{Tr}_r\{\rho_0 b_k^\dagger\} = 0, \tag{6.187}$$

and therefore,

$$\langle F(t)\rangle = \langle F^\dagger(t)\rangle = 0. \tag{6.188}$$

On the other hand,

$$\langle b_k^\dagger b_1\rangle = \langle N_k\rangle\,\delta_{kl}, \qquad \langle b_k b_l^\dagger\rangle = \langle N_k + 1\rangle\,\delta_{kl}, \tag{6.189}$$

and

$$\begin{aligned}\langle F^\dagger(t)F(t')\rangle &= \sum_k |\lambda_k|^2 e^{i\omega_k(t-t')}\langle N_k\rangle,\\ &\simeq \lambda^2(\Omega)\langle n\rangle \int_{-\infty}^{\infty} d\Omega e^{i\Omega(t-t')}\\ &= 2\pi\lambda^2(\Omega)\langle n\rangle\,\delta(t-t'),\\ &= 2\gamma\langle n\rangle\,\delta(t-t') = \langle F^\dagger(t')F(t)\rangle,\end{aligned} \tag{6.190}$$

in which $\langle n\rangle$ is the average number of (reservoir) photons at the temperature T that are nearly resonant with the radiation mode. In the same fashion, it is found that

$$\langle F(t)F^\dagger(t')\rangle = \langle F(t')F^\dagger(t)\rangle = 2\gamma[\langle n\rangle + 1]\,\delta(t-t'), \tag{6.191}$$

$$\langle F(t)F(t')\rangle = \langle F^\dagger(t)F^\dagger(t')\rangle = 0. \tag{6.192}$$

Returning to Eq. (6.185), the solutions are

$$a_{\mathrm{H}}(t) = a(0)e^{-(i\omega+\gamma)t} + \int_0^t F(t')e^{-(i\omega+\gamma)(t-t')}\,dt', \tag{6.193}$$

$$a_{\mathrm{H}}^\dagger(t) = a^\dagger(0)e^{(i\omega-\gamma)t} + \int_0^t F^\dagger(t')e^{(i\omega-\gamma)(t-t')}\,dt', \tag{6.194}$$

and with these relations we find

$$\begin{aligned}
\frac{d}{dt} a_H a_H^\dagger &= \dot{a}_H a_H^\dagger + a_H \dot{a}_H^\dagger, \\
&= [-(i\omega + \gamma)a_H + F(t)]a_H^\dagger + a_H[(i\omega - \gamma)a_H^\dagger + F^\dagger(t)], \\
&= -2\gamma a_H a_H^\dagger + F(t)a_H^\dagger + a_H F^\dagger(t), \\
&= -2\gamma a_H a_H^\dagger + F(t)a^\dagger(0)e^{(i\omega-\gamma)t} \\
&\quad + \int_0^t F(t)F^\dagger(t')e^{(i\omega-\gamma)(t-t')}\,dt' + a(0)e^{-(i\omega+\gamma)t}F^\dagger(t) \\
&\quad + \int_0^t F(t')F^\dagger(t)e^{-(i\omega+\gamma)(t-t')}\,dt'.
\end{aligned} \tag{6.195}$$

Therefore, the rate of change of the ensemble average is

$$\begin{aligned}
\frac{d}{dt}\langle a_H a_H^\dagger \rangle &= -2\gamma\langle a_H a_H^\dagger\rangle + 2\gamma[\langle n\rangle + 1]\int_0^t \delta(t - t')e^{(i\omega-\gamma)(t-t')}\,dt' \\
&\quad + 2\gamma[\langle n\rangle + 1]\int_0^t \delta(t - t')e^{-(i\omega+\gamma)(t-t')}\,dt', \\
&= -2\gamma\langle a_H a_H^\dagger\rangle + 2\gamma[\langle n\rangle + 1].
\end{aligned} \tag{6.196}$$

Here, too, in the final step, the upper limit on the integral was extended to infinity and the principal value was ignored; in effect, we used the relation shown in Eq. (4.217) with $f(x) = 1$. Following the same procedure,

$$\frac{d}{dt}\langle a_H^\dagger a_H\rangle = -2\gamma\langle a_H^\dagger a_H\rangle + 2\gamma\langle n\rangle. \tag{6.197}$$

Combining Eqs. (6.196) and (6.197),

$$\frac{d}{dt}\langle [a_H, a_H^\dagger]\rangle = -2\gamma\langle [a_H, a_H^\dagger]\rangle + 2\gamma, \tag{6.198}$$

which has a solution

$$\langle [a_H, a_H^\dagger]\rangle = \mathrm{C}e^{-2\gamma t} + 1, \tag{6.199}$$

where C is a constant evaluated from the initial conditions $a_H(0) = a$, $a_H^\dagger(0) = a^\dagger$, and the commutation rule $[a, a^\dagger] = 1$. Therefore, at $t = 0$

$$\langle [a_H, a_H^\dagger]\rangle = \langle [a, a^\dagger]\rangle = 1 = \mathrm{C} + 1, \tag{6.200}$$

or C = 0. We then have the result that

$$\langle [a_{\mathrm{H}}(t), a_{\mathrm{H}}^{\dagger}(t)] \rangle = 1 \tag{6.201}$$

for all values of time. This relation, which is consistent with Eq. (6.175), underlines the importance of the fluctuating terms $F(t)$ and $F^{\dagger}(t)$ in Eq. (6.185). Were it not for their presence, the two equations would read

$$\dot{a}_{\mathrm{H}}(t) = -(i\omega + \gamma)a_{\mathrm{H}}(t), \qquad a_{\mathrm{H}}^{\dagger}(t) = (i\omega - \gamma)a_{\mathrm{H}}^{\dagger}(t), \tag{6.202}$$

with solutions

$$a_{\mathrm{H}}(t) = ae^{-(\omega+\gamma)t}, \qquad a_{\mathrm{H}}^{\dagger}(t) = a^{\dagger}e^{(i\omega-\gamma)t}, \tag{6.203}$$

and

$$[a_{\mathrm{H}}(t), a_{\mathrm{H}}^{\dagger}(t)] = [a, a^{\dagger}]e^{-2\gamma t} = e^{-2\gamma t}. \tag{6.204}$$

The last relation would be in clear violation of a fundamental principle of quantum mechanics which requires that in both the Schrödinger and Heisenberg representations,

$$[a, a^{\dagger}] = [a_{\mathrm{H}}(t), a_{\mathrm{H}}^{\dagger}(t)] = 1. \tag{6.205}$$

What has been shown is that in situations where damping terms arise, they cannot be arbitrarily incorporated into the boson operators without violating the basic commutation rules. When fluctuating terms are included, the commutation rules are restored but only in the statistical sense since it is the ensemble average of the commutator and not the commutator itself that becomes independent of time.

The results of this section may be connected with those of the previous section. As a consequence of Eq. (6.188), we have from Eqs. (6.193) and (6.194),

$$\langle a_{\mathrm{H}}(t) \rangle = \langle a(0) \rangle e^{-(i\omega+\gamma)t} = \langle a \rangle, \tag{6.206}$$
$$\langle a_{\mathrm{H}}^{\dagger}(t) \rangle = \langle a^{\dagger}(0) \rangle e^{(i\omega-\gamma)t} = \langle a^{\dagger} \rangle.$$

Comparing these equations with Eqs. (6.169) and (6.170) it is seen that

$$\gamma = \hbar(\mathrm{A}_{21} - A_{12}). \tag{6.207}$$

Similarly, the solution to Eq. (6.197) gives

$$\langle a_{\mathrm{H}}^{\dagger} a_{\mathrm{H}} \rangle = \langle a^{\dagger} a \rangle = \langle a^{\dagger} a \rangle_0 e^{-2\gamma t} + \langle n \rangle (1 - e^{-2\gamma t}) \tag{6.208}$$

which may be compare with Eq. (6.171) to yield

$$\langle n \rangle = \frac{\mathrm{A}_{12}}{\mathrm{A}_{21} - \mathrm{A}_{12}}. \tag{6.209}$$

6.8 Radiation Mode Coupled to an Atomic Reservoir

As another example of damping induced in a dynamical system by a reservoir, we consider a radiation field contained in a cavity and a beam of two-level atoms that traverses the cavity. The specific assumption are the following:

1. The time spent by each atom in the cavity is of the order of the radiative lifetime τ of the atomic excited state $|e\rangle$.

2. The atomic level populations are distributed according to the Boltzmann distribution and are not altered by interaction with the radiation field; in other words, the atoms are in thermal equilibrium at all times. Hence, the atomic density operator is $\rho_A(t) = \rho_0$ where ρ_0 is given by Eq. (5.28).

3. The radiation field consists of a single mode of frequency ω which is resonant with the transition frequency ω_0 of the two-level atoms.

4. The coupling between the atoms and the field is sufficiently weak to permit the use of low-order perturbation theory.

The computation begins with the Hamiltonian for a single field mode interacting with a single atom; at a later stage, it will be necessary to sum over all atoms. According to (5.113)

$$\mathscr{H}_0 = \mathscr{H}_A + \mathscr{H}_F = \frac{1}{2}\hbar\omega_0\sigma_z + \hbar\omega a^\dagger a,$$

$$\mathscr{H}_{AF} = \hbar[ga\sigma^+ + g^* a^\dagger \sigma^-].$$

We now proceed along the general lines of Section 6.1 for a dynamical system interacting with a reservoir. The temporal development of the density operator ρ associated with the total system is computed by successive approximations followed by a tracing operation over the atomic variables to yield the density operator ρ_F for the field alone. Hence, we write

$$\rho_F(t) = \mathrm{Tr}_A\{\rho(t)\} = \mathrm{Tr}_A\{\rho^{(0)}(t) + \rho^{(1)}(t) + \rho^{(2)}(t) + \cdots\}, \qquad (6.212)$$

and carry the computations to second order, which is sufficient to account for the main effects.

The various orders of approximation to the density operator are given by Eq. (2.169) and the equations that immediately follow. These are written in the interaction representation. In view of the resonance condition $\omega = \omega_0$ (Assumption 3), however, and the relations (4.89) for the field operators and Eq. (5.23) for the atomic operators, we have

$$\tilde{a}(t)\tilde{\sigma}^+(t) = a\sigma^+ e^{i(\omega_0 - \omega)(t - t_0)} = a\sigma^+,$$
$$\tilde{a}^\dagger(t)\tilde{\sigma}^-(t) = a^\dagger\sigma^- e^{-i(\omega_0 - \omega)(t - t_0)} = a^\dagger\sigma^-. \tag{6.213}$$

Therefore, for all values of t,

$$\tilde{\mathscr{H}}_{AF}(t) = e^{i\mathscr{H}_0(t - t_0)/\hbar}\mathscr{H}_{AF}e^{-i\mathscr{H}_0(t - t_0)/\hbar} = \mathscr{H}_{AF}. \tag{6.214}$$

The atom-field interaction is assumed to begin at $t = t_0$. We then have, in zero order,

$$\tilde{\rho}^{(0)}(t) = \rho(t_0) = \rho_A(t_0)\rho_F(t_0) = \rho_0\rho_F(t_0) = \rho_F(t_0)\rho_0. \tag{6.215}$$

Note that the atomic density operator ρ_0 commutes with the field operators $\rho_F(t_0)$, a, and $a^\dagger$. However, $\rho_F(t_0)$ does *not* commute with a and $a^\dagger$, so that the order in which products of these operators are written is not arbitrary.

The first-order term is

$$\tilde{\rho}^{(1)}(t) = -\frac{i}{\hbar}\int_{t_0}^{t} dt_1 [\tilde{\mathscr{H}}_{AF}(t_1), \rho(t_0)],$$
$$= -\frac{i}{\hbar}\int_{t_0}^{t} dt_1 [\mathscr{H}_{AF}, \rho_0\rho_F(t_0)] = \rho^{(1)}(t), \tag{6.216}$$

in which the last equality comes about again because of the resonance condition and the specific form of the interaction Hamiltonian $\mathscr{H}_{AF}$. The time-independent commutator may be conveniently evaluated by writing the operators in matrix form. Thus,

$$\mathscr{H}_{AF} = \hbar\begin{pmatrix} 0 & ga \\ g^*a^\dagger & 0 \end{pmatrix} \qquad \rho_0\rho_F(t_0) = \begin{pmatrix} \rho_e & 0 \\ 0 & \rho_g \end{pmatrix}\rho_F(t_0), \tag{6.217}$$

where $\rho_e \equiv \rho_{ee}$, $\rho_g \equiv \rho_{gg}$, and

$$\rho_e = \frac{1}{Z}e^{-\beta\hbar\omega_0/2}, \qquad \rho_g = \frac{1}{Z}e^{\beta\hbar\omega_0/2}. \tag{6.218}$$

Carrying out the matrix multiplications, we obtain

$$\rho^{(1)}(t) = -i(t - t_0)\begin{pmatrix} m_{11} & m_{12} \\ m_{21} & m_{22} \end{pmatrix}, \tag{6.219}$$

$$m_{11} = 0, \qquad m_{12} = g[a\rho_F(t_0)\rho_g - \rho_F(t_0)a\rho_e], \tag{6.220}$$
$$m_{21} = g^*[a^\dagger\rho_F(t_0)\rho_e - \rho_F(t_0)a^\dagger\rho_g], \qquad m_{22} = 0.$$

Proceeding to the second-order term,

$$\rho^{(2)}(t) = -\frac{i}{\hbar}\int_{t_0}^{t} dt_1 [\mathscr{H}_{\mathrm{AF}}, \rho^{(1)}(t_1)].$$

$$= -(t-t_0)^2\begin{pmatrix} n_{11} & n_{12} \\ n_{21} & n_{22} \end{pmatrix}, \tag{6.221}$$

$$n_{11} = \frac{|g|^2}{2}[aa^\dagger \rho_{\mathrm{F}}(t_0)\rho_{\mathrm{e}} - 2a\rho_{\mathrm{F}}(t_0)a^\dagger \rho_{\mathrm{g}} + \rho_{\mathrm{F}}(t_0)aa^\dagger \rho_{\mathrm{e}}],$$

$$n_{22} = \frac{|g|^2}{2}[a^\dagger a\rho_{\mathrm{F}}(t_0)\rho_{\mathrm{g}} - 2a^\dagger \rho_{\mathrm{F}}(t_0)a\rho_{\mathrm{e}} + \rho_{\mathrm{F}}(t_0)a^\dagger a\rho_{\mathrm{g}}], \tag{6.222}$$

$$n_{12} = n_{21} = 0.$$

We now set $\tau = t - t_0$ and collect the results to second order:

$$\rho(t) - \rho^{(0)}(t) = \rho(t) - \rho_{\mathrm{F}}(t_0)\rho_0 = \rho^{(1)}(t) + \rho^{(2)}(t),$$

$$= -i\tau\begin{pmatrix} m_{11} & m_{12} \\ m_{21} & m_{22} \end{pmatrix} - \tau^2\begin{pmatrix} n_{11} & n_{12} \\ n_{21} & n_{22} \end{pmatrix},$$

$$= -i\tau(m_{12}\sigma^+ + m_{21}\sigma^-) - \tau^2(n_{11}\sigma^+\sigma^- + n_{22}\sigma^-\sigma^+). \tag{6.223}$$

Since $\rho(t)$ is the density operator for the total system (atoms plus field), it is necessary to trace over the atomic variables to obtain the density operator for the field alone. The quantity of greater interest, however, the *change* in the density operator of the field due to interactions with the atoms that traverse the cavity. Therefore, let

$$\delta\rho_{\mathrm{F}}(\tau) = \mathrm{Tr}_{\mathrm{A}}\{\rho(t) - \rho_{\mathrm{F}}(t_0)\rho_0\} = \mathrm{Tr}_{\mathrm{A}}\{\rho^{(1)}(t) + \rho^{(2)}(t)\},$$

$$= \langle \mathrm{g}|\rho^{(1)}(t) + \rho^{(2)}(t)|\mathrm{g}\rangle + \langle \mathrm{e}|\rho^{(1)}(t) + \rho^{(2)}(t)|\mathrm{e}\rangle. \tag{6.224}$$

Inserting Eq. (6.223) one finds

$$\delta\rho_{\mathrm{F}}(\tau) = \langle \mathrm{g}|\rho^{(2)}(t)|\mathrm{g}\rangle + \langle \mathrm{e}|\rho^{(2)}(t)|\mathrm{e}\rangle = -\tau^2(n_{11} + n_{22}),$$

$$= -\frac{|g|^2}{2}\tau^2\{[a^\dagger a\rho_{\mathrm{F}}(t_0) - 2a\rho_{\mathrm{F}}(t_0)a^\dagger + \rho_{\mathrm{F}}(t_0)a^\dagger a]\rho_{\mathrm{g}}$$

$$+[aa^\dagger\rho_{\mathrm{F}}(t_0) - 2a^\dagger\rho_{\mathrm{F}}(t_0)a + \rho_{\mathrm{F}}(t_0)aa^\dagger]\rho_{\mathrm{e}}\},$$

$$= [\delta\rho_{\mathrm{F}}(\tau)]_{\mathrm{g}} + [\delta\rho_{\mathrm{F}}(\tau)]_{\mathrm{e}}. \tag{6.225}$$

Thus, for example, if an atom in the excited state $|\mathrm{e}\rangle$ enters the cavity and leaves after a time τ, we have $\rho_{\mathrm{g}} = 0$, $\rho_{\mathrm{e}} = 1$, and

$$\delta\rho_{\mathrm{F}}(\tau) = [\delta\rho_{\mathrm{F}}(\tau)]_{\mathrm{e}}. \tag{6.226}$$

If τ is the radiative lifetime of $|e\rangle$, the atom relaxes by emitting a photon during its passage through the cavity. The photon is added to the field, thereby altering the density operator of the field by $[\delta\rho_F(\tau)]_e$. An analogous change in the density operator, $[\delta\rho_F(\tau)]_g$, occurs when an atom in the ground state $|g\rangle$ is injected into the cavity and absorbs a photon from the field.

Thus far, the effects on the radiation field due to the passage of single atoms have been considered. The more relevant situation is one in which a beam of two-level atoms traverses the cavity. Hence, we now consider the cumulative effects of a beam that injects r_g atoms per second in the state $|g\rangle$ and r_e atoms per second in the state $|e\rangle$ into the cavity. If $r = r_g + r_e$, and the atoms are assumed to remain in thermal equilibrium at all times (Assumption 2), r_g and r_e are related by

$$r_g = r\rho_g, \qquad r_e = r\rho_e, \tag{6.227}$$

where ρ_g and ρ_e are given by Eq. (6.218).

To describe macroscopic effects on the radiation field due to the traversal of the atomic beam, it is necessary to adopt a coarser time scale. This means that we will be interested in changes that occur over a time Δt that is much greater than τ, the time of passage of an atom through the cavity. The assumption of weak coupling between the field and the atomic system (Assumption 4), however, implies that ρ_F, the density operator of the field, is a slowly varying function of the time. Therefore, with respect to the slow temporal variations in ρ_F, it will be assumed that Δt is small enough to be regarded as an infinitesimal despite the assumption that $\Delta t \gg \tau$. To obtain the change in ρ_F over the longer time Δt, we still may use Eq. (6.225) but with $\rho_F(t_0)$ replaced by $\rho_F(t)$. That is, while ρ_F on the right side of Eq. (6.225) may be regarded as a constant over the short time interval τ, it is no longer independent of time over intervals of longer duration. With these considerations in mind, we define the *coarse-grained* time derivative of $\rho_F(t)$ (Section 6.1)

$$\begin{aligned}\frac{d\rho_F(t)}{dt} &\simeq \frac{\Delta\rho_F}{\Delta t} = r_g(\delta\rho_F) + r_e(\delta\rho_F),\\ &= -\frac{|g|^2}{2}\tau^2 r\{[a^\dagger a\rho_F(t) - 2a\rho_F(t)a^\dagger + \rho_F(t)a^\dagger a]\rho_g\\ &\quad + [aa^\dagger\rho_F(t) - 2a^\dagger\rho_F(t)a + \rho_F(t)aa^\dagger]\rho_e\}.\end{aligned} \tag{6.228}$$

This is the fundamental equation governing the change in the statistics of the radiation field due to its one-photon interactions with the atomic system. In Eq. (6.161) an expression was derived for the total (free motion plus reservoir interaction) rate of change of ρ_s, the density operator for a general dynamical system. For the present case, ρ_s is equated to ρ_F, the density operator of the radiation field. But since the free motion is not contained in

Eq. (6.228), we ignore the first term (the free motion) in Eq. (6.161) in order to compare the two expressions. We then have

$$\frac{|g|^2}{2}\tau^2 r\rho_g = \hbar A_{21}, \qquad \frac{|g|^2}{2}\tau^2 r\rho_e = \hbar A_{12}. \tag{6.229}$$

Let use now compute the matrix elements of $d\rho_F/dt$ in the photon number basis. We have

$$\begin{aligned}\langle n|aa^\dagger \rho_F(t)|n'\rangle &= \sum_l \langle n|aa^\dagger|l\rangle\langle l|\rho_F(t)|n'\rangle,\\ &= (n+1)\langle n|\rho_F(t)|n'\rangle.\end{aligned} \tag{6.230a}$$

Similarly,

$$\langle n|a^\dagger \rho_F(t) a|n'\rangle = \sqrt{nn'}\,\langle n-1|\rho_F(t)|n'-1\rangle, \tag{6.230b}$$

$$\langle n|\rho_F(t) aa^\dagger|n'\rangle = (n'+1)\langle n|\rho_F(t)|n'\rangle, \tag{6.230c}$$

$$\langle n|a^\dagger a\rho_F(t)|n'\rangle = n\langle n|\rho_F(t)|n'\rangle, \tag{6.230d}$$

$$\langle n|a\rho_F(t)a^\dagger|n'\rangle = \sqrt{(n+1)(n'+1)}\,\langle n+1|\rho_F(t)|n'+1\rangle, \tag{6.230e}$$

$$\langle n|\rho_F(t)a^\dagger a|n'\rangle = n'\langle n|\rho_F(t)|n'\rangle. \tag{6.230f}$$

With these relations,

$$\begin{aligned}\left\langle n\left|\frac{d\rho_F(t)}{dt}\right|n'\right\rangle = -\frac{|g|^2}{2}\tau^2 r\{&[(n+n')\langle n|\rho_F(t)|n'\rangle\\ &-2\sqrt{(n+1)(n'+1)}\,\langle n+1|\rho_F(t)|n'+1\rangle]\rho_g\\ &+[\{(n+1)+(n'+1)\}\langle n|\rho_F(t)|n'\rangle\\ &-2\sqrt{nn'}\,\langle n-1|\rho_F(t)|n'-1\rangle]\rho_e\},\end{aligned} \tag{6.231}$$

and

$$\begin{aligned}\left\langle n\left|\frac{d\rho_F(t)}{dt}\right|n\right\rangle = -|g|^2\tau^2 r\{&[n\langle n|\rho_F(t)|n\rangle - (n+1)\langle n+1|\rho_F(t)|n+1\rangle]\rho_g\\ &+[(n+1)\langle n|\rho_F(t)|n\rangle - n\langle n-1|\rho_F(t)|n-1\rangle]\rho_e\}.\end{aligned} \tag{6.232}$$

A diagrammatic representation of the four terms is shown in Fig. 6.6. The two terms proportional to ρ_g in Eq. (6.232) are associated with atoms in the ground state $|g\rangle$; such atoms will absorb photons in the course of a transition $|g\rangle \to |e\rangle$ and will therefore decrease the photon number of the field. Since the matrix element $\langle n|\rho_F(t)|n\rangle$ is interpretable as the probability of finding n photons in the field, the term $-|g|^2\tau^2 rn\langle n|\rho_F(t)|n\rangle\rho_g$ represents a transition rate from a state with n photons to a state with $(n-1)$ photons, that is

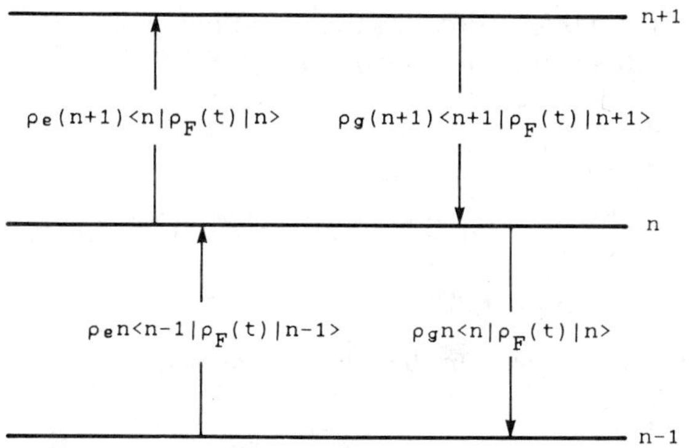

FIGURE 6.6 Diagram to represent the change in the statistics of a radiation field due to its interaction with a two-level atomic system.

$|n\rangle \to |n-1\rangle$. Such transitions reduce the magnitude of $\langle n|\rho_F(t)|n\rangle$. The term $|g|^2\tau^2 r(n+1)\langle n+1|\rho_F(t)|n+1\rangle\rho_g$ also represents a loss rate of photons from the field. The loss, however, is from a state with $n+1$ photons; hence, the field makes a transition to a state with n photons, or, $|n+1\rangle \to |n\rangle$. Transitions of this type *increase* the magnitude of $\langle n|\rho_F(t)|n\rangle$. The first term in Eq. (6.232) therefore appears with a negative sign and the second term with a positive sign. Similar interpretations apply to the terms proportional to ρ_e. Both terms increase the photon number of the field owing to the atomic transitions $|e\rangle \to |g\rangle$ but $-|g|^2\tau^2 r(n+1)\langle n|\rho_F(t)|n\rangle\rho_e$ is associated with transitions $|n\rangle \to |n+1\rangle$ and $|g|^2\tau^2 rn\langle n-1|\rho_F(t)|n-1\rangle\rho_e$ is associated with transitions $|n-1\rangle \to |n\rangle$. Consequently, the magnitude of $\langle n|\rho_F(t)|n\rangle$ is reduced by the third term and enhanced by the fourth term in Eq. (6.232).

We now shall examine the equilibrium situation where

$$\left\langle n\left|\frac{d\rho_F(t)}{dt}\right|n\right\rangle = 0. \tag{6.233}$$

Although this condition ensures equilibrium for the state $|n\rangle$, the states $|n+1\rangle$ and $|n-1\rangle$ are not necessarily in equilibrium since there can be a net flow of photon probability between $|n+1\rangle$ and $|n-1\rangle$ without disturbing the equilibrium of $|n\rangle$. We encountered a similar situation in the discussion of the random walk problem (Section 1.7). To make sure that all the states are in equilibrium, it is necessary to invoke the principle of detailed balance, which requires that the net flow of probability between adjacent pairs of states vanishes:

$$\begin{aligned} \rho_e n\langle n-1|\rho_F(t)|n-1\rangle - \rho_g n\langle n|\rho_F(t)|n\rangle &= 0, \\ \rho_e(n+1)\langle n|\rho_F(t)|n\rangle - \rho_g(n+1)\langle n+1|\rho_F(t)|n+1\rangle &= 0. \end{aligned} \tag{6.234}$$

Since the value of n is arbitrary, the two equations express the same relation; the general requirement for the steady state is

$$\begin{aligned}\langle n|\rho_F(t)|n\rangle &= \frac{\rho_e}{\rho_g}\langle n-1|\rho_F(t)|n-1\rangle \\ &= e^{-\beta\hbar\omega}\langle n-1|\rho_F(t)|n-1\rangle = e^{-2\beta\hbar\omega}\langle n-2|\rho_F(t)|n-2\rangle, \\ &\vdots \\ &= e^{-n\beta\hbar\omega}\langle 0|\rho_F(t)|0\rangle,\end{aligned} \tag{6.235}$$

in which the atomic system has been assumed to remain in thermal equilibrium. Also,

$$\begin{aligned}\operatorname{Tr}\rho_F(t) &= \sum_n \langle n|\rho_F(t)|n\rangle = \langle 0|\rho_F(t)|0\rangle \sum_n e^{-n\beta\hbar\omega} \\ &= \langle 0|\rho_F(t)|0\rangle \frac{1}{1-e^{-\beta\hbar\omega}} = 1.\end{aligned} \tag{6.236}$$

Thus,

$$\langle 0|\rho_F(t)|0\rangle = 1 - e^{-\beta\hbar\omega}, \tag{6.237}$$

and

$$\langle n|\rho_F(t)|n\rangle = e^{-n\beta\hbar\omega}(1 - e^{\beta\hbar\omega}), \tag{6.238}$$

which is nothing more than the probability of finding n photons in the radiation mode at thermal equilibrium, as in Eq. (4.213). This is not surprising since the radiation field interacts with a reservoir (the atomic system) that is maintained at thermal equilibrium. The field, therefore, ultimately must achieve a steady state at the same temperature as the atomic system.

It is often of interest to investigate the manner in which the expectation values of the field operators $\langle a\rangle$, $\langle a^\dagger\rangle$ and $\langle a^\dagger a\rangle$ develop in time. Recalling the invariance of the trace to cyclic permutations of the operators and the commutation rules of the field operators,

$$\begin{aligned}\frac{d\langle a\rangle}{dt} &= \frac{d}{dt}\operatorname{Tr}\{a\rho_F\} = \operatorname{Tr}\left\{a\frac{d\rho_F}{dt}\right\} \\ &= -\operatorname{Tr}\{[a^\dagger a\rho_F a - 2a\rho_F a^\dagger a + \rho_F a^\dagger a a]\hbar A_{21} \\ &\quad + [aa^\dagger\rho_F a - 2a^\dagger\rho_F aa + \rho_F aa^\dagger a]\hbar A_{12}\}, \\ &= -\operatorname{Tr}\{\rho_F[aa^\dagger a - 2a^\dagger aa + a^\dagger aa]\hbar A_{21} \\ &\quad + \rho_F[aaa^\dagger - 2aaa^\dagger + aa^\dagger a]\hbar A_{12}\} \\ &= -\operatorname{Tr}\{\rho_F a A_{21} - \rho_F a A_{12}\} \\ &= -\hbar(A_{21} - A_{12})\langle a\rangle = -\gamma\langle a\rangle,\end{aligned} \tag{6.239}$$

where γ is given by Eq. (6.207). Similarly, with $\langle n \rangle$ as in Eq. (6.209),

$$\frac{d\langle a^\dagger \rangle}{dt} = -\hbar(A_{21} - A_{12})\langle a^\dagger \rangle = -\gamma\langle a^\dagger \rangle,$$
$$\frac{d\langle a^\dagger a \rangle}{dt} = -2\hbar(A_{21} - A_{12})\langle a^\dagger a \rangle + 2\hbar A_{12} = -2\gamma\langle a^\dagger a \rangle + 2\gamma\langle n \rangle. \tag{6.240}$$

These equations indicate the manner in which the statistics of a light beam are altered by lowest order (single photon) processes during traversal of an absorbing medium.

The results pertaining to a field mode coupled to a two-level atomic reservoir may be used to demonstrate that the function $P(\alpha)$ in the diagonal representation of the density operator (Eq. (4.232)) in a coherent state basis satisfies the Fokker-Planck equation.

In the notation of Eq. (6.238), Eq. (6.228) for the density operator associated with the field mode is

$$\frac{d\rho_F}{dt} = -\hbar A_{21}[a^\dagger a\rho_F - 2a\rho_F a^\dagger + \rho_F a^\dagger a] - \hbar A_{12}[aa^\dagger\rho_F - 2a^\dagger\rho_F a + \rho_F aa^\dagger]. \tag{6.241}$$

We now insert the diagonal representation for ρ into the right side of the equation:

$$\begin{aligned}\frac{d\rho_F}{dt} &= -\hbar A_{21}\int d^2\alpha P(\alpha)[a^\dagger a|\alpha\rangle\langle\alpha| - 2a|\alpha\rangle\langle\alpha|a^\dagger + |\alpha\rangle\langle\alpha|a^\dagger a]\\ &\quad -\hbar A_{12}\int d^2\alpha P(\alpha)[aa^\dagger|\alpha\rangle\langle\alpha| - 2a^\dagger|\alpha\rangle\langle\alpha|a + |\alpha\rangle\langle\alpha|aa^\dagger],\\ &= -\hbar A_{21}\int d^2\alpha P(\alpha)\left(\alpha\frac{\partial}{\partial\alpha} + \alpha^*\frac{\partial}{\partial\alpha^*}\right)|\alpha\rangle\langle\alpha|\\ &\quad + \hbar A_{12}\int d^2\alpha P(\alpha)\left(2\frac{\partial^2}{\partial\alpha\partial\alpha^*} + \alpha\frac{\partial}{\partial\alpha} + \alpha^*\frac{\partial}{\partial\alpha^*}\right)|\alpha\rangle\langle\alpha|,\end{aligned} \tag{6.242}$$

in which the second equation is based on the relations in Eqs. (4.201) to (4.208). An integration by parts with $P(\alpha) = 0$ at $\alpha = \pm\infty$ yields

$$\begin{aligned}\int d^2\alpha P\alpha\frac{\partial}{\partial\alpha}|\alpha\rangle\langle\alpha| &= -\int d^2\alpha\frac{\partial}{\partial\alpha}(\alpha P)|\alpha\rangle\langle\alpha|,\\ \int d^2\alpha P\alpha^*\frac{\partial}{\partial\alpha^*}|\alpha\rangle\langle\alpha| &= -\int d^2\alpha\frac{\partial}{\partial\alpha^*}(\alpha^* P)|\alpha\rangle\langle\alpha|,\\ \int d^2\alpha P\frac{\partial^2}{\partial\alpha\partial\alpha^*}|\alpha\rangle\langle\alpha| &= \int d^2\alpha\frac{\partial^2}{\partial\alpha\partial\alpha^*}P|\alpha\rangle\langle\alpha|.\end{aligned} \tag{6.243}$$

With these equations,

$$\frac{d\rho_F}{dt} = \hbar(A_{21} - A_{12}) \int d^2\alpha \left(\frac{\partial}{\partial\alpha}(\alpha P) + \frac{\partial}{\partial\alpha^*}(\alpha^* P) \right) |\alpha\rangle\langle\alpha| + 2\hbar A_{12} \int d^2\alpha \frac{\partial^2}{\partial\alpha\partial\alpha^*} P|\alpha\rangle\langle\alpha|. \tag{6.244}$$

But

$$\frac{d\rho_F}{dt} = \int d^2\alpha \dot{P}(\alpha)|\alpha\rangle\langle\alpha|, \tag{6.245}$$

and since α is arbitrary,

$$\dot{P}(\alpha) = \hbar(A_{21} - A_{12}) \left(\frac{\partial}{\partial\alpha}(\alpha P) + \frac{\partial}{\partial\alpha^*}(\alpha^* P) \right) + 2\hbar A_{12} \frac{\partial^2}{\partial\alpha\partial\alpha^*} P. \tag{6.246}$$

This is the Fokker-Planck equation for the distribution function in the diagonal representation of the density operator associated with a radiation mode interacting with a two-level atomic reservoir [29–31].

An alternative form of (6.246) is obtained by writing $\alpha = x + iy$. Then

$$\frac{\partial}{\partial\alpha} = \frac{1}{2}\left(\frac{\partial}{\partial x} - i\frac{\partial}{\partial y} \right), \qquad \frac{\partial}{\partial\alpha^*} = \frac{1}{2}\left(\frac{\partial}{\partial x} + i\frac{\partial}{\partial y} \right), \qquad \frac{\partial^2}{\partial\alpha\partial\alpha^*} = \frac{1}{4}\left(\frac{\partial^2}{\partial x^2} + \frac{\partial^2}{\partial y^2} \right), \tag{6.247}$$

$$\begin{aligned} \frac{\partial}{\partial\alpha}(\alpha P) &= \frac{1}{2}\left[\frac{\partial}{\partial x}(xP) + \frac{\partial}{\partial y}(yP) + i\left(y\frac{\partial P}{\partial x} - x\frac{\partial P}{\partial y} \right) \right], \\ \frac{\partial}{\partial\alpha^*}(\alpha^* P) &= \frac{1}{2}\left[\frac{\partial}{\partial x}(xP) + \frac{\partial}{\partial y}(yP) - i\left(y\frac{\partial P}{\partial x} - x\frac{\partial P}{\partial y} \right) \right]. \end{aligned} \tag{6.248}$$

Thus,

$$\dot{P}(\alpha, t) = \hbar(A_{21} - A_{12}) \left[\frac{\partial}{\partial x}(xP) + \frac{\partial}{\partial y}(yP) \right] + \frac{\hbar A_{12}}{2} \left[\frac{\partial^2 P}{\partial x^2} + \frac{\partial^2 P}{\partial y^2} \right]. \tag{6.249}$$

Much of present-day laser theory is based on the Fokker-Planck equation.

6.9 Vacuum Fluctuations and Spontaneous Emission

Spontaneous emission was discussed in Section 5.6; an alternative approach is based on the general formalism of a dynamical system interacting with a reservoir (Section 6.1). The dynamical system in this case consists of a two-level atom in a cavity and the reservoir consists of fluctuating quantized modes of a radiation field in the cavity. It further is assumed that the radiation modes are vacant, that is, the radiation field as a whole is in the vacuum state. Consistent with the general notion of a reservoir, the radiation field remains in the vacuum state for all time.

The total Hamiltonian is given by Eq. (6.1), which for the present case is identical with Eq. (5.113):

$$\mathscr{H}_s = \mathscr{H}_A = \frac{\hbar}{2}\omega_0\sigma_z, \qquad \mathscr{H}_r = \mathscr{H}_F = \hbar\sum_j \omega_j a_j^\dagger a_j, \tag{6.250}$$

$$V = \mathscr{H}_{AF} = \hbar\left[\sigma^+ \sum_j g_j a_j + \sigma^- \sum_j g_j^* a_j^\dagger\right], \tag{6.251}$$

with

$$[a_j, a_k^\dagger] = \delta_{jk}, \qquad [a_j, a_k] = [a_j^\dagger, a_k^\dagger] = 0, \tag{6.252}$$

$$[\sigma^\pm, a_j] = [\sigma^\pm, a_j^\dagger] = 0. \tag{6.253}$$

The coupling constants g_j and g_j^*, defined by Eq. (5.109), provide a measure for the strength of interaction between the atom and the field modes.

Since the interaction $\mathscr{H}_{AF}$ evidently conforms to Eq. (6.4), and ρ_A corresponds to ρ_s—the density operator for the dynamical system (the atom)—we may refer to Eq. (6.5) to obtain the equation of motion for ρ_A. Because the first term in Eq. (6.5) vanishes due to the fluctuations in the reservoir, we have

$$\dot{\tilde{\rho}}_A = -\frac{1}{\hbar^2}\int_0^t dt' \mathrm{Tr}_F\{[\tilde{\mathscr{H}}_{AF}(t), [\tilde{\mathscr{H}}_{AF}(t'), \tilde{\rho}(t')]]\}, \tag{6.254}$$

where

$$\begin{aligned}\tilde{\mathscr{H}}_{AF}(t) &= \hbar\left[\tilde{\sigma}^+ \sum_j g_j \tilde{a}_j + \tilde{\sigma}^- \sum_j g_j^* \tilde{a}_j^\dagger\right] \\ &= \hbar\left[\sigma^+ \sum_j g_j a_j e^{i(\omega_0-\omega_j)t} + \sigma^- \sum_j g_j^* a_j^\dagger e^{-i(\omega_0-\omega_j)t}\right].\end{aligned} \tag{6.255}$$

In the second equation, we referred to Eq. (4.89) and Eq. (5.23). Previously, an approximation for $\tilde{\rho}(t')$—the density operator for the total system—was written as in Eq. (6.11). We invoke the same approximation in the present case

$$\tilde{\rho}(t') = \tilde{\rho}_A(t)\rho_F, \tag{6.256}$$

and for ρ_F, in view of our assumption that the field in the cavity is in the vacuum state for all time, we take

$$\rho_r = \rho_F = |0\rangle\langle 0|. \tag{6.257}$$

Let us now consider the four terms in the integrand of Eq. (6.254). One such term is

$$\begin{aligned}
&\mathrm{Tr_F}\{\tilde{\mathscr{H}}_{\mathrm{AF}}(t)\tilde{\mathscr{H}}_{\mathrm{AF}}(t')\tilde{\rho}(t')\} \\
&\quad = Tr_{\mathrm{F}}\left\{\hbar\left[\sigma^+\sum_j g_j a_j e^{i(\omega_0-\omega_j)t} + \sigma^-\sum_j g_j^* a_j^\dagger e^{-i(\omega_0-\omega_j)t}\right]\right. \\
&\qquad \left.\times\,\hbar\left[\sigma^+\sum_k g_k a_k e^{i(\omega_0-\omega_k)t'} + \sigma^-\sum_k g_k^* a_k^\dagger e^{-i(\omega_0-\omega_k)t'}\right]\tilde{\rho}_{\mathrm{A}}(t)|0\rangle\langle 0|\right\}.
\end{aligned} \tag{6.258}$$

Recalling that

$$(\sigma^+)^2 = (\sigma^-)^2 = 0 \tag{6.259}$$

and that field operators commute with atomic operators, we have

$$\begin{aligned}
&\mathrm{Tr_F}\{\tilde{\mathscr{H}}_{\mathrm{AF}}(t)\tilde{\mathscr{H}}_{\mathrm{AF}}(t')\tilde{\rho}(t')\} \\
&\quad = \hbar^2\left[\sigma^+\sigma^-\tilde{\rho}_{\mathrm{A}}(t)\sum_{jk} g_j g_k^* e^{i[(\omega_0-\omega_j)t-(\omega_0-\omega_k)t']}\mathrm{Tr_F}\{a_j a_k^\dagger|0\rangle\langle 0|\}\right. \\
&\qquad \left. + \sigma^-\sigma^+\tilde{\rho}_{\mathrm{A}}(t)\sum_{jk} g_j^* g_k e^{-i[(\omega_0-\omega_j)t-(\omega_0-\omega_k)t']}\mathrm{Tr_F}\{a_j^\dagger a_k|0\rangle\langle 0|\}\right].
\end{aligned} \tag{6.260}$$

Since

$$\mathrm{Tr_F}\{a_j^\dagger a_k|0\rangle\langle 0|\} = 0, \qquad \mathrm{Tr_F}\{a_j a_k^\dagger|0\rangle\langle 0|\} = \delta_{jk}, \tag{6.261}$$

the second term in Eq. (6.260) is zero and

$$\mathrm{Tr_F}\{\tilde{\mathscr{H}}_{\mathrm{AF}}(t)\tilde{\mathscr{H}}_{\mathrm{AF}}(t')\tilde{\rho}(t')\} = \hbar^2\sigma^+\sigma^-\tilde{\rho}_{\mathrm{A}}(t)\sum_j |g_j|^2 e^{i(\omega_0-\omega_j)(t-t')}. \tag{6.262}$$

Another term in the integrand of Eq. (6.254) is

$$\begin{aligned}
&\mathrm{Tr_F}\{\tilde{\mathscr{H}}_{\mathrm{AF}}(t)\tilde{\rho}(t')\tilde{\mathscr{H}}_{\mathrm{AF}}(t')\} \\
&\quad = \hbar^2\left[\sigma^+\tilde{\rho}_{\mathrm{A}}(t)\sigma^+\sum_{jk} g_j g_k e^{i[(\omega_0-\omega_j)t+(\omega_0-\omega_k)t']}\mathrm{Tr_F}\{a_j|0\rangle\langle 0|a_k\}\right. \\
&\qquad + \sigma^+\tilde{\rho}_{\mathrm{A}}(t)\sigma^-\sum_{jk} g_j g_k^* e^{i[(\omega_0-\omega_j)t-(\omega_0-\omega_k)t']}\mathrm{Tr_F}\{a_j|0\rangle\langle 0|a_k^\dagger\} \\
&\qquad + \sigma^-\tilde{\rho}_{\mathrm{A}}(t)\sigma^+\sum_{jk} g_j^* g_k e^{-i[(\omega_0-\omega_j)t-(\omega_0-\omega_k)t']}\mathrm{Tr_F}\{a_j^\dagger|0\rangle\langle 0|a_k\} \\
&\qquad \left. + \sigma^-\tilde{\rho}_{\mathrm{A}}(t)\sigma^-\sum_{jk} g_j^* g_k^* e^{-i[(\omega_0-\omega_j)t+(\omega_0-\omega_k)t']} \times \mathrm{Tr_F}\{a_j^\dagger|0\rangle\langle 0|a_k^\dagger\}\right].
\end{aligned} \tag{6.263}$$

But

$$\mathrm{Tr_F}\{a_j|0\rangle\langle 0|a_k\} = \mathrm{Tr_F}\{a_j|0\rangle\langle 0|a_k^\dagger\} = \mathrm{Tr_F}\{a_j^\dagger|0\rangle\langle 0|a_k^\dagger\} = 0.$$
$$\mathrm{Tr_F}\{a_j^\dagger|0\rangle\langle 0|a_k\} = \delta_{jk}. \tag{6.264}$$

Hence,

$$\mathrm{Tr_F}\{\tilde{\mathscr{H}}_{\mathrm{AF}}(t)\tilde{\rho}(t')\tilde{\mathscr{H}}_{\mathrm{AF}}(t')\} = \hbar^2\sigma^-\tilde{\rho}_{\mathrm{A}}(t)\sigma^+\sum_j|\mathrm{g_j}|^2e^{-i(\omega_0-\omega_j)(t-t')}. \tag{6.265}$$

Following the same procedure, the remaining two terms in the integrand of Eq. (6.254) are

$$\mathrm{Tr_F}\{\tilde{\mathscr{H}}_{\mathrm{AF}}(t')\tilde{\rho}(t')\tilde{\mathscr{H}}_{\mathrm{AF}}(t)\} = \hbar^2\sigma^-\tilde{\rho}_{\mathrm{A}}(t)\sigma^+\sum_j|\mathrm{g_j}|^2e^{i(\omega_0-\omega_j)(t-t')}, \tag{6.266}$$

$$\mathrm{Tr_F}\{\tilde{\rho}(t')\tilde{\mathscr{H}}_{\mathrm{AF}}(t')\tilde{\mathscr{H}}_{\mathrm{AF}}(t)\} = \hbar^2\tilde{\rho}_{\mathrm{A}}(t)\sigma^+\sigma^-\sum_j|\mathrm{g}_j|^2e^{-i(\omega_0-\omega_j)(t-t')}. \tag{6.267}$$

As was done previously, the summation over the modes may be replaced by an integral, the frequencies ω_j by a continuous variable ω, and $|\mathrm{g_j}|^2$ by a continuous function $g^2(\omega)$ which has the dimensions of ω. In this way,

$$\sum_j|\mathrm{g_j}|^2e^{\pm i(\omega_0-\omega_j)(t-t')} \to \int_{-\infty}^{\infty} g^2(\omega)e^{\pm i(\omega_0-\omega)(t-t')}\,d\omega. \tag{6.268}$$

The terms in Eq. (6.254) now may be evaluated. Thus, with $\tau = t - t'$,

$$\int_0^t dt'\mathrm{Tr_F}\{\tilde{\mathscr{H}}_{\mathrm{AF}}(t)\tilde{\mathscr{H}}_{\mathrm{AF}}(t')\tilde{\rho}(t')\} = \hbar^2\sigma^+\sigma^-\tilde{\rho}_{\mathrm{A}}(t)\int_{-\infty}^{\infty} g^2(\omega)\,d\omega\int_0^t e^{\pm i(\omega_0-\omega)\tau}\,d\tau. \tag{6.269}$$

The integral over ω in Eq. (6.269) is

$$\int_{-\infty}^{\infty} g^2(\omega)e^{\pm i\omega\tau}\,d\omega \tag{6.270}$$

which is just the Fourier transform of $g^2(\omega)$. It may be assumed that in the region around ω_0 the coupling constants do not vary a great deal and that they fall off smoothly as ω departs from ω_0. The Fourier transform of $g^2(\omega)$ then will be a narrow pulse peaked at $\tau = 0$, and if t long compared to the width of the pulse, the upper limit on the τ-integral may be extended to infinity. Thus,

$$\int_0^t e^{\pm i(\omega_0-\omega)\tau}\,d\tau \to \int_0^{\infty} e^{\pm i(\omega_0-\omega)\tau}\,d\tau = \pm iP\left(\frac{1}{\omega_0-\omega}\right) + \pi\delta(\omega_0-\omega), \tag{6.271}$$

in which we invoked the relation (Eq. (2.287)). Inserting this expression into Eq. (6.269),

$$\int_{-\infty}^{\infty} g^2(\omega)\,d\omega \int_0^t e^{\pm i(\omega_0-\omega)\tau}\,d\tau = \pm iP\int_{-\infty}^{\infty} \frac{g^2(\omega)}{\omega_0-\omega}\,d\omega + \pi g^2(\omega)$$
$$\equiv \pm i\,\Delta\omega + \frac{\Gamma}{2}, \tag{6.272}$$

where

$$\Delta\omega = P\int_{-\infty}^{\infty} \frac{g^2(\omega)}{\omega_0-\omega}\,d\omega,$$
$$\frac{\Gamma}{2} = \pi g^2(\omega). \tag{6.273}$$

With Eqs. (6.262), (6.265)–(6.267), and (6.272), the dynamic Eq. (6.254) is transformed to

$$\dot{\tilde{\rho}}_{\mathrm{A}} = \sigma^-\tilde{\rho}_{\mathrm{A}}(t)\sigma^+\Gamma - \left(\frac{\Gamma}{2} + i\,\Delta\omega\right)\sigma^+\sigma^-\tilde{\rho}_{\mathrm{A}}(\mathrm{t}) - \left(\frac{\Gamma}{2} - i\,\Delta\omega\right)\tilde{\rho}_{\mathrm{A}}(t)\sigma^+\sigma^-,$$
$$\simeq \sigma^-\tilde{\rho}_{\mathrm{A}}(t)\sigma^+\Gamma - \frac{\Gamma}{2}[\sigma^+\sigma^-\tilde{\rho}_{\mathrm{A}}(t) + \tilde{\rho}_{\mathrm{A}}(t)\sigma^+\sigma^-], \tag{6.274}$$

in which the last expression neglects the small frequency shift. The matrix elements now may be evaluated with the help of Eq. (5.9). For the diagonal elements,

$$(\dot{\tilde{\rho}}_{\mathrm{A}})_{\mathrm{ee}} = -(\dot{\tilde{\rho}}_{\mathrm{A}})_{\mathrm{gg}} = \Gamma\langle \mathrm{e}|\sigma^-\tilde{\rho}_{\mathrm{A}}(t)\sigma^+|\mathrm{e}\rangle$$
$$-\frac{\Gamma}{2}[\langle \mathrm{e}|\sigma^+\sigma^-\tilde{\rho}_{\mathrm{A}}(t)|\mathrm{e}\rangle + \langle \mathrm{e}|\tilde{\rho}_{\mathrm{A}}(t)\sigma^+\sigma^-|\mathrm{e}\rangle]$$
$$= -\frac{\Gamma}{2}[\langle \mathrm{e}|\sigma^+\sigma^-\tilde{\rho}_{\mathrm{A}}(t)|\mathrm{e}\rangle + \langle \mathrm{e}|\tilde{\rho}_{\mathrm{A}}(t)\sigma^+\sigma^-|\mathrm{e}\rangle]$$
$$= -\Gamma\langle \mathrm{e}|\tilde{\rho}_{\mathrm{A}}(t)|\mathrm{e}\rangle, \tag{6.275}$$

with solutions

$$\langle \mathrm{e}|\tilde{\rho}_{\mathrm{A}}(t)|\mathrm{e}\rangle = \langle \mathrm{e}|\rho_{\mathrm{A}}(t)|\mathrm{e}\rangle = e^{-\Gamma t}\langle \mathrm{e}|\rho_{\mathrm{A}}(0)|\mathrm{e}\rangle, \tag{6.276a}$$
$$\langle \mathrm{g}|\tilde{\rho}_{\mathrm{A}}(t)|\mathrm{g}\rangle = \langle \mathrm{g}|\rho_{\mathrm{A}}(t)|\mathrm{g}\rangle = 1 - \langle \mathrm{e}|\rho_{\mathrm{A}}(t)|\mathrm{e}\rangle$$
$$= \langle \mathrm{g}|\rho_{\mathrm{A}}(0)|\mathrm{g}\rangle + (1 - e^{-}\Gamma^{t})\langle \mathrm{e}|\rho_{\mathrm{A}}(0)|\mathrm{e}\rangle. \tag{6.276b}$$

For the off-diagonal matrix elements,

$$(\dot{\tilde{\rho}}_A)_{eg} = \Gamma\langle e|\sigma^- \rho_A(t)\sigma^+|g\rangle - \frac{\Gamma}{2}[\langle e|\sigma^+\sigma^- \tilde{\rho}_A(t)|g\rangle + \langle e|\tilde{\rho}_A(t)\sigma^+\sigma^-|g\rangle]$$

$$= -\frac{\Gamma}{2}\langle e|\sigma^+\sigma^- \tilde{\rho}_A(t)|g\rangle = -\frac{\Gamma}{2}\langle e|\tilde{\rho}_A(t)|g\rangle, \tag{6.277a}$$

$$(\dot{\tilde{\rho}}_A)_{ge} = -\frac{\Gamma}{2}\langle g|\tilde{\rho}_A(t)\sigma^+\sigma^-|e\rangle = -\frac{\Gamma}{2}\langle g|\tilde{\rho}_A(t)|e\rangle. \tag{6.277b}$$

Thus,

$$\begin{aligned}\langle e|\tilde{\rho}_A(t)|g\rangle &= e^{-\Gamma t/2}\langle e|\rho_A(0)|g\rangle,\\ \langle g|\tilde{\rho}_A(t)|e\rangle &= e^{-\Gamma t/2}\langle g|\rho_A(0)|e\rangle.\end{aligned} \tag{6.278}$$

Equations (6.276) indicate that if the atomic population is initially in the upper state $|e\rangle$ while the lower state $|g\rangle$ is vacant, the population in $|e\rangle$ will be depleted exponentially at a rate Γ, as we have found previously in Eq. (5.174). At the same time the population in $|g\rangle$ will grow, and as $t \to \infty$, the entire population will have been shifted to $|g\rangle$. On the other hand, if $|e\rangle$ is initially vacant and $|g\rangle$ is occupied, the population will remain in $|g\rangle$ for all time. It is further observed that according to Eq. (6.278), the off-diagonal density matrix elements also decay exponentially but at the slower rate of $\Gamma/2$. Thus, the interaction of an excited atom with the vacuum state of the radiation field provides an alternative picture of spontaneous emission fully consistent with the Schrödinger derivation but with emphasis on the density matrix elements.

One also may connect the results of this section with those in Section 6.3. Comparing Eq. (6.73) with Eq. (6.275), it is seen that if the reservoir is the vacuum state of the field there can be no transitions $|g\rangle \to |e\rangle$; hence, $W_{eg} = 0$ and $1/T_1$ becomes Γ. Similarly, a comparison of the off-diagonal equations in Eq. (6.69) with Eq. (6.277) leads to the identification $1/T_2$ with $\Gamma/2$. Therefore, when damping occurs entirely as a result of spontaneous emission,

$$\frac{1}{T_2} = \frac{1}{2T_1}. \tag{6.279}$$

The more general relation is

$$\frac{1}{T_2} = \frac{1}{T_2'} + \frac{1}{T_2^*} \tag{6.280}$$

in which T_2 is the total transverse damping constant, T_2^* is the contribution from inhomogeneous broadening, and T_2' is associated with all other damping mechanisms including the contribution from T_1. In practice, the computation of these constants from realistic models may be quite difficult; the use of phenomenological results then may be the only recourse.

References

[1] H. Haken, *In Handbuch der Physik, v. XXV/2c.* (S. Flugge, ed.), Springer-Verlag, Berlin, 1970.

[2] W. H. Louisell, *Quantum Statistical Properties of Radiation.* J. Wiley, New York, 1973.

[3] K. Blum, *Density Matrix Theory and Applications.* Plenum Press, New York, 1981.

[4] I. R. Sensitzky, *Phys. Rev.* **119**, 670(1960).

[5] I. R. Sensitzky, *Phys. Rev.* **124**, 642(1961).

[6] I. R. Sensitzky, *Phys. Rev.* **131**, 2827(1963).

[7] M. Lax, *In Dynamical Processes in Solid State Optics.* (R. Kubo and H. Kamimura, eds.), W. A. Benjamin, New York, 1967.

[8] B. R. Mollow and M. M. Miller, *Ann.* of *Phys.* **52**, 464(1969).

[9] G. S. Agarwal, *In Springer Tracts in Modern Physics, v. 70.* (G. Hohlder, ed.), Springer-Verlag, Berlin, 1974.

[10] K. J. McNeil and D. F. Walls, *J. Phys.* A**7**, 617(1974).

[11] S. Sachdev, *Phys. Rev.* A**29**, 2627(1984).

[12] H. C. Torrey, *Phys. Rev.* **76,** 1059(1949).

[13] R. G. Brewer, *In Nonlinear Optics.* (P. G. Harper and B. S. Wherrett, eds.), Academic Press, New York, 1977.

[14] L. Allen and J. H. Eberly, *Optical Resonance and Two-Level Atoms.* J. Wiley, New York, 1975.

[15] F. T. Hioe, *Phys. Rev.* A**30**, 2100(1984).

[16] A. L. Schawlow, *Rev. Mod. Phys.* **54**, 697(1982).

[17] M. M. Salour, *Rev. Mod. Phys.* **50**, 667(1978).

[18] S. R. Hartmann, *In Quantum Optics, Proc. Int'l School of Physics Enrico Fermi XLII.* (R. J. Galuber, ed.), Academic Press, New York, 1969.

[19] A. Schenzle and R. G. Brewer, *Phys. Rev.* A**14**, 1756(1976).

[20] D. T. Pegg, *In Laser Physics*, (D. F. Walls and J. D. Harvey, eds.), Academic Press, New York, 1980.

[21] W. H. Hesselink and D. A. Wiersma, *In Spectroscopy and Excitation Dynamics of Condensed Molecular Systems.* (V. M. Agranovich and R. M. Hochstrasser, eds.), North-Holland, Amsterdam, 1983.

[22] M. J. Burns, W. K. Liu and A. H. Zewail, *In Spectroscopy and Excitation Dynamics of Condensed Molecular Systems.* (V. M. Agranovich and R. M. Hochstrasser, eds.), North-Holland, Amsterdam, 1983.

[23] S. L. McCall and E. L. Hahn, *Phys. Rev. Lett.* **18**, 908(1967).

[24] S. L. McCall and E. L. Hahn, *Phys. Rev.* **183**, 457(1969).

[25] A. Schenzle, *In Quantum Optics.* (C. A. Engelbrecht, ed.), Springer-Verlag, Berlin, 1982.

[26] B. R. Mollow, *Phys. Rev.* A**2**, 76(1970).

[27] B. R. Mollow, *In Progress in Optics, XIX.* (E. Wolf, ed.), North-Holland, Amsterdam, 1981.

[28] F. Casagrande and L. A. Lugiato, *In Quantum Optics.* (C. A. Engelbrecht, ed.), Springer-Verlag, Berlin, 1982.

[29] H. Haken, *In Quantum Optics.* (C. A. Engelbrecht, ed.), Springer-Verlag, Berlin, 1982.

[30] P. Zoller and J. Cooper, *Phys. Rev.* **A28**, 2310(1983).

[31] M. Scully, *In Quantum Optics, Int'l School of Physics Enrico Fermi.* (R. J. Glauber, ed.) Academic Press, New York, 1969.

VII Nonlinear and Multiphoton Processes

Practically all physical experiments involve the application of an external disturbance or force $f(t')$ on a system initially in equilibrium in order to elicit a response, $y(t)$. Such is the case, for example, when an electric field induces an electric dipole moment in an atom or a molecule. To satisfy the requirement of causality, which prevents the system from responding before the imposition of the disturbance, it will be assumed that at a particular time t, the response $y(t)$ may depend on $f(t')$ for all earlier values of t' up to, but not later than, the time t. A relation such as that between $y(t)$ and $f(t')$ where t' is not necessarily equal to t is said to be *nonlocal*. One might also visualize a nonlocal spatial dependence of the response on the disturbance; that is, where the response at a point $\mathbf{r}$ depends not only on $f(\mathbf{r})$ but also on $f(\mathbf{r}')$ where $f(\mathbf{r}')$ is the disturbance at other points $\mathbf{r}'$. We shall be concerned with interactions in the dipole approximation, however, in which case the region of interaction is small compared with a wavelength. The spatial dependence then may be regarded as local, i.e., the response at a point depends only on the force acting on the same point and no other.

If the disturbance is sufficiently weak, the relation between $y(t)$ and $f(t')$ is linear. As the disturbance grows in magnitude, the assumption of linear dependence becomes less tenable and ultimately must be replaced by a nonlinear dependence. The response then no longer depends on the first power of $f(t')$ but rather on higher powers or, if several disturbances are acting simultaneously, on their product. It was the development of laser sources producing beams of high intensity that ushered in the era of nonlinear effects in optics [1–12].

7.1 Polarization and Susceptibility

We begin with the linear case to introduce the general formalism. The assumption of local spatial dependence permits us to omit the spatial dependence from the notation. (Spatial dependence that may arise from propagation effects is considered separately.) The general relation between $y(t)$ and $f(t')$ then may be written

$$\langle y(t)\rangle = \int_{-\infty}^{t} K(t, t')f(t')\,dt', \tag{7.1}$$

in which $\langle y(t)\rangle$ is an ensemble average to allow for possible fluctuations in the response. $K(t, t')$ is known as the *response function* (also the *after-effect* or *Green's* function) [13]; it is an intrinsic property of the physical system and is independent of the specific form of the applied force $f(t)$. The lower limit has been put at $-\infty$ to allow $f(t)$ to build up gradually (adiabatically) from the distant past. In this way, one avoids the production of transients associated with the sudden onset of a disturbance at a finite time. The formal procedure for evaluating the integral at the lower limit is facilitated by the insertion of a convergence factor $\exp(\varepsilon t)$ in the integrand with $\varepsilon > 0$; at the end of the calculation, the limit $\varepsilon \to 0$ is evaluated (see Section 2.10 for an analogous procedure). Finally, in order to ensure that $\langle y(t)\rangle$ is real when $f(t)$ is real, it will be necessary for $K(t, t')$ to be real.

We shall be interested in situations in which the fluctuations in the response are stationary. The response function $K(t, t')$ then depends on the difference $t' - t$ rather than on t and t' separately. Specifically, it is assumed that $f(t)$ is a component of an electric field $\mathbf{E}(t)$ and the response $\langle y(t)\rangle$ is a component of the polarization $\mathbf{P}(t)$ arising from a redistribution of charges within the medium exposed to the electric field. Equation (7.1) then assumes the form

$$\mathbf{P}(t) = \int_{-\infty}^{t} K(t' - t)\mathbf{E}(t')\,dt, \tag{7.2}$$

where the response function $K(t' - t)$ now must be interpreted as a second rank tensor to allow for the possibility that the two vectors $\mathbf{P}(t)$ and $\mathbf{E}(t)$ are not colinear as would be the case in a nonisotropic medium. In component form, with $\alpha, \mu = x, y, z$, and $\tau = t' - t$,

$$P_\mu(t) = \sum_\alpha \int_{-\infty}^{t} K_{\mu\alpha}(t' - t)E_\alpha(t')\,dt' \tag{7.3a}$$

$$= \sum_\alpha \int_{-\infty}^{0} K_{\mu\alpha}(\tau)E_\alpha(t + \tau)\,d\tau. \tag{7.3b}$$

To proceed further, it is necessary to evaluate $K_{\mu\alpha}(\tau)$—or quantities closely related to it—from the microscopic radiative interactions occuring in the medium. This is a quantum mechanical problem involving the eigenstates and eigenvalues of the atoms or molecules of which the medium is composed. But the electric field, in many important applications, need not be quantized. This semiclassical approach, adopted here, turns out to be amply justified and usually more convenient for computations.

It is assumed that at an initial time t_0 (later to be taken as $-\infty$), the system is in an eigenstate $\psi(t_0)$ of the Hamiltonian $\mathscr{H}_0$ and that a disturbance, represented by a perturbation V, is applied adiabatically at $t = t_0$. For $t > t_0$, the total Hamiltonian is

$$\mathscr{H} = \mathscr{H}_0 + V(t). \tag{7.4}$$

We now refer to Eqs. (2.176) and (2.178) with the identification

$$\langle y(t)\rangle = P_\mu(t) = \langle A(t)\rangle. \tag{7.5}$$

At $t = t_0$ when $V = 0$, an operator A representing a physical quantity (an observable) has an expectation value

$$\langle A\rangle_0 = \langle \psi(t_0)|A|\psi(t_0)\rangle, \tag{7.6}$$

which also may be regarded as an ensemble average since at $t = t_0$ the system is in a pure state. As time progresses, however, the statistical average $\langle A(t)\rangle$ will depart from $\langle A\rangle_0$ as a consequence of the onset of the perturbation V. The difference

$$\langle A(t)\rangle - \langle A\rangle_0 \equiv \Delta\langle A(t)\rangle \tag{7.7}$$

then will be the response of the system to the perturbation, but since $\langle A\rangle_0$ is a constant it need not be carried along in the computation. We therefore shall refer to $\langle A(t)\rangle$ as the response.

In the case of a linear response, which implies a weak force, Eq. (2.176) or (2.178) for the time development of $\langle A(t)\rangle$ may be truncated at the first order term. Further, it will be assumed that at $t = -\infty$ the system is in thermal equilibrium. Then, with

$$\rho(t_0) = \rho(-\infty) = \rho_0, \tag{7.8}$$

where ρ_0 is the thermal density operator defined by Eq. (2.226), we have

$$\begin{aligned}\langle A(t)\rangle &= -\frac{i}{\hbar}\int_{-\infty}^{t} dt_1 \mathrm{Tr}\{\rho_0[\tilde{A}(t), \tilde{V}(t_1)]\},\\ &= -\frac{i}{\hbar}\int_{-\infty}^{t} dt_1 \langle[\tilde{A}(t), \tilde{V}(t_1)]\rangle,\end{aligned} \tag{7.9}$$

or

$$\langle A(t)\rangle = -\frac{i}{\hbar}\int_{-\infty}^{t} dt_1 \langle [A, \tilde{V}(t_1 - t)]\rangle$$
$$= -\frac{i}{\hbar}\int_{-\infty}^{t} dt_1 Tr\{\rho_0[A, \tilde{V}(t_1 - t)]\}. \qquad (7.10)$$

For the interaction of an electric field $\mathbf{E}(t)$ with a medium consisting of independent subsystems, $\langle A(t)\rangle$ becomes the polarization $\mathbf{P}(t)$ where

$$\mathbf{P}(t) = N\langle \mathbf{d}(t)\rangle, \qquad (7.11)$$

and $\langle \mathbf{d}(t)\rangle$ is the ensemble average of the elementary dipole moments induced by the field; N is the number of dipoles in a unit volume and the interaction term in the dipole approximation is

$$V(t) = -\mathbf{d}\cdot\mathbf{E}(t) = -\sum_{\alpha} d_\alpha E_\alpha(t), \qquad (\alpha = x, y, z). \qquad (7.12)$$

The components of the polarization, to first order, now may be written in accordance with the general form of Eq. (2.176):

$$P_\mu^{(1)}(t) = \frac{i}{\hbar}N\sum_\alpha \int_{-\infty}^{t} dt_1 \langle [\tilde{d}_\mu(t), \tilde{d}_\alpha(t_1)]\rangle E_\alpha(t_1). \qquad (7.13)$$

Since the fluctuations have been assumed to be stationary, we may shift the time axis in the correlation functions, so that,

$$P_\mu^{(1)}(t) = \frac{i}{\hbar}N\sum_\alpha \int_{-\infty}^{t} dt_1 \langle [d_\mu, \tilde{d}_\alpha(t_1 - t)]\rangle E_\alpha(t_1) \qquad (7.14a)$$
$$= \frac{i}{\hbar}N\sum_\alpha \int_{-\infty}^{0} d\tau \langle [d_\mu, \tilde{d}_\alpha(\tau)]\rangle E_\alpha(t + \tau), \qquad (7.14b)$$

with $\tau = t_1 - t$. Comparing this expression with Eq. (7.3), it is seen that the first-order response function is expressible in the form

$$K_{\mu\alpha}^{(1)}(t) = \frac{i}{\hbar}N\langle [d_\mu, \tilde{d}_\alpha(\tau)]\rangle = \frac{i}{\hbar}N\langle [\tilde{d}_\mu(-\tau), d_\alpha]\rangle \qquad (7.15)$$

Instead of concentrating on the time-development of the system's response to an external perturbation, it is often preferable to examine the frequency spectrum. If $E_\alpha(\omega)$ is the Fourier transform of $E_\alpha(t)$ then

$$E_\alpha(t + \tau) = \int_{-\infty}^{\infty} E_\alpha(\omega)e^{-i\omega(t+\tau)}\,d\omega, \qquad (7.16)$$

and, by substitution in Eq. (7.3),

$$P_{\mu}^{(1)}(t) = \sum_{\alpha} \int_{-\infty}^{\infty} E_{\alpha}(\omega)e^{-i\omega t}\, d\omega \int_{-\infty}^{0} K_{\mu\alpha}^{(1)}(\tau)e^{-i\omega\tau}\, d\tau. \tag{7.17}$$

We then may write

$$P_{\mu}^{(1)}(t) = \varepsilon_0 \sum_{\alpha} \int_{-\infty}^{\infty} \chi_{\mu\alpha}^{(1)}(\omega)E_{\alpha}(\omega)e^{-i\omega t}\, d\omega, \tag{7.18}$$

where

$$\chi_{\mu\alpha}^{(1)}(\omega) = \frac{1}{\varepsilon_0} \int_{-\infty}^{0} K_{\mu\alpha}^{(1)}(\tau)e^{-i\omega\tau}\, d\tau \tag{7.19}$$

is defined as the dimensionless, frequency-dependent linear or first order), complex *susceptibility function.* We now may insert Eq. (7.15) into (7.19) to obtain a relation between the susceptibility and the dipole moment correlation functions:

$$\chi_{\mu\alpha}^{(1)}(\omega) = \frac{i}{\varepsilon_0 \hbar} N \int_{-\infty}^{0} \langle [d_{\mu}, \tilde{d}_{\alpha}(\tau)] \rangle\, e^{-i\omega\tau}\, d\tau, \tag{7.20a}$$

$$= \frac{i}{\varepsilon_0 \hbar} N \int_{0}^{\infty} \langle [\tilde{d}_{\mu}(\tau), d_{\alpha}] \rangle e^{i\omega\tau}\, d\tau. \tag{7.20b}$$

In the second expression, τ was replaced by $-\tau$ followed by a shift in the time axis of the correlation functions. Note that the first-order susceptibility function is a tensor of rank two.

For the Fourier transform of the polarization we have

$$\begin{aligned} P_{\mu}^{(1)}(\omega) &= \frac{1}{2\pi} \int_{-\infty}^{\infty} P_{\mu}^{(1)}(t)e^{i\omega t}\, dt, \\ &= \frac{\varepsilon_0}{2\pi} \sum_{\alpha} \int_{-\infty}^{\infty} \chi_{\mu\alpha}^{(1)}(\omega')E_{\alpha}(\omega')\, d\omega' \int_{-\infty}^{\infty} e^{i(\omega-\omega')t}\, dt, \\ &= \varepsilon_0 \sum_{\alpha} \int_{-\infty}^{\infty} \chi_{\mu\alpha}^{(1)}(\omega')E_{\alpha}(\omega')\, \delta(\omega - \omega')\, d\omega', \\ &= \varepsilon_0 \sum_{\alpha} \chi_{\mu\alpha}^{(1)}(\omega)E_{\alpha}(\omega), \end{aligned} \tag{7.21}$$

which indicates that in the linear case a field of frequency ω induces an oscillating polarization at the same frequency and of a magnitude governed by the susceptibility function.

At low power densities, the assumption of a linear relation between the polarization and the electric field is sufficient for the description of optical processes. As indicated previously, with increasing power densities currently available from laser sources, it is necessary to consider higher order, nonlinear interactions. Let us, therefore, extend the formalism to second order; in place of Eq. (7.3b) we have

$$P_\mu^{(2)}(t) = \sum_{\alpha\beta} \int_{-\infty}^{0} d\tau_1 \int_{-\infty}^{\tau_1} d\tau_2 K_{\mu\alpha\beta}^{(2)}(\tau_1, \tau_2) E_\alpha(t + \tau_1) E_\beta(t + \tau_2), \tag{7.22}$$

where each of the indices μ, α, β stands for x, y, or z and $\tau_1 = t_1 - t, \tau_2 = t_2 - t$. Likewise, extension of Eq. (7.14b) leads to

$$P_\mu^{(2)}(t) = -\frac{N}{\hbar^2} \sum_{\alpha\beta} \int_{-\infty}^{0} d\tau_1 \int_{-\infty}^{\tau_1} d\tau_2 \langle[[d_\mu, \tilde{d}_\alpha(\tau_1)], \tilde{d}_\beta(\tau_2)]\rangle \times E_\alpha(t + \tau_1) E_\beta(t + \tau_2). \tag{7.23}$$

This implies that one may set

$$K_{\mu\alpha\beta}^{(2)}(\tau_1, \tau_2) = -\frac{N}{\hbar^2} \langle[[d_\mu, d_\alpha(\tau_1)], d_\beta(\tau_2)]\rangle. \tag{7.24}$$

Now let

$$E_\alpha(t + \tau_1) = \int_{-\infty}^{\infty} E_\alpha(\omega_1) e^{-i\omega_1(t+\tau_1)}\, d\omega_1,$$
$$E_\beta(t + \tau_2) = \int_{-\infty}^{\infty} E_\beta(\omega_2) e^{-i\omega_2(t+\tau_2)}\, d\omega_2. \tag{7.25}$$

Then

$$P_\mu^{(2)}(t) = -\frac{N}{\hbar^2} \sum_{\alpha\beta} \int_{-\infty}^{\infty} d\omega_1 \int_{-\infty}^{\infty} d\omega_2 e^{-i(\omega_1+\omega_2)t} E_\alpha(\omega_1) E_\beta(\omega_2) \times \int_{-\infty}^{0} d\tau_1 \int_{-\infty}^{\tau_1} d\tau_2 \langle[[d_\mu, \tilde{d}_\alpha(\tau_1)], \tilde{d}_\beta(\tau_2)]\rangle e^{-i(\omega_1\tau_1+\omega_2\tau_2)}. \tag{7.26}$$

By extension of Eq. (7.18), the second order susceptibility function—a tensor of rank three—is defined by

$$\chi_{\mu\alpha\beta}^{(2)}(\omega_1, \omega_2) = \frac{1}{\varepsilon_0} \int_{-\infty}^{0} d\tau_1 \int_{-\infty}^{\tau_1} d\tau_2 K_{\mu\alpha\beta}^{(2)}(\tau_1, \tau_2) e^{-i(\omega_1\tau_1+\omega_2\tau_2)}$$
$$= -\frac{N}{\varepsilon_0 \hbar^2} \int_{-\infty}^{0} d\tau_1 \int_{-\infty}^{\tau_1} d\tau_2 \langle[[d_\mu, \tilde{d}_\alpha(\tau_1)], \tilde{d}_\beta(\tau_2)]\rangle e^{-i(\omega_1\tau_1+\omega_2\tau_2)}. \tag{7.27}$$

Therefore, the second-order polarizability, expressed in terms of the susceptibility, is

$$P_\mu^{(2)}(t) = \varepsilon_0 \sum_{\alpha\beta} \int_{-\infty}^{\infty} d\omega_1 \int_{-\infty}^{\infty} d\omega_2 \chi_{\mu\alpha\beta}^{(2)}(\omega_1, \omega_2) E_\alpha(\omega_1) E^\beta(\omega_2) e^{-i(\omega_1+\omega_2)t} \tag{7.28}$$

or, in terms of the Fourier transform,

$$\begin{aligned} P_\mu^{(2)}(\omega) &= \frac{1}{2\pi} \int_{-\infty}^{\infty} P_\mu^{(2)}(t) e^{i\omega t}\, dt \\ &= \frac{\varepsilon_0}{2\pi} \sum_{\alpha\beta} \int_{-\infty}^{\infty} \chi_{\mu\alpha\beta}^{(2)}(\omega_1, \omega_2) E_\alpha(\omega_1) E_\beta(\omega_2)\, d\omega_1\, d\omega_2 \int_{-\infty}^{\infty} e^{i(\omega-\omega_1-\omega_2)t}\, dt \\ &= \varepsilon_0 \sum_{\alpha\beta} \int_{-\infty}^{\infty} \chi_{\mu\alpha\beta}^{(2)}(\omega_1, \omega_2) E_\alpha(\omega_1) E_\beta(\omega_2)\, \delta(\omega - \omega_1 - \omega_2)\, d\omega_1\, d\omega_2 \\ &= \varepsilon_0 \sum_{\alpha\beta} \chi_{\mu\alpha\beta}^{(2)}(\omega_1, \omega_2) E_\alpha(\omega_1) E_\beta(\omega_2). \end{aligned} \tag{7.29}$$

The δ-function imposes the requirement $\omega = \omega_1 + \omega_2$ and is reflected in the manner in which the argument of the susceptibility function is written (more about that later). It is seen, then, that Eq. (7.29) describes the interaction of two fields with frequencies ω_1 and ω_2 to produce a third field at the sum of the two frequencies. This is an example of a general coherent 3-wave mixing process in which the three fields interchange energy among themselves in a nonlinear medium. If the sign of ω_2 is changed to $-\omega_2$ the effect would be to change $E_\beta(\omega_2)$ to $E_\beta^*(\omega_2)$; therefore, Eq. (7.29), in principle, also encompasses a mixing process in which $\omega = \omega_1 - \omega_2$. Further discussion of such matters is contained in Section 7.3.

In third order,

$$\begin{aligned} P_\mu^{(3)}(t) = \varepsilon_0 \sum_{\alpha\beta\gamma} \int_{-\infty}^{\infty} d\omega_1 \int_{-\infty}^{\infty} d\omega_2 \int_{-\infty}^{\infty} d\omega_3 \chi_{\mu\alpha\beta\gamma}^{(3)}(\omega_1, \omega_2, \omega_3) \\ \times E_\alpha(\omega_1) E_\beta(\omega_2) E_\gamma(\omega_3) e^{-i(\omega_1+\omega_2+\omega_3)t}, \end{aligned} \tag{7.30}$$

$$\begin{aligned} \chi_{\mu\alpha\beta\gamma}^{(3)}(\omega_1, \omega_2, \omega_3) = -\frac{N}{\varepsilon_0}\left(\frac{i}{\hbar}\right)^3 \int_{-\infty}^{0} d\tau_1 \int_{-\infty}^{\tau_1} d\tau_2 \int_{-\infty}^{\tau_2} d\tau_3 \\ \times \langle [[[d_\mu, \tilde{d}_\alpha(\tau_1)], \tilde{d}_\beta(\tau_2)], \tilde{d}_\gamma(\tau_3)] \rangle e^{-i(\omega_1\tau_1+\omega_2\tau_2+\omega_3\tau_3)}, \end{aligned} \tag{7.31}$$

and the Fourier transform of Eq. (7.30) is

$$P_\mu^{(3)}(\omega) = \varepsilon_0 \sum_{\alpha\beta\gamma} \chi_{\mu\alpha\beta\gamma}^{(3)}(\omega_1, \omega_2, \omega_3) E_\alpha(\omega_1) E_\beta(\omega_2) E_\gamma(\omega_3). \tag{7.32}$$

The third-order susceptibility function $\chi^{(3)}(\omega_1, \omega_2, \omega_3)$ is associated with a 4-wave mixing process wherein three input fields at ω_1, ω_2, and ω_3 (not necessarily all different) interact coherently in a nonlinear medium to produce a field at ω where $\omega = \omega_1 + \omega_2 + \omega_3$. With suitable changes in sign, however, as indicated previously, $\chi^{(3)}$ characterizes a nonlinear mixing process where ω is any algebraic combination of the input fields. Clearly, these relations can be extended to higher orders although in practice, orders higher than the third rarely are needed.

Equations such as Eq. (7.31) and (7.32) are the fundamental equations governing the polarization induced in a medium by the application of an external field or combination of fields. The characteristics of the medium, insofar as its response to the applied fields is concerned, are contained in the susceptibility function. Although the latter often may be regarded as an experimental quantity, the quantum theoretical formulation of the susceptibility function provides additional insight into the interactions between the medium and the external fields. Let us, therefore, indicate how such a computation is implemented.

In terms of the density operator, Eq. (7.11) is written

$$\mathbf{P}(t) = N\,\mathrm{Tr}\{\rho(t)\mathbf{d}(t)\}. \tag{7.33}$$

Since the temporal development of the density operator may be expressed in the form of a series, as shown by Eq. (2.157) and (2.158d), similarly, one may expand the polarization into the series

$$\mathbf{P}(t) = \mathbf{P}^{(0)}(t) + \mathbf{P}^{(1)}(t) + \mathbf{P}^{(2)}(t) + \cdots, \tag{7.34}$$

where the nth order term is

$$\begin{aligned}\mathbf{P}^{(n)}(t) &= N\,\mathrm{Tr}\{\rho^{(n)}(t)\mathbf{d}(t)\} = N\sum_k \langle k|\rho^{(n)}(t)\mathbf{d}(t)|^k\rangle \\ &= N\sum_{kl}\langle k|\rho^{(n)}(t)|l\rangle\langle l|\mathbf{d}(t)|k\rangle = N\sum_{kl}\rho_{kl}^{(n)}(t)\mathbf{d}_{lk}(t).\end{aligned} \tag{7.35}$$

For a single component of the polarization vector,

$$P_\mu^{(n)}(t) = N\sum_{kl}\rho_{kl}^{(n)}(t)d_{lk}^{\mu}(t), \qquad \mu = x, y, z. \tag{7.36}$$

The basis set employed in these expressions is the same as the one in Section 2.7. It also is assumed that $\rho^{(0)}(t)$ is the thermal density operator ρ_0 (Section 2.8), which is diagonal in the basis set we have chosen. Diagonal elements of the dipole operator, however, are zero because of parity restrictions; hence, $\mathbf{P}^{(0)}(t)$ is identically zero (unless there are permanent electric dipoles) and the lowest value of n in the series (Eq. (7.34)) that gives a nonvanishing polarization is unity.

This formalism is of sufficient generality to encompass a large range of linear and nonlinear phenomena such as harmonic generation, production of sum and difference frequencies, stimulated Raman processes, parametric amplification, and many other effects discovered since the advent of laser sources.

Examination of the density matrix in first order (Section 6.3) indicates that the leading term $\mathbf{P}^{(1)}(t)$ corresponds to linear interactions, that is, interactions in which the polarization is proportional to the applied field and has the same frequency. Similarly, from the form of the second-order density matrix it is seen that $P^{(2)}(t)$ is the lowest order nonlinear polarization; it is proportional to the product of two fields, or to the square of a single field, and may contain frequencies not found in the incident field(s). The nth order polarization $\mathbf{P}^{(n)}(t)$ is proportional to n fields (or a combination of fields corresponding to an nth order product) and may contain numerous frequency combinations, although, as stated earlier, in most current experiments $n \leq 3$.

Once the temporal behavior of the induced polarization has been established it can serve as a source of electromagnetic radiation and from here on the computation proceeds in a purely classical fashion on the basis of Maxwell's equations. Specifically, through the wave equation

$$\nabla^2 \mathbf{E}(\mathbf{r}, t) - \mu_0 \varepsilon_0 \frac{\partial^2}{\partial t^2} \mathbf{E}(\mathbf{r}, t) = \mu_0 \frac{\partial^2}{\partial t^2} \mathbf{P}(\mathbf{r}, t) \tag{7.37}$$

for propagation in a medium devoid of free charge and magnetization. When the refractive index is n, the free-space velocity is replaced by c/n. Introducing

$$\mathbf{P}(\mathbf{r}, t) = \mathbf{P}^{(1)}(\mathbf{r}, t) + \mathbf{P}_{\mathrm{NL}}(\mathbf{r}, t), \tag{7.38}$$

$$\mathbf{P}^{(1)}(\mathbf{r}, t) = \varepsilon_0 \chi^{(1)} \mathbf{E}(\mathbf{r}, t), \qquad \varepsilon = \varepsilon_0(1 + \chi^{(1)}), \tag{7.39}$$

where $\mathbf{P}^{(1)}(\mathbf{r}, t)$ is the linear polarization, $\mathbf{P}_{\mathrm{NL}}(\mathbf{r}, t)$ is the nonlinear polarization, and $\chi^{(1)}$ is the first-order or linear susceptibility. Equation (7.37) transforms to

$$\nabla^2 \mathbf{E}(\mathbf{r}, t) - \mu_0 \varepsilon \frac{\partial^2}{\partial t^2} \mathbf{E}(\mathbf{r}, t) = \mu_0 \frac{\partial^2}{\partial t^2} \mathbf{P}_{\mathrm{NL}}(\mathbf{r}, t). \tag{7.40}$$

It is customary to incorporate the *slowly-varying-amplitude* approximation [1, 2]; Eq. (7.40) then leads to a set of coupled, first-order differential equations for the amplitudes of the interacting fields (see Section 7.7). When these are solved, with appropriate boundary conditions, one obtains the characteristics of the emitted radiation and the alterations in the spectrum as the beam propagates through the medium.

7.2 First-Order Susceptibility

We now discuss the first-order (linear) polarization and susceptibility computed on the basis of Eq. (7.36) for the interaction of a medium with a monochromatic field. The matrix element $\rho_{kl}^{(1)}(t)$ was evaluated in Eq. (6.59);

hence, the first-order polarization is

$$P_{\mu}^{(1)}(t) = N \sum_{kl} \rho_{kl}^{(1)}(t)\, d_{lk}^{\mu}(t)$$

$$= \frac{N}{2\hbar} \sum_{kl} (\rho_{ll}^{0} - \rho_{kk}^{0}) \sum_{\alpha} \frac{d_{kl}^{\alpha} d_{lk}^{\mu} E_{\alpha}(\omega) e^{-i\omega t}}{\omega_{kl} - \omega - i\Gamma_{kl}}. \tag{7.41}$$

Let us now write

$$P_{\mu}^{(1)}(t) = P_{\mu}^{(1)}(\omega) e^{-i\omega t}. \tag{7.42}$$

Comparing the two relations, it is seen that

$$P_{\mu}^{(1)}(\omega) = \frac{N}{\hbar} \sum_{kl} (\rho_{ll}^{0} - \rho_{kk}^{0}) \sum_{\alpha} \frac{d_{kl}^{\alpha} d_{lk}^{\mu} E_{\alpha}(\omega)}{\omega_{kl} - \omega - i\Gamma_{kl}}, \tag{7.43a}$$

$$= \varepsilon_0 \sum_{\alpha} \chi_{\mu\alpha}^{(1)}(\omega) E_{\alpha}(\omega), \tag{7.43b}$$

where

$$\chi_{\mu\alpha}^{(1)}(\omega) = \frac{N}{\varepsilon_0 \hbar} \sum_{kl} \frac{d_{kl}^{\alpha} d_{lk}^{\mu}}{\omega_{kl} - \omega - i\Gamma_{kl}} [\rho_{ll}^{0} - \rho_{kk}^{0}]. \tag{7.44}$$

Equation (7.43b) is identical with Eq. (7.21); therefore, $\chi_{\mu\alpha}^{(1)}(\omega)$, written as in Eq. (7.44), is the first-order susceptibility function expressed in terms of atomic parameters. After interchanging the summation indices in the first term of the sum and setting $\Gamma_{kl} = \Gamma_{lk}$,

$$\chi_{\mu\alpha}^{(1)}(\omega) = \frac{N}{\varepsilon_0 \hbar} \sum_{kl} \rho_{kk}^{0} \left[\frac{d_{kl}^{\mu} d_{lk}^{\alpha}}{\omega_{lk} - \omega - i\Gamma_{lk}} + \frac{d_{kl}^{\alpha} d_{lk}^{\mu}}{\omega_{lk} + \omega + i\Gamma_{lk}} \right]. \tag{7.45}$$

The susceptibility function is obviously complex with singularities in the lower half of the complex plane, i.e., when

$$\omega = \pm\omega_{lk} - i\Gamma_{lk}. \tag{7.46}$$

When the dipole matrix elements are real, as is usually the case, we have

$$\chi_{\mu\alpha}^{*(1)}(\omega) = \chi_{\mu\alpha}^{(1)}(-\omega) \tag{7.47}$$

or, upon separation into real and imaginary components,

$$\chi_{\mu\alpha}^{(1)}(\omega) = \chi_{\mu\alpha}^{\prime(1)}(\omega) + i\chi_{\mu\alpha}^{\prime\prime(1)}(\omega) \tag{7.48}$$

in which

$$\chi_{\mu\alpha}^{\prime(1)}(\omega) = \chi_{\mu\alpha}^{\prime(1)}(-\omega), \qquad \chi_{\mu\alpha}^{\prime\prime(1)}(\omega) = -\chi_{\mu\alpha}^{\prime\prime(1)}(-\omega). \tag{7.49}$$

When the system is sufficiently far from resonance to permit the neglect of damping constants, i.e., a lossless medium (but the matrix elements of the

Hermitian dipole moment operator are not necessarily real),

$$\chi_{\mu\alpha}^{*(1)}(\omega) = \chi_{\alpha\mu}^{(1)}(\omega), \tag{7.50a}$$

which, together with Eq. (7.47) yields

$$\chi_{\alpha\mu}^{(1)}(\omega) = \chi_{\mu\alpha}^{(1)}(-\omega). \tag{7.50b}$$

We conclude that $\chi_{\mu\alpha}^{(1)}(\omega)$ (far from resonance) is invariant under the simultaneous interchange of α with μ and ω with $-\omega$. This is known as *overall permutation symmetry*; Eq. (7.45) then may be written more compactly in the form

$$\chi_{\mu\alpha}^{(1)}(\omega) = \frac{N}{\varepsilon_0 \hbar} P_0 \sum_{kl} \rho_{kk}^0 \frac{d_{kl}^{\mu} d_{lk}^{\alpha}}{\omega_{lk} - \omega}, \tag{7.51}$$

in which the operator P_0 signifies that the expression which follows is to be summed over the two possible permutations of the pairs (α, ω) and $(\mu, -\omega)$.

If, in addition, the eigenfunctions are real so that all the matrix elements in Eq. (7.45) are real, the net result is

$$\chi_{\mu\alpha}^{*(1)}(\omega) = \chi_{\mu\alpha}^{(1)}(\omega) = \chi_{\mu\alpha}^{(1)}(-\omega) = \chi_{\alpha\mu}^{(1)}(\omega). \tag{7.52}$$

Under these circumstances, it is seen that $\chi_{\mu\alpha}^{(1)}(\omega)$ is a real, symmetric function of ω.

Let us recall, from Section 5.2, that the possibility of choosing real eigenfunctions requires a Hamiltonian that is invariant under a transformation by the time reversal operator T defined by Eq. (5.47). When the Hamiltonian is real this requirement is satisfied. Because of such considerations, the relations expressed by Eq. (7.52) are said to be a consequence of time reversal invariance. It will be assumed henceforth, unless stated otherwise, that real eigenfunctions have been chosen.

For an isotropic, two-level system, we may set $\mu = \alpha$ and

$$|d_{\text{eg}}^{\alpha}|^2 = |d_{\text{ge}}^{\alpha}|^2, \qquad \omega_{\text{eg}} = \omega_0 = -\omega_{\text{ge}}, \qquad \Gamma_{\text{eg}} = \Gamma_{\text{ge}} = \Gamma. \tag{7.53}$$

The expression for the susceptibility, from Eq. (7.45), then becomes

$$\chi_{\alpha\alpha}^{(1)}(\omega) = \frac{N}{\varepsilon_0 \hbar} |d_{\text{ge}}^{\alpha}|^2 [\rho_{\text{gg}}^0 - \rho_{\text{ee}}^0] \left(\frac{1}{\omega_0 - \omega - i\Gamma} + \frac{1}{\omega_0 + \omega + i\Gamma} \right). \tag{7.54}$$

Near resonance the major contribution comes from the two terms containing the denominator $\omega_0 - \omega - i\Gamma$; therefore, to a good approximation

$$\chi_{\alpha\alpha}^{(1)}(\omega) = \frac{N}{\varepsilon_0 \hbar} |d_{\text{ge}}^{\alpha}|^2 \frac{[\rho_{\text{gg}}^0 - \rho_{\text{ee}}^0]}{\omega_0 - \omega - i\Gamma}. \tag{7.55}$$

If $\hbar\omega_0$ is much larger than kT practically all the atoms will be in the ground state $|g\rangle$, in which case $\rho^0_{gg} \simeq 1$, $\rho^0_{ee} \simeq 0$, and

$$\chi^{(1)}_{\alpha\alpha}(\omega) = \frac{N}{\varepsilon_0 \hbar}|d^\alpha_{ge}|^2 \frac{\omega_0 - \omega + i\Gamma}{(\omega_0 - \omega)^2 + \Gamma^2}. \tag{7.56}$$

Omitting the superscript, the real and imaginary parts of the first-order susceptibility are

$$\chi'_{\alpha\alpha}(\omega) = \frac{N}{\varepsilon_0 \hbar}|d^\alpha_{ge}|^2 \frac{\omega_0 - \omega}{(\omega_0 - \omega)^2 + \Gamma^2}, \tag{7.57a}$$

$$\chi''_{\alpha\alpha}(\omega) = \frac{N}{\varepsilon_0 \hbar}|d^\alpha_{ge}|^2 \frac{\Gamma}{(\omega_0 - \omega)^2 + \Gamma^2} \tag{7.57b}$$

or, with $\Gamma = 1/T_2$,

$$\chi'_{\alpha\alpha}(\omega) = \frac{N}{\varepsilon_0 \hbar}|d^\alpha_{ge}|^2 \frac{T_2^2(\omega_0 - \omega)}{1 + T_2^2(\omega_0 - \omega)^2}, \tag{7.58a}$$

$$\chi''_{\alpha\alpha}(\omega) = \frac{N}{\varepsilon_0 \hbar}|d^\alpha_{ge}|^2 \frac{T_2}{1 + T_2^2(\omega_0 - \omega)^2}. \tag{7.58b}$$

These equations are precisely of the same form as Eq. (3.133) for the magnetic case and therefore have the same interpretation, namely, $\chi'_{\alpha\alpha}(\omega)$ corresponds to dispersion and $\chi''_{\alpha\alpha}(\omega)$ corresponds to absorption as shown in Fig. 3.7. It should come as no surprise that the imaginary component of the susceptibility is associated with energy dissipation since the same result can be shown classically. Thus, according to Maxwell's equations, the power per unit volume expended by a field $\mathbf{E}(t)$ to orient the dipoles that result in the polarization $\mathbf{P}(t)$ is

$$\frac{dU}{dt} = \mathbf{E}(t) \cdot \frac{\partial \mathbf{P}(t)}{\partial t}. \tag{7.59}$$

When

$$\mathbf{E}(t) = \mathbf{E}e^{-i\omega t}, \qquad \mathbf{P}(t) = \varepsilon_0 \chi(\omega)\mathbf{E}e^{-i\omega t} = \mathbf{P}e^{-i\omega t},$$
$$\chi(\omega) = \chi_{\alpha\alpha}(\omega) = \chi'(\omega) + i\chi''(\omega), \tag{7.60a}$$

we have

$$\frac{\partial \mathbf{P}(t)}{\partial t} = -i\omega \mathbf{P}e^{-i\omega t} = -i\omega\varepsilon_0[\chi'(\omega) + \chi''(\omega)]\mathbf{E}. \tag{7.60b}$$

To obtain the absorbed power per unit volume averaged over a cycle, it is noted that if

$$x = \mathrm{Re}(Ae^{-i\omega t}), \qquad y = \mathrm{Re}(Be^{-i\omega t}), \tag{7.61}$$

the cycle average $\langle xy \rangle$ is defined by

$$\langle xy \rangle = \frac{1}{T}\int_0^T xy\,dt = \frac{1}{2}\,\mathrm{Re}(AB^*) = \frac{1}{2}\,\mathrm{Re}(A^*B), \qquad T = \frac{2\pi}{\omega}. \tag{7.62}$$

Then

$$\begin{aligned}\left\langle \frac{dU}{dt} \right\rangle &= \frac{1}{2}\,\mathrm{Re}[\mathbf{E}^* \cdot (-i\omega\mathbf{P})] \\ &= \frac{1}{2}\varepsilon_0 \omega |E|^2 \chi''(\omega).\end{aligned} \tag{7.63}$$

The linear absorption coefficient α is defined by

$$\alpha = \frac{\left\langle \dfrac{dU}{dt} \right\rangle}{I}, \tag{7.64}$$

where

$$I = \mathrm{c}U = \tfrac{1}{2}\varepsilon_0 \mathrm{c}|E|^2 \tag{7.65}$$

is the energy flux. Accordingly,

$$\alpha = \frac{\omega}{\mathrm{c}}\chi''(\omega). \tag{7.66}$$

Returning to Eq. (7.60), it is seen that the first-order (linear) polarization $P_\alpha(t)$ produced by a field $E_\alpha(\omega)$ in an isotropic medium is

$$\begin{aligned}P_\alpha(t) &= \varepsilon_0 E_\alpha(\omega)\chi_{\alpha\alpha}(\omega)e^{-i\omega t} \\ &= \varepsilon_0 E_\alpha(\omega)[\chi'_{\alpha\alpha}(\omega) + i\chi''_{\alpha\alpha}(\omega)][\cos\omega t - i\sin\omega t],\end{aligned} \tag{7.67}$$

with a real part

$$\begin{aligned}\mathrm{Re}\,P_\alpha(t) &= \varepsilon_0 E_\alpha(\omega)[\chi'_{\alpha\alpha}(\omega)\cos\omega t + \chi''_{\alpha\alpha}(\omega)\sin\omega t] \\ &= E_\alpha(\omega)\frac{N}{\hbar}|d^\alpha_{\mathrm{ge}}|^2 \frac{1}{(\omega_0 - \omega)^2 + \Gamma^2}[(\omega_0 - \omega)\cos\omega t + \Gamma\sin\omega t].\end{aligned} \tag{7.68}$$

The relation between the imaginary part of the first-order susceptibility and energy dissipation may be expressed in another way. Referring to Eq. (7.20b),

$$\begin{aligned}\chi''_{\mu\alpha}(\omega) &= \frac{1}{2i}[\chi_{\mu\alpha}(\omega) - \chi^*_{\mu\alpha}(\omega)], \\ &= \frac{N}{2\varepsilon_0\hbar}\int_0^\infty \langle[\tilde{d}_\mu(\tau), d_\alpha]\rangle(e^{i\omega\tau} - e^{-i\omega\tau})\,d\tau.\end{aligned} \tag{7.69}$$

But after replacing τ by $-\tau$ and shifting the time axis in the correlation functions in the commutator

$$\int_0^{\infty} \langle[\tilde{d}_\mu(\tau), d_\alpha]\rangle e^{-i\omega\tau}\, d\tau = -\int_{-\infty}^{0} \langle[\tilde{d}_\alpha(\tau), d_\mu]\rangle e^{i\omega\tau}\, d\tau. \tag{7.70}$$

Therefore, for the diagonal elements of Eq. (7.69) we have

$$\chi''_{\alpha\alpha}(\omega) = \frac{N}{2\varepsilon_0 \hbar} \int_{-\infty}^{\infty} \langle[\tilde{d}_\alpha(\tau), d_\alpha]\rangle e^{i\omega\tau}\, d\tau. \tag{7.71}$$

Under thermal equilibrium, Eqs. (2.260) to (2.262) enable us to convert Eq. (7.71) to

$$\begin{aligned}
\chi''_{\alpha\alpha}(\omega) &= \frac{N}{2\varepsilon_0 \hbar}(1 - e^{-\beta\hbar\omega}) \int_{-\infty}^{\infty} \langle \tilde{d}_\alpha(\tau)\, d_\alpha\rangle e^{i\omega\tau}\, d\tau, \\
&= \frac{N}{2\varepsilon_0 \hbar}(e^{\beta\hbar\omega} - 1) \int_{-\infty}^{\infty} \langle d_\alpha \tilde{d}_\alpha(\tau)\rangle e^{i\omega\tau}\, d\tau, \\
&= \frac{N}{2\varepsilon_0 \hbar} \tan h\left(\frac{\beta\hbar\omega}{2}\right) \int_{-\infty}^{\infty} (\langle \tilde{d}_\alpha(\tau)\, d_\alpha\rangle + \langle d_\alpha \tilde{d}_\alpha(\tau)\rangle) e^{i\omega\tau}\, d\tau.
\end{aligned} \tag{7.72}$$

Equation (7.72) in one or another of its several forms is known as the *fluctuation-dissipation theorem.* It relates the imaginary part of the susceptibility to the correlation function associated with the parameter of the system that responds to the external force. In the present case, the parameter is the dipole moment and the external force is exerted by the applied radiation field. Thus, the three quantities—imaginary part of the susceptibility, energy dissipation, and the system's correlation functions—are intimately related. It is also worth noting that the correlation functions are defined with reference to the equilibrium state despite the fact that the dissipative process only arises when the system is driven away from equilibrium by the external force.

The real and imaginary parts of the susceptibility satisfy an important set of relations whose derivation requires the frequency to be regarded as a complex quantity. Consider the contour integral

$$\oint \frac{\chi(\omega_1)}{\omega_1 - \omega}\, d\omega_1 \tag{7.73}$$

over the contour shown in Fig. 7.1. If the integrand has no singularities within the contour,

$$\begin{aligned}
\oint \frac{\chi(\omega_1)}{\omega_1 - \omega}\, d\omega_1 &= \int_{C'} \frac{\chi(\omega_1)}{\omega_1 - \omega}\, d\omega_1 + \int_{-R}^{\omega-\varepsilon} \frac{\chi(\omega_1)}{\omega_1 - \omega}\, d\omega_1 \\
&\quad + \int_{\omega+\varepsilon}^{R} \frac{\chi(\omega_1)}{\omega_1 - \omega}\, d\omega_1 + \int_{C} \frac{\chi(\omega_1)}{\omega_1 - \omega}\, d\omega_1 = 0.
\end{aligned} \tag{7.74}$$

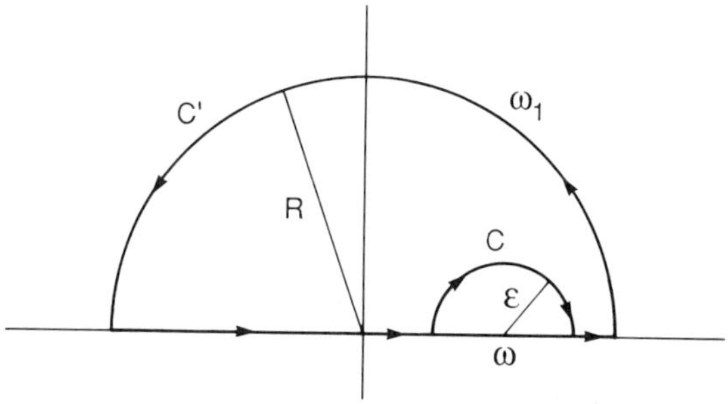

FIGURE 7.1 Integration contour for the Kramers-Kronig relations.

Assuming that $|\chi(\omega_1)|$ approaches zero faster than $1/|\omega_1|$ as $\omega_1 \to \infty$, we have

$$\lim_{R\to\infty} \int_{C'} \frac{\chi(\omega_1)}{\omega_1 - \omega} d\omega_1 = 0. \tag{7.75}$$

The integral over C may be evaluated by writing

$$\omega_1 = \omega + \varepsilon e^{i\phi}, \tag{7.76}$$

in which case

$$\lim_{\varepsilon\to 0} \int_{C} \frac{\chi(\omega_1)}{\omega_1 - \omega} d\omega_1 = -i\pi\chi(\omega). \tag{7.77}$$

Combining Eqs. (7.74) and (7.77) we obtain the principle value integral

$$\lim_{R\to\infty,\varepsilon\to 0} \left(\int_{-R}^{\omega-\varepsilon} \frac{\chi(\omega_1)}{\omega_1 - \omega} d\omega_1 + \int_{\omega+\varepsilon}^{R} \frac{\chi(\omega_1)}{\omega_1 - \omega} d\omega_1 \right) \equiv P \int_{-\infty}^{\infty} \frac{\chi(\omega_1)}{\omega_1 - \omega} d\omega_1$$
$$= i\pi\chi(\omega). \tag{7.78}$$

Therefore, with $\chi(\omega) = \chi'(\omega) + i\chi''(\omega)$,

$$\chi'(\omega) = \frac{1}{\pi} P \int_{-\infty}^{\infty} \frac{\chi''(\omega_1)}{\omega_1 - \omega} d\omega_1,$$
$$\chi''(\omega) = -\frac{1}{\pi} P \int_{-\infty}^{\infty} \frac{\chi'(\omega_1)}{\omega_1 - \omega} d\omega_1. \tag{7.79}$$

These are the *Kramers-Kronig relations*, also known as *Hilbert transforms* or *dispersion relations*. Since the principle value of each integral corresponds to an integration along the real axis, with the singular point omitted, the frequencies in Eq. (7.79) are real.

It is often desirable to cast the Kramers-Kronig relations into a form in which only positive frequencies appear. Thus,

$$\begin{aligned}\chi'(\omega) &= \frac{1}{\pi} P \int_{-\infty}^{0} \frac{\chi''(\omega_1)}{\omega_1 - \omega} d\omega_1 + \frac{1}{\pi} P \int_{0}^{\infty} \frac{\chi''(\omega_1)}{\omega_1 - \omega} d\omega_1 \\ &= \frac{1}{\pi} P \int_{0}^{\infty} \frac{-\chi''(-\omega_1)}{\omega_1 + \omega} d\omega_1 + \frac{1}{\pi} P \int_{0}^{\infty} \frac{\chi''(\omega_1)}{\omega_1 - \omega} d\omega_1.\end{aligned} \tag{7.80}$$

Since $\chi'(-\omega) = \chi'(\omega)$ and $\chi''(-\omega) = -\chi''(\omega)$ we have

$$\chi'(\omega) = \frac{2}{\pi} P \int_{0}^{\infty} \frac{\omega_1 \chi''(\omega_1)}{\omega_1^2 - \omega^2} d\omega_1. \tag{7.81}$$

A similar development for $\chi''(\omega)$ yields

$$\chi''(\omega) = -\frac{2}{\pi} P \int_{0}^{\infty} \frac{\omega \chi'(\omega_1)}{\omega_1^2 - \omega^2} d\omega_1. \tag{7.82}$$

Parallel expressions may be written in terms of the dielectric constant ε since

$$\varepsilon = \varepsilon_0(1 + \chi) = \varepsilon' + i\varepsilon''. \tag{7.83}$$

The Kramers-Kronig relations are quite general; they hold whenever there is a causal relation between the response (or output) of a system and the external disturbance (or input), such as the case discussed at the beginning of the previous section.

7.3 Second- and Third-Order Susceptibility

It already has been noted that a second-order interaction with two frequencies ω_1 and ω_2 is capable of polarizing the medium at frequencies $2\omega_1$, $2\omega_2$, 0, $\omega_1 - \omega_2$, and $\omega_1 + \omega_2$. We will derive expressions for the polarization and the susceptibility at $\omega = \omega_1 + \omega_2$ which includes the other frequencies as special cases.

For the general case, we have

$$P_\mu^{(2)}(t) = N \sum_{kl} \rho_{kl}^{(2)}(t)\, d_{lk}^{\mu}(t). \tag{7.84}$$

It already has been noted that a second-order interaction with two fre- $\exp[-i(\omega_1 + \omega_2)t]$, i.e.,

$$P_\mu^{(2)}(t) = P_\mu^{(2)}(\omega) e^{-i(\omega_1 + \omega_2)t}, \tag{7.85}$$

the required expression for $\rho_{kl}^{(2)}(t)$ is the one given by Eq. (6.68). Thus,

$$P_{\mu}^{(2)}(\omega_1+\omega_2)=\frac{N}{\hbar^2}\sum_{klr}\sum_{\alpha\beta}\frac{E_{\alpha}(\omega_1)E_{\beta}(\omega_2)}{\omega_{kl}-\omega_1-\omega_2-i\Gamma_{kl}}$$
$$\times\left[\frac{d_{kr}^{\alpha}d_{rl}^{\beta}d_{lk}^{\mu}}{\omega_{rl}-\omega_2-i\Gamma_{rl}}(\rho_{ll}^{0}-\rho_{rr}^{0})-\frac{d_{kr}^{\alpha}d_{rl}^{\beta}d_{lk}^{\mu}}{\omega_{kr}-\omega_1-i\Gamma_{kr}}(\rho_{rr}^{0}-\rho_{kk}^{0})\right.$$
$$\left.+\frac{d_{kr}^{\beta}d_{rl}^{\alpha}d_{lk}^{\mu}}{\omega_{rl}-\omega_1-i\Gamma_{rl}}(\rho_{ll}^{0}-\rho_{rr}^{0})-\frac{d_{kr}^{\beta}d_{rl}^{\alpha}d_{lk}^{\mu}}{\omega_{kr}-\omega_2-i\Gamma_{kr}}(\rho_{rr}^{0}-\rho_{kk}^{0})\right]. \quad (7.86)$$

A digression on notation is now in order. Since polarizations of order higher than the first involve more than one frequency and the frequencies may combine in various ways, the notation must be explicit in this regard. In the most common format one writes

$$\mathbf{P}^{(n)}(\omega)=\varepsilon_0\chi^{(n)}(-\omega;\theta_1\omega_1,\theta_2\omega_2,\ldots,\theta_n\omega_n)\times\mathbf{E}_{\theta_1}(\omega_1)\mathbf{E}_{\theta_2}(\omega_2)\ldots\mathbf{E}_{\theta_n}(\omega_n), \quad (7.87)$$

where

$$\theta_j=\pm1,\qquad \omega=\theta_1\omega_1+\theta_2\omega_2+\cdots+\theta_n\omega_n, \quad (7.88)$$

$$\mathbf{E}_{\theta_j}(\omega_j)=\begin{cases}\mathbf{E}(\omega_j) & \text{if } \theta_j=+1,\\ \mathbf{E}^*(\omega_j) & \text{if } \theta_j=-1.\end{cases} \quad (7.89)$$

As an example, a component of the susceptibility in second order for the coupling of three waves, ω_1, ω_2, and ω such that $\omega=\omega_1+\omega_2$ (3-wave mixing) is written

$$\chi_{\mu\alpha\beta}^{(2)}(-\omega;\omega_1,\omega_2) \quad (7.90)$$

in which the frequency of the polarization (ω), with a minus sign, is placed first in the argument followed by the frequencies (ω_1, ω_2) whose combination produces the resultant polarization. The algebraic sum of all the frequencies in the argument equals zero.

Equation (7.86) now may be written

$$P_{\mu}^{(2)}(\omega)=\varepsilon_0\sum_{\alpha\beta}\chi_{\mu\alpha\beta}^{(2)}(-\omega;\omega_1,\omega_2)E_{\alpha}(\omega_1)\,E_{\beta}(\omega_2) \quad (7.91)$$

where

$$\chi_{\mu\alpha\beta}^{(2)}(-\omega;\omega_1,\omega_2)=\frac{N}{\varepsilon_0\hbar^2}\sum_{klr}\frac{1}{\omega_{kl}-\omega_1-\omega_2-i\Gamma_{kl}}$$
$$\times\left[\frac{d_{kr}^{\alpha}d_{rl}^{\beta}d_{lk}^{\mu}}{\omega_{rl}-\omega_2-i\Gamma_{rl}}(\rho_{ll}^{0}-\rho_{rr}^{0})-\frac{d_{kr}^{\alpha}d_{rl}^{\beta}d_{lk}^{\mu}}{\omega_{kr}-\omega_1-i\Gamma_{kr}}(\rho_{rr}^{0}-\rho_{kk}^{0})\right.$$
$$\left.+\frac{d_{kr}^{\beta}d_{rl}^{\alpha}d_{lk}^{\mu}}{\omega_{rl}-\omega_1-i\Gamma_{rl}}(\rho_{ll}^{0}-\rho_{rr}^{0})-\frac{d_{kr}^{\beta}d_{rl}^{\alpha}d_{lk}^{\mu}}{\omega_{kr}-\omega_2-i\Gamma_{kr}}(\rho_{rr}^{0}-\rho_{kk}^{0})\right]. \quad (7.92)$$

Since we have the freedom of interchanging the summation indices, all the terms in the susceptibility function may be referred to ρ^0_{kk}. That is, l and k are interchanged in the terms proportional to ρ^0_{ll} and r and k are interchanged in the terms proportional to ρ^0_{rr}. It is noted that a frequency such as ω_{kl} is the same as $-\omega_{lk}$ but $\Gamma_{kl} = \Gamma_{lk}$. Also, having assumed that the atomic wave functions are real, a matrix element such as d^α_{kl} is the same as d^α_{lk}. The products of the matrix elements, of course, may be written in any order. Then

$$\chi^{(2)}_{\mu\alpha\beta}(-\omega; \omega_1, \omega_2) = \frac{N}{\varepsilon_0 \hbar^2} \sum_{klr} (U_1 + U_2 + \cdots + U_8), \tag{7.93}$$

with

$$\begin{aligned} U_1 &= \frac{1}{\omega_{kl} - \omega_1 - \omega_2 - i\Gamma_{kl}} \frac{d^\alpha_{kr} d^\beta_{rl} d^\mu_{lk}}{\omega_{rl} - \omega_2 - i\Gamma_{rl}} \rho^0_{ll} \\ &= \frac{d^\mu_{kl} d^\alpha_{lr} d^\beta_{rk}}{(\omega + \omega_{kl} + i\Gamma_{kl})(\omega_2 + \omega_{kr} + i\Gamma_{kr})} \rho^0_{kk}, \end{aligned} \tag{7.94a}$$

$$\begin{aligned} U_2 &= \frac{1}{\omega_{kl} - \omega_1 - \omega_2 - i\Gamma_{kl}} \frac{d^\alpha_{kr} d^\beta_{rl} d^\mu_{lk}}{\omega_{rl} - \omega_2 - i\Gamma_{rl}} \rho^0_{rr} \\ &= -\frac{d^\beta_{kl} d^\mu_{lr} d^\alpha_{rk}}{(\omega + \omega_{lr} + i\Gamma_{lr})(\omega_2 + \omega_{lk} + i\Gamma_{lk})} \rho^0_{kk}, \end{aligned} \tag{7.94b}$$

$$\begin{aligned} U_3 &= -\frac{1}{\omega_{kl} - \omega_1 - \omega_2 - i\Gamma_{kl}} \frac{d^\alpha_{kr} d^\beta_{rl} d^\mu_{lk}}{\omega_{kr} - \omega_1 - i\Gamma_{kr}} \rho^0_{rr} \\ &= -\frac{d^\beta_{kl} d^\mu_{lr} d^\alpha_{rk}}{(\omega + \omega_{lr} + i\Gamma_{lr})(\omega_1 + \omega_{kr} + i\Gamma_{kr})} \rho^0_{kk}, \end{aligned} \tag{7.94c}$$

$$\begin{aligned} U_4 &= \frac{1}{\omega_{kl} - \omega_1 - \omega_2 - i\Gamma_{kl}} \frac{d^\alpha_{kr} d^\beta_{rl} d^\mu_{lk}}{\omega_{kr} - \omega_1 - i\Gamma_{kr}} \rho^0_{kk} \\ &= \frac{d^\alpha_{kl} d^\beta_{lr} d^\mu_{rk}}{(\omega + \omega_{rk} + i\Gamma_{rk})(\omega_1 + \omega_{lk} + i\Gamma_{lk})} \rho^0_{kk}, \end{aligned} \tag{7.94d}$$

$$\begin{aligned} U_5 &= \frac{1}{\omega_{kl} - \omega_1 - \omega_2 - i\Gamma_{kl}} \frac{d^\beta_{kr} d^\alpha_{rl} d^\mu_{lk}}{\omega_{rl} - \omega_1 - i\Gamma_{rl}} \rho^0_{ll} \\ &= \frac{d^\mu_{kl} d^\beta_{lr} d^\alpha_{rk}}{(\omega + \omega_{kl} + i\Gamma_{kl})(\omega_1 + \omega_{kr} + i\Gamma_{kr})} \rho^0_{kk}, \end{aligned} \tag{7.94e}$$

$$\begin{aligned} U_6 &= -\frac{1}{\omega_{kl} - \omega_1 - \omega_2 - i\Gamma_{kl}} \frac{d^\beta_{kr} d^\alpha_{rl} d^\mu_{lk}}{\omega_{rl} - \omega_1 - i\Gamma_{rl}} \rho^0_{rr} \\ &= -\frac{d^\alpha_{kl} d^\mu_{lr} d^\beta_{rk}}{(\omega + \omega_{lr} + i\Gamma_{lr})(\omega_1 + \omega_{lk} + i\Gamma_{lk})} \rho^0_{kk}, \end{aligned} \tag{7.94f}$$

$$U_7 = -\frac{1}{\omega_{kl} - \omega_1 - \omega_2 - i\Gamma_{kl}} \frac{d^{\beta}_{kr} d^{\alpha}_{rl} d^{\mu}_{lk}}{\omega_{kr} - \omega_2 - i\Gamma_{kr}} \rho^0_{rr}$$

$$= -\frac{d^{\alpha}_{kl} d^{\mu}_{lr} d^{\beta}_{rk}}{(\omega + \omega_{lr} + i\Gamma_{lr})(\omega_2 + \omega_{kr} + i\Gamma_{kr})} \rho^0_{kk}, \qquad (7.94g)$$

$$U_8 = \frac{1}{\omega_{kl} - \omega_1 - \omega_2 - i\Gamma_{kl}} \frac{d^{\beta}_{kr} d^{\alpha}_{rl} d^{\mu}_{lk}}{\omega_{kr} - \omega_2 - i\Gamma_{kl}} \rho^0_{kk}$$

$$= \frac{d^{\beta}_{kl} d^{\alpha}_{lr} d^{\mu}_{rk}}{(\omega + \omega_{rk} + i\Gamma_{rk})(\omega_2 + \omega_{lk} + i\Gamma_{lk})} \rho^0_{kk}. \qquad (7.94h)$$

Feynman diagrams for the computation of density matrix elements were described in Section 4.11. It would be highly desirable to extend the diagrammatic approach to compute the terms in the susceptibility function directly, without the prior computation of the density matrix elements. This, indeed, turns out to be possible [14]. Referring to Fig. 7.2, which corresponds to Fig. 4.10 for one term in the density matrix element, the lowest horizontal line connecting $|k\rangle$ with $\langle k|$ identifies ρ^0_{kk}. Progressing in time (upward on the diagram) beyond the first interaction point (on the bra side) associated with $-\omega_2$, the horizontal line connecting $|k\rangle$ with $\langle l|$ identifies the factor $-1/(\omega_{kl} - \omega_2 - i\Gamma_{kl})$. The negative sign in the numerator is associated with the location of the interaction point on the bra side. Continuing above the next interaction point (on the ket side) associated with $-\omega_1$, the horizontal line connecting $|r\rangle$ with $\langle l|$ identifies the factor $1/(\omega_{rl} - \omega - i\Gamma_{rl})$ in which $\omega = \omega_1 + \omega_2$. Here the numerator carries a positive sign owing to the location of the interaction point on the ket side. We note that each horizontal line generates a factor that contains all the frequencies (including their signs) below the line.

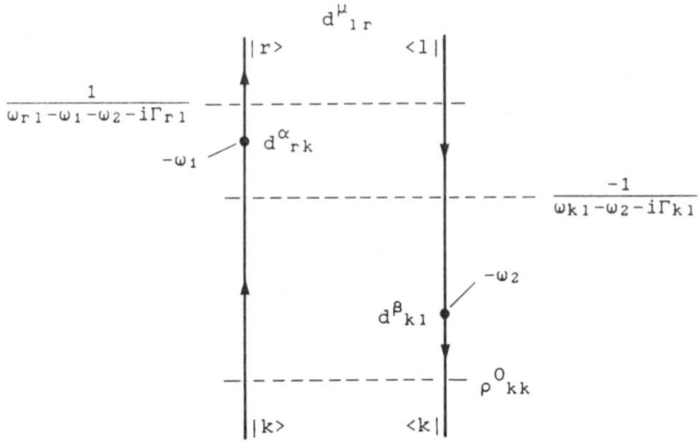

FIGURE 7.2 Steps in the development of the expression for U_2 (Eq. (7.94b)) from the associated Feynman diagram.

To obtain the matrix elements, we follow the arrows—up on the ket side and down on the bra side, bearing in mind that there is a direct association between α and ω_1; β and ω_2; and between μ and $\omega(=\omega_1 + \omega_2)$. We arrive first at $-\omega_1$, which lies between $|r\rangle$ and $|k\rangle$; this gives the matrix element d^{α}_{rk}. Crossing to the bra side at $\langle l|$ from the ket side at $|r\rangle$ gives rise to d^{μ}_{lr}. Finally, arriving at $-\omega_2$ between $\langle k|$ and $\langle l|$ gives d^{β}_{kl}. In this manner, we have generated the term U_2. Diagrams for $U_1, \ldots, U_8$ are shown in Fig. 7.3. An example of a diagram for $\omega = \omega_1 - \omega_2$ is shown in Fig. 7.4.

An interchange of the summation indices r and l in U_1 (Eq. (7.94a)) yields the result

$$U_1^*(\omega_1, \omega_2) = U_8(-\omega_1, \omega_2), \tag{7.95}$$

and a similar match-up exists for all the terms $U_1, \ldots, U_8$. Hence, we have the extension of Eq. (7.47) to second order:

$$\chi^{*(2)}_{\mu\alpha\beta}(-\omega; \omega_1, \omega_2) = \chi^{(2)}_{\mu\alpha\beta}(\omega; -\omega_1, -\omega_2). \tag{7.96}$$

A type of symmetry that appears exclusively in the nonlinear case is known as *intrinsic permutation symmetry*. In second order, it states that the susceptibility is invariant under an interchange between (α, ω_1) and (β, ω_2); i.e.,

$$\chi^{(2)}_{\mu\alpha\beta}(-\omega; \omega_1, \omega_2) = \chi^{(2)}_{\mu\beta\alpha}(-\omega; \omega_2, \omega_1). \tag{7.97}$$

This relation is verified easily by noting that each of the pairs (U_1, U_5), (U_2, U_6), (U_3, U_7) and (U_4, U_8) involves an interchange between (α, ω_1) and (β, ω_2). Obviously, since the linear case contains only one frequency, there can be no intrinsic symmetry. The expression for the susceptibility given by Eq. (7.93) then may be written

$$\chi^{(2)}_{\mu\alpha\beta}(-\omega; \omega_1, \omega_2) = \frac{N}{\varepsilon_0 \hbar^2} \sum_{P_i} \sum_{klr} (U_1 + U_2 + U_3 + U_4), \tag{7.98}$$

in which the sum over P_i is understood as a sum over the permutations of (α, ω_1) and (β, ω_2).

For the nonresonant case (damping terms excluded), the susceptibility function is real; hence, Eq. (7.96) becomes

$$\chi^{(2)}_{\mu\alpha\beta}(-\omega; \omega_1, \omega_2) = \chi^{(2)}_{\mu\alpha\beta}(\omega; -\omega_1, -\omega_2), \tag{7.99}$$

which indicates that for a lossless medium, $\chi^{(2)}$ is invariant under a change of sign of all frequencies. Overall symmetry in lossless media was defined previously by Eq. (7.50b) for first-order susceptibility; in second order, overall symmetry implies that the susceptibility is invariant under a permutation among the pairs $(\mu, -\omega)$, (α, ω_1), and (β, ω_2). Thus,

$$\begin{aligned}\chi^{(2)}_{\mu\alpha\beta}(-\omega; \omega_1, \omega_2) &= \chi^{(2)}_{\alpha\mu\beta}(\omega_1; -\omega, \omega_2)\\ &= \chi^{(2)}_{\alpha\mu\beta}(-\omega_1; \omega, -\omega_2).\end{aligned} \tag{7.100}$$

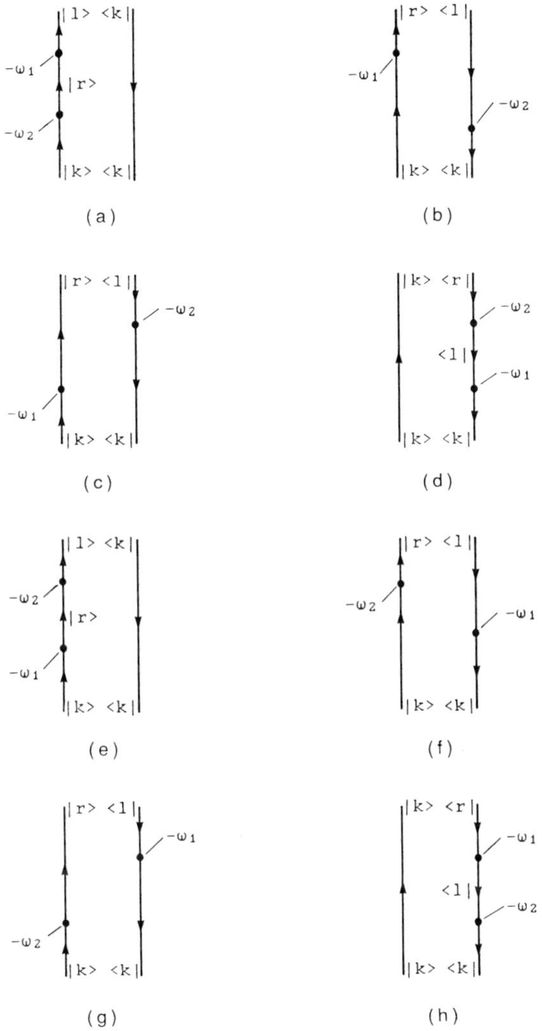

FIGURE 7.3 Diagrams for terms belonging to $\chi^{(2)}(-\omega; \omega_1, \omega_2)$. (a) U_1 (Eq. (7.94a))$\cdots$(h) U_8 (Eq. (7.94h)).

In the same fashion,

$$\begin{aligned}\chi^{(2)}_{\mu\alpha\beta}(-\omega; \omega_1, \omega_2) &= \chi^{(2)}_{\alpha\beta\mu}(-\omega_1; -\omega_2, \omega) \\ &= \chi^{(2)}_{\alpha\mu\beta}(-\omega_1; \omega, -\omega_2) = \chi^{(2)}_{\mu\beta\alpha}(-\omega; \omega_2, \omega_1) \\ &= \chi^{(2)}_{\beta\alpha\mu}(-\omega_2; -\omega_1, \omega) = \chi^{(2)}_{\beta\mu\alpha}(-\omega_2; \omega, -\omega_1). \quad (7.101)\end{aligned}$$

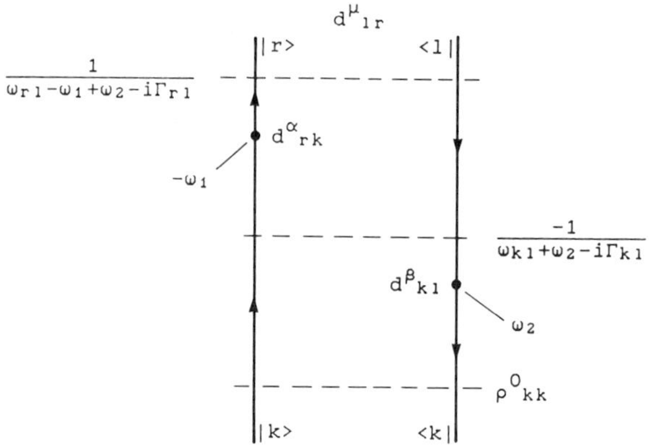

FIGURE 7.4 Diagram for

$$\rho^0_{kk}\left(\frac{-1}{\omega_{kl}+\omega_2-i\Gamma_{kl}}\right)\left(\frac{1}{\omega_{rl}-\omega_1+\omega_2-i\Gamma_{rl}}\right)d^\alpha_{rk}\,d^\mu_{lr}\,d^\beta_{kl}$$

belonging to $\chi^{(2)}(-\omega;\omega_1,-\omega_2)$ with $\omega=\omega_1-\omega_2$.

Physically, these relations indicate that in addition to the polarization at the frequency ω produced by the two frequencies ω_1 and ω_2, the same susceptibility function is responsible for a polarization at the frequency ω_1 as a result of ω_2 combining with ω; similarly, a polarization at ω_2 can be achieved by mixing ω_1 with ω.

The symmetry of the medium through which the electromagnetic fields propagate imposes certain restrictions on the susceptibility functions. To demonstrate this feature consider a coordinate transformation $\mathbf{r}'=R\mathbf{r}$ where R is an orthogonal matrix with matrix elements a_{ij}. Under the coordinate transformation R, a tensor such as $\chi^{(2)}_{\mu\alpha\beta}$ transforms according to the rule

$$\chi'^{(2)}_{\mu\alpha\beta}=\sum_{u'\alpha'\beta'}a_{\mu\mu'}a_{\alpha\alpha'}a_{\beta\beta'}\chi^{(2)}_{\mu'\alpha'\beta'}. \tag{7.102}$$

When R belongs to the point group of the medium, the susceptibility tensors must be identical in the two coordinate systems. Thus,

$$\chi'^{(2)}_{\mu\alpha\beta}=\chi^{(2)}_{\mu'\alpha'\beta'}. \tag{7.103}$$

But this relation generally cannot be satisfied unless one imposes restrictions on the coefficients in Eq. (7.102). To take a specific example, let R be the inversion matrix

$$R=\begin{pmatrix}-1&0&0\\0&-1&0\\0&0&-1\end{pmatrix} \tag{7.104}$$

or

$$a_{\mu\mu'} = -\delta_{\mu\mu'} \qquad a_{\alpha\alpha'} = -\delta_{\alpha\alpha'} \qquad a_{\beta\beta'} = -\delta_{\beta\beta'} \tag{7.105}$$

Equation (7.102) then becomes

$$\chi'^{(2)}_{\mu\alpha\beta} = (-1)^3 \chi^{(2)}_{\mu\alpha\beta}, \tag{7.106}$$

which is clearly impossible. Thus, the second-order susceptibility tensor vanishes in a medium with inversion symmetry as, for example, in the case of crystals that have a center of symmetry. The first nonlinear susceptibility is then $\chi^{(3)}$, and all even order susceptibilities are zero. An externally applied dc field, however, can destroy inversion symmetry. For other symmetry classes, the general effect is to reduce the number of nonvanishing, independent components of the susceptibility function.

One also make take note of the fact that atomic systems have inversion symmetry and the wave functions therefore have a definite parity. Then, in a product such as $d^{\mu}_{kl} d^{\alpha}_{lr} d^{\beta}_{rk}$, the wave functions $|k\rangle$ and $|l\rangle$ must have opposite parity in order for the matrix element of the dipole operator (which has odd parity) not to vanish. In the same way, $|l\rangle$ and $|r\rangle$ must have opposite parity, which means that $|k\rangle$ and $|r\rangle$ have the same parity. But in that case, $d^{\beta}_{rk} = 0$, and so the second-order susceptibility vanishes. This argument is simply another manifestation of the fact, previously established, that the second-order susceptibility, as well as all higher even orders, vanish in systems with inversion symmetry.

An additional restriction arises from the requirement of phase matching. That is, if ω_1 and ω_2 combine to produce the frequency $\omega = \omega_1 + \omega_2$, it is necessary to satisfy the condition

$$\mathbf{k}(\omega) = \mathbf{k}(\omega_1) + \mathbf{k}(\omega_2) \tag{7.107}$$

for efficient conversion. Otherwise, the relative phases will fluctuate during the propagation and, in fact, may be reversed, resulting in a loss of energy transfer to the wave at ω. Since the photon momentum is $\hbar\mathbf{k}$, Eq. (7.107) may be interpreted as the condition for the conservation of momentum among the three photons. For maximum efficiency, the three waves should be colinear, in which case

$$|\mathbf{k}(\omega)| = |\mathbf{k}(\omega_1)| + |\mathbf{k}(\omega_2)| \tag{7.108}$$

or, in terms of the refractive indices,

$$\omega n = \omega_1 n_1 + \omega_2 n_2 \tag{7.109}$$

Clearly, the direct calculation of $\chi^{(2)}$ is difficult—one must know the relation between the applied field and the effective field in the medium, the transition frequencies, and the matrix elements. When an incident frequency or a combination of both frequencies falls in the vicinity of a resonance transition, however, one or more denominators become very small. A good approximation to the susceptibility is obtained then by keeping only the near-resonant terms with denominators largely dominated by the damping constants.

Third-order polarization (four-wave mixing) can occur in a medium with or without a center of symmetry; consequently, third-order polarization is the lowest order nonlinearity in a medium with inversion symmetry. The susceptibility function is represented by eight basic Feynman diagrams (Fig. 7.5); each diagram contains three interaction points, and for each diagram there are six permutations of three input frequencies ω_1, ω_2, and ω_3. Figure 7.6 shows the permutations for the diagram in Fig. 7.5b. The susceptibility at a frequency ω arising from a specific combination of ω_1, ω_2, and ω_3 (e.g., $\omega = \omega_1 + \omega_2 - \omega_3$) consequently consists of 48 terms. Moreover, there are eight possible combinations of the three frequencies resulting in 384 possible terms that may contribute to a third-order polarization. Again, under circumstances where resonances occur, the number of terms often may be substantially reduced. Fig. 7.7 illustrates a contribution to $\chi^{(3)}(-\omega; \omega_1, \omega_2, -\omega_3)$ from the arrangement of frequencies in Fig. 7.6e. The corresponding

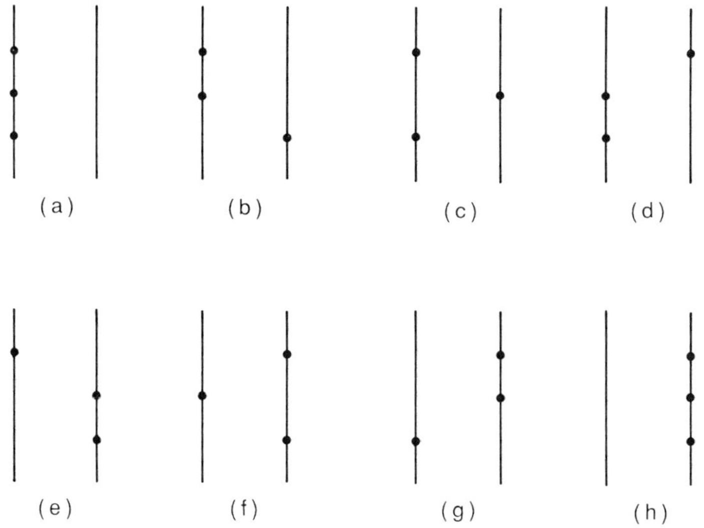

FIGURE 7.5 Basic Feynman diagrams for the third-order susceptibility. Each diagram contains three interaction points.

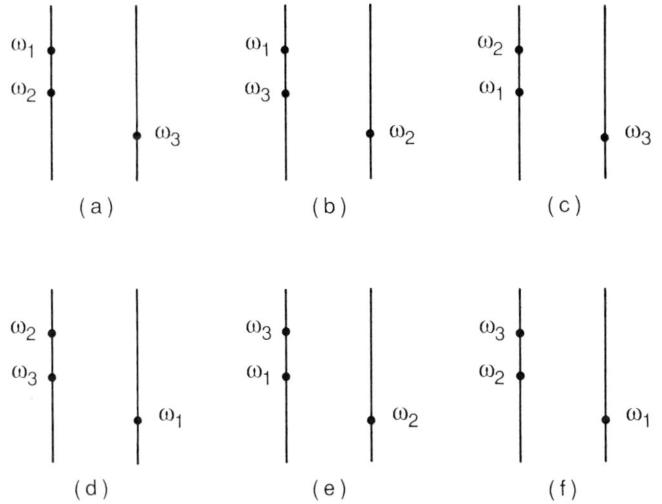

FIGURE 7.6 The six permutations of three input frequencies ω_1, ω_2, and ω_3 for the basic diagram shown in Fig. 7.5b.

polarization is

$$P_\mu^{(3)}(t) = P_\mu^{(3)}(\omega)e^{-i(\omega_1+\omega_2-\omega_3)} \tag{7.110}$$

where

$$P_\mu^{(3)}(\omega) = \varepsilon_0 \sum_{\alpha\beta\gamma} \chi_{\mu\alpha\beta\gamma}^{(3)}(-\omega; \omega_1, \omega_2, -\omega_3)E_\alpha(\omega_1)E_\beta(\omega_2)E_\gamma^*(\omega_3). \tag{7.111}$$

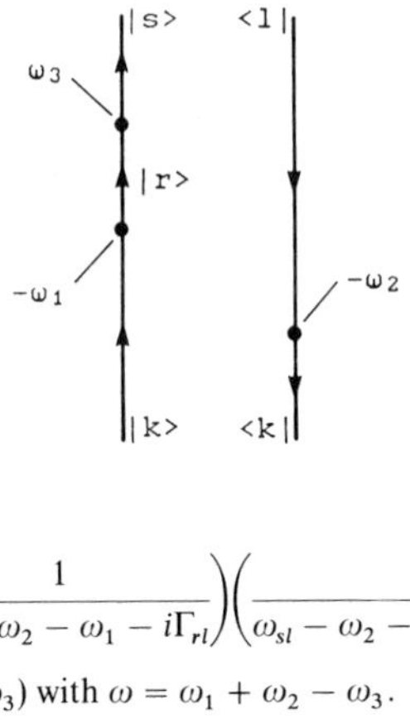

FIGURE 7.7 Diagram for

$$\rho_{kk}^0\left(\frac{-1}{\omega_{kl}-\omega_2-i\Gamma_{kl}}\right)\left(\frac{1}{\omega_{rl}-\omega_2-\omega_1-i\Gamma_{rl}}\right)\left(\frac{1}{\omega_{sl}-\omega_2-\omega_1+\omega_3-i\Gamma_{sl}}\right)d_{rk}^\alpha d_{sr}^\gamma d_{ls}^\mu d_{kl}^\beta$$

belonging to $\chi^{(3)}(-\omega; \omega_1, \omega_2, -\omega_3)$ with $\omega = \omega_1 + \omega_2 - \omega_3$.

7.4 Two-Photon Absorption and Emission

Ordinary absorption and emission are one-photon processes since in every transition, the field either loses or gains a single photon while the atom undergoes a corresponding change of state to keep the total energy (atom plus field) constant. Two-photon processes, including scattering, two-photon absorption, and two-photon emission involve a change of two photons. In a general scattering event, the field loses an incident photon and gains a scattered photon, which may or may not differ from the incident photon in energy, direction of propagation, or polarization. In two-photon absorption or emission, two photons are lost or gained, respectively, by the field with a compensating energy change in the atom, molecule, crystal, etc., with which the field interacts.

A two-photon process may occur in two sequential one-photon steps. Thus, an atom may reach an excited state by absorbing a single photon. During the lifetime of the excited state, the atom may absorb another photon to reach a higher excited state. The essential physics of such processes is entirely contained in the physics of one-photon (linear) processes, as described in previous chapters. New physics is encountered when two photons are simultaneously absorbed or emitted as a single, nonlinear indivisible event in which only the beginning and end of the event are observable experimentally [15]. Although for theoretical purposes intermediate steps might be conceptualized, they are not observable experimentally. This is very similar to Raman scattering (Section 5.11) where intermediate states appear in the formalism but one does not actually observe absorption followed by emission (or the converse) as separate events. Nor do we observe an intermediate atomic state in a nonlinear two-photon absorption or emission. These concepts are extended readily to multiphoton processes.

It is recalled that in a system with inversion symmetry (e.g., an atom) a one-photon (dipole) transition requires the initial and final states to have opposite parity. For a two-photon transition, as soon will become evident, the selection rule requires the two states to have the same parity. As a consequence, states that cannot be populated by one-photon transitions, because of parity restrictions, may become accessible by two-photon (or N-photon, N even) transitions. Spectroscopic investigations have benefited considerably as a result of this feature. In a typical two-photon absorption experiment, one measures the loss in transmission of one light beam of frequency ω_1 in the presence of a second—usually a strong laser beam—of frequency ω_2.

Another property that distinguishes the multiphoton, nonlinear process from the single-photon, linear process involves the statistics of light beams.

The rates of nonlinear interactions are influenced by the photon distributions in the participating light beams and by their correlations. In turn, the nonlinear interaction couples different photon modes which leads to the annihilation of photons in certain modes and the creation of photons in other modes. Hence, the photon distributions or the statistics of the emerging light beams and their correlations are altered in the nonlinear process. We return to this subject in Section 7.6.

In a scattering process, one photon is absorbed and another emitted; in two-photon absorption or emission, both photons are either absorbed or emitted. Hence, the cross sections for two-photon absorption and emission may be derived by following the pattern employed for the derivation of the Kramers-Heisenberg formulas.

Consider, first, the absorption process with the dipole interaction Hamiltonian (Eq. (5.252)). For the initial and final states we take

$$|\mathrm{I}\rangle = |g, n_1, n_2\rangle, \qquad |\mathrm{F}\rangle = |e, n_1 - 1, n_2 - 1\rangle, \tag{7.112}$$

with energies

$$\begin{aligned} E_\mathrm{I} &= E_\mathrm{g} + n_1\hbar\omega_1 + n_2\hbar\omega_2, \\ E_\mathrm{F} &= E_\mathrm{e} + (n_1 - 1)\hbar\omega_1 + (n_2 - 1)\hbar\omega_2. \end{aligned} \tag{7.113}$$

The requirement $E_\mathrm{I} = E_\mathrm{F}$ to conserve energy yields

$$E_\mathrm{e} - E_\mathrm{g} = \hbar(\omega_1 + \omega_2). \tag{7.114}$$

The general matrix element (Eq. (5.256)) for a two-photon process requires construction of intermediate states that are connected to the initial and final states by one-photon processes. Hence, the analog to Eq. (5.257) for two-photon absorption is written

$$|T\rangle = \begin{cases} |T_1\rangle = |t_1, n_1 - 1, n_2\rangle, \\ |T_2\rangle = |t_2, n_1, n_2 - 1\rangle, \end{cases} \tag{7.115}$$

in which $|T_1\rangle$ is a state of the total system after a photon of frequency ω_1 has been absorbed. $|T_2\rangle$ is a state after absorption of an ω_2-photon. Diagrams for the two paths $|\mathrm{I}\rangle \to |T_1\rangle \to |\mathrm{F}\rangle$ and $|\mathrm{I}\rangle \to |T_2\rangle \to |\mathrm{F}\rangle$ are shown in Fig. 7.8a. The energies are

$$\begin{aligned} E_{T_1} &= E_{t_1} + (n_1 - 1)\hbar\omega_1 + n_2\hbar\omega_2, \\ E_{T_2} &= E_{t_2} + n_1\hbar\omega_1 + (n_2 - 1)\hbar\omega_2, \end{aligned} \tag{7.116}$$

$$\begin{aligned} E_\mathrm{I} - E_{T_1} &= E_\mathrm{g} - E_{t_1} + \hbar\omega_1, \\ E_\mathrm{I} - E_{T_2} &= E_\mathrm{g} - E_{t_2} + \hbar\omega_2. \end{aligned} \tag{7.117}$$

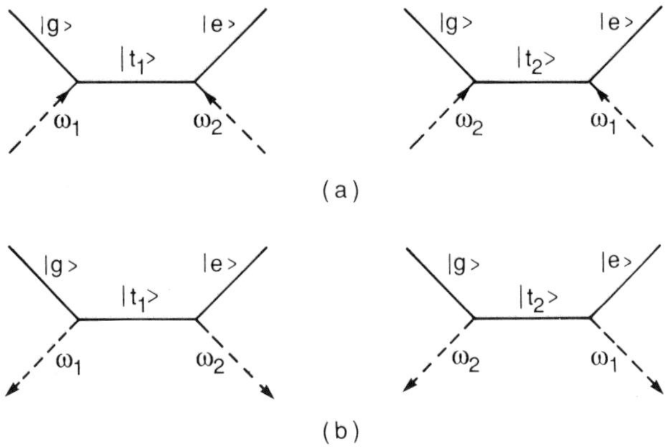

FIGURE 7.8 Diagrams for two-photon absorption and emission. (a) In absorption the two paths are $|\mathrm{I}\rangle = |\mathrm{g}, n_1, n_2\rangle \to |T_1\rangle = |t_1, n_1 - 1, n_2\rangle \to |\mathrm{F}\rangle = |\mathrm{e}, n_1 - 1, n_2 - 1\rangle$ and $|\mathrm{I}\rangle = |\mathrm{g}, n_1, n_2\rangle \to |T_2\rangle = |t_2, n_1, n_2 - 1\rangle \to |\mathrm{F}\rangle = |\mathrm{e}, n_1 - 1, n_2 - 1\rangle$. (b) In emission the paths are $|\mathrm{I}\rangle = |\mathrm{e}, n_1, n_2\rangle \to |T_1\rangle = |t_1, n_1 + 1, n_2\rangle \to |\mathrm{F}\rangle = |\mathrm{g}, n_1 + 1, n_2 + 1\rangle$ and $|\mathrm{I}\rangle = |\mathrm{e}, n_1, n_2\rangle \to |T_2\rangle = |t_2, n_1, n_2 + 1\rangle \to |\mathrm{F}\rangle = |\mathrm{g}, n_1 + 1, n_2 + 1\rangle$. n_1 and n_2 refer to the number of photons of frequency ω_1 and ω_2, respectively. Note that the intermediate states $|t_1\rangle$ and $|t_2\rangle$ for emission are not the same as $|t_1\rangle$ and $|t_2\rangle$ for absorption.

For absorption, the relevant part of the interaction operator is $\mathscr{H}^{(+)}_{\mathrm{AF}}$, which contains the annihilation operator a. Therefore, referring to Eq. (5.260),

$$\langle \mathrm{F}|\mathscr{H}_{\mathrm{AF}}|T_1\rangle = \langle \mathrm{e}, n_1 - 1, n_2 - 1|\mathscr{H}^{(+)}_{\mathrm{AF}}|t_1, n_1 - 1, n_2\rangle$$

$$= -i\sqrt{\frac{n_2 \hbar\omega_2}{2\varepsilon_0 V}}\langle \mathrm{e}|\mathbf{d}\cdot\hat{\boldsymbol{\varepsilon}}_2|t_1\rangle, \tag{7.118a}$$

$$\langle T_1|\mathscr{H}_{\mathrm{AF}}|\mathrm{I}\rangle = \langle t_1, n_1 - 1, n_2|\mathscr{H}^{(+)}_{\mathrm{AF}}|\mathrm{g}, n_1, n_2\rangle$$

$$= -i\sqrt{\frac{n_1 \hbar\omega_1}{2\varepsilon_0 V}}\langle t_1|\mathbf{d}\cdot\hat{\boldsymbol{\varepsilon}}_1|\mathrm{g}\rangle, \tag{7.118b}$$

$$\langle \mathrm{F}|\mathscr{H}_{\mathrm{AF}}|T_2\rangle = \langle \mathrm{e}, n_1 - 1, n_2 - 1|\mathscr{H}^{(+)}_{\mathrm{AF}}|t_2, n_1, n_2 - 1\rangle$$

$$= -i\sqrt{\frac{n_1 \hbar\omega_1}{2\varepsilon_0 V}}\langle \mathrm{e}|\mathbf{d}\cdot\hat{\boldsymbol{\varepsilon}}_1|t_2\rangle, \tag{7.118c}$$

$$\langle T_2|\mathscr{H}_{\mathrm{AF}}|\mathrm{I}\rangle = \langle t_2, n_1, n_2 - 1|\mathscr{H}^{(+)}_{\mathrm{AF}}|\mathrm{g}, n_1, n_2\rangle$$

$$= -i\sqrt{\frac{n_2 \hbar\omega_2}{2\varepsilon_0 V}}\langle t_2|\mathbf{d}\cdot\hat{\boldsymbol{\varepsilon}}_2|\mathrm{g}\rangle. \tag{7.118d}$$

With these expressions, the total matrix element for two-photon absorption is

$$\begin{aligned} M_{\text{tpa}} &= M_1 + M_2 \\ &= \sum_{T_1} \frac{\langle \text{F}|\mathcal{H}_{\text{AF}}|T_1\rangle\langle T_1|\mathcal{H}_{\text{AF}}|\text{I}\rangle}{E_{\text{I}} - E_{\text{T}_1}} + \sum_{T_2} \frac{\langle \text{F}|\mathcal{H}_{\text{AF}}|T_2\rangle\langle T_2|\mathcal{H}_{\text{AF}}|\text{I}\rangle}{E_{\text{I}} - E_{\text{T}_2}} \\ &= -\frac{\hbar}{2\varepsilon_0 V}\sqrt{n_1 n_2 \omega_1 \omega_2}\left[\sum_{t_1} \frac{\langle \text{e}|\mathbf{d}\cdot\hat{\varepsilon}_2|t_1\rangle\langle t_1|\mathbf{d}\cdot\hat{\varepsilon}_1|\text{g}\rangle}{E_{\text{g}} - E_{t_1} + \hbar\omega_1}\right. \\ &\quad \left. + \sum_{t_2} \frac{\langle \text{e}|\mathbf{d}\cdot\hat{\varepsilon}_1|t_2\rangle\langle t_2|\mathbf{d}\cdot\hat{\varepsilon}_2|\text{g}\rangle}{E_{\text{g}} - E_{t_2} + \hbar\omega_2}\right], \\ &= \frac{1}{2\varepsilon_0 V}\sqrt{n_1 n_2 \omega_1 \omega^2}\sum_t \left[\frac{\langle \text{e}|\mathbf{d}\cdot\hat{\varepsilon}_2|t\rangle\langle t|\mathbf{d}\cdot\hat{\varepsilon}_1|\text{g}\rangle}{\omega_{\text{tg}} - \omega_1}\right. \\ &\quad \left. + \frac{\langle \text{e}|\mathbf{d}\cdot\hat{\varepsilon}_1|t\rangle\langle t|\mathbf{d}\cdot\hat{\varepsilon}_2|\text{g}\rangle}{\omega_{\text{tg}} - \omega_2}\right], \end{aligned} \tag{7.119}$$

where, in the last expression, the summation indices t_1 and t_2 have been replaced by a general summation index t. Finally, the transition probability per unit time for two-photon absorption is obtained from the Golden Rule

$$W_{\text{tpa}} = \frac{2\pi}{\hbar}|M_{\text{tpa}}|^2\,\delta(E_{\text{I}} - E_{\text{F}}). \tag{7.120}$$

If the sum in Eq. (7.119) contains resonance terms where $\omega_{\text{tg}} \simeq \omega_1$ or $\omega_{\text{tg}} \simeq \omega_2$, it is necessary to include damping constants to avoid the singularities. Such terms, when they exist, dominate the sum and strongly enhance the probability for the occurrence of the process. In practical calculations, as in one-photon absorption, the δ-function is replaced by a density of states. Evidently, W_{tpa} is proportional to $n_1 n_2$ or, equivalently, the product of the intensities of the two beams. If only one beam is employed, $\omega_1 = \omega_2 = \omega$, $E_{\text{e}} - E_{\text{g}} = 2\omega$, and the transition rate is proportional to the square of the intensity whereas the one-photon absorption rate is proportional to the first power of the intensity.

As mentioned earlier, the states associated with a nonvanishing dipole matrix element must have opposite parity. That is, the parities of $|t_1\rangle$ and $|t_2\rangle$ must be opposite to the parities of both $|\text{g}\rangle$ and $|\text{e}\rangle$ which means, then, that $|\text{g}\rangle$ and $|\text{e}\rangle$ must have the same parity. This selection rule is the same as the one for Raman scattering. Note, also, that M_{tpa} depends on the polarization vectors ε_1 and ε_2, which implies that the matrix element may vanish for certain orientations of the vectors even when the parity selection rule is satisfied. The same formalism is applicable to two-photon ionization in which the transition occurs from a bound atomic state to a continuum state. Because the final state lies in the continuum, the sum of the photon energies must exceed the ionization limit but is otherwise not restricted as it is in the case of two-photon absorption between two bound states.

Turning now to the emission process (Fig. 7.8b), the initial, final, and intermediate states are

$$|\mathrm{I}\rangle = |\mathrm{e}, n_1, n_2\rangle, \qquad |\mathrm{F}\rangle = |\mathrm{g}, n_1 + 1, n_2 + 1\rangle, \tag{7.121}$$

$$|\mathrm{T}\rangle = \begin{cases} |T_1\rangle = |t_1, n_1 + 1, n_2\rangle, \\ |T_2\rangle = |t_2, n_1, n_2 + 1\rangle, \end{cases} \tag{7.122}$$

with energies

$$\begin{aligned} E_\mathrm{I} &= E_\mathrm{e} + n_1\hbar\omega_1 + n_2\hbar\omega_2, \\ E_\mathrm{F} &= E_\mathrm{g} + (n_1 + 1)\hbar\omega_1 + (n_2 + 1)\hbar\omega_2, \\ E_{T_1} &= E_{t_1} + (n_1 + 1)\hbar\omega_1 + n_2\hbar\omega_2, \\ E_{T_2} &= E_{t_2} + n_1\hbar\omega_1 + (n_2 + 1)\hbar\omega_2, \\ E_\mathrm{I} - E_{T_1} &= E_\mathrm{e} - E_{t_1} - \hbar\omega_1, \\ E_\mathrm{I} - E_{T_2} &= E_\mathrm{e} - E_{t_2} - \hbar\omega_2. \end{aligned} \tag{7.123}$$

Note, however, that the intermediate states $|t_1\rangle$ and $|t_2\rangle$, as well as the energies, are not the same for emission and absorption. The interaction matrix elements are

$$\begin{aligned} \langle \mathrm{F}|\mathscr{H}_{\mathrm{AF}}|T_1\rangle &= \langle \mathrm{g}, n_1 + 1, n_2 + 1|\mathscr{H}^{(-)}_{\mathrm{AF}}|t_1, n_1 + 1, n_2\rangle \\ &= i\sqrt{\frac{(n_2 + 1)\hbar\omega_2}{2\varepsilon_0 V}}\langle \mathrm{g}|\mathbf{d}\cdot\hat{\varepsilon}_2|t_1\rangle, \end{aligned} \tag{7.124a}$$

$$\begin{aligned} \langle T_1|\mathscr{H}_{\mathrm{AF}}|\mathrm{I}\rangle &= \langle t_1, n_1 + 1, n_2|\mathscr{H}^{(-)}_{\mathrm{AF}}|\mathrm{e}, n_1, n_2\rangle \\ &= i\sqrt{\frac{(n_1 + 1)}{2\varepsilon_0 V}}\langle t_1|\mathbf{d}\cdot\hat{\varepsilon}_1|\mathrm{e}\rangle, \end{aligned} \tag{7.124b}$$

$$\begin{aligned} \langle \mathrm{F}|\mathscr{H}_{\mathrm{AF}}|T_2\rangle &= \langle \mathrm{g}, n_1 + 1, n_2 + 1|\mathscr{H}^{(-)}_{\mathrm{AF}}|t_2, n_1, n_2 + 1\rangle \\ &= i\sqrt{\frac{(n_1 + 1)\hbar\omega_1}{2\varepsilon_0 V}}\langle \mathrm{g}|\mathbf{d}\cdot\hat{\varepsilon}_1|t_2\rangle, \end{aligned} \tag{7.124c}$$

$$\begin{aligned} \langle T_2|\mathscr{H}_{\mathrm{AF}}|\mathrm{I}\rangle &= \langle t_2, n_1, n_2 + 1|\mathscr{H}^{(-)}_{\mathrm{AF}}|\mathrm{e}, n_1, n_2\rangle \\ &= i\sqrt{\frac{(n_2 + 1)\hbar\omega_j}{2\varepsilon_0 V}}\langle t_2|\mathbf{d}\cdot\hat{\varepsilon}_2|\mathrm{e}\rangle. \end{aligned} \tag{7.124d}$$

Hence, the matrix element for two-photon emission is

$$\begin{aligned}
M_{\text{tpe}} &= M_1 + M_2 \\
&= \sum_{T_1} \frac{\langle \mathrm{F}|\mathscr{H}_{\mathrm{AF}}|T_1\rangle\langle T_1|\mathscr{H}_{\mathrm{AF}}|\mathrm{I}\rangle}{E_{\mathrm{I}} - E_{\mathrm{T}_1}} + \sum_{T_2} \frac{\langle \mathrm{F}|\mathscr{H}_{\mathrm{AF}}|T_2\rangle\langle T_2|\mathscr{H}_{\mathrm{AF}}|\mathrm{I}\rangle}{E_{\mathrm{I}} - E_{\mathrm{T}_2}} \\
&= \frac{1}{2\varepsilon_0 V}\sqrt{(n_1+1)(n_2+1)}\sum_t \left[\frac{\langle \mathrm{g}|\mathbf{d}\cdot\hat{\varepsilon}_2|t\rangle\langle t|\mathbf{d}\cdot\hat{\varepsilon}_1|\mathrm{e}\rangle}{\omega_{\mathrm{te}} - \omega_1}\right. \\
&\quad \left. + \frac{\langle \mathrm{g}|\mathbf{d}\cdot\hat{\varepsilon}_1|t\rangle\langle t|\mathbf{d}\cdot\hat{\varepsilon}_2|\mathrm{e}\rangle}{\omega_{\mathrm{te}} - \omega_2}\right],
\end{aligned} \tag{7.125}$$

and the probability per unit time for two-photon emission is

$$W_{\text{tpe}} = \frac{2\pi}{\hbar}|M_{\text{tpe}}|^2\,\delta(E_{\mathrm{I}} - E_{\mathrm{F}}). \tag{7.126}$$

For this case, too, the parities of $|\mathrm{g}\rangle$ and $|\mathrm{e}\rangle$ must be the same, and damping constants are to be included near resonances. Note that two-photon emission shares with one-photon emission the feature that it may be subdivided into a spontaneous part ($n_1 = n_2 = 0$) and a stimulated part ($n_1, n_2 \neq 0$).

A two-photon process may be represented by means of an effective Hamiltonian

$$\mathscr{H}_{\text{tpa}} = M_{\text{tpa}} a_1 a_2 \sigma^+, \tag{7.127a}$$

$$\mathscr{H}_{\text{tpe}} = M_{\text{tpe}} a_1^\dagger a_2^\dagger \sigma^-, \tag{7.127b}$$

since then

$$\langle \mathrm{F}|\mathscr{H}_{\text{tpa}}|\mathrm{I}\rangle = M_{\text{tpa}}\langle \mathrm{e}, n_1 - 1, n_2 - 1|a_1 a_2 \sigma^+|\mathrm{g}, n_1, n_2\rangle = M_{\text{tpa}}, \tag{7.128a}$$

$$\langle \mathrm{F}|\mathscr{H}_{\text{tpe}}|\mathrm{I}\rangle = M_{\text{tpe}}\langle \mathrm{g}, n_1 + 1, n_2 + 1|a_1^\dagger a_2^\dagger \sigma^-|\mathrm{e}, n_1, n_2\rangle = M_{\text{tpe}}. \tag{7.128b}$$

Although there are little, if any, computational advantages from the effective Hamiltonian, it is often useful for theoretical purposes because it explicitly displays the details of the particular process. The formalism is extended easily to higher order multiphoton processes.

Two photon absorption and ionization are special cases of the more general processes of N-photon absorption and ionization [16–21]. With each additional photon that is absorbed or ionized, it is necessary to advance one order higher in perturbation theory. For transitions between bound states, the sum of the photon energies must be equal to the energy difference between the two states. Here too, N-photon absorption and ionization can occur from multiple beams or from a single beam. The general N-photon ionization rate is usually written

$$W^{(N)} = \sigma_N I^N (s^{-1}) \tag{7.129}$$

where I is the photon flux (photons/cm^{2}s) and σ_N is the generalized cross section (cm^{2n}s^{n-1}). In a later section, it will be shown that the statistics of the light beam cannot be ignored. The cross section σ_N must be regarded as a product of $\sigma_0^{(N)}$, the cross section computed from perturbation theory (including the line shape) and $G^{(N)}$, the statistical factor associated with the light beam.

Finally, we note that two-photon absorption and emission also may be described in terms of a semiclassical, third-order susceptibility function (Section 7.7).

7.5 Stimulated Raman Processes

The general theory of Raman scattering in the dipole approximation is contained in the Kramers-Heisenberg formula (Section 5.9). In particular, we refer to Eq. (5.271) for the transition probability per unit time, W_{FI}, for a transition from $|\mathrm{I}\rangle$ to $|\mathrm{F}\rangle$ in which an incident photon is lost and a scattered photon is gained. It was shown that W_{FI} is proportional to $n_{\mathrm{i}}(n_{\mathrm{s}}+1)$ where n_{i} is the number of incident photons of frequency ω_{i}, and n_{s} is the number of scattered photons of frequency ω_{s}. The term proportional to n_{i} alone represents spontaneous scattering since the process can start without the presence of scattered photons. But the term proportional to the product $n_{\mathrm{i}}n_{\mathrm{s}}$ can occur only when $n_{\mathrm{s}} \neq 0$; therefore, it represents stimulated scattering. The derivation leading to Eq. (5.271) employed a quantized electromagnetic field. Quantization is essential to obtain the spontaneous part of the scattering; the stimulated scattering, however, may equally well be formulated in terms of classical fields. In this format, stimulated scattering is treated as a coherent third-order process [22–24].

Consider two frequencies $\omega_1 > \omega_2$ such that

$$\omega_1 - \omega_2 = \omega_{\mathrm{eg}} \tag{7.130}$$

where $\hbar\omega_{\mathrm{eg}} = E_{\mathrm{e}} - E_{\mathrm{g}}$ is the energy of excitation in the medium. Of the many types of excitations, vibrational excitation is one of the most common; therefore, we shall refer to ω_{eg} as a vibrational frequency with the understanding that other excitations (e.g., electronic, rotational, phonon) are possible. As we have seen previously, Raman scattering may be either Stokes Raman or anti-Stokes Raman scattering (Fig. 5.12). The frequencies are related by

$$\begin{aligned} \omega_{\mathrm{st}} &= \omega_1 - \omega_{\mathrm{eg}} = \omega_2 && \text{(Stokes)} \\ \omega_{\mathrm{as}} &= \omega_1 + \omega_{\mathrm{eg}} = 2\omega_1 - \omega_2 && \text{(anti-Stokes).} \end{aligned} \tag{7.131}$$

For stimulated Stokes scattering at low intensities, it is necessary to propagate two beams, at ω_1 and ω_2, through a medium containing Raman-active atoms or molecules. With laser-generated high-intensity beams it is sufficient to employ a single beam at the incident (pump) frequency ω_1. Spontaneous Raman scattering produces Stokes-shifted photons at a frequency ω_2 and an excitation of the medium at an energy $\hbar\omega_{eg}$. That is, a pump photon (ω_1) is absorbed and a (lower energy) scattered photon (ω_2) is emitted together with an excitation at ω_{eg}. We may regard the stimulated process as one in which light at the frequency ω_1 interacts with light at the frequency $\omega_1 - \omega_2$ by difference-frequency mixing to produce light at ω_2. But since $\omega_1 - \omega_2 = \omega_{eg}$, the process also may be viewed as a difference-frequency mixing of ω_1 and the excitation frequency ω_{eg}. Still another point of view regards stimulated Stokes scattering as a parametric process in which the pump frequency ω_1 splits into a signal frequency ω_2 (Stokes) and an idler frequency ω_{eg}, the latter being an internal excitation of the medium. In this sense, it is seen that as the waves propagate through the medium the Stokes wave will grow in intensity at the expense of the pump wave. It also should be mentioned that since there is no phase associated with ω_{eg}, the phase matching condition (Eq. (7.107) does not apply to stimulated Stokes scattering.

The three frequencies interact to produce a third-order polarization

$$\mathbf{P}^{(3)}(\omega_{st}) = \varepsilon_0 \chi^{(3)}(-\omega_{st}; \omega_2, \omega_1, -\omega_1)\mathbf{E}(\omega_2)\mathbf{E}(\omega_1)\mathbf{E}^*(\omega_1). \tag{7.132}$$

Clearly, the most important contributors to $\chi^{(3)}$ are those terms that contain the Raman resonance denominator $\omega_{eg} - \omega_1 + \omega_2$. It is therefore a good approximation to confine the computation to the resonance terms and to neglect all others. Of the eight basic Feynman diagrams for the third-order susceptibility, only four are capable of exhibiting a Raman resonance. These diagrams, assuming only $|g\rangle$ is populated, are shown in Fig. 7.9, and their contributions to the stimulated Stokes Raman susceptibility are

$$U_1^{st} = \rho_{gg}^0 \left(\frac{-1}{\omega_{gs} + \omega_1}\right)\left(\frac{-1}{\omega_{ge} + \omega_1 - \omega_2 - i\Gamma_{ge}}\right)\left(\frac{-1}{\omega_{gr} - \omega_2}\right) d_{rg}^{\mu} d_{er}^{\beta} d_{se}^{\alpha} d_{gs}^{\gamma}$$

$$= \frac{d_{gs}^{\gamma} d_{se}^{\alpha} d_{er}^{\beta} d_{rg}^{\mu}}{(\omega_{sg} - \omega_1)(\omega_{eg} - \omega_1 + \omega_2 + i\Gamma_{eg})(\omega_{rg} + \omega_2)} \rho_{gg}^0, \tag{7.133a}$$

$$U_2^{st} = \rho_{gg}^0 \left(\frac{-1}{\omega_{gs} - \omega_2}\right)\left(\frac{-1}{\omega_{ge} - \omega_2 + \omega_1 - i\Gamma_{ge}}\right)\left(\frac{-1}{\omega_{gr} - \omega_2}\right) d_{rg}^{\mu} d_{er}^{\beta} d_{se}^{\gamma} d_{gs}^{\alpha}$$

$$= \frac{d_{gs}^{\alpha} d_{se}^{\gamma} d_{er}^{\beta} d_{rg}^{\mu}}{(\omega_{sg} + \omega_2)(\omega_{eg} + \omega_2 - \omega_1 + i\Gamma_{eg})(\omega_{rg} + \omega_2)} \rho_{gg}^0, \tag{7.133b}$$

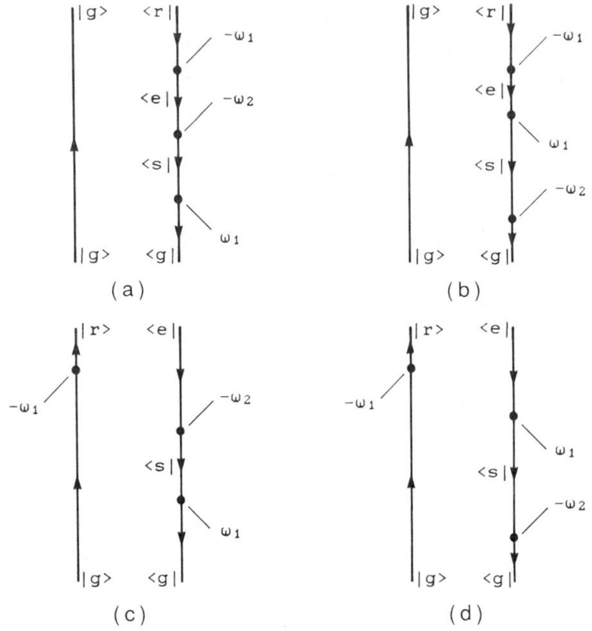

FIGURE 7.9 The four major contributors to $\chi^{(3)}(-\omega_{st}; \omega_2, \omega_1, -\omega_1)$ for the stimulated Stokes Raman process are those that contain the resonance denominator $\omega_{eg} - \omega_1 + \omega_2$. The diagrams (a)$\cdots$(d) correspond to $U_1^{st} \cdots U_4^{st}$ shown in Eq. (7.133).

$$U_3^{st} = \rho_{gg}^0 \left(\frac{-1}{\omega_{gs} + \omega_1}\right)\left(\frac{-1}{\omega_{ge} + \omega_1 - \omega_2 - i\Gamma_{ge}}\right)\left(\frac{1}{\omega_{re} - \omega_2}\right) d_{rg}^{\beta} d_{er}^{\mu} d_{se}^{\alpha} d_{gs}^{\gamma}$$

$$= \frac{d_{gs}^{\gamma} d_{se}^{\alpha} d_{er}^{\mu} d_{rg}^{\beta}}{(\omega_{sg} - \omega_1)(\omega_{eg} - \omega_1 + \omega_2 + i\Gamma_{eg})(\omega_{rg} - \omega_1)} \rho_{gg}^0, \tag{7.133c}$$

$$U_4^{st} = \rho_{gg}^0 \left(\frac{-1}{\omega_{gs} - \omega_2}\right)\left(\frac{-1}{\omega_{ge} - \omega_2 + \omega_1 - i\Gamma_{ge}}\right)\left(\frac{1}{\omega_{re} - \omega_2}\right) d_{rg}^{\beta} d_{er}^{\mu} d_{se}^{\gamma} d_{gs}^{\alpha}$$

$$= \frac{d_{gs}^{\alpha} d_{se}^{\gamma} d_{er}^{\mu} d_{rg}^{\beta}}{(\omega_{sg} + \omega_2)(\omega_{eg} + \omega_2 - \omega_1 + i\Gamma_{eg})(\omega_{rg} - \omega_1)} \rho_{gg}^0. \tag{7.133d}$$

In these expressions, we employed the associations α with ω_2, β with $+\omega_1$, and γ with $-\omega_1$. Note also that $\omega_{re} - \omega_2 = \omega_{rg} - \omega_1$ as a consequence of $\omega_{eg} = \omega_1 - \omega_2$. Near resonance (i.e., when $\omega_{eg} - \omega_1 + \omega_2 \simeq 0$), the damping constant Γ_{eg} becomes dominant and its retention is therefore required. But the other factors in the denominators are not near resonance (barring a fortuitous arrangement of energy levels); hence, all other damping constants may be omitted. The approximation to the susceptibility, consisting entirely of resonant terms, is

$$\chi^{(3)}_{\mu\alpha\beta\gamma}(-\omega_{st}; \omega_2, \omega_1, -\omega_1) = \frac{N}{\varepsilon_0 \hbar^3} \sum_{rs} (U_1^{st} + \cdots + U_4^{st}). \tag{7.134}$$

We note that in accordance with the convention defined by Eqs. (7.87) to (7.89), the pump field $\mathbf{E}(\omega_1)$ appears in the polarization associated with $\chi^{(3)}$ in the form of a product $\mathbf{E}(\omega_1)\mathbf{E}^*(\omega_1)$. Consequently, the phase of the pump field cancels and phase matching does not apply, as was indicated previously. The gain in the Stokes wave and the corresponding loss in the pump wave is proportional to the imaginary part of $\chi^{(3)}$; the real part of the susceptibility is associated with dispersion, that is, with a change of refractive index at the Stokes frequency. This is known as the optical Kerr effect.

Whereas the intensity of spontaneous anti-Stokes Raman scattering is low owing to the small thermal population in the excited state, stimulated anti-Stokes scattering does not suffer from the same limitation. To see how this comes about, we note again that when a pump wave at the higher frequency ω_1 produces a gain in the Stokes wave at the lower frequency ω_2, there is a concurrent buildup of population in the excited state at the energy $\hbar\omega_{eg}$. Then by sum-frequency mixing, an anti-Stokes wave at the frequency $\omega_{as} = \omega_1 + \omega_{eg} = 2\omega_1 - \omega_2$ is produced. One of the most important techniques based on the stimulated anti-Stokes Raman process, with numerous spectroscopic and nonspectroscopic applications, is known as Coherent Anti-Stokes Raman Scattering (CARS), illustrated schematically and in terms of an energy level diagram in Fig. 7.10. In this case, *two* laser beams of frequencies ω_1 and ω_2 ($\omega_1 > \omega_2$) propagate through the Raman-active medium which, again, contains molecules with an excitation frequency $\omega_{eg} = \omega_1 - \omega_2$ (Fig. 7.10).

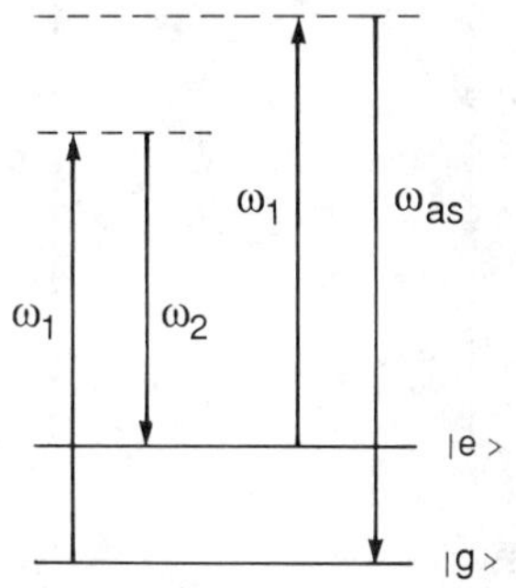

FIGURE 7.10 Energy level diagram for Coherent Anti-Stokes Raman scattering (CARS). Dashed lines indicate virtual levels.

These frequencies may combine in various ways; in particular we are interested in the CARS (anti-Stokes) frequency $\omega_{as} = 2\omega_1 - \omega_2$, as given by Eq. (7.131b).

Here, too, we have a third-order process with a polarization of the form

$$\mathbf{P}^{(3)}(\omega) = \varepsilon_0 \chi^{(3)}(-\omega_{as}: \omega_1, \omega_1, -\omega_2) E(\omega_1) E(\omega_1) E^*(\omega_2). \qquad (7.135)$$

One way to interpret this process is through a sequence of (virtual) absorption and emission steps (Fig. 7.10). That is, a pump photon is absorbed, a Stokes photon is emitted, and the medium acquires an excitation $\hbar\omega_{eg}$. Subsequently, a pump photon is absorbed from the excited state, an anti-Stokes photon is emitted, and the medium returns to its ground state. In another interpretation, the excitation at ω_{eg} is regarded as a photon that mixes with photons at ω_1 and ω_2. According to this view, the pump and Stokes waves combine to produce a wave at the difference frequency $\omega_{eg} = \omega_1 - \omega_2$. The pump at ω_1 then mixes with ω_{eg} to produce the sum frequency $\omega = \omega_1 + \omega_{eg} = 2\omega_1 - \omega_2$.

As we have done previously for the case of the stimulated Stokes Raman process, we will confine our attention to those terms in the susceptibility that contain the Raman resonance denominator $\omega_{eg} - \omega_1 + \omega_2$. Here, too, there

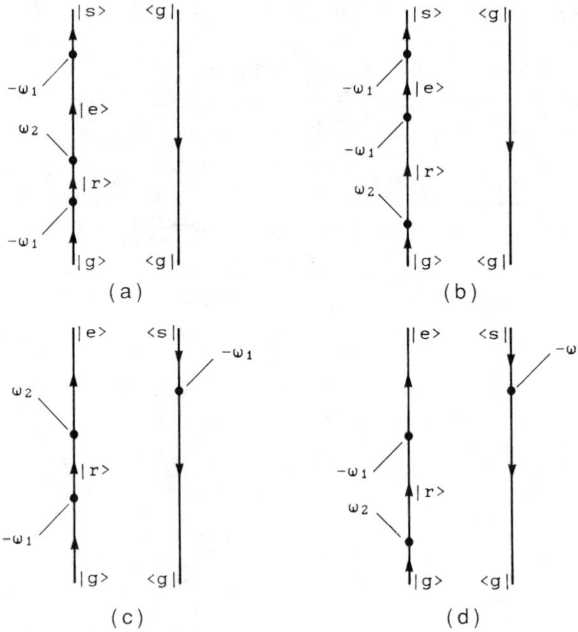

FIGURE 7.11 The four major contributors to $\chi^{(3)}(-\omega_{as}; \omega_1, \omega_1, -\omega_2)$ for the CARS process are those that contain the resonance denominator $\omega_{eg} - \omega_1 + \omega_2$, as in the stimulated Stokes Raman process. The diagrams (a)$\cdots$(d) correspond to $U_1^{as} \cdots U_4^{as}$ shown in Eq. (9.136).

are only four diagrams (Fig. 7.11). If only $|g\rangle$ has a nonzero population and all other states are vacant, the contributors to the CARS susceptibility, neglecting damping constants in the nonresonant denominators, are

$$U_1^{as} = \rho_{gg}^0\left(\frac{1}{\omega_{rg} - \omega_1}\right)\left(\frac{1}{\omega_{eg} - \omega_1 + \omega_2 - i\Gamma_{eg}}\right)\left(\frac{1}{\omega_{sg} - 2\omega_1 + \omega_2}\right)$$
$$\times\, d_{rg}^{\alpha} d_{er}^{\beta} d_{se}^{\alpha} d_{gs}^{\mu}$$
$$= \frac{d_{gs}^{\mu} d_{se}^{\alpha} d_{er}^{\beta} d_{rg}^{\alpha}}{(\omega_{rg} - \omega_1)(\omega_{eg} - \omega_1 + \omega_2 - i\Gamma_{eg})(\omega_{sg} - 2\omega_1 + \omega_2)}\rho_{gg}^0, \quad (7.136a)$$

$$U_2^{as} = \rho_{gg}^0\left(\frac{1}{\omega_{rg} + \omega_2}\right)\left(\frac{1}{\omega_{eg} + \omega_2 - \omega_1 - i\Gamma_{eg}}\right)\left(\frac{1}{\omega_{sg} - 2\omega_1 + \omega_2}\right)$$
$$\times\, d_{rg}^{\beta} d_{er}^{\alpha} d_{se}^{\alpha} d_{gs}^{\mu}$$
$$= \frac{d_{gs}^{\mu} d_{se}^{\alpha} d_{er}^{\alpha} d_{rg}^{\beta}}{(\omega_{rg} + \omega_2)(\omega_{eg} + \omega_2 - \omega_1 - i\Gamma_{eg})(\omega_{sg} - 2\omega_1 + \omega_2)}\rho_{gg}^0, \quad (7.136b)$$

$$U_3^{as} = \rho_{gg}^0\left(\frac{1}{\omega_{rg} - \omega_1}\right)\left(\frac{1}{\omega_{eg} - \omega_1 + \omega_2 - i\Gamma_{eg}}\right)\left(\frac{-1}{\omega_{es} - 2\omega_1 + \omega_2}\right)$$
$$\times\, d_{rg}^{\alpha} d_{er}^{\beta} d_{se}^{\mu} d_{gs}^{\alpha}$$
$$= \frac{d_{gs}^{\alpha} d_{se}^{\mu} d_{er}^{\beta} d_{rg}^{\alpha}}{(\omega_{rg} - \omega_1)(\omega_{eg} - \omega_1 + \omega_2 - i\Gamma_{eg})(\omega_{se} + 2\omega_1 - \omega_2)}\rho_{gg}^0, \quad (7.136c)$$

$$U_4^{as} = \rho_{gg}^0\left(\frac{1}{\omega_{rg} + \omega_2}\right)\left(\frac{1}{\omega_{eg} + \omega_2 - \omega_1 - i\Gamma_{eg}}\right)\left(\frac{-1}{\omega_{es} - 2\omega_1 + \omega_2}\right)$$
$$\times\, d_{rg}^{\beta} d_{er}^{\alpha} d_{se}^{\mu} d_{gs}^{\alpha}$$
$$= \frac{d_{gs}^{\alpha} d_{se}^{\mu} d_{er}^{\alpha} d_{rg}^{\beta}}{(\omega_{rg} + \omega_2)(\omega_{eg} + \omega_2 - \omega_1 - i\Gamma_{eg})(\omega_{se} + 2\omega_1 - \omega_2)}\rho_{gg}^0. \quad (7.136d)$$

In this approximation, the CARS susceptibility is

$$\chi_{\mu\alpha\alpha\beta}^{(3)}(-\omega_{as}; \omega_1, \omega_1, -\omega_2) = \frac{N}{\varepsilon_0 \hbar^3}\sum_{rs}(U_1^{as} + \cdots + U_4^{as}). \quad (7.137)$$

In constrast to the stimulated Stokes case, a CARS process does not change the state of the medium, as is evident from the energy level diagram in Fig. 7.10. Therefore, the three waves involved in the stimulated anti-Stokes case must satisfy the phase matching condition

$$\mathbf{k}_{as} = 2\mathbf{k}_p - \mathbf{k}_s \quad (7.138)$$

in order to produce gain at ω_{as}.

7.6 Statistics of Two-Photon Absorption

The previous calculations of the probabilities for two-photon processes (Section 7.4) were performed on the assumption that the radiation field is describable in terms of photon-number states. Our present objective is to demonstrate that the statistics of the light beam can have a decided influence on the probabilities (or cross sections) of a two-photon process. For this purpose we consider a more general format in which the radiation field is described by an general state $|R\rangle$ which will be made more explicit at a later stage. We now shall compute the matrix element for two-photon absorption under this more general assumption.

If $|R_g\rangle$ and $|R_e\rangle$ are the initial and final states of the field, respectively, let

$$|\mathrm{I}\rangle = |\mathrm{g}, R_g\rangle, \qquad |\mathrm{F}\rangle = |\mathrm{e}, R_e\rangle, \tag{7.139}$$

where $|\mathrm{g}\rangle$ and $|\mathrm{e}\rangle$ are the initial and final atomic states. The corresponding energies are

$$E_\mathrm{I} = E_g + E(R_g), \qquad E_\mathrm{F} = E_e + E(R_e). \tag{7.140}$$

Since $E_\mathrm{I} = E_\mathrm{F}$,

$$E_e - E_g = E(R_g) - E(R_e) = \hbar(\omega_1 + \omega_2). \tag{7.141}$$

The computation proceeds along the same lines as in Section 7.4 but without the commitment to photon-number states. After the photon of frequency ω_1 has been absorbed, let the atomic state be $|t_1\rangle$, the state of the field $|R_1\rangle$, and the state of the total system $|T_1\rangle = |t_1, R_1\rangle$. Similarly, $|T_2\rangle = |t_2, R_2\rangle$ is the state of the system after the absorption of an ω_2-photon. The energies are

$$\begin{aligned} E_{T_1} &= E_{t_1} + E(R_1) = E_{t_1} + E(R_g) - \hbar\omega_1, \\ E_{T_2} &= E_{t_2} + E(R_2) = E_{t_2} + E(R_g) - \hbar\omega_2. \end{aligned} \tag{7.142}$$

For a transition from $|T_1\rangle$ to $|\mathrm{F}\rangle$ it is necessary to absorb an ω_2-photon while the transition $|T_2\rangle \to |\mathrm{F}\rangle$ requires that an ω_1-photon be absorbed.

With these preliminaries and the dipole interaction Hamiltonian as in (4.329),

$$\langle \mathrm{F}|\mathscr{H}_{\mathrm{AF}}|T_1\rangle = -\langle \mathrm{F}|\mathbf{d}\cdot\mathbf{E}_2|T_1\rangle = -\langle \mathrm{F}|\mathbf{d}\cdot\mathbf{E}_2^{(+)}|T_1\rangle \tag{7.143}$$

where $\mathbf{E}_2^{(+)}$ is the annihilation part of the electric field operator for ω_2-photons. Writing

$$\mathbf{E}_2^{(+)} = \hat{\boldsymbol{\varepsilon}} E^{(+)}, \tag{7.144}$$

the matrix element may be factored into an atomic part and a radiation part

$$\langle \mathrm{F}|\mathscr{H}_{\mathrm{AF}}|T_1\rangle = -\langle \mathrm{e}|\mathbf{d}\cdot\hat{\varepsilon}_2|t_1\rangle\langle R_{\mathrm{e}}|E_2^{(+)}|R_1\rangle. \tag{7.145}$$

The other matrix elements are

$$\begin{aligned}\langle T_1|\mathscr{H}_{\mathrm{AF}}|\mathrm{I}\rangle &= -\langle t_1|\mathbf{d}\cdot\hat{\varepsilon}_1|\mathrm{g}\rangle\langle R_1|E_1^{(+)}|R_{\mathrm{g}}\rangle,\\ \langle \mathrm{F}|\mathscr{H}_{\mathrm{AF}}|T_2\rangle &= -\langle \mathrm{e}|\mathbf{d}\cdot\hat{\varepsilon}_1|t_2\rangle\langle R_{\mathrm{e}}|E_1^{(+)}|R_2\rangle,\\ \langle T_2|\mathscr{H}_{\mathrm{AF}}|\mathrm{I}\rangle &= -\langle t_2|\mathbf{d}\cdot\hat{\varepsilon}_2|\mathrm{g}\rangle\langle R_2|E_2^{(+)}|R_{\mathrm{g}}\rangle.\end{aligned} \tag{7.146}$$

For the complete matrix element for two-photon absorption we have

$$\begin{aligned}M_{\mathrm{tpa}} &= \sum_{T_1}\frac{\langle \mathrm{F}|\mathscr{H}_{\mathrm{AF}}|T_1\rangle\langle T_1|\mathscr{H}_{\mathrm{AF}}|\mathrm{I}\rangle}{E_{\mathrm{I}}-E_{\mathrm{T}_1}} + \sum_{T_2}\frac{\langle \mathrm{F}|\mathscr{H}_{\mathrm{AF}}|T_2\rangle\langle T_2|\mathscr{H}_{\mathrm{AF}}|\mathrm{I}\rangle}{E_{\mathrm{I}}-E_{\mathrm{T}_2}}\\ &= \sum_{t_1}\sum_{R_1}\frac{\langle \mathrm{e}|\mathbf{d}\cdot\hat{\varepsilon}_2|t_1\rangle\langle t_1|\mathbf{d}\cdot\hat{\varepsilon}_1|\mathrm{g}\rangle}{E_{\mathrm{g}}-E_{t_1}-\hbar\omega_1}\langle \mathrm{R}_{\mathrm{e}}|E_2^{(+)}|R_1\rangle\langle R_1|E_1^{(+)}|R_{\mathrm{g}}\rangle\\ &\quad+\sum_{t_2}\sum_{R_2}\frac{\langle \mathrm{e}|\mathbf{d}\cdot\hat{\varepsilon}_1|t_2\rangle\langle t_2|\mathbf{d}\cdot\hat{\varepsilon}_2|\mathrm{g}\rangle}{E_{\mathrm{g}}-E_{t_2}-\hbar\omega_2}\langle R_{\mathrm{e}}|E_1^{(+)}|R_2\rangle\langle R_2|E_2^{(+)}|R_{\mathrm{g}}\rangle.\end{aligned} \tag{7.147}$$

The states contained in the sums over R_1 and R_2 may or may not be complete. If not, additional states can be included to complete each set, since all nonvanishing matrix elements are already accounted for in the two sums. Thus,

$$\begin{aligned}\sum_{R_1}\langle R_{\mathrm{e}}|E_2^{(+)}|R_1\rangle\langle R_1|E_1^{(+)}|R_{\mathrm{g}}\rangle &= \langle R_{\mathrm{e}}|E_2^{(+)}E_1^{(+)}|R_{\mathrm{g}}\rangle\\ \sum_{R_2}\langle R_{\mathrm{e}}|E_1^{(+)}|R_2\rangle\langle R_2|E_2^{(+)}|R_{\mathrm{g}}\rangle &= \langle R_{\mathrm{e}}|E_1^{(+)}E_2^{(+)}|R_{\mathrm{g}}\rangle\end{aligned} \tag{7.148}$$

But since $E_1^{(+)}$ and $E_2^{(+)}$ are proportional to the annihilation operators a_1 and a_2, respectively, and the latter commute, as shown in Eq. (4.96),

$$\langle R_{\mathrm{e}}|E_2^{(+)}E_1^{(+)}|R_{\mathrm{g}}\rangle = \langle R_{\mathrm{e}}|E_1^{(+)}E_2^{(+)}|R_{\mathrm{g}}\rangle, \tag{7.149}$$

and the two-photon absorption matrix element becomes

$$\begin{aligned}M_{\mathrm{tpa}} = \langle R_{\mathrm{e}}|E_2^{(+)}E_1^{(+)}|R_{\mathrm{g}}\rangle\sum_t\Bigg[&\frac{\langle \mathrm{e}|\mathbf{d}\cdot\hat{\varepsilon}_2|t\rangle\langle t|\mathbf{d}\cdot\hat{\varepsilon}_1|\mathrm{g}\rangle}{E_{\mathrm{g}}-E_t+\hbar\omega_1}\\ &+\frac{\langle \mathrm{e}|\mathbf{d}\cdot\hat{\varepsilon}_1|t\rangle\langle t|\mathbf{d}\cdot\hat{\varepsilon}_2|\mathrm{g}\rangle}{E_{\mathrm{g}}-E_t+\hbar\omega_2}\Bigg].\end{aligned} \tag{7.150}$$

The transition rate is then computed according to Eq. (7.120).

In most cases, the relevant quantity is the transition rate to all possible final states. Then

$$\begin{aligned}\sum_{R_e}|\langle R_e|E_2^{(+)}E_1^{(+)}|R_g\rangle|^2 &= \sum_{R_e}\langle R_g|E_1^{(-)}E_2^{(-)}|R_e\rangle\langle R_e|E_2^{(+)}E_1^{(+)}|R_g\rangle,\\ &= \langle R_g|E_1^{(-)}E_2^{(-)}E_2^{(+)}E_1^{(+)}|R_g\rangle,\\ &= \mathrm{Tr}\{\rho E_1^{(-)}E_2^{(-)}E_2^{(+)}E_1^{(+)}\}. \qquad (7.151)\end{aligned}$$

It is evident from this relation that the transition rate for two-photon absorption depends on the statistical properties of the light beams [15, 25]. Specifically, since $E^{(-)}$ is proportional to the creation operator $a^\dagger$ and $E^{(+)}$ is proportional to the annihilation operator a, the transition rate is proportional to the normally ordered second-order correlation function of the electric fields.

We note that Eq. (7.150) may be derived from an effective Hamiltonian

$$\mathscr{H}_{\text{eff}} = d_{\text{tpa}}^{(2)}[\sigma^+ E_1^{(+)}E_2^{(+)} + \sigma^- E_1^{(-)}E_2^{(-)}], \qquad (7.152)$$

where σ^+ and σ^- are the atomic raising and lowering operators, respectively, and $d^{(2)}$ is the second-order electric dipole matrix element:

$$d_{\text{tpa}}^{(2)} = \sum_t \left[\frac{\langle \text{e}|\mathbf{d}\cdot\hat{\varepsilon}_2|t\rangle\langle t|\mathbf{d}\cdot\hat{\varepsilon}_1|\text{g}\rangle}{E_g - E_t + \hbar\omega_1} + \frac{\langle \text{e}|\mathbf{d}\cdot\hat{\varepsilon}_1|t\rangle\langle t|\mathbf{d}\cdot\hat{\varepsilon}_2|\text{g}\rangle}{E_g - E_t + \hbar\omega_2}\right]. \qquad (7.153)$$

When $|\text{I}\rangle$ and $|\text{F}\rangle$ are given by Eq. (7.139), only the first term in the square braces of Eq. (7.152) contributes to the matrix element $\langle \text{F}|\mathscr{H}_{\text{eff}}|\text{I}\rangle$, which then immediately yields Eq. (7.150). Alternatively, the electric field operators in Eq. (7.152) may be replaced by their definitions (Eq. (4.119)) in terms of annihilation and creation operators. In the dipole approximation, for the two-mode case

$$\begin{aligned}\mathscr{H}_{\text{eff}} &= -k_{\text{tpa}}[\sigma^+ a_1 a_2 + \sigma^- a_1^\dagger a_2^\dagger]\\ &= -k_{\text{tpa}}\begin{pmatrix} 0 & a_1 a_2 \\ a_1^\dagger a_2^\dagger & 0 \end{pmatrix}, \qquad (7.154)\end{aligned}$$

where

$$k_{\text{tpa}} = \frac{\hbar}{2\varepsilon_0 V} d_{\text{tpa}}^{(2)} \sqrt{\omega_1\omega_2}. \qquad (7.155)$$

Further insight into the statistics of the field may be obtained by means of a computation analogous to the one in Section 6.8. In the present case, we will couple the two radiation modes to an atomic reservoir consisting of identical two-level atoms initially in their ground states. Following the procedure leading to $\rho^{(1)}(t)$, we write

$$\rho^{(1)}(t) = \frac{-i\tau}{\hbar}[\mathscr{H}_{\mathrm{eff}}, \rho_{\mathrm{A}}(t_0)\rho_{\mathrm{F}}] \tag{7.156}$$

in which $\tau = t - t_0$ and the atomic density matrix is taken to be

$$\rho_{\mathrm{A}}(t_0) = \begin{pmatrix} 0 & 0 \\ 0 & 1 \end{pmatrix}. \tag{7.157}$$

It is further assumed that ρ_{F}, the density opeator for the field, is a slowly varying function of time. Employing the matrix form of $\mathscr{H}_{\mathrm{eff}}$ and $\rho_{\mathrm{A}}(t_0)$,

$$\rho^{(1)}(t) = K_{\mathrm{tpa}}\begin{pmatrix} 0 & a_1 a_2 \rho_{\mathrm{F}} \\ -\rho_{\mathrm{F}} a_1^\dagger a_2^\dagger & 0 \end{pmatrix}, \tag{7.158a}$$

$$\rho^{(2)}(t) = K_{\mathrm{tpa}}^2\begin{pmatrix} n_{11} & 0 \\ 0 & n_{22} \end{pmatrix}, \tag{7.158b}$$

where

$$K_{\mathrm{tpa}} = \frac{i\tau k_{\mathrm{tpa}}}{\hbar} \tag{7.159}$$

$$\begin{aligned} n_{11} &= -2a_1 a_2 \rho_{\mathrm{F}} a_1^\dagger a_2^\dagger, \\ n_{22} &= a_1^\dagger a_2^\dagger a_1 a_2 \rho_{\mathrm{F}} + \rho_{\mathrm{F}} a_1^\dagger a_2^\dagger a_1 a_2. \end{aligned} \tag{7.160}$$

Since $\rho^{(1)}(t)$ has no diagonal elements, the rate of change of ρ_{F} to second order is given by the trace of $\rho^{(2)}(t)$:

$$\begin{aligned} \frac{d\rho_{\mathrm{F}}(t)}{dt} &= K_{\mathrm{tpa}}^2 r_{\mathrm{g}}(a_1^\dagger a_2^\dagger a_1 a_2 \rho_{\mathrm{F}} - 2a_1 a_2 \rho_{\mathrm{F}} a_1^\dagger a_2^\dagger + \rho_{\mathrm{F}} a_1^\dagger a_2^\dagger a_1 a_2), \\ &= K_{\mathrm{tpa}}^2 r_{\mathrm{g}}\{[a_1^\dagger a_2^\dagger, a_1 a_2 \rho_{\mathrm{F}}] + [\rho_{\mathrm{F}} a_1^\dagger a_2^\dagger, a_1 a_2]\}. \end{aligned} \tag{7.161}$$

in which r_{g} is the rate of injection of atoms in their ground states.

With population ρ_{g} in $|\mathrm{g}\rangle$ and ρ_{e} in $|\mathrm{e}\rangle$, the initial atomic density matrix is

$$\rho_{\mathrm{A}}(t_0) = \begin{pmatrix} \rho_{\mathrm{e}} & 0 \\ 0 & \rho_{\mathrm{g}} \end{pmatrix}, \tag{7.162}$$

and in place of (7.161), the time rate of change of $\rho_{\mathrm{F}}(t)$ becomes

$$\begin{aligned} \frac{d\rho_{\mathrm{F}}(t)}{dt} = K_{\mathrm{tpa}}^2 r[&(a_1^\dagger a_2^\dagger a_1 a_2 \rho_{\mathrm{F}} - 2a_1 a_2 \rho_{\mathrm{F}} a_1^\dagger a_2^\dagger + \rho_{\mathrm{F}} a_1^\dagger a_2^\dagger a_1 a_2)\rho_{\mathrm{g}} \\ &+ (a_1 a_2 a_1^\dagger a_2^\dagger \rho_{\mathrm{F}} - 2a_1^\dagger a_2^\dagger \rho_{\mathrm{F}} a_1 a_2 + \rho_{\mathrm{F}} a_1 a_2 a_1^\dagger a_2^\dagger)\rho_{\mathrm{e}}], \end{aligned} \tag{7.163}$$

where r is defined by Eq. (6.227). By means of this relation, which is the analog of Eq. (6.228) for the one-photon process together with the cyclic property of the trace and the commutation rules for the annihilation and creation operators, one obtains

$$\frac{d\langle a_1\rangle}{dt} = \mathrm{Tr}\left\{a_1\frac{d\rho_F}{dt}\right\} = K_{\mathrm{tpa}}^2 r[\langle a_1 a_2^\dagger a_2\rangle(\rho_g - \rho_e) - \langle a_1\rangle\rho_e], \quad (7.164)$$

$$\frac{d\langle a_1^\dagger a_1\rangle}{dt} = \mathrm{Tr}\left\{a_1^\dagger a_1\frac{d\rho_F}{dt}\right\} = 2K_{\mathrm{tpa}}^2 r[\langle a_1^\dagger a_1 a_2^\dagger a_2\rangle(\rho_g - \rho_e) - \langle a_1^\dagger a_1 + a_2^\dagger a_2 + 1\rangle\rho_e], \quad (7.165)$$

With analogous relations for $\langle a_2\rangle$ and $\langle a_2^\dagger a_2\rangle$.

When two-photon absorption occurs from a single mode, we have $\omega_1 = \omega_2 = \omega$, $a_1 = a_2 = a$, and $a_1^\dagger = a_2^\dagger = a^\dagger$. The total rate of two-photon absorption is then given by

$$\frac{d\langle a^\dagger a\rangle}{dt} = 2K_{\mathrm{tpa}}^2 r[\langle a^\dagger a^\dagger aa\rangle(\rho_g - \rho_e) - (2\langle a^\dagger a\rangle + 1)\rho_e], \quad (7.166)$$

which reduces to

$$\frac{d\langle a^\dagger a\rangle}{dt} = 2K_{\mathrm{tpa}}^2 r\langle a^\dagger a^\dagger aa\rangle \quad (7.167)$$

when $\rho_e = 0$ and $\rho_g = 1$. For a coherent beam, according to Eq. (4.143),

$$\langle a^\dagger a^\dagger aa\rangle_c = \langle\alpha|a^\dagger a^\dagger aa|\alpha\rangle = |\alpha|^4 = \langle a^\dagger a\rangle_c^2, \quad (7.168)$$

and for a chaotic beam, according to Eq. (4.224),

$$\langle a^\dagger a^\dagger aa\rangle_{\mathrm{ch}} = \langle a^\dagger aa^\dagger a\rangle_{\mathrm{ch}} - \langle a^\dagger a\rangle_{\mathrm{ch}} = 2\langle a^\dagger a\rangle_{\mathrm{ch}}^2. \quad (7.169)$$

Thus, for the same average intensity, a chaotic beam is twice as effective as a coherent beam in a two-photon absorption process. For single-photon absorption, on the other hand, the transition rate depends only on the average intensity and is independent of the statistics. The difference between the two cases resides in the fact that the two photons must be absorbed in a very short time for a two-photon transition to occur. The transition rate is therefore sensitive to the statistical properties of the light beam. Since a chaotic beam is bunched, whereas the coherent beam is Poissonian, the former contains more pairs of photons with very short time intervals between them. Stated more generally, the N-photon absorption rate depends on the Nth order correlation function. Chaotic light produces a higher rate of transition, by a factor of $N!$, compared with purely coherent light.

7.7 Examples of Three- and Four-Wave Processes

Nonlinear wave-mixing processes became a reality with the advent of lasers capable of producing high-intensity coherent radiation. As indicated previously, such processes are conveniently treated by semiclassical theory; that is, the fields retain their classical form but the susceptibilities, which reflect the microscopic features of the medium, are treated on the basis of quantum mechanics. Given the fields and the susceptibility, one computes the polarization at the particular frequency generated by the mixing process. The polarization is then inserted as a source term into the wave equation from which one derives the physical characteristics of the process. One also should bear in mind that nonlinear processes and the statistics of the light beams are mutually related. A nonlinear process alters the statistics of the input beams and, in turn, the statistics of the input beams affect the nonlinear process. This feature was illustrated in the previous previous section for two-photon absorption.

In second order, two waves at frequencies ω_1 and ω_2 combine to produce a third wave at the frequency ω that is some combination of the input frequencies; hence, the name three-wave process. Similarly, in third order, a four-wave process is one in which three waves at frequencies ω_1, ω_2, and ω_3, combine to produce various combinations of the inputs. One also may distinguish between a resonant process—one in which some combination of the frequencies corresponds to a possible transition in the medium—and a nonresonant process, where such relations do not exist [26–28].

Examples of three-wave processes, assuming the medium is not centrosymmetric, are the following:

1. *Sum-frequency mixing*, also known as *up-conversion*, $\omega = \omega_1 + \omega_2$ (Fig. 7.12).

$$P_\mu^{(2)}(\omega) = \varepsilon_0 \sum_{\alpha\beta} \chi_{\mu\alpha\beta}^{(2)}(-\omega; \omega_1, \omega_2) E_\alpha(\omega_1) E_\beta(\omega_2). \tag{7.170}$$

Second harmonics are generated when $\omega_1 = \omega_2$. For colinear propagation, the phase matching condition Eq. (7.107) reduces to $n(\omega) = n(\omega_1)$, i.e., the phase velocity of the second harmonic must be the same as that of the fundamental for efficient second harmonic generation.

2. *Difference-frequency mixing*, also known as *down-conversion*, $\omega = \omega_3 = \omega_1 - \omega_2$.

$$P_\mu^{(2)}(\omega_3) = \varepsilon_0 \sum_{\alpha\beta} \chi_{\mu\alpha\beta}^{(2)}(-\omega_3; \omega_1, -\omega_2) E_\alpha(\omega_1) E_\beta^*(\omega_2). \tag{7.171}$$

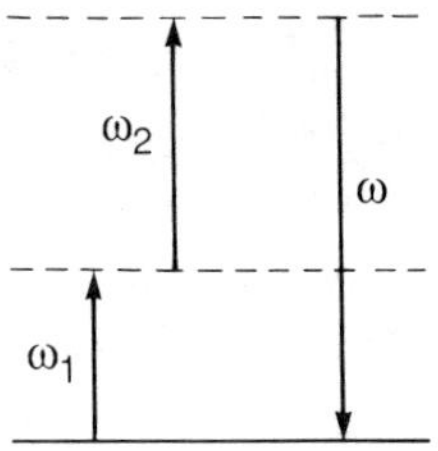

FIGURE 7.12 Schematic diagram for sum-frequency mixing in second order.

We may use this case to illustrate the application of the wave equation (Eq. (7.40)). For propagation in the z direction,

$$\nabla^2 = \frac{\partial^2}{\partial z^2}, \qquad \mathbf{E}(\mathbf{r}, t) = \mathbf{E}(z, t), \qquad \mathbf{P}_{\mathrm{NL}}(\mathbf{r}, t) = \mathbf{P}_{\mathrm{NL}}(z, t). \tag{7.172}$$

The simplest situation is one in which the fields are (transverse) plane waves,

$$\mathbf{E}_i(z, t) = \mathbf{E}_i(k_i, \omega_i) = \mathscr{E}_i(z) e^{i(k_i z - \omega_i t)}, \qquad i = 1, 2, 3, \tag{7.173}$$

$$\frac{\partial^2}{\partial z^2}\mathbf{E}_i(z, t) = \left[\frac{\partial^2}{\partial z^2}\mathscr{E}_i(z) + 2ik_i\frac{\partial}{\partial z}\mathscr{E}_i(z) - k_i^2\mathscr{E}_i(z)\right] e^{i(k_i z - \omega_i t)}$$

$$\simeq \left[2ik\frac{\partial}{\partial z}\mathscr{E}_i(z) - k_i^2\mathscr{E}_i(z)\right] e^{i(k_i z - \omega_i t)}, \tag{7.174}$$

where, in the last expression, the slowly-varying-amplitude approximation

$$\left|\frac{\partial^2}{\partial z^2}\mathscr{E}_i(z)\right| \ll \left|k_i\frac{\partial}{\partial z}\mathscr{E}_i(z)\right| \tag{7.175}$$

has been invoked. With $\mu_0\varepsilon = 1/v^2$ where v is the phase velocity, $k_i = \omega_i/v$, $\mu_0\varepsilon\omega_i^2 = k_i^2$, the wave Eq. (7.40) becomes

$$\frac{\partial^2}{\partial z^2}\mathbf{E}_i(z, t) - \mu_0\varepsilon\frac{\partial^2}{\partial t^2}\mathbf{E}_i(z, t) = 2ik_i\frac{\partial}{\partial z}\mathscr{E}_i(z) e^{i(k_i z - \omega_i t)}$$

$$= \mu_0\frac{\partial^2}{\partial t^2}\mathbf{P}_{\mathrm{NL}}(z, t). \tag{7.176}$$

Now let

$$\mathbf{P}_{\mathrm{NL}}(z, t) = \mathbf{P}^{(2)}(k_3, \omega_3) = \mathbf{P}_3 e^{i(k_3 z - \omega_3 t)}, \tag{7.177}$$

$$\frac{\partial^2}{\partial t^2}\mathbf{P}_{\mathrm{NL}}(z, t) = \frac{\partial^2}{\partial t^2}\mathbf{P}^{(2)}(k_3, \omega_3) = -\omega_3^2 P^{(2)}(k_3, \omega_3). \tag{7.178}$$

Also, from Eqs. (7.171) and (7.173),

$$\mathbf{P}^{(2)}(k_3, \omega_3) = \varepsilon_0 \chi^{(2)}(-\omega_3; \omega_1, -\omega_2)\mathbf{E}_1(k_1, \omega_1)\mathbf{E}_2^*(k_2, \omega_2)$$
$$= \varepsilon_0 \chi^{(2)}(-\omega_3; \omega_1, -\omega_2)\mathscr{E}_1(z)\mathscr{E}_2^*(z)e^{i[(k_1 - k_2)z - (\omega_1 - \omega_2)t]}. \quad (7.179)$$

Substituting these relations into Eq. (7.176) yields the propagation equation

$$2ik_3 \frac{\partial}{\partial z}\mathscr{E}_3(z) = -\varepsilon_0 \mu_0 \omega_3^2 \chi^{(2)}(-\omega_3; \omega_1, -\omega_2)\mathscr{E}_1(z)\mathscr{E}_2^*(z)e^{i(k_1 - k_2 - k_3)z} \quad (7.180)$$

3. *Parametric processes.* In a parametric amplifier, an intense (laser) pump wave at the frequency ω_p and a weak signal wave at the frequency ω_s, with $\omega_p > \omega_s$, propagate through a nonlinear medium. Amplification of the incident signal wave can occur, and, in the course of this process, an idler wave at the frequency ω_i is generated. This process may be regarded as a special case of difference-frequency mixing,

$$P_\mu^{(2)}(\omega_i) = \varepsilon_0 \sum_{\alpha\beta} \chi^{(2)}(-\omega_i; \omega_p, -\omega_s)E_\alpha(\omega_p)E_\beta^*(\omega_s), \quad (7.181)$$

in which reference has been made to Eq. (7.171) with the identification $\omega_3 = \omega_i$, $\omega_1 = \omega_p$, $\omega_2 = \omega_s$. This expression indicates that a polarization at ω_i is created which, in turn, radiates a wave at ω_i. Since $\omega_p = \omega_s + \omega_i$ is the condition for the conservation of energy, the signal and idler waves grow in intensity at the expense of the pump wave. The phase-matching condition of Eq. (7.109) for colinear propagation takes the form

$$\omega_p n_p = \omega_s n_s + \omega_i n_i, \quad (7.182)$$

which means that the three waves must travel at the same phase velocity. The refractive indices n_p, n_s, and n_i generally depend on crystal orientation, temperature, and possibly other parameters, which means that the amplifier is tunable.

In addition to an amplifier, a parametric oscillator is also possible. That is, with no signal input, the wave at ω_p splits into two waves at ω_s and ω_i governed by the same two conditions—energy conservation and phase matching (or momentum conservation)—that select the unique values of ω_s and ω_i for a given pump frequency. One often refers to the amplifier as stimulated parametric scattering and to the oscillator as spontaneous parametric scattering. A degenerate parametric process is one in which $\omega_s = \omega_i$.

4. *Rectification,* $\omega = 0$.

$$P_\mu^{(2)}(\omega) = \varepsilon_0 \sum_{\alpha\beta} \chi_{\mu\alpha\beta}^{(2)}(-\omega; \omega_1, -\omega_1)E_\alpha(\omega_1)E_\beta^*(\omega_1). \quad (7.183)$$

The optical beam induces a dc polarization proportional to the intensity of the light.

5. *Pockels effect*, $\omega = \omega_1$ $(\omega_2 = 0)$.

$$P_\mu^{(2)}(\omega) = \varepsilon_0 \sum_{\alpha\beta} \chi_{\mu\alpha\beta}^{(2)}(-\omega; \omega_1, 0) E_\alpha(\omega_1) E_\beta(0). \tag{7.184}$$

For a wave of frequency ω_1 propagating in a crystal that has no center of inversion, the index of refraction will be altered when the crystal is subjected to a static electric field. The change in the index is proportional to the first power of the field.

Some examples of four-wave processes follow.

1. *Sum-frequency mixing*, $\omega = \omega_1 + \omega_2 + \omega_3$.

$$P_\mu^{(3)}(\omega) = \varepsilon_0 \sum_{\alpha\beta\gamma} \chi_{\mu\alpha\beta\gamma}^{(3)}(-\omega; \omega_1, \omega_2, \omega_3) E_\alpha(\omega_1) E_\beta(\omega_2) E_\gamma(\omega_3). \tag{7.185}$$

Fig. 7.13 illustrates some of the possible resonant and nonresonant mixing processes. Third harmonics are generated when $\omega_1 = \omega_2 = \omega_3$.

2. *Difference-frequency mixing*, $\omega = \omega_1 + \omega_2 - \omega_3$ (Fig. 7.14).

$$P_\mu^{(3)}(\omega) = \varepsilon_0 \sum_{\alpha\beta\gamma} \chi_{\mu\alpha\beta\gamma}^{(3)}(-\omega; \omega_1, \omega_2, -\omega_3) E_\alpha(\omega_1) E_\beta(\omega_2) E_\gamma^*(\omega_3). \tag{7.186}$$

This type of mixing includes the nonresonant case (Fig. 7.14a), resonance by two-photon absorption (Fig. 7.14b), and Raman resonance (Fig. 7.14c).

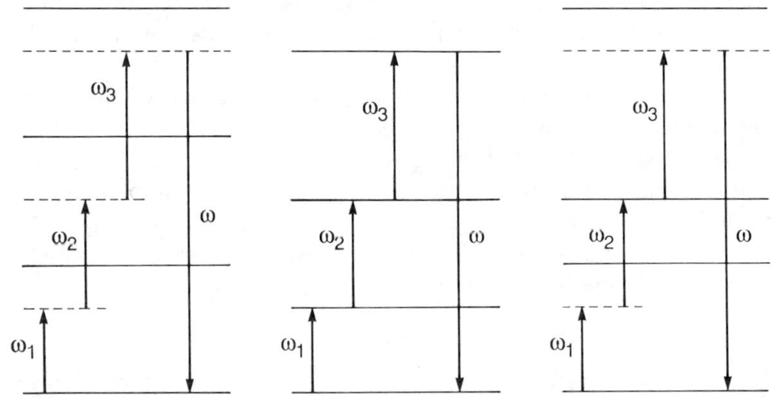

FIGURE 7.13 Examples of resonant and nonresonant mixing processes.

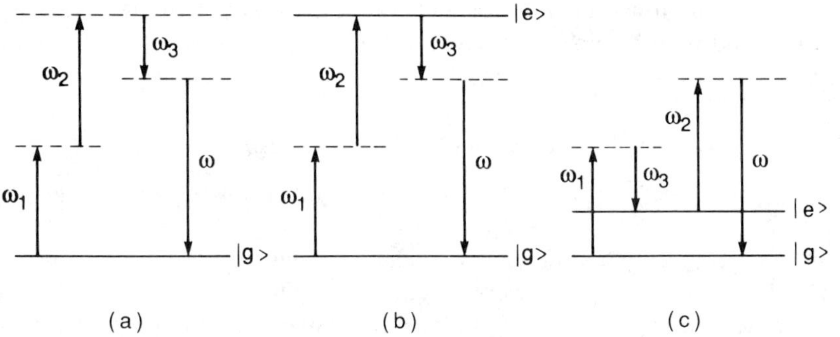

FIGURE 7.14 Examples of difference-frequency mixing in which $\omega = \omega_1 + \omega_2 - \omega_3$. (a) Nonresonant case; (b) resonance by two-photon absorption and (c) Raman resonance.

3. *DC Kerr effect*, $\omega = \omega_1(\omega_2 = \omega_3 = 0)$.

$$P_\mu^{(3)}(\omega) = \varepsilon_0 \sum_{\alpha\beta\gamma} \chi^{(3)}_{\mu\alpha\beta\gamma}(-\omega; \omega_1, 0, 0) E_\alpha(\omega_1) E_\beta(0) E_\gamma(0). \qquad (7.187)$$

Crystals with centers of symmetry, or isotropic materials such as liquids, when placed in a dc electric field acquire the characteristics of a uniaxial crystal with the optic axis in the direction of the applied field. The effect of the field is to replace the original refractive index n at the frequency ω by two indices n_1 and n_2 associated with propagation parallel and perpendicular to the applied field; the difference $n_1 - n_2$ is proportional to the square of the dc field. The Kerr effect is an electro-optic effect closely related to the Pockels effect. We note, however, that the latter applies to a material with no center of symmetry and therefore is describable in terms of a second-order susceptibility function that leads to a linear dependence on the field. The Kerr effect, on the other hand, applies to media with centers of symmetry; hence, the lowest order susceptibility function is in third order, which leads to a quadratic dependence on the static field.

4. *Self-induced, intensity-dependent, refractive index*, $\omega = \omega_1(\omega_1 = \omega_2 = \omega_3)$.

$$P_\mu^{(3)}(\omega) = \varepsilon_0 \sum_{\alpha\beta\gamma} \chi^{(3)}_{\mu\alpha\beta\gamma}(-\omega; \omega_1, -\omega_1, \omega_1) E_\alpha(\omega_1) E^*_\beta(\omega_1) E_\gamma(\omega_1). \qquad (7.188)$$

The refractive index of an intense beam at ω_1 is modified by a term proportional to the intensity at the same frequency, ω_1; that is, the beam modifies its own index of refraction. This effect can lead to self-focusing. For example, if a circular beam is more intense at the center than at the periphery—as in the case of a beam with a Gaussian transverse profile—the refractive index will be greater (and the phase velocity smaller) at the center than at the periphery.

Since rays bend toward regions of higher index, the wave front will converge inwards and become more concave. Such a profile corresponds to the effect of a positive lens.

5. *Optical Kerr effect* and two-photon absorption, $\omega = \omega_1, (\omega_2 = \omega_3)$.

$$P_\mu^{(3)}(\omega) = \varepsilon_0 \sum_{\alpha\beta\gamma} \chi^{(3)}_{\mu\alpha\beta\gamma}(-\omega; \omega_1, \omega_2, -\omega_2) E_\alpha(\omega_1) E_\beta(\omega_2) E_\gamma^*(\omega_2). \tag{7.189}$$

This is a special case of difference-frequency mixing. The real part of the susceptibility corresponds to dispersion and the imaginary part corresponds to two-photon absorption. That is, for the dispersive effect, an intense beam at ω_2 gives rise to a change in the refractive index at ω_1, proportional to the intensity of the beam at ω_2. The absorption at ω_1 is also proportional to the intensity at ω_2. The most important contributions to the susceptibility occur when $\omega_1 + \omega_2$ corresponds to a transition between two states $|g\rangle$ and $|e\rangle$. At exact resonance $\chi^{(3)}$ is imaginary.

The last two effects indicate that an intense light beam modifies its own refractive index as well as that of beams at other frequencies. These phenomena are most pronounced in media containing molecules with anisotropic polarizabilities since a strong electric field will cause such molecules to reorient themselves.

7.8 Dressed States

For a two-level atom in the presence of a radiation field consisting of a single mode, the two states of the *uncoupled* system (atom plus field) $|\mathrm{I}\rangle = |\mathrm{g}, n\rangle$ and $|\mathrm{F}\rangle = |\mathrm{e}, n-1\rangle$, are eigenstates of $\mathscr{H}_\mathrm{A} + \mathscr{H}_\mathrm{F}$—the Hamiltonian in the absence of an interaction between the atom and the field. The energies are $E_\mathrm{I} = -1/2\hbar\omega_0 + n\hbar\omega$ and $E_\mathrm{F} = 1/2\hbar\omega_0 + (n-1)\hbar\omega$ with

$$E_\mathrm{F} - E_\mathrm{I} = \hbar(\omega_0 - \omega) \equiv \hbar\,\Delta, \tag{7.190}$$

as in Eq. (5.128). When $\omega = \omega_0$, the two states are degenerate with a common energy $(n - 1/2)\hbar\omega$. When the total Hamiltonian is $\mathscr{H} = \mathscr{H}_\mathrm{A} + \mathscr{H}_\mathrm{F} + \mathscr{H}_\mathrm{AF}$, however, that is, when the atom and field interact, $|\mathrm{I}\rangle$ and $|\mathrm{F}\rangle$ are no longer eigenstates of $\mathscr{H}$. The effect of "turning on" the interaction Hamiltonian $\mathscr{H}_\mathrm{AF}$ is to couple the two states and to alter their energies.

To investigate the eigenstates and eigenvalues of the total Hamiltonian, we shall require that the superposition state

$$|\Psi\rangle = c_\mathrm{I}|\mathrm{I}\rangle + c_\mathrm{F}|\mathrm{F}\rangle \tag{7.191}$$

satisfy

$$\mathscr{H}|\Psi\rangle = (\mathscr{H}_A + \mathscr{H}_F + \mathscr{H}_{AF})|\Psi\rangle = E|\Psi\rangle. \tag{7.192}$$

Since $|I\rangle$ and $|F\rangle$ are orthonormal, the matrix elements of $\mathscr{H}$ obey the relations

$$c_I\langle I|\mathscr{H}|I\rangle + c_F\langle I|\mathscr{H}|F\rangle = Ec_I, \tag{7.193a}$$

$$c_I\langle F|\mathscr{H}|I\rangle + c_F\langle F|\mathscr{H}|F\rangle = Ec_F, \tag{7.193b}$$

or

$$c_I(E_I - E) + c_F\hbar g^*\sqrt{n} = 0, \tag{7.194a}$$

$$c_I\hbar g\sqrt{n} + c_F(E_F - E) = 0, \tag{7.194b}$$

in which Eq. (5.132a) has been used. Upon solving the secular equation, one obtains the eigenvalues

$$\begin{aligned} E_\pm(n) &= \tfrac{1}{2}(E_I + E_F) \pm \tfrac{1}{2}\sqrt{(E_I + E_F)^2 - 4(E_I E_F - \hbar^2|g|^2 n)} \\ &= (n - \tfrac{1}{2})\hbar\omega \pm \tfrac{1}{2}\hbar\sqrt{\Delta^2 + 4|g|^2 n} = (n - \tfrac{1}{2})\hbar\omega \pm \tfrac{1}{2}\hbar\Omega, \end{aligned} \tag{7.195}$$

where

$$\Omega \equiv \sqrt{\Delta^2 + 4|g|^2 n}, \qquad \Delta \equiv \omega_0 - \omega, \tag{7.196}$$

is just the Rabi frequency. With the definitions

$$\sin 2\theta = \frac{2|g|\sqrt{n}}{\Omega} \equiv \frac{\Omega_0}{\Omega}, \qquad \cos 2\theta = \frac{\Delta}{\Omega}, \tag{7.197}$$

and with $E = E_+(n)$, we obtain from Eq. (7.194a)

$$\frac{1}{2}c_I(\Delta + \Omega) = c_F g^*\sqrt{n}. \tag{7.198}$$

Squaring both sides and imposing the normalization condition on the coefficients, it is found that

$$|c_I|^2 = \sin^2\theta, \qquad |c_F|^2 = \cos^2\theta, \tag{7.199}$$

or

$$|\Psi\rangle = |n\rangle_+ = \sin\theta|I\rangle + \cos\theta|F\rangle. \tag{7.200}$$

For the eigenvalue $E_-(n)$,

$$|c_I|^2 = \cos^2\theta, \qquad |c_F|^2 = \sin^2\theta, \tag{7.201}$$

and the corresponding (orthogonal) eigenfunction is

$$|\Psi\rangle = |n\rangle_- = \cos\theta|I\rangle - \sin\theta|F\rangle. \tag{7.202}$$

In place of $|I\rangle$ and $|F\rangle$ we now have the states $|n\rangle_+$ and $|n\rangle_-$, called *dressed states* [29–32], with energies $E_+(n)$ and $E_-(n)$, respectively. They are mixtures of $|I\rangle = |g, n\rangle$ and $|F\rangle = |e, n-1\rangle$ and are eigenstates of the total system consisting of an atom immersed in a radiation field with the atom-field interaction included. At resonance,

$$\Omega = \Omega_0, \qquad E_+(n) - E_-(n) = \hbar\Omega_0, \qquad \theta = \frac{\pi}{2}, \tag{7.203}$$

$$|n\rangle_\pm = \frac{1}{\sqrt{2}}[|I\rangle \pm |F\rangle]. \tag{7.204}$$

Thus, even at resonance, the states $|n\rangle_+$ and $|n\rangle_-$ are not degenerate.

We have shown that it is possible to construct linear combinations of $|I\rangle$ and $|F\rangle$ that are eigenstates of $\mathscr{H}$ with energies $E_+(n)$ and $E_-(n)$ that are not degenerate even when $\omega = \omega_0$, because the application of $\mathscr{H}_{AF}$ removes the degeneracy inherent in the uncoupled system. An immediate consequence of the separation between $|I\rangle$ and $|F\rangle$ is that, in a transition to a third state, the spectrum consists of two lines, known as an *Autler-Townes* doublet [33]. Clearly, a multimode field will produce a band of shifted energies. It nevertheless should be kept in mind that the photon-atom coupling is intrinsically small ($\alpha \ll 1$); the shifts in energy therefore are small compared with the separation between atomic states, unless the intensity of the radiation is very high.

Let us now consider another pair of degenerate states $|I'\rangle = |g, n+1\rangle$ and $|F'\rangle = |e, n\rangle$ which differ from $|I\rangle$ and $|F\rangle$ by the addition of an extra photon; their common energy at resonance is then $[(n+1) - 1/2]\hbar\omega$. As in the previous case, a pair of dressed states $|n+1\rangle_\pm$ may be constructed with energies

$$E_\pm(n+1) = [(n+1) - \tfrac{1}{2}]\hbar\omega \pm \tfrac{1}{2}\hbar\Omega. \tag{7.205}$$

Hence, the states $|n+1\rangle_\pm$ may be regarded as excited states relative to $|n\rangle_\pm$. An energy level diagram for the two systems is shown in Fig. 7.15. It is seen that the energies of the four possible transitions are

$$\begin{aligned} E_+(n+1) - E_+(n) &= \hbar\omega, \\ E_+(n+1) - E_-(n) &= \hbar(\omega + \Omega), \\ E_-(n+1) - E_+(n) &= \hbar(\omega - \Omega), \\ E_-(n+1) - E_-(n) &= \hbar\omega. \end{aligned} \tag{7.206}$$

If the system initially resides in the excited states $|n+1\rangle_\pm$, the subsequent spontaneous decay to the lower lying states $|n\rangle_\pm$ produces a fluorescence

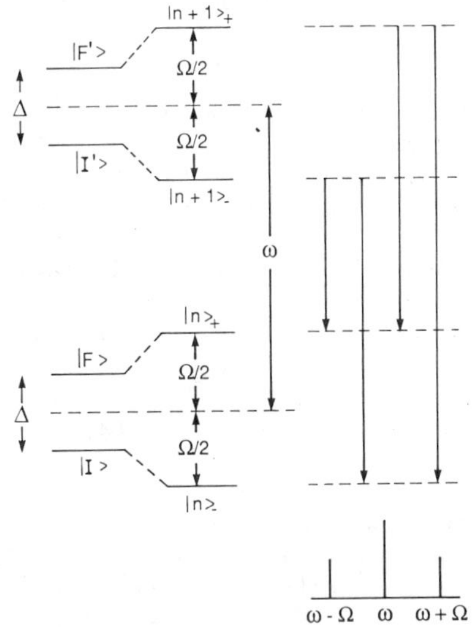

FIGURE 7.15 Energy level diagram for the demonstration of the dynamic Stark effect on the basis of dressed states.

spectrum. Whereas the low-intensity fluorescence spectrum consists of a single (broadened) line, we now find a fluorescence spectrum that, according to Eq. (7.206), contains three lines: a central line at the frequency ω and two satellites at $\omega \pm \Omega$ (Fig. 7.15). Since the central component originates from two transitions, it is more intense: at resonance the peak heights are in the ratio of 1:3:1. The appearance of a triplet in the fluorescence spectrum is known as the *dynamic Stark effect* [34–40]. Clearly, the triplet will not be resolved unless the Rabi frequency is greater than the natural linewidth.

In absorption, the central component is missing. The reason for this may be understood from the following consideration. Since

$$E_+(n+1) - E_-(n+1) = E_+(n) - E_-(n) = \hbar\Omega, \tag{7.207}$$

the equilibrium population in $|n+1\rangle_+$ is the same as the population in $|n\rangle_+$. But as we have seen previously (Section 6.5), a transition from a lower level to an upper level cannot occur when the populations are equal; hence, the transition $|n\rangle_+ \to |n+1\rangle_+$ is missing. The same argument applies to the transition $|n\rangle_- \to |n+1\rangle_-$. This restriction does not apply to the transition $|n\rangle_- \to |n+1\rangle_+$, because the equilibrium population in $|n\rangle_-$ exceeds that of $|n+1\rangle_+$. An absorption line is seen therefore at the frequency $\omega + \Omega$. The transition $|n\rangle_+ \to |n+1\rangle_-$ at the frequency $\omega - \Omega$, however, is an *amplifying transition* because the equilibrium population of $|n\rangle_+$ is smaller than that of $|n+1\rangle_-$.

References

[1] N. Bloembergen, *Nonlinear Optics.* W. A. Benjamin, New York, 1965.

[2] P. N. Butcher, *Nonlinear Optical Phenomena.* Bulletin 200, Engineering Experimental Station, Ohio State University, Columbus, Ohio, 1965.

[3] J. Ducuing, *In Proc. Int'l School of Physics XLII.* (R. J. Glauber, ed.), Academic Press, New York, 1969.

[4] N. Bloembergen, *In Quantum Optics.* (S. M. Kay and A. Maitland, eds.), Academic Press, New York, 1970.

[5] C. Flytzanis, *In Nonlinear Optics, v.1, Part A.* (H. Rabin and C. L. Tang, eds.), Academic Press, New York, 1975.

[6] Y. R. Shen, *Rev. Mod. Phys.* **48**, 1(1976).

[7] V. S. Letokhov and V. P. Chebotaev, *Nonlinear Laser Spectroscopy.* Springer-Verlag, Berlin, 1977.

[8] D. C. Hanna, M. A. Yuratich and D. Cotter, *Nonlinear Optics of Free Atoms and Molecules.* Springer-Verlag, Berlin, 1979.

[9] N. Bloembergen, *Rev, Mod. Phys.* **54**, 685(1982).

[10] M. Levenson, *Introduction to Nonlinear Laser Spectroscopy.* Academic Press, New York, 1982.

[11] J. F. Reintjes, *Nonlinear Optical Parametric Processes in Liquids and Gases.* Academic Press, New York, 1984.

[12] Y. R. Shen, *The Principles of Nonlinear Optics.* J. Wiley, New York, 1984.

[13] B. J. Berne and G. D. Harp, *Adv. in Chem. Phys.* **17**, 63(1970).

[14] Y. Prior, *IEEE J. Quant. Elect.* **QE-20**, 37(1984).

[15] B. R. Mollow, *Phys. Rev.* **175**, 1555(1968).

[16] A. L'Huillier, L. Jonson and G. Wendin, *Int'l J. Quant. Chem.* **31**, 833(1987).

[17] A. T. Georges and P. Lambropoulos, *Adv. Electr.* and *Electr. Phys.* (L. Marton and C. Marton, eds.), **54**, 191(1980).

[18] S. H. Lin, Y. Fujimura, H. J. Neusser and E. W. Schlag, *Multiphoton Spectroscopy of Molecules.* Academic Press, New York, 1984.

[19] P. Lambropoulos, *Adv. Atomic Mol. Phys.* **12**, 87(1976).

[20] B. R. Mollow, *Phys. Rev.* **A12**, 1919(1975).

[21] G. S. Agarwal, *Phys. Rev.* **A1**, 1445(1970).

[22] A. Penzkofer, A. Laubereau and W. Kaiser, *Prog. Quant. Electr.* **6**, 56(1979).

[23] N. Bloembergen, *Am. J. Phys.* **35**, 989(1967).

[24] S. A. J. Druet and J-P. E. Taran, *Prog. Quant. Electr.* **7**, 1(1981).

[25] M. Schubert and B. Wilhelmi, *In Progress in Optics XVII.* (E. Wolf, ed.), North-Holland, Amsterdam, 1980.

[26] M. C. Gower, *IEEE J. Quant. Electr.* **QE21**, 182(1985).

[27] C. L. Tang, *In Quantum Electronics: A Treatise v.1 Nonlinear Optics.* (H. Rabin and C. L. Tang, eds.), Academic Press, New York, 1975.

[28] M. Ducloy, *In* Methods of Laser Spectroscopy. (Y. Prior, A. Ben-Reuven and M. Rosenbluth, eds.), Plenum, New York, 1986.

[29] C. Cohen-Tannoudji, *In Laser Spectroscopy.* (S. Haroche, J. C. Pebay-Peyroula, T. W. Hansch and S. E. Harris, eds.), Springer-Verlag, Berlin, 1975.

[30] C. Cohen-Tannoudji and S. Reynaud, *In Multiphoton Processes* (J. H. Eberly and P. Lambropoulos, eds.), J. Wiley, New York, 1978.

[31] C. Cohen-Tannoudji and S. Reynaud, *J. Phys. B: Atom. Mol. Phys.* **10**, 345(1977).

[32] N. Lu, P. R. Berman, Y. S. Bai, J. E. Golub and T. W. Mossberg, *Phys. Rev.* **A34**, 319(1986).

[33] S. H. Autler and C. H. Townes, *Phys. Rev.* **100**, 703(1955).
[34] E. Courtens and A. Szoke, *Phys. Rev.* **A15**, 1588(1977).
[35] C. Cohen-Tannoudji and S. Reynaud, *J. Phys. B: Atom. Mol. Phys.* **10**, 365(1977).
[36] S. Feneuille, *Rep. Prog. Phys.* **40**, 1257(1977).
[37] B. R. Mollow, *Phys. Rev.* **188**, 1969(1969).
[38] B. R. Mollow, *Phys. Rev.* **A5**, 2217(1972).
[39] B. R. Mollow, *Phys. Rev.* **A5**, 1522(1972).
[40] B. R. Mollow, *Phys. Rev.* **A8**, 1949(1973).

General References

Stochastic Processes and Statistical Mechanics.

C. W. Gardiner, *Handbook of Stochastic Methods.* Springer-Verlag, Berlin, 1983.

M. Kac and J. Logan, *In Studies in Statistical Mechanics* (E. W. Montroll and J. L. Lebowitz, eds.), North-Holland, Amsterdam, 1979.

R. Kubo, M. Toda, and N. Hashitsume, *Statistical Physics II, Nonequilibrium Statistical Mechanics.* Springer-Verlag, Berlin, 1985.

D. K. C. MacDonald, *Noise and Fluctuations: An Introduction.* J. Wiley, New York, 1962.

D. A. McQuarrie, *Statistical Mechanics.* Harper and Row, New York, 1973.

A. Papoulis, *Probability, Random Variables and Stochastic Processes.* McGraw-Hill, New York, 1965.

R. K. Pathria, *Statistical Mechanics.* Pergamon, New York, 1972.

L. E. Reichl, *A Modern Course in Statistical Physics.* Texas Press, Austin, 1980.

F. Reif, *Fundamental of Statistical and Thermal Physics.* McGraw-Hill, New York, 1965.

M. Toda, R. Kubo, and N. Saito, *Statistical Physics I, Equilibrium Statistical Mechanics.* Springer-Verlag, Berlin, 1983.

N. G. Van Kampen, *Stochastic Processes in Physics and Chemistry.* North-Holland, Amsterdam, 1981.

Quantum Mechanics and Atomic Physics.

C. Cohen-Tannoudji, *Quantum Mechanics.* J. Wiley, New York, 1977.

R. D. Cowan, *The Theory of Atomic Structure and Spectra.* Univ. Calif. Press, Berkeley, 1981.

L. D. Landau and E. M. Lifshitz, *Quantum Mechanics, Nonrelativistic Theory.* Pergamon Press, London, 1958.

A. Messiah, *Quantum, Mechanics.* North-Holland, Amsterdam, 1962.

J. J. Sakurai, *Advanced Quantum Mechanics.* Addison-Wesley, Reading, Mass., 1967.

M. Weissbluth, *Atoms and Molecules.* Academic Press, New York, 1978.

Theory of Radiation.

L. Allen and J. H. Eberly, *Optical Resonance and Two-Level Atoms.* J. Wiley, New York, 1975.

P. R. Fontana, *Atomic Radiative Processes.* Academic Press, New York, 1982.

H. Haken, *Light. V. 1, Waves, Photons, Atoms.* North-Holland, Amsterdam, 1981.

H. Haken, *Light. V. 2, Laser Light Dynamics.* North-Holland, Amsterdam, 1985.

W. Heitler, *The Quantum Theory of Radiation.* Oxford Press, Oxford, 1954.

R. Loudon, *The Quantum Theory of Light Second Edition.* Clarendon Press, Oxford, 1983.

M. H. Mittleman, *Theory of Laser-Atom Interactions.* Plenum Press, New York, 1982.

M. Sargent III, M. O. Scully, and W. E. Lamb, Jr., *Laser Physics.* Addison-Wesley, Reading, Mass., 1974.

Quantum Electronics and Nonlinear Optics.

N. Bloembergen, *Nonlinear Optics.* W. A. Benjamin, New York, 1965.

W. Demtroder, *Laser Spectroscopy.* Springer-Verlag, Berlin, 1982.

V. M. Fain and Ya. I. Khanin, *Quantum Electronics.* MIT Press, Cambridge, 1969.

D. C. Hanna, M. A. Yuratich and D. Cotter, *Nonlinear Optics of Free Atoms and Molecules.* Springer-Verlag, Berlin, 1979.

J. R. Klauder and E. C. G. Sudershan, *Fundamentals of Quantum Optics.* W. A. Benjamin, New York, 1968.

V. S. Letokhov and V. P. Chebotaev, *Nonlinear Laser Spectroscopy.* Springer-Verlag, Berlin, 1977.

M. Levenson, *Introduction to Nonlinear Laser Spectroscopy.* Academic Press, New York, 1982.

D. Marcuse, *Principles of Quantum Electronics.* Academic Press, New York, 1980.

H. M. Nussenzveig, *Introduction to Quantum Optics.* Gordon and Breach, New York, 1973.

R. H. Pantell and H. E. Puthoff, *Fundamentals of Quantum Electronics.* J. Wiley, New York, 1969.

J. F. Reintjes, *Nonlinear Optical Parametric Processes in Liquids and Gases.* Academic Press, New York, 1984.

M. Schubert and B. Wilhelmi, *Nonlinear Optics and Quantum Electronics.* J. Wiley, New York, 1986.

Y. R. Shen, *The Principles of Nonlinear Optics.* J. Wiley, New York, 1984.

A. Yariv, *Quantum Electronics Second Edition.* J. Wiley, New York, 1975.

Index

A

B

C